CONCEPTUAL ECOLOGY AND INVASION BIOLOGY: RECIPROCAL APPROACHES TO NATURE

INVADING NATURE -
SPRINGER SERIES IN INVASION ECOLOGY
Volume 1

Series Editor: JAMES A. DRAKE
University of Tennessee,
Knoxville, TN, U.S.A.

Conceptual Ecology and Invasion Biology: Reciprocal Approaches to Nature

Edited by

MARC WILLIAM CADOTTE
University of Tennessee, Knoxville, U.S.A.

SEAN M. McMAHON
University of Tennessee, Knoxville, U.S.A.

and

TADASHI FUKAMI
University of Hawaii at Manoa, Honolulu, U.S.A.

Springer

A C.I.P. Catalogue record for this book is available from the Library of Congress.

ISBN-10 1-4020-4158-6 (PB)
ISBN-13 978-1-4020-4158-7 (PB)
ISBN-10 1-4020-4157-8 (HB)
ISBN-13 978-1-4020-4157-0 (HB)
ISBN-10 1-4020-4925-0 (e-book)
ISBN-13 978-1-4020-4925-5 (e-book)

Published by Springer,
P.O. Box 17, 3300 AA Dordrecht, The Netherlands.

www.springer.com

Printed on acid-free paper

Cover design by Jeff McMahon

Printed in the Netherlands.

Contents

Contributors

Malika L. Aïnouche

Department of Evolutionary Ecology, Genome Evolution and Speciation lab, UMR CNRS 6553, University of Rennes1, Campus de Beaulieu 35 042 Rennes cedex, France.

Ottar N. Bjørnstad

Department of Entomology and Biology, The Pennsylvania State University, University Park, PA 16802, USA.

Yvonne M. Buckley

The Ecology Centre, Hines Building, University of Queensland, St Lucia, Brisbane, QLD 4072, Australia

Marc W. Cadotte

Department of Ecology and Evolutionary Biology, 569 Dabney Hall, University of Tennessee, Knoxville TN, 37996, USA.

Ragan M. Callaway

Division of Biological Sciences, The University of Montana, Missoula, MT 59812, USA.

Hal Caswell

Woods Hole Oceanographic Institution, Woods Hole, MA, 02543, USA.

James S. Clark

Nicholas School of the Environment and Earth Sciences, A221 LSRC, Box 90228, Duke University, Durham, NC, 27708, USA.

Stephane Caut

Ecologie, Systématique & Evolution, Bât. 362 - Université Paris-Sud XI, F-91405 Orsay Cedex, France.

Franck Courchamp

Ecologie, Systématique & Evolution, Bât. 362 - Université Paris-Sud XI, F-91405 Orsay Cedex, France.

Mark A. Davis

Department of Biology, Macalester College, Saint Paul, MN 55105, USA.

Peter M. Dowling

NSW Agriculture, Australian Centre for Weed Research and CRC for Weed Management System, Agricultural and Veterinary Centre, Forest Road, Orange, NSW 2800, NSW 2200, Australia.

Nicholas K. Dulvy

Ecosystem Research Group, Centre for Environment, Fisheries and Aquaculture Science, Pakefield Road, Lowestoft, Suffolk, NR33 OHT, UK

Richard P. Duncan

Ecology and Entomology Group, P.O. Box 84, Lincoln University, Canterbury, New Zealand; and Landcare Research, P.O. Box 69, Lincoln 8152, New Zealand.

David M. Forsyth

Arthur Rylah Institute for Environmental Research, Department of Sustainability and Environment, 123 Brown Street, Heidelberg, Victoria 3084, Australia.

Robert P. Freckleton

Department of Zoology, University of Oxford, South Parks Road, Oxford OX1 3PS, UK.

Tadashi Fukami

Department of Zoology, University of Hawaii at Manoa, 2538 McCarthy Mall, Edmondson 152, Honolulu, HI 96822, USA

Steven D. Gaines

Department of Ecology, Evolution and Marine Biology, University of California, Santa Barbara, CA, USA.

Gregory S. Gilbert

Environmental Studies, University of California, Santa Cruz, CA 95064, USA.

Karin C. Harding

Department of Marine Ecology, Göteborg University, Box 461, 405 30 Göteborg, Sweden.

Robert D. Holt

Department of Zoology, University of Florida, 111 Bartram Hall, PO Box 118525, Gainesville, Florida, USA.

Yoh Iwasa

Department of Biology, Faculty of Sciences, Kyushu University, Fukuoka 812-8581, Japan.

Michio Kondoh

Department of Environmental Solution Technology, Faculty of Science and Technology, Ryukoku University, 1-5 Yokoya, Seta Oe-cho, Otsu 520-2194, Japan; and Center for Ecological Research, Kyoto University, Kamitanakami, Otsu 520-2113, Japan.

Richard Law

Department of Biology, University of York, PO Box 373, York YO10 5YW, UK.

Mark A. Lewis

Department of Mathematical and Statistical Sciences, University of Alberta, 632 Central Academic Building, Edmonton, AB, T6G 2G1, Canada.

Jon Lovett-Doust

Department of Biology, University of Windsor, Windsor, Ontario, N9B 3P4, Canada.

Lesley Lovett-Doust

Department of Biology, University of Windsor, Windsor, Ontario, N9B 3P4, Canada.

Sean M. McMahon

Department of Ecology and Evolutionary Biology, 569 Dabney Hall, University of Tennessee, Knoxville TN, 37996, USA.

John M. McNamara

Department of Mathematics, University of Bristol, University Walk, Bristol, BS8 1TW, UK.

Jessica Metcalf

Department of Biology, Imperial College London, Silwood Park Campus, Ascot, Berkshire, SL5 7PY, UK.

Helen Murphy

Department of Biology, University of Windsor, Windsor, Ontario, N9B 3P4, Canada. Present address: CSIRO Sustainable Ecosystems, PO Box 780, Atherton QLD 4883, Australia.

David John Murrell

NERC Centre for Population Biology, Imperial College London, Silwood Park Campus, Ascot, Berkshire, SL5 7PY, UK.

Michael G. Neubert

Woods Hole Oceanographic Institution, Woods Hole, MA, 02543, USA.

Ingrid M. Parker

Ecology and Evolutionary Biology, University of California, Santa Cruz, CA 95064, USA.

Akiko Satake

Center for Ecological Research, Kyoto University, Otsu 520-2113, Japan.

Dov F. Sax

Institute of Ecology, University of Georgia, Athens, Georgia 30602, USA.

Kristina A. Schierenbeck

Department of Biological Sciences, California State University Chico, Chico, CA, USA.

Katriona Shea

Biology Department, Pennsylvania State University, 208 Mueller Lab, University Park, PA 16802-5301, USA.

Jonathan B. Shurin

Department of Zoology, University of British Columbia, 6270 University Blvd., Vancouver, BC, V6T 1Z4, Canada.

Scott A. Smith

Department of Biology, McGill University, 1205 Avenue Dr. Penfield, Montreal, Quebec, H3A 1B1, Canada.

Andrea S. Thorpe

Division of Biological Sciences, The University of Montana, Missoula, MT 59812, USA.

Jeremy VanDerWal

Department of Biology, University of Windsor, Windsor, Ontario, N9B 3P4, Canada.

Diego P. Vázquez

Instituto Argentino de Investigaciones de las Zonas Aridas, Centro Regional de Investigaciones Cientificas y Tecnologicas, Av. Ruiz Leal s/n, (5500) Mendoza, Argentina.

Philip H. Warren

Department of Animal and Plant Sciences, University of Sheffield, Sheffield S10 2TN, UK.

Anita Weatherby

The Ponds Conservation Trust: Policy and Research, c/o BMS, Oxford Brookes University, Gipsy Lane, Headington, Oxford OX3 0BP, UK.

Foreword

"When I took this plot of land five years ago," he said, "I was disappointed to find that, instead of its being one with the rich sandy soil of the district, it was practically a pocket of clay, which had to be put under treatment before it would yield the result I desired. I have been fairly successful, and now have more than two hundred subtropical shrubs flourishing out there in the grounds. It has been an experiment in acclimatisation, with which I am well pleased. Many of the plants came from Australia and New Zealand. In most cases I have bought the seed, and reared them entirely by myself".

Alfred Russel Wallace
from an interview by E. H. Rann

Q: What are the characteristics of a good invader?
A: It depends.

Q: Depends on what?
A: That depends too!

Biological invasion has become one of those rare themes so profound in its implications and scope that it cuts a broad path across the academic disciplines. From genes to ecosystems and economics to law, species incursions have confronted the global community with a most transcendent Gordian knot. Ours is a peculiar knot however, as invaders colonize ecological systems that have been fashioned from strands of complexity with origins in evolutionary time. A half-century has passed since the publication of Elton's delightful if not foreboding treatise. Of course Elton's assessment was correct and indeed the problem is even greater today. Biological invasions have become a pivotal force in directing the future expression of nature. While considerable progress has been made, generality remains elusive, prediction hit-or-miss, and invasions are proceeding largely unchecked.

Adopting an integrative approach, *Conceptual Ecology and Invasions Biology: Reciprocal Approaches to Nature*, offers essential insights into the topology of our ecological knot. The editors have brought together a superb cast of scientists with a singular aim, orchestrating the diversity of approaches to biological invasions into a more coherent and synthetic whole. What emerges from this effort is compelling framework that captures the intricacy and nuance of the processes, dynamics and mechanisms at play when species invade nature. While no one would argue that this framework is complete, it is clear that any solution to the problem of biological invasions is founded herein.

This book is the inaugural volume in Springer's *Invading Nature* series, and I am pleased to have had the opportunity to introduce both.

James Drake, Series Editor

Preface

Human activities affect natural systems in every corner of the globe, and one of the most widespread and potentially disruptive of these is the introduction of non-indigenous species into new environments. These introductions can have profound consequences for resident populations, community dynamics, and ecosystem function. Over and above the need to understand how a non-indigenous species can affect a native community, introductions offer ecologists the potential to learn about how communities are put together, how species contribute to ecosystem function, or how populations evolve in the face of novel environments and novel species. Just like the railroad spike that pierced Phineas Gage's skull taught physicians much about the function of the cerebral cortex, environmental perturbations can teach us about natural systems. Invasions can inform ecological theory by serving as natural experiments.

On the other hand, modern-day ecologists use sophisticated conceptual tools to aid in the understanding of the functioning of natural systems. These tools, we feel, have been largely under-utilized in understanding species invasions. We see here the potential for a reciprocal approach to gain a deeper understanding of how ecological systems are put together.

We have assembled an excellent group of authors that, each in their own way, attempt to use their tools to explore this reciprocal relationship. The authors in this volume were invited because of their recent publications, which show their use of novel approaches to either understanding the causes and consequences of species invasions, or else use conceptual tools to understand how natural systems function. We asked authors, regardless of their individual approaches, to think about this reciprocal relationship between invasions and theory. The chapters here use examples from numerous organisms, ecological systems, and geographic locations, and are evenly split between theory and natural examples.

This volume could not have been possible without the involvement of others. We are grateful first and foremost to the authors for producing such fine work for us. We wish to thank the chapter reviewers, who, as per usual in science, selflessly reviewed chapters, and greatly improved the quality of this book. We offer a special thanks to the series editor and our mentor, Jim Drake, who allowed us to focus our creative energies on to this project, and to agree to make this the premiere volume of this book series. Finally, Suzanne Mekking and Martine van Bezooijen, at Springer, formed a supremely encouraging and well-organized publishing team, instrumental in making this book a reality.

M.W.C.
S.M.M.
Knoxville, Tennessee

T.F.
Lincoln, New Zealand

July, 2005

Introduction, history and terminology

Tracking the tractable: using invasions to guide the exploration of conceptual ecology

S. M. McMahon, M. W. Cadotte and T. Fukami

INTRODUCTION

Our understanding of the complex functioning of human physiology, neurology, and behavior has long been guided by the study of anomalies in these systems. When faced with a seemingly intractably complex system, it is at times easier to fathom its mechanisms by disrupting or disturbing it rather than trying to delicately explore the 'natural' entity. It has recently become clear to ecologists this tact of using the disruption or distortion of a system to better understand its normal workings can be applied to the study of the behavior, spread, and impact of non-indigenous species (NIS) on natural communities (Sax *et al.* 2005). The power of this view of NIS is that systems are inadvertently being perturbed all over the globe. What we can learn from the perturbations will not only advance the science of ecology, but also will allow ecologists to restore and rebuild these very systems. However, not all disruptions of a system are necessarily informative. Further, if the system as it normally functions holds some intrinsic value,

M.W. Cadotte et al. (eds.), Conceptual ecology and invasion biology, 3–14.

as do invaded ecological systems, it is in the interest of the researcher to work to both understand invasions but also to correct or mitigate the damaging impact of non-native populations.

NIS pose a wide variety of threats to their non-native habitats through immediate and long-term influences on species composition and ecosystem functioning (Parker *et al.* 1999). Because the negative impacts of NIS are first, directly, and primarily ecological, ecologists and evolutionary biologists constitute the forward line in meeting the challenges posed by the introduction and spread of NIS. The many pathways through which an introduced population can affect non-native habitats and communities, however, are difficult to determine and predict. Which tools can the ecologist apply to the problem of NIS? Any solution must develop from a solid understanding of the following: 1) how populations intrinsically behave; 2) how populations influence the biotic and abiotic systems of which they are components; 3) how, conversely, an environment can influence a non-indigenous population; and 4) how NIS populations can change, phenotypically and genotypically, in new environments. Although all of these questions fall solidly within the tradition of ecological research, they demand, with a startling immediacy, capable answers culled from unresolved, ambivalent and even contentious ecological debate. While ecologists are at once well positioned to address the key elements that have arisen from the importance of NIS, our answers are likely to carry with them some of the struggles born of classic theoretical and conceptual debates in ecology and evolutionary biology. Here, however, the study of NIS may not necessarily suffer for these challenges.

The study of NIS, when approached from the realm of conceptual ecology, can act as interesting tests of competing ecological theories. This potential has long been recognized and more recently been promoted (Lodge 1993, Davis *et al.* 2001). Yet, not all ecological theory is applicable or appropriate for this type of test. Nor will the answers gained from all applications of ecological theory assist in improving our understanding, intervention, or prediction about actual colonizations of NIS. One of the key goals of this book is to better refine what ecologists believe the study of NIS can do for the field of ecology and, conversely, how conceptual ecology can advance our ability to explain and address the challenges posed by NIS. In this chapter we will organize some of the ways in which the study of NIS can advance conceptual ecology and vise versa, focusing on recent literature in the field and the ways in which the chapters of this book fit into that scheme. This chapter will also discuss how this book establishes guidelines not only for what we may learn from NIS, but for what we are unlikely to learn. Again, with imminent challenges and limited resources, understanding which scientific approaches, theoretical and applied, are intractable or impractical is as important to advancing this field as knowing which approaches show promise when confronting this complex problem.

TRACTABLE, INTRACTABLE, AND PRACTICAL STUDIES OF BIOLOGICAL INVASIONS

Although studies on the dynamics of invasive organisms demonstrate many research methodologies, three approaches are crucial to the advancement of our understanding of invasions; they are mathematical models, experiments (either manipulative or observational), and reviews or meta-analyses. The strength derived from exploring invasion dynamics through mathematical modeling is that scientists with expertise in a variety of theoretical applications can tackle the challenges of understanding biological invasions through novel techniques (Shea and Chesson 2002). Modeling suggests both possible dynamics, probable findings, and, perhaps more importantly, the limits to certain conceptual approaches to invasions (Harding *et al.* Chapter 17). It is broadly acknowledged that in order to better understand invasions, experimental approaches are crucial. There has been a growth of strong experiments in invasions, ranging from laboratory microcosms (Dick and Platvoet 2000), to field mesocosms (Dukes 2002), to large-scale manipulative field experiments (Tilman 1997). Many of the strengths of inferences derived from these studies, in order to be broadly appreciated, need to be reincorporated into mathematical models. Finally, meta-analyses and reviews provide a generalized assessment of the theoretical and empirical work that has collected over several years. The synthesis derived from these analyses can, as with the models, serve to inform, direct, or delimit the questions researchers ask about invasions (Cadotte *et al.* 2005). The dialectic derived from the complementary approaches of supposition, experimentation, and assessment offers the most efficient way to study invasions and draw broader understanding of ecological processes through the study of invasions. All three of these approaches are featured in this book, and in the following introduction to the themes of this book, we pay special attention to how this dialectic can guide important research and thinking on invasions.

The behavior of populations

Even in models of populations that do not include abiotic or interspecific environmental interactions, intraspecific population dynamics can illustrate important properties of invading populations through patterns of intrinsic population growth and spread (see Cadotte, *et al.* Chapter 21). Although a number of scales may ultimately prove important when assessing the arrival, establishment, growth, and impact of an NIS (Lloret *et al.* 2004, Hamilton *et al.* 2005), the population scale has long provided some of the clearest insights into the basic questions of invasion dynamics (Petrovskii *et al.* 2005, Thomson 2005). All of the features that figure prominently in population biology, from the role of carrying capacity to mortality and reproduction, provide a baseline understanding of how an NIS can be expected to behave when colonizing a new area. A key distinction when exploring these intrinsic properties lies in the different expectations of stochastic versus deterministic processes. Quite a few chapters in this book implicitly and explicitly address this important distinction.

Buckley and Metcalf (Chapter 5) review the role of density dependence in determining the population densities that can be reached by an invasive population and the influences of those densities on invasive success. They show how density dependence can delimit demographic parameters such as intrinsic rate of increase and dispersal ability. Using microbial microcosms, Warren *et al.* (Chapter 16) show that chance events play a key role in the establishment of an invader. Initial demographic stochasticity, propagule pressure, and the initial rate of increase all contribute to whether a population is able to establish in a community, but not necessarily persist in that community. Species-specific effects were also found to be important in both establishment and persistence, which indicates that careful characterization of the demographic qualities of an invading species may be important to understanding its potential impact.

What population-scale patterns do lead to persistence over time? Murrel (Chapter 7) takes a 'plant-eye view' in using simulations to show how the density and aggregation of a population can create a 'neighborhood' effect that determines ultimate population growth. This approach could be quite important in plant invasions, as intra-specific competition can determine ultimate viability and spread of a population. After the population is established, its dispersal becomes a crucial issue in anticipating the extent and impact of an invasion (Puth and Post 2005). Lewis *et al.* (Chapter 8) demonstrate a novel approach to modeling dispersal through two dimensions. Using kernel methods to incorporate long-distance dispersal into models and two-dimensional models to correct for biases in more commonly used one-dimensional models, a more accurate prediction of the dispersal behavior of an invading population can be made.

Metapopulation theory also emphasizes the role of dispersal in maintaining NIS persistence over time. Harding *et al.* (Chapter 17) show that for single NIS colonizing patchy environments, colonization success into empty patches must be greater than extinctions. This implies that species with strong dispersal may be predisposed to colonize. Combined with buffering life history traits, such as plants that can self-pollinate or vegetatively reproduce, a good dispersal mechanism can lead to successful colonization of a heterogeneous landscape.

Environment influencing NIS populations

Intrinsic population behaviors are clearly crucial to any understanding of the establishment and spread of NIS, but as with many ecological studies of populations, the behaviors of the population can depend on the biotic and abiotic components of the environment (Lodge 1993, Dukes and Mooney 1999, Holway *et al.* 2002). Incorporating the additional level of community complexity to population dynamics, however, can be daunting when determining the impact of NIS (Davis *et al.* 2000). The study of the establishment and spread of NIS can provide insight into how these more complex relationships develop, lending important case studies to the analysis of ecological communities. The complex and complicated dynamics of the environment also mean that these additional components can move problems from less realistic but tractable problems to intractable ones. Therefore, it is important to use studies of environmental influences on NIS to determine which approaches may be appropriate and which an inefficient use of time and resources.

A classic concept of how an environment can dictate whether or not a species can persist is that of the ecological niche. The idea that an environment is parsed into a finite set of niches has been an important one for the development of invasions biology (Cadotte Chapter 2). Vazquez (Chapter 14) uses a series of meta-analyses to explore how niche breadth and invasion success relate. When looking at fundamental niches (where species requirements are independent of community interactions) versus realized niches (where the suitability of an environment takes into account competition), the realized niche ultimately determines the success of an NIS. Vazquez points out that realized niches, however, are impossible to measure and warns against pursuing straightforward investigation into niche-based predictions of potential NIS success. This marks a powerful recommendation for invasions biology as a traditional line of inquiry into predicting the establishment of NIS has been the susceptibility of different habitats to invasions (Holway 1998, Levine and D'Antonio 1999).

In other circumstances, invasions can shed light on conceptual topics that are difficult to assess in natural communities. Duncan and Forsyth (Chapter 18) use historical records of bird invasions to test classic hypotheses of competition and community structure. Invasion records offer a unique way of discovering if invasions of bird species into certain types of environments are more likely to persist. They find that priority effects do occur, and that depending on circumstances, stochastic processes and deterministic processes (such as competition) can determine ultimate community composition. Sax and Gaines (Chapter 20) suggest that species rich communities are likely to be more susceptible to invasions (because conditions are favorable for many species). They argue, however, that this is true only up to a critical threshold (such as highly speciose tropical rain forests) at which point Elton's (Elton 1958) model, in which species rich communities have less available niche space unexploited by the best competitor, become applicable. Both Sax and Gaines (Chapter 20) and Smith and Shurin (Chapter 19) explain that heterogeneity within regional scales can mean that multiple processes (stochastic and deterministic) and multiple outcomes (invasion resistance and susceptibility) can be found in any one habitat. This indicates that even in unsaturated communities (where presumably there is ample niche space for well-dispersed invaders), invasion success is contingent on other factors. This role of contingency and complexity is becoming a common theme in both invasions literature and the empirical studies of community composition and assembly, and therefore points to an important area of reciprocal research.

Another example of the challenges inherent in complex community dynamics is illustrated in Harding *et al.* (Chapter 17). Although, as mentioned above, they show that metapopulations without competition rely on a simple strategy of colonization and buffers from extinction, adding even one more species into the model can radically change the number of population parameters and life-history traits that may ultimately dictate the successful establishment and persistence of an NIS. This complexity of potential outcomes indicates that metapopulation theory, as applied to multiple species living in subdivided habitats might be well applied to case studies, experiments, and management problems that offer specific species characteristics and robust estimation of population parameters. With such a meta-analysis, it may be possible to determine whether a metapopulation model can accurately describe and predict certain types of invasions.

Using microbial microcosms, Warren *et al.* (Chapter 16) reinforce this conclusion. In a series of ambitious experiments where different species are introduced into established communities, they show that predicting invasions can depend on a number of characteristics of the species involved as well as the community characteristics. Specifically, the intrinsic rate of increase (a simple population parameter) should be thought of as having as much to do with the interaction (contact experience) between an invader and a community as intrinsic properties of the invading population.

Demographic parameters (like the intrinsic rate of increase) themselves are not constant. These parameters can change stochastically due to both intrinsic demographic factors, such as random changes in birth or death rates. They can also change due to environmental stochasticity. Freckleton *et al.* (Chapter 6) distinguish demographic versus environmental stochasticity at different stages of an invasion (arrival, establishment, and spread). They show with mathematical models that stochastic processes influence these stages of the invasion in different ways and with different consequences. Stochasticity may be more important in phases with low densities (arrival and establishment), while density dependent behaviors may hold a greater influence on the spread of already established populations. Again, models like this can indicate when and how to proceed with an experimental program to test stochastic and deterministic components of invasion success.

Although direct agonistic interactions between species, such as competition and predation, have long been held as central to understanding the population structure of communities, indirect effects and mutualisms have recently been thrust into the forefront of thinking on a number of conceptual themes in ecology (Wootton 1994, Lortie *et al.* 2004), such as richness-productivity, population viability analysis, and invasions (Lach 2003). Vazquez (Chapter 14) shows in his meta-analyses of niche breath and invasions that the presence of mutualists and other niche categories that may be tricky to measure can have an important influence on the ultimate viability of a population in a new habitat. Thorpe and Callaway (Chapter 15) extend the discussion of mutualisms to the level of indirect mutualist interactions and feedback loops. Good experimental evidence has shown that presence of soil pathogens decrease NIS plant success in native range and absence of those pathogens increases NIS success in new range. Further, mutualist interactions can more easily create positive feedbacks in a new range, enhancing plant success. This conceptual approach to invasions might be well suited for a mathematical study, as feedbacks are both common in nature, yet difficult to tease apart from other community effects.

In addition to using analyses of environmental contexts to predict whether an NIS will successfully establish and persist in a novel environment, Gilbert and Parker (Chapter 13) and Satake *et al.* (Chapter 12) address how the environment can prevent an invasion or even reverse an already persistent NIS. Gilbert and Parker argue that the understudied role that pathogens play in plant population regulation can have important consequences in understanding, predicting, and mitigating or reversing high-density persistence of NIS. Satake *et al.* (Chapter 12) model 'classic' mast seeding, and show that when seed set is negatively correlated in time but positively correlated across space a system is most resistant to invading seed predators.

Phenotypic and genotypic change

A new and exciting theme in invasions biology addresses how populations of NIS may change in phenotypic expression (Schweitzer and Larson 1999) or genotype (Sakai *et al.* 2001) due to the novel environments they experience in their new range: this includes population changes inherent in a colonizing event such as founder effects or drift (Palumbi 2002, Cox 2004). Just as mutations in a virus within human populations can mean the difference between rarity and a pandemic, so too changes in NIS populations can turn a rare or isolated population into a pest. This can also be one of the major factors in creating create time lags between the arrival phase of an NIS and an outbreak (Mack *et al.* 2000, Byers and Goldwasser 2001, Memmott *et al.* 2004). New methods in genetics, life-history evolution, and coevolutionary theory can provide insight into and draw inference from the growing examples of NIS populations that have changed fundamentally since their arrival in a new range.

Evolutionary change in NIS populations can follow many possible pathways. Schierenbeck and Aïnouche (Chapter 9) emphasize that pre-adapted generalist genotypes will facilitate invasion success. Substantial genetic variation can provide the requisite genetic material for rapid evolutionary response to the novel environment, and Shierenbeck and Aïnouche (Chapter 9) point out that there are many pathways to this variation (heterozygosity, polyploidy, and perhaps now even extra-genomic genetic information (Lalle *et al.* 2005)). However, general genetic diversity may not alone be enough to influence invasion success (Herben *et al.* 2004). More importantly, how labile life-history traits are (phenology, clonal or vegetative reproduction in plants, mating behaviors, etc.) may mark the important genetic distinction. Buckley and Metcalf (Chapter 5) describe how life-history evolution due to unique forces at work in novel environments (different life-history strategies) may benefit NIS populations.

While life-history evolution may be responsible for rare large outbreaks, this would be difficult to test. Thorpe and Callaway (Chapter 15) explain that evolutionary response of microbial communities can happen so fast that this may mute or mitigate an NIS from long term dominance through positive soil feedback pathways. Because of a potentially rapid response and the short generation times of microbial communities, it may be possible to test soil systems for rapid evolutionary response to encouraging or mitigating the successful establishment of an NIS. NIS colonization, additionally, may provide ideal systems to research the stability of microbial communities as well as their relationship to the plants with which they interact through belowground pathways (Wardle 2001). Pathogens would also fit this model, and Gilbert and Parker (Chapter 13) describe the coevolution of plant-pathogen systems and maternally inherited induced resistance. The evolutionary response of an NIS to the biotic community may be crucial not only in the establishment, but in the resulting effect on the native biota. Kondoh (Chapter 10) models how contact experience, the previous evolutionary filtering from prior interactions between an NIS and related novel competitors, predators, and prey, can result in successful invasion. He describes how the contact experience, in many forms, can influence the specific components of contact between the NIS and novel biota as information retrieval (detecting the other species), processing (recognizing characteristics of that novel species), and behavioral response (being able to prey upon, escape, or compete with that species).

NIS populations influencing populations and environment

It should be noted that even identifying an invasive species can be problematic (Colautti and MacIsaac 2004, Murphy *et al.* Chapter 4). But regardless of definition, there is a general understanding among biologists that ultimately any invasive will in some key way be defined by its influence on the biotic and abiotic components of its new range. There may be millions of NIS propagules entering complex environments, but our concern for their range expansion is ultimately a function of their impact on other species in their new range, the ways in which they affect ecosystem processes, and how these changes may persist or increase over time.

Historically documented invasions can also provide insights into the pathways through which the context of and invading organisms can influence both persistence and the concomitant influence on native biota. Courchamp and Caut (Chapter 11) used models incorporating direct and indirect interactions to explain invasions, as well as show how the invasions themselves illustrate the complex and sometimes counterintuitive behaviors of interacting populations. Complex dynamics between populations may also mean that after an invasion it takes many generations for the effects of the new interactions to show their impact (Freckleton *et al.* Chapter 6).

Although documented changes in native biota or ecosystem function may be clear after an invasive becomes a damaging pest, predicting the ultimate impact of an NIS on native habitats is difficult. A number of factors, intrinsic to the invading population and due to the novel environment, can elicit complex responses to the non-native populations. Indirect effects on the environment are difficult to detect, let alone predict, yet may be among the more common and serious responses to NIS. Invasive plants can alter the soil nutrient feedback loops and render long term changes to the soil environment (Thorpe and Callaway Chapter 15). Mutualisms, such as pollination services, can be disrupted by invasives. Though the converse may occur. Non-native bees have been shown to increase pollination efficiency of some plants (Richardson *et al.* 2000), although the effect on native pollinator communities adds additional uncertainty to the diagnosis of these effects. Generalism versus specialism can be crucial to the extent of influence a non-native species may have in pathogen-host relationships (Gilbert and Parker, Chapter 13), pollinator systems (Vazquez, Chapter 14), and soil mutualisms (Thorpe and Callaway Chapter 15). The symmetry of these mutualist relationships is of further consequence (Vazquez, Chapter 14) when inferring the potential impact an NIS may have on novel systems.

Whether direct or indirect, the stochastic nature of population interactions can radically affect the ultimate consequences of a persistent NIS (Freckleton *et al.* Chapter 5, Warren *et al.* Chapter 16) especially as systems may contain non-linear dynamics. A final level of complexity results from the many applications of our inference to political, social, and economic issues. It is broadly accepted that NIS can affect the inherent richness of native biota, specific endangered species, the ecosystem function of native habitats, as well as biota of economic importance. In their review of how NIS influence novel communities, Murphy *et al.* (Chapter 4) found a correlation between level of invasive on the SE scale and community impact, there was no correlation with economic impact. This may trouble our ability to apply a level of community change due to an invasive to some system of social or economic 'importance' of the invader.

CONCLUSION

Invasions biology is faced with many of the same struggles that have long faced mainstream ecology. Organisms show an array of intrinsic qualities that are both indicative of their population dynamics as well as responsible for those dynamics. Yet populations of organisms respond and influence their biotic and abiotic contexts, and they adapt phenotypically and genotypically to them. The problem of filtering that which is broadly important to populations from that which is idiosyncratic is highlighted in the study of NIS. We certainly need to develop a general understanding of the qualities of populations and habitats that can lead to invasion persistence and impact, yet we often find that the causes of any specific invasion can be unique to that specific species or habitat. A major goal of this book is to carefully assess when simplicity is informative (e.g., a population-level analysis of density dependence), when complexity is instrumental (such as when abiotic factors influence NIS persistence), and when generalization is not likely possible (such as the factors that make up the realized niche of an NIS).

One of the greatest challenges facing the study of invasions biology arises from species-specific characteristics. As is often shown in this book, a better understanding of the interface between intrinsic population properties and the stochastic context of a population is likely to provide important insights into NIS establishment and expansion. Frustrating many straightforward paths to invasion prediction is the combination of stochastic and deterministic processes that influence populations behaviors (Dennis *et al.* 2003). Mathematical models may help indicate which parameters under which assumptions can show a definitive influence on population dynamics. Thus, when entering an observational or manipulative experimental program, the theoretical literature can prove an important guide to developing hypotheses and testing them.

Reflecting recent conceptual advances in general ecological theory, often ignored processes such as mutualisms, intra-specific competition, and below ground processes are also likely to influence the potential impact of NIS (Callaway and Aschehoug 2000, Richardson *et al.* 2000, Parker and Haubensak 2002, Callaway *et al.* 2004) and mark potential pathways to bridge what appears to be idiosyncratic to more generalizable patterns. In these areas, the study of NIS may be well suited to inform our greater understanding of such complex and often masked mechanisms in general. Thus, not only is there a reciprocal interest in understanding traditional problems in ecology and invasions biology, these two approaches to natural systems should interact and inform each other as they advance in techniques, concepts, and understanding.

REFERENCES

Byers, J. E. and L. Goldwasser. 2001. Exposing the mechanism and timing of impact of nonindigenous species on native species. Ecology **82**, 1330-1343.

Cadotte, M. W., B. Murray and J. Lovett-Doust. 2005. Using database analyses to test widespread ecological generalities in invasions biology. Biological Invasions **in press**.

Callaway, R. M. and E. T. Aschehoug. 2000. Invasive plants versus their new and old neighbors: A mechanism for exotic invasion. Science **290**, 521-523.

Callaway, R. M., G. C. Thelen, A. Rodriguez and W. E. Holben. 2004. Soil biota and exotic plant invasion. Nature **427**, 731-733.

Colautti, R. I. and H. J. MacIsaac. 2004. A neutral terminology to define 'invasive' species. Diversity and Distributions **10**, 135-141.

Davis, M. A., J. P. Grime and K. Thompson. 2000. Fluctuating resources in plant communities: a general theory of invasibility. Journal of Ecology **88**, 528-534.

Davis, M. A., K. Thompson and J. P. Grime. 2001. Charles S. Elton and the dissociation of invasion ecology from the rest of ecology. Diversity and Distributions **7**, 97-102.

Dennis, B., R. A. Besharnais, J. M. Cushing, S. M. Henson and R. F. Constantino. 2003. Can noise induce chaos? OIKOS **102**, 329-339.

Dick, J. T. A. and D. Platvoet. 2000. Invading predatory crustacean Dikerogammarus villosus eliminates bath native and exotic species. Proceedings of the Royal Society of London Series B-Biological Sciences **267**, 977-983.

Dukes, J. S. 2002. Species composition and diversity affect grassland susceptibility and response to invasion. Ecological Applications **12**, 602-617.

Dukes, J. S. and H. A. Mooney. 1999. Does global change increase the success of biological invaders? Trends in Ecology & Evolution **14**, 135-139.

Elton, C. S. 1958. The ecology of invasions by animals and plants. Methuen and Co. LTD., London, UK.

Hamilton, M. A., B. R. Murray, M. W. Cadotte, G. C. Mose, A. C. Baker, C. J. Harris and D. Licari. 2005. Life-history correlates of plant invasiveness at regional and continental scales. Ecology Letters **8**, 1066-1074.

Herben, T., B. Mandak, K. Bimova and Z. Munzbergova. 2004. Invasibility and species richness of a community: A neutral model and a survey of published data. Ecology **85**, 3223-3233.

Holway, D. A. 1998. Factors governing rate of invasion: a natural experiment using Argentine ants. Oecologia **115**, 206-212.

Holway, D. A., A. V. Suarez and T. J. Case. 2002. Role of abiotic factors in governing susceptibility to invasion: A test with argentine ants. Ecology **83**, 1610-1619.

Lach, L. 2003. Invasive ants: Unwanted partners in ant-plant interactions? Annals of the Missouri Botanical Garden **90**, 91-108.

Levine, J. M. and C. M. D'Antonio. 1999. Elton revisited: A review of evidence linking diversity and invasibility. OIKOS **87**, 15-26.

Lloret, F., F. Medail, G. Brundu and P. E. Hulme. 2004. Local and regional abundance of exotic plant species on Mediterranean islands: are species traits important? Global Ecology and Biogeography **13**, 37-45.

Lodge, D. M. 1993. Biological Invasions – Lessons for Ecology. Trends in Ecology & Evolution **8**, 133-137.

Lortie, C. J., R. W. Brooker, P. Choler, Z. Kikvidze, R. Michalet, F. I. Pugnaire and R. M. Callaway. 2004. Rethinking plant community theory. Oikos **107**, 433-438.

Mack, R. N., D. Simberloff, W. M. Lonsdale, H. Evans, M. Clout and F. A. Bazzaz. 2000. Biotic invasions: Causes, epidemiology, global consequences and control. Ecological Applications **10**, 689-710.

Memmott, J., N. M. Waser and M. V. Price. 2004. Tolerance of pollination networks to species extinctions. Proceedings of the Royal Society of London Series B-Biological Sciences **271**, 2605-2611.

Parker, I. M. and K. A. Haubensak. 2002. Comparative pollinator limitation of two non-native shrubs: do mutualisms influence invasions? Oecologia **130**, 250-258.

Parker, I. M., D. Simberloff, W. M. Lonsdale, K. Goodell, M. Wonham, P. M. Kareiva, M. H. Williamson, B. Von Holle, P. B. Moyle, J. E. Byers and L. Goldwasser. 1999. Impact: toward a framework for understanding the ecological effects of invaders. Biological Invasions **1**.

Petrovskii, S., A. Morozov and B. L. Li. 2005. Regimes of biological invasion in a predator-prey system with the Allee effect. Bulletin of Mathematical Biology **67**, 637-661.

Puth, L. M. and D. M. Post. 2005. Studying invasion: have we missed the boat? Ecology Letters **8**, 715-721.

Richardson, D. M., N. Allsopp, C. M. D'Antonio, S. J. Milton and M. Rejmanek. 2000. Plant invasions – the role of mutualisms. Biological Reviews **75**, 65-93.

Sakai, A. K., F. W. Allendorf, J. S. Holt, D. M. Lodge, J. Molofsky, K. A. With, S. Baughman, R. J. Cabin, J. E. Cohen, N. C. Ellstrand, D. E. McCauley, P. O'Neil, I. M. Parker, J. N. Thompson and S. G. Weller. 2001. The population biology of invasive species. Annual Review of Ecology and Systematics **32**, 305-332.

Sax, D. F., J. J. Stachowicz and S. D. Gaines, editors. 2005. Species Invasions: Insights into Ecology, Evolution and Biogeography. Sinauer Associates, Sunderland, MA.

Schweitzer, J. A. and K. C. Larson. 1999. Greater morphological plasticity of exotic honeysuckle species may make them better invaders than native species. Journal of the Torrey Botanical Society **126**, 15-23.

Shea, K. and P. Chesson. 2002. Community ecology theory as a framework for biological invasions. Trends in Ecology & Evolution **17**, 170-176.

Thomson, D. M. 2005. Matrix models as a tool for understanding invasive plant and native plant interactions. Conservation Biology **19**, 917-928.

Tilman, D. 1997. Community invasibility, recruitment limitation and grassland biodiversity. Ecology **78**, 81-92.

Wootton, J. T. 1994. The Nature and Consequences of Indirect Effects in Ecological Communities. Annual Review of Ecology and Systematics **25**, 443-466.

Chapter two

Darwin to Elton: early ecology and the problem of invasive species

M. W. Cadotte

INTRODUCTION

A number of prominent modern authors have perceived a chasm separating mainstream ecology and invasion biology (e.g., Davis *et al.* 2001). Introspection into the nature of invasion biology could be seen as part of a larger enterprise of prescriptive philosophy within ecology (e.g., Egler 1951; Peters 1991; Lawton 1999). These types of introspection are not science *per se*, that is they are not trying to answer questions through deduction and empiricism. Rather, introspection is about the epistemology of science, or which ideas and concepts are valuable and what knowledge is important — basically HOW science is done. Epistemology is no trivial matter, what questions we ask and how we come to conclusions are under the purview of epistemology. This chapter is about the epistemology of ecological research and invasion biology.

Recent calls for ecology and invasion biology to reintegrate (Johnstone 1986; Huston 1994; Thompson *et al.* 1995, Davis *et al.* 2001) are based upon a perceived dissociation instigated by Elton's (1958) magisterial book on invasion biology. These calls, I believe, rightly draw attention to the potential fertile ground that lay between ecology and invasion biology. However, if Elton caused the dissociation

M.W. Cadotte et al. (eds.), Conceptual ecology and invasion biology, 15–33.

(Davis *et al.* 2001) then what was the prior state of the relationship between ecology and non-indigenous species (NIS)? I will examine how early ecologists dealt with non-native species in their writings, especially at the time when ecology became its own field of science (late 1800's to early 1900's).

ECOLOGICAL SCIENCE AT THE BEGINNING

Even though many great thinkers thought about why species occur where they do (e.g., Aristotle, Paley, Agassiz, de Candolle) it wasn't until the mid 1800's that these types of investigations moved beyond metaphysical accounts and began to use, what could be referred to as modern scientific epistemology. Early 'ecology' was typified by natural history observations (Egerton 1976), but the contribution of early workers cannot be undervalued, simply because scientific epistemology was not an independent branch of knowledge. That said, the new science that was to emerge in the mid to late 1800's would eclipse, in a matter of decades, the previous 2000 years of thought on the origin and distribution of species. Before there could be a science of ecology, the ultimate causes of species diversity and distributions need to be articulated in a way that would remove non-naturalistic explanations. What follows through the rest of this essay is not meant to be a exhaustive summary of early ecology — for that see Egerton 1976 and McIntosh 1976 — rather I will use selected examples to examine how early ecology dealt with NIS.

Early, pre-Darwinian ecology was imbued with notions of 'balance of nature' (Egerton 1976), and so changes in the abundance and distribution of species were not prominent in early writings. The science of species distributions was largely conceived in the 18th and 19th centuries in the work of Buffon, de Candolle, and Darwin, and strong explanatory science did not emerge until Charles Darwin's *The Origin of the Species* in 1859. Though some of Darwin's contemporaries did not recognize *The Origin* as 'scientific', on the grounds that it was not inductive enough (Owens 1860; Sedgwick 1860), Darwin's conceptual leaps and the linking of formerly disparate pieces of information resemble modern science more than anything that predates Darwin.

In trying to explain patterns of species distributions and abundance, Darwin was faced with the problem of NIS. How did Darwin assimilate such an apparent fact, that species readily move from their native range? Darwin cleverly used non-indigenous species as a device to test or illustrate his theory of natural selection and descent with modification. NIS are species locally coexisting with other species, yet independent of these local assemblages with their own evolutionary histories. Therefore Darwin saw the presence of NIS as efficacious examples of natural selection in already distinct forms.

Darwin notes that an important aspect of the struggle for existence is that species naturally have a propensity to increase at high rates, so the struggle is an issue of population processes. He mentions that, when domestic animals are released, their populations often rapidly increase. Furthermore "so is it with plants" that there are "introduced plants which have become common throughout whole islands in a period of less than ten years." (p. 118). The explanation for this is "that the conditions of life have been very favourable, and that there has consequently been less destruction of the young and old" (p. 118). Essentially, what better way to see the potential high rates of increase than in species arriving in places where they didn't exist before?

Isolation is a fundamental process in ecology and evolutionary biology and Darwin reminds the reader of this, that on isolated islands "let it be remembered how powerful the influence of a single introduced tree or mammal" (p. 131). Adaptation is not absolute, it is relative to species that exist together and that barriers to immigration that isolate habitats "into which new and better adapted forms could not freely enter" and that "had the area been open to immigration, these same places would hence been seized on by intruders" (p. 131). There is then always room for improvement, and "we may safely concluded that the natives might have been modified with advantage, so as to have better resisted such intruders" (p. 132). For example, " The endemic productions of New Zealand, are perfect one compared with another; but they are now rapidly yielding before the advancing legions of plants and animals introduced from Europe" (p. 229).

Whereas species differences are slight compared to those among genera, Darwin expected, as Alphonse de Candolle (1855) suggested, "that floras gain by naturalisation ... proportionally ... far more in new genera than in new species" (Darwin 1859, p. 158, see Daehler 2001).

INVASIVE SPECIES AND ECOLOGY BEFORE THE 20TH CENTURY

By the early 20th century, ecology was a science and exotic species were a concern, but these two things did not seem to influence one another. For one thing, modern calls for ecologists to work towards saving the natural world (e.g., Peters 1991; Shrader-Frechette 2001; Ehrlich 2002) simply did not happen until the mid 20th century (see Davis, Chapter 3). Incipient modernism was surfacing, and modern civilization and her science was quickly being regarded as the only valid knowledge-acquiring method and was neutral to, or above morality and ethics. In America this period saw unprecedented change in higher education, and the pre-existing emphasis on theology and philosophy was being replaced with science and psychology curricula with many institutions, departments and endowments created promoting the training of professional scientists at postsecondary institutions (Croker 2001; Menand 2002). This was the time period when great American thinkers, such as William James and John Dewey, would cull out the last remaining notions of supernatural explanation in American Acedemia.

But all was not well, people were noticing that modern civilization was having undesirable impacts on the natural world. In 1793, a devastating yellow fever outbreak in Philadelphia, likely imported from the West Indies, had people fearing the accedental importation of pathogenic NIS (Egerton 1976). Ralph

Waldo Emerson had already been pleading to the American public to rethink its unhealthy relationship with nature (Emerson 1959). The impact of certain invaders was becoming apparent. Eighteenth century botanists, such as John Bartram, were starting to comment on the fact that some introduced plants were having negative impacts, and were difficult to control (see Mack 2003). Hooker (1867) noticed that European species were quickly replacing natives in New Zealand.

Some of the earliest concerns about exotic species were in terms of the maintenance of the 'natural', based on a morality of aesthetics (e.g., Olmstead 1888; Sargent 1888; both cited and discussed by Simberloff 2003). Soon after a more (distinctly American) pragmatic approach recognized the economic, agricultural and ecological effects by non-indigenous species (e.g., Fitch 1861, Howard 1897a). In his address to the American Association for the Advancement of Science, Leland Howard says: "Potentially cosmopolitan forms, ... have by (human) agency become dispersed over nearly all of the civilized parts of the globe, while thousands of other species have been carried thousands of miles from their native homes, and have established themselves and flourished, often with a new vigor, in a new soil and with a novel environment" (Howard 1897a, p. 4).

Howard published more than 10 articles on non-native insect pests and the realized or potential impacts of introduced species (e.g., Howard 189?, 1893, 1895, 1897a, b, 1898, 1899). Howard's peer and friend, Stephen Forbes — the father of American ecology — also became professionally dedicated to the effects that insect pests posed for agriculture (Croker 2001). Forbes wrote many articles on the biology and impacts of insect pests (e.g., Forbes 1883, 1886, 1887a, 1897, 1898). By the turn of the century we have legitimate, non-aesthetic concerns about the importation of novel species. Of course, much of this concern was in terms of agricultural and economic costs. During this time ecology was developing into its own science (Egerton 1976, McIntosh 1976), and how did this science of ecology address these concerns about invasive species?

During this period the development of the science of ecology was largely a European phenomenon, without a uniquely American school until shortly after the turn of the century (McIntosh 1976). Of course there was Forbes, who wrote several papers before the 20th century, which we should call 'ecological', especially his famous *Lake as a Microcosm* talk delivered to the Peoria Scientific Association (Forbes 1887b). Here he stressed the interaction among species and a balance of nature view, but we cannot call this a 'school' as little subsequent work was done, spawning little research and no textbooks. Forbes and his contemporaries (Henry Cowles and Leland Howard) could be said to have started applied ecology. It is curious, then, that on the one hand Forbes worked extensively with insect pests and exotic agricultural diseases, and yet maintained a closed-system, balance of nature approach in his ecological writings. The fact that a single individual, recognizing the importance of pests and understanding ecological systems, would keep these ideas dissociated is prophetic for the rest of early ecology.

Ecology grew out of the apparent need to separate causes in whole-organism biology. Whereas Darwin's theory explained the ultimate causes and patterns of why we find species where we do (e.g., why cacti are diverse in desert regions and the purpose of the panda's thumb), it did not include a strong framework to account for proximate causes and patterns (e.g., cyclic population dynamics and patterns of abundances). Early forays into the ultimate causes of species distributions featured geographical and geological causes of species distributions (e.g., Wallace 1876). Wallace (1876) observed that biotas were large regional entities, often with fairly discrete boundaries. Common was the idea of barriers to movement, with all species being able to move to new locals, but rarely doing so (e.g., Heilprin 1887). Writings of de Candolle and Darwin certainly address issues of proximate patterns, but it took a separate field to create their own proximate explanations.

The biogeographer Oscar Drude was able to separate these causes and patterns in his *Manual de Geographie Botanique* (1896). He noted that there were two forces of change in species ranges and abundance: geological and biological. Drude rightly thought that all ultimate change happened in "cours de l'évolution geologique" (p. 81) — the course of geological evolution. This included adaptation, as well as constant change, but he also saw the roles of the organism-environment interface, dispersal, and competition. The distinction between geological and biological causes was important for Drude's understanding of rapid changes in species distributions. Drude classifies limits to species ranges as either geographical, that is mountains, oceans, and the like, arising from geological processes, and biological, by which he means chance dispersal and the agency of humans. ("Les limits des aires sont purement géographiques, ou bien résultent de l'action combinée des conditions de vie" p. 87). Drude says that without the aid of humans, geographical barriers limit species spread, but that rare chance events still play an important role for range expansion. With human aid, species will spread as far as the external environment allows. Drude tells the reader that moving to a new location, which must be climatically similar to that in a species' native range, is only half the equation, and once a plant has arrived, "elle lutte avec succes contre une plante indigène tout-à-fait adaptée au climat" (p. 82) — it has to be successful in the struggle against indigenous plants perfectly adapted to the climate. Given this limitation, plants still come to "couvrir de grandes surfaces" (p. 85), and in covering these large areas, displace other species. He goes even further to say that some species do not dominate one another, and competition is indeterminate. In these species the ranges are set up by whoever arrives there first.

Drude's account sounds very modern. His biogeography incorporates geography, chance, community assembly, and local competition. He also has a system of proximate causes and patterns that readily accounts for the dynamics of invasive species. However, in reading the chapter on range expansion in Drude (1896), one gets the impression that he is trying to explain the fact of NIS spread in light of late-nineteenth century knowledge of biogeography, rather than trying to understand invasions and specific invaders. This is not a criticism, as he was developing a system of knowledge about the natural world, and his epistemology called for explaining extant patterns within the existing framework.

Another European ecologist, Eugene Warming, wrote an influential ecology book before the turn of the century. The English translation of his 1895 *Oecology of Plants* (1909) was concerned with population and community ecology, and he tried to develop a system of ecological understanding. Warming was very much interested in the process of ecological change, and this change included the arrival of new species. He believed that ecological processes, including change, were realized at the community level. For example he believed that successional change resulted from competition among community types, affected by changes in the environment. Essentially, Warming's change is assemblages of species moving among equilibria, but what was his mechanism for these changes?

Warming says that there are two types of community change, first is from 'changes in soil' and the other is change 'without new soil'. In the first type of change, disturbances, both natural and human caused, are important and abrupt. Slower types of change include increases or decreases in water content or by the accumulation of soil. The second type of change happens through migrations and struggles among individual species.

This second type of community change, Warming says, is atypical, "the struggles in question are rare" and that countries "whose vegetation is not appreciably disturbed by man or animals, and which has been exposed for long ages past to immigration from adjoining countries, in this case a certain equilibrium must have been attained" (p. 364). This equilibrium view means that the problem of invasive species is one primarily associated with disturbances — which is largely true. However his view may have some difficultly explaining why some species invade intact communities. He concedes this point on the next page, "some changes in the nature of forests may have been simply due to the entry of new species" (p. 365).

Warming does make an emotional plea to be weary of importing new species, because of the competition to which they may subject native forms: "a new ... community which did not naturally occur before the soil had been drawn into the service of Man; and the species which now occur in vast numbers, and form a community with its own special stamp and economy" (p. 82-83).

Warming's system makes predictions as to what types of communities ought to be more invaded, and our modern understanding of the invasion process would also predict that intact communities are less invadable than disturbed ones. However, both Warming and Drude were not explicitly trying to understand invasions; rather they were trying to fit the fact that we see invaders into their respective ecological systems. For them there was no real association between developing ecological understanding and addressing the problem of invasive species. They felt that, if their research programmes explained the general phenomenon of NIS, then their ecology was robust. The relationship between ecology and invaders was limited and one-way. At most, invaders were tests of early ecologist's ideas or at worst phenomena needing explanation.

THE NATURE OF CHANGE: CLEMENTS AND THE AMERICAN SCHOOL

Shortly after the turn of the century, an American school, embodied in Frederic Clements, come to the fore (McIntosh 1976). I say that Clements embodied the school because he did all the things that make a successful school of thought. He interacted with a number of students and colleagues, he wrote numerous successful textbooks, and most importantly he emphasized a strong heuristic paradigm in much of his writing — namely, the nature of ecological change.

Clements' first two books really laid the groundwork for this new school. His first, *Research Methods in Ecology* (1905), detailed how the science of ecology should be done, what techniques and research topics were most fruitful. This was soon followed by a second book, *Plant Physiology and Ecology* (1907), which had much overlap with his 1905 book, but reinforced which areas of research were most important and the paradigm of plant ecology. Specifically, these books had an unprecedented focus on ecological change, or what became known as succession. Intuitively, such a paradigm seems like NIS would be of central concern.

Both books define invasion almost identically, as "The movement of one or more plants from one area into another and their establishment in the latter (1907, p. 270, also a very similar quote in 1905, p. 210). He notes "Invasion is a regular occurrence between adjacent formations, but it also takes place into remote ones, as a result of long carriage by wind, water, birds, railroads, or vessels." (1907, p. 270). He tells us that dispersal is usually constrained by barriers, but that humans, as an agent of dispersal, are able to transcend these barriers. Invasions are important for a number of reasons. First, invasions can significantly affect species distributions. Secondly, ecological change depends on invasions. Clements defines succession as a "series of invasions ...in the same spot" (1905, p. 239; 1907, p. 273).

Of course, several factors work against the success of an invasion. First of all there are the well-known geographical barriers to migration (mountains, oceans, etc.), but more importantly Clements details other types of barriers. There is an abiotic barrier — the species must find itself in an environment in which it can survive. Many biological forces work against species establishment, including predation, disease, competition, and humans (trying to control weeds). Also, Clements finds that whether the habitat is full or not is of paramount importance to invasion success, as competition from full communities will be too severe for the colonizer to succeed.

This view of ecological succession is common in the writings of Clements' colleagues and collaborators. Shantz (1905) saw invasions as the process of succession, and succession as the process of changing formations or community types. The idea of species invasions as the force behind succession is also dominant in Weaver and Clements (1938). Humanity's role is examined in greater detail, and they notice that anthropogenic effects on community succession can be severe, as we both disturb habitats and bring in new individuals. They note that the motor of succession is not disturbance, but rather biotic processes

— migration and competition — and that climax communities often resist these biotic factors. Clements and Shelford (1938) clarified who are called invaders. Specifically, invaders are those species that did not previously exist in the habitat they are invading. Previously Clements did not explicitly stipulate whether the invaders had to be separate species. Here Clements and Shelford (1938) make no distinction about where species come from, in terms of their native or introduced status. Earlier, Clements (1907) did create terms to reflect species that migrate from nearby versus those originating further away. Yet none of these works explicitly examines NIS, just ecological processes of idealized species. Again like the other early ecologists, Clements and colleagues were interested in establishing a strong programme that *could potentially* explain the occurrence of NIS.

EXPLAINING SPECIES OCCURRENCES WITH THE NICHE

I continue by taking a brief tour of ecology up to the 1950's. During this period primary publications (i.e., academic journal articles) were becoming increasingly important for the development and dissemination of new ideas. One of the most important conceptual developments in early ecology was the idea of the niche. Through understanding a species niche, one could potentially understand and predict where and when species is found, and which species are likely to replace one another.

We could probably trace the niche concept all the way back to Darwin, but the idea of examining the distribution of species through a species' suite of environmental needs is commonly attributed to Grinnell (1917). Grinnell's niche concept has been an important theme in the development of ecology. His premise was that patterns of species occurrence are best understood through their abiotic and biotic requirements. Hutchinson (1957) famously used this niche concept to advanced the notion of a multidimensional niche space defining where each species could potentially live. Hutchinson's concept could potentially explain where species could live and therefore where they could invade, as each species potentially occupies a unique region of the nth dimensional niche hyperspace. The definition of 'niche' has been recognized as problematic (Simberloff 1978; Conner and Simberloff 1979) and open to various interpretations (Leibold 1995; Chase and Leibold 2003). An early reinterpretation was provided by Elton (1927), who saw the niche as a species 'status' in the community. This included the role an organism played and most importantly its relationship to other species and especially the trophic status of a species. Elton's invasions book (1958) actually used both concepts of the niche to understand when and where NIS invade and which communities were more likely to be invaded. Only recently had there been an attempt at reconciling these two major conceptualizations of the niche concept (Leibold 1995, Chase and Leibold 2003).

Since Grinnell's (1917) paper, research on competition has been intimately associated with the niche concept. Elton (1958), like Warming (1905), thought species-rich communities ought to be less invadable because available niches should be filled and competition for niche space more intense. Competition as a major structuring force surfaced before Elton (e.g., Gause 1934, D'Ancona 1954), with the foundation in what Gause (1934) called the 'competitive exclusion principle' (see also Hardin 1960), probably first articulated by Grinnell (1925). The central premise of this principle is that "No two species of identical or even closely similar biological predilections can long occupy the same niche ... at the same time." (Grinnell 1925, p. 621). In fact Grinnell (1925) used this reasoning to argue against the purposeful introduction by the state of Claifornia of the Hungarian partridge for hunting. Grinnell's concern was that the Hungarian partridge shared its niche with the native California quail, and that the latter would likely be competitively excluded. This prediction was based on the observation that species having large native ranges usually exclude smaller-ranged competitors. Grinnell applied ecological reasoning to predicting the outcome of a species introduction.

Gause (1934) saw the value that invasions have for understanding general ecological processes. To study competition Gause commented on the importance of experimentation and that in natural experiments "the most frequent ones consist in the transportation of animals into countries new to them, which commonly leads to a great number of highly interesting processes of the struggle for existence." (p. 13). One example highlighted by Gause is the introduction of the European carp into eastern North America in the 1870's and how it "has now in many streams and lakes multiplied to such an extent that several native species are found in greatly diminished number." (p. 21).

Much understanding of the nature of species interactions, and especially those that stimulated Gause (1934), was advanced by the theoretical models of Lotka (e.g., 1924) and Volterra (e.g., 1927). The work of Lotka and Volterra not only revolutionized how ecologists thought about species interactions, both competition and predation, but also showed how models could be used to understand applied problems. The famous instance is that the son-in-law, Umberto D'Ancona, asked his father-in-law, Vito Volterra, if it were possible to explain mathematically why it should be that with the cessation of fishing in Fiume, near Trieste during WWI, predatory sharks and rays came to constitute a higher mean proportion of the catch biomass (e.g., D'Ancona 1954; Hallam 1986). Of course, using his predator-prey models, Volterra was able to explain this increase in predator proportion. Volterra assumed that the harvesting prior to WWI was indiscriminant and resulted in a fixed proportion of individuals removed, so that the periodic equilibrium between predator and prey densities shifted to have a center with higher prey and lower predator abundance when there is density independent mortality.

In the same vein, D'Ancona (1954) showed how Lotka-Volterra-type formulations could be used to understand the dynamics of species invasions. Using these tyes of equations, D'Ancona (1954) showed that a new invader will either move to an equilibrium of zero density (i.e., unsuccessful) or change the equilibrium dynamics of the extant competitors, which essentially alters the densities of the other species.

It is worth mentioning that the Lotka-Volterra equations have been criticized, with a number of shortcomings recognized, namely they are difficult to apply in specific circumstances and the interaction coefficients are problematic to compute (Heck 1976, Hallam 1986). That said, they do offer great use as heuristic devices and as a mechanism to produce qualitative hypotheses (Heck 1976). These types of models have found use in understanding and predicting the consequences of invasions (e.g., Courchamp *et al.* 1999; Courchamp and Caut, Chapter X).

UNDERSTANDING SPATIAL SPREAD

One major development, prior to Elton's book, bridging ecology and invasions biology was Skellam's (1951) use of spatial spread models. The major thrust of this research was the application of models describing the diffusion of particles in a physical system to the spread of a reproducing population over a two-dimensional landscape (Skellam 1951, Kiersteadt and Slobodkin 1953). Skellam (1951) applied these physical models to understanding and potentially predicting population spread across a landscape. By the 1950's, the rapid spread of NIS from their points of origin had been documented and to Skellam were in need of mechanistic explanation. Like Darwin, Skellam used NIS to show how his theory could be used to explain spatial spread of a population. What better way to study the dynamics of a spreading population then with newly arriving species. The 1905 introduction and spread of the muskrat, *Ondatra zibethica*, in central Europe was an illustrative example for Skellam. Skellam (1951) was able to add population growth terms, either exponential or logistic, to spread models. Adding exponential growth, the change in density is:

$$\frac{\partial n}{\partial t} = rn + D \frac{\partial^2 n}{\partial x^2}$$

where n is the population density and location x with growth rate r and diffusivity D. Following Kot (2001) the velocity of spread is:

$$\frac{x}{t} = 2\sqrt{rD}$$

Adding logistic growth, the change in density is:

$$\frac{\partial n}{\partial t} = rn\left(1 - \frac{n}{K}\right) + D\frac{\partial^2 n}{\partial x^2}$$

Again following Kot (2001) the velocity of spread is:

$$\frac{dx}{dt} = 2\sqrt{rD}$$

Skellam's spatial spread models have been used to understand the spread of natural populations (e.g., Ovaskainen 2004) and especially for the spread of populations of invasive species (e.g., Hastings 1996, Kot *et al.* 1996; Buchan and Padilla 1999).

GENERAL TRENDS IN ECOLOGY UP TO ELTON

Simultaneously with Clements' earlier work, another plant ecologist, Volney Spalding (1909), published a monograph on the distribution of desert plants. He showed that competition and migration are the most important factors structuring desert communities. His studies showed that change is constantly taking place and invasion is a constant process. This invasion process, he noted, is by both native and exotic species, and the exotics are no different, except that they are just part of the local flora owing to "the hand of Man" (p. 55). Therefore the same process, succession, occurs regardless of which species and where they came from.

Animal ecology had a later start than plant ecology and seems to have followed its lead (McIntosh 1976), but a quick flurry of texts made up for plant ecology's head start. Arthur Pearse's *Animal Ecology* (1926) definitely took its lead from plant ecology on the nature of change: "succession is usually due to one of two causes: 1) the environment is changing, or 2) the increasing group of animals is better able to take advantage of existing conditions" (p. 101). There is no mention of where the better-adapted animals come from, much less of the overarching importance of dispersal for succession.

Similarly, Charles Elton's first book, *Animal Ecology* (1927), notes that animal dispersal is important, with some moving voluntarily and others through chance. Yet there is no mention of NIS or their impacts. Moreover, Royal Chapman's fittingly titled *Animal Ecology* (1931) was the first to examine explicitly the effects of migration on animal communities and the causes of stability. He expounded on two types of resistance to invasions and population explosions. The first was physical resistance, where the abiotic environment must be optimal for the population in question. The second he called biotic resistance, in which populations are kept in check through competition, predation and parasites. More diverse communities would therefore be more resistant to invaders. As with the other authors, there is no explicit discussion of where invaders come from and their impacts.

There was little advance in the field of plant ecology leading up to Elton's invasion book (1958). For example, McDougall's *Plant Ecology* (1941) reads like an updated version of Clements. To McDougall, succession has both abiotic and biotic causes and the biotic mechanism is invasion, which "means the movement of plants from one area into another and their colonization in the latter" (p. 207). Sounds familiar.

In Good's *The Geography of Flowering Plants* (1947), he says more about NIS invaders than his predecessors. He notes that, wherever humans have gone, so have NIS. He makes some predictions as to why we ought not to worry about NIS. When it comes to competition "the struggle of [exotic] versus native is much less in favour of the former" (p. 304). Furthermore, NIS are relegated to disturbed ground because of this competitive inferiority. Where NIS do spread and become abundant, as do a number of freshwater species, they "tend after a time to settle down to a more reasonable scale of existence" (p. 305). Good's view is interesting, as he does use prediction based on ecological reasoning to address invasive species. However, we must question whether he was actually addressing what was being observed with invasive NIS, or if he was simply trying to fit what should happen into his ecological worldview.

INCIPIENT INVASIONS BIOLOGY

It really wasn't until the mid-1930s that studies, which we can refer to as invasion biology, began to surface. As late as 1932 (Storer 1932), NIS invaders were not considered to be a major source of concern for maintaining the natural environment. Tracy Storer (1932) writing about factors affecting the wildlife of California listed the major factors as agriculture, grazing by cattle, manipulation of hydrology, removal of predation, deforestation, and the changing of fire regimes. Further he noted that government and professional organizations were actively managing wildlife by regulating hunting means and seasons, protecting reserves and refuges, removing predators, and propagating and releasing young of particular species. He showed no concern about NIS or their control.

H. H. Allen (1936) gave serious and academic attention to the problem of NIS in New Zealand. Whereas authors like Hooker (1867) and Tansley (1935) thought that invaders as a group change a flora with uniform success, Allen (1936) showed that the NIS were not a horde of competitive invaders. In fact relatively few NIS (48 of 603 species) became problem species, and these few problem species represented a diverse array of ecologies. Further, Allen thought that if humans were to cease clearing and grazing land, native communities would return and replace the novel communities.

Similarly, Egler (1942) thought NIS invaders were not a uniform group changing the ecology of Hawaii. Rather he saw NIS as a diverse group, which made predicting the future of community types nearly impossible. He showed that many factors affect the success of NIS, and whether these factors were grazing, anthropogenic disturbances or species interactions, the fate of any one NIS remained foggy. Like Allen (1936), Egler (1942) thought that after grazing and disturbances were removed native communities would be much better suited to compete against NIS.

The late 1940s and early 1950s heralded the arrival of studies of the ecology of invasions. These types of studies spanned the biology of NIS in their new environments to species interactions to control. Stewart and Hall (1949) depicted the biology of the NIS, cheatgrass, and how it was affecting the native communities of southern Idaho. Baker (1948) used the invasions of *Melandrium album* and *M. dioicum* in Great Britain in order to study the dynamics of invasion and range expansion as well as large-scale competition. Finally, the control of problematic NIS was gaining momentum because of famous early instances of successful biological control. These examples would include not only the control of *Opuntia* cacti in Australia by *Cactoblastis cactorum* by 1926 (Groves 1989) but also the control of *Hypericum perforatum* by *Chrysomela* beetles in western United States (Huffaker and Holloway 1949; Huffaker 1951).

WHAT HISTORY TELLS ABOUT ECOLOGY AND INVASION BIOLOGY

I am painting the picture that the new science of ecology in the early 20th century was developing its epistemological domain, and the question is: did they try to use their new science to understand invasive exotics? I think they did not. These early ecologists seemed to incorporate NIS because they yielded real-world observations, which obeyed the very natural laws these authors were exploring. Furthermore, NIS were, in some cases, used to showcase their ideas about natural laws or as natural experiments. By and large the early ecologists did not use their conceptual ideas to understand and predict the probable outcomes of species invasions. They did not often implement Gause's (1934) suggestion that invasions constitute natural experiments to understand ecology better.

Of course this review does not scour the literature for every recorded idea or every discussion of NIS invader. Rather, I have attempted to find the prominent modes of thinking in ecology and especially with respect to NIS invaders. I have relied extensively on texts as records of these trends, and a valid criticism is that general texts do not capture all the diversity of thinking within a field of study, both subtle and stark. However, Egler (1951), in his criticism of the epistemology of American plant ecology during the 1940s, found that the major texts did largely represent the schools of thought and philosophies of American ecology.

Davis *et al.* (2001) associate Elton's (1958) invasions book with the cause of the dissociation between ecology and invasions biology. If we think of ecology as a science of case studies (e.g., Shrader-Frechette and McCoy 1994), then perhaps there would be no dissociation, and by studying individual cases of species invasions we can then form larger pictures of what is happening in nature (Shrader-Frechette 2001). Yet much of ecology is based on the precept that there are fundamental operations in nature. These fundamental operations are approximated through theory construction, experimentation and observation. The dissociation seen by Davis *et al.* (2001) is that these scientific exercises (theory, experiment and observation) are what both invasions biologists and ecologists do, but they do it largely in isolation from one another. Also, these two scientific enterprises, though driven by sometimes separate goals, could benefit from use of each other's ideas, techniques and observations.

However, after reviewing much of the early ecology literature, I find this dissociation seems to have a much longer pedigree, long before Elton. Many early ecologists do not seem to apply their science to applied problems. That is not to say that many were not concerned about human impacts on the environment, but that they used their science little in these concerns.

For one thing there was no moral call for ecologists to solve the problems of the natural world. The modern concern for the environment appeared in the work of naturalists and ecologists in the later 1950's and early 1960's (e.g., Elton 1958, Carson 1962); it is now commonplace. The societal impetus was not there. Funding agencies simply did not exist, much less support research programmes purporting to address applied problems.

More fundamentally, the foundations of ecology, like the foundations of any scientific field seemed laid by those seeking general explanations. That is not to say that there were not authors looking at very specific problems, but ecology textbooks relay the general ideas that seem best suited to explaining natural phenomena. Many of these early authors address issues intimately linked to understanding species invasions (e.g., dispersal, succession, change, etc.). Yet their considerations of non-native species involved using these species as efficacious tests of their ideas.

CONCLUSION

Here I have, ever so briefly, reviewed some of the major works of the science of ecology in the hundred or so years before Elton published *The Ecology of Invasions by Animals and Plants* in 1958. In reviewing these I have focused particularly on the way in which the development of ideas in ecology pertain to understanding the phenomenon of species invasions. Coupled with this historical examination, I have considered the notion that Elton's work caused dissociation between ecology and invasion biology (Davis *et al.* 2001). To be sure, there was no independent field of invasion biology in the early years of ecology, beyond agricultural and economic concerns posed by some non-indigenous species. The current field of invasion biology owes much to Elton's book. This review shows that early ecologists did address species invasions because they we a real and evident phenomena, and any general account on the functioning of biological communities would need to be able to explain why some species reach new lands and survive and grow. Some of these early ecologists saw invaders as a way to test theory. Yet the concern and energy devoted to understanding invaders went little beyond explaining and use as example, and it really wasn't until the mid to late 1950's that we see anything resembling invasion biology.

The current trend to unify mainstream ecology and invasion biology is proceeding (as the current volume attests), exploring new ground and pushing the limits of what has come before. Rather than look at invasion biology as a separate field, it may be better to consider it a necessary condition to generating an accurate view of nature, one that ecology must be able to use. Further, the plethora of case studies of invaders should be invaluable to the enterprise of trying to synthesize natural phenomena into a holistic framework. Perhaps concepts in ecology are adequate in their current development, but learning from the ways in which invaders respond to new environments and change the communities they are invading can tell us more about nature than most other exercises.

ACKNOWLEDGEMENTS

I am truly standing on the shoulders of giants, and I hope that my perception of the history of ecology and invasion biology is an accurate one (lest those giants turn in their graves). I am indebted to Jim Drake, Dan Simberloff, Mark Davis, Sean McMahon, Tad Fukami, for reading and earlier version of this chapter. Of course any inaccuracies are no fault of theirs.

REFERENCES

Allan, H. H. 1936. Indigene versus alien in the New Zealand plant world. Ecology **17**, 187-193.

Baker, H. G. 1948. Stages in invasion and replacement demonstrated by species of Melandrium. Journal of Ecology **36**, 96-119.

Buchan, L. A. J. and D. K. Padilla. 1999. Estimating the probability of long-distance overland dispersal of invading aquatic species. Ecological Applications **9**, 254-265.

Candolle, de, A. L. P. 1855. Géographie Botanique raissoné. V. Masson, Paris.

Carson, R. 1962. Silent Spring. Houghton Mifflin, Boston, MA.

Chapman, R. N. 1931. Animal Ecology. McGraw-Hill Book Company, Inc., New York, NY.

Chase, J. M. and M. A. Leibold. 2003. Ecological Niches. The University of Chicago Press, Chicago, IL.

Clements, F. E. 1905. Research Methods in Ecology. University Publishing Co., Lincoln, NE.

Clements, F. E. 1907. Plant Physiology and Ecology. Henry Holt and Company, New York, NY.

Clements, F. E. and V. E. Shelford. 1938. Bio-ecology. John Wiley & Sons Inc., New York, NY.

Conner, E. F. and D. S. Simberloff. 1979. The assembly of species communities. Chance or competition? Ecology **60**, 1132-1140.

Courchamp, F., M. Langlais and G. Sugihara. 1999. Cats protecting birds: modeling the mesopredator release effect. Journal of Animal Ecology **68**, 282-292.

Croker, R. A. 2001. Stephen Forbes and the Rise of American Ecology. Smithsonian Institution Press, Washington, DC.

Daehler, C. C. 2001. Darwin's naturalization hypothesis revisited. American Naturalist **158**, 324-330.

D'Ancona, U. 1954. The Struggle for Existence. E. J. Brill, Leiden, The Netherlands.

Darwin, C. 1859. The Origin of the Species. J. Murray, London.

Davis, M. A., K. Thompson and J. P. Grime. 2001. Charles S. Elton and the dissociation of invasion ecology from the rest of ecology. Diversity and Distributions **7**, 97-102.

Drude, O. 1896. Manuel de Geographie Botanique. Librairie Des Sciences Naturelles, Paris.

Egerton, F. N. 1976. Ecological studies and observations before 1900. Pages 311-351 in B. J. Taylor and T. J. White, editors. Issues and Ideas in America. University of Oklahoma Press, Oklahoma, OK.

Egler, F. E. 1942. Indigene versus alien in the development of arid Hawaiian vegetation. Ecology **23**, 14-23.

Egler, F. E. 1951. A commentary on American plant ecology, based on the textbooks of 1947-1949. Ecology **32**, 673-695.

Ehrlich, P. R. 2002. Human natures, nature conservation, and environmental ethics. Bioscience **52**, 31-43.

Elton, C. S. 1927. Animal Ecology. The MacMillan Company, New York, NY.

Elton, C. S. 1958. The Ecology of Invasions by Animals and Plants. Methuen, London.

Emerson, R. W. 1959. The Early Lectures, 1833-1836, edited by S. E. Whicher and R. E. Spiller. Harvard University Press, Cambridge, MA.

Fitch, A. 1861. Sixth report on the noxious and other insects of the state of New York. New York State Agricultural Society Transactions **20**, 745-868.

Forbes, S. A. 1883. Insects affecting corn. Illinois State Entomology Office, Circulation 1-21.

Forbes, S. A. 1886, The clinch-bug in Illinois. Illinois State Entomology Office, Circulation 8pp (September).

Forbes, S. A. 1887a. Arsenical poisons for the codling moth: record and discussion of experiments for 1885 and 1886. Transactions of the Illinois State Horticultural Society **20**, 109-119.

Forbes, S. A. 1887b. The lake as a microcosm. Bulletin of the Scientific Association (Peoria, IL) **1887**, 77-87.

Forbes, S. A. 1897. On the principle corn insects and methods of collecting them. Illinois Agriculturalist pp. 36-48.

Forbes, S. A. 1898. The season's campaign against the San José and other scale insects in Illinois. Transactions of the Illinois State Horticultural Society **31**, 105-119.

Gause, G. F. 1934 (1964). The Struggle for Existence. Hafner Publishing Company, New York, NY.

Good, R. 1947. The Geography of the Flowering Plants. Longmans, Green and Co., New York, NY

Grinnell, J. 1917. The niche relationships of the California thrasher. Auk **34**, 427-433.

Grinnell, J. 1925. Risks incurred in the introduction of alien game birds. Science **61**, 621-623.

Groves, R. H. 1989. Ecological control of invasive terrestrial plants. Pages 437-461 in J. A. Drake, H. A. Mooney, F. di Castri, R. H. Groves, F. J. Kruger, M. Rejmanek and M. Williamson, editors. Biological Invasions: A Global Perspective. John Wiley and Sons, New York, NY.

Hallam, T. G. 1986. Community dynamics in a homogeneous environment. Pages 242-285 in T. G. Hallam and S. A. Levin, editors. Mathematical Ecology. Springer-Verlag, Berlin.

Hardin, G. 1960. The competitive exclusion principle. Science **131**, 1292-1297.

Hastings, A. 1996. Models of spatial spread: is the theory complete? Ecology **77**, 1675-1679.

Heck, K. L. 1976. Some critical considerations of the theory of species packing. Evolutionary Theory **1**, 247-258.

Heilprin, A. 1887. The Geographical and Geological distribution of Animals. D. Appleton & Co., New York, NY.

Hooker, J. D. 1867. Handbook of the New Zealand Flora. Volume 2. Sir Joseph Dalton Hooker, London.

Howard, L. O. 189?. Insects affecting the cotton plant. Bulletin of the Office of Experimental Stations **33**, 317-350.

Howard, L. O. 1893. The fly weevil. Address at the Farmers' Institute, Manassas, Va.

Howard, L. O. 1895. The shade-tree insect problem in the eastern United States. Yearbook of the U. S. Department of Agriculture 361-384.

Howard, L. O. 1897a. The spread of land species by the agency of Man; with especial reference to insects. Proceedings of the American Association for the Advancement of Science **46**, 3-26.

Howard, L. O. 1897b. Danger of importing insect pests. Yearbook of Department of Agriculture 529-552.

Howard, L. O. 1898. The principal insects affecting the tobacco plant. Yearbook of Department of Agriculture 121-150.

Howard, L. O. 1899. Three insect enemies of shade trees. U. S. Department of Agriculture Farmers' Bulletin **99**, 5-29.

Huffaker, C. B. 1951. The return of native perennial bunchgrass following the removal of Klamath weed (*Hypericum perforatum* L.) by imported beetles. Ecology **32**, 443-458.

Huffaker, C. B. and J. K. Holloway. 1949. Changes in range plant population structure associated with feeding of imported enemies of Klamath weed (*Hypericum perforatum* L.). Ecology **30**, 167-175.

Huston, M. A. 1994. Biological Diversity. Cambridge University Press, Cambridge.

Hutchinson, G. E. 1957. Concluding remarks. Cold Spring Harbor Symposia on Quantitative Biology **22**, 415-427.

Johnstone, I. M. 1986. Plant invasions windows: a time-based classification of invasion potential. Biological Review **61**, 369-394.

Kierstead, S. E. and L. B. Slobodkin. 1953. The size of water masses containing plankton blooms. Journal of Marine Research **12**, 141-147.

Kot, M. 2001. Elements of Mathematical Ecology. Cambridge University Press, New York, NY.

Kot, M., M. A. Lewis and P. van den Driessch. 1996. Dispersal data and the spread of invading organisms. Ecology **77**, 2027-2042.

Lawton, J. H. Are there general laws in ecology? Oikos **84**, 177-192.

Leibold, M. A. 1995. The niche concept revisited: mechanistic models and community context. Ecology **76**, 1371-1382.

Lotka, A. J. 1924. Elements of Physical Biology. Williams and Wilkins, Baltimore, MD.

Mack, R. N. 2003. Plant naturalizations and invasions in the eastern United States: 1634-1860. Annals of the Missouri Botanical Garden **90**, 77-90.

McDougall, W. B. 1941. Plant Ecology. Lea and Febiger, Philadelphia, PA.

McIntosh, R. P. 1976. Ecology since 1900. Pages 353-372 in B. J. Taylor and T. J. White, editors. Issues and Ideas in America. University of Oklahoma Press, Oklahoma, OK.

Menand, L. 2002. The Metaphysical Club: A Story of Ideas in America. Farrar Straus & Giroux, New York, NY.

Olmsted, F. L. 1888. Letter to the editor: Foreign plants and American scenery. Garden and Forest **1**, 418-419.

Owens, R. 1860 (2001). Darwin on the origin of species. Pages 267-270 in P. Appleman, editor. Darwin, A Norton Critical Edition, third edition. W. W. Norton & Company, New York, NY.

Pearse, A. S. 1926. Animal Ecology. McGraw-Hill Book Company, Inc., New York, NY.

Peters, R. H. 1991. A Critique for Ecology. Cambridge University Press, New York, NY.

Sargent, C. S. 1888. Editorial. Garden and Forest **1**, 266.

Sedgwick, A. 1860 (2001). Objections to Mr. Darwin's theory of the origin of species. Pages 265-267 in P. Appleman, editor. Darwin, A Norton Critical Edition, third edition. W. W. Norton & Company, New York, NY.

Shantz, H, L. 1906. A study of the vegetation of the Mesa region east of Pike's Peak. Botanical Gazette **42**, 179-207.

Shrader-Frechette, K. 2001. Non-indigenous species and ecological explanation. Biology and Philosophy **16**, 507-519.

Shrader-Frechette, K. and E. D. McCoy. 1994. Applied ecology and the logic of case studies. Philosophy of Science **61**, 228-249.

Simberloff, D. S. 1978. Using island biogeographic distributions to determine if colonization is stochastic. American Naturalist **112**, 713-726.

Simberloff, D. 2003. Confronting introduced species: a form of xenophobia? Biological Invasions **5**, 179-192.

Skellam, J. G. 1951. Random dispersal in theoretical populations. Biometrika **38**, 196-218.

Spalding, V. M. 1909. Distribution and Movements of Desert Plants. Carnegie Institute of Washington, Washington, DC.

Stewart, G. and A. C. Hull. 1949. Cheatgrass (*Bromus tectorum* L.): an ecological intruder in southern Idaho. Ecology **30**, 58-74.

Storer, T. I. 1932. Factors influencing wild life in California, past and present. Ecology **13**, 315-327.

Tansley, A. G. 1935. The use and abuse of vegetational concepts and terms. Ecology **16**, 284-307.

Thompson, K., J. G. Hodgson and T. C. G. Rich. 1995. Native and alien plant invasive plants: more of the same? Ecography **18**, 390-402.

Volterra, V. 1927. (1978). Variations and fluctuations in the numbers of coexisting animal species. Pages 65-236 in F. M. Scudo and J. R. Ziegler, editors. The Golden Age of Theoretical Ecology: 1923-1940. Springer-Verlag, New York, NY.

Wallace, A. R. 1876 (1962). The Geographical Distribution of Animals. Hafner Publishing Company, New York, NY.

Warming, E. 1909. Oecology of Plants: An Introduction to the Study of Plant Communities. Clarendon Press, Oxford, UK.

Weaver, J. E. and F. E. Clements. 1938. Plant Ecology. McGraw-Hill book Company, New York, NY.

Chapter three

Invasion biology 1958-2005: the pursuit of science and conservation

M. A. Davis

INTRODUCTION

The history of invasion biology would be a dream dissertation topic for some history of science graduate student. The list of researchers who have weighed in on the topic at one time or another reads like a Who's Who of late 20th century ecology. Controversy and disagreements, occasionally resulting in lively exchanges among researchers, have created an intellectually dynamic and sometimes emotionally charged atmosphere in recent years. Authors of both technical and popular articles have often used evocative language and imagery in their writing. Some of the issues have attracted the interests of scholars outside the biological sciences, particularly philosophers. And, for the past several decades, invasion ecology research has been conducted within a larger social milieu of contentious environmental values and politics. No doubt for all these reasons, invasion ecology has captured the attention of national and international media outlets, which, having their own agendas, have emphasized and presented particular research findings and perspectives for their audiences.

However, an examination of the field's history should be of interest not just to historians of science but to ecologists as well. After all, a look backwards is

M.W. Cadotte et al. (eds.), Conceptual ecology and invasion biology, 35–64.

almost always a good idea before charting new paths. How have research questions changed over time? What have we learned? What continues to challenge our understanding? If we can answer these questions, we can more effectively formulate future priorities. A historical review cannot guarantee answers to these questions, but neither can the answers be obtained without such a review. In researching and writing this chapter, I have strived to provide a historical review of invasion biology since 1958 (the publication date Charles Elton's invasion classic, *The Ecology of Invasions by Animals and Plants*) in order to answer the above questions, and, based on whatever insights could be gained from these efforts, to offer a few suggestions as to how invasion biology might profitably proceed from this point forward.

THE 1960S

In 1964, The International Union of Biological Sciences held its first Biological Sciences Symposium, part of what was to be a series of such meetings designed to focus attention on biological topics having international significance. The objective of the first symposium (held in Asilomar, California) was to bring together geneticists, ecologists, taxonomists, and applied scientists working in the area of pest control and to present and discuss facts and ideas "about the kinds of evolutionary change which take place when organisms are introduced into new territories" (Waddington 1965). The proceedings were published in a *The Genetics of Colonizing Species* (Baker and Stebbins 1965) now regarded as a classic. As would be expected given the organizing group, the book represented a distinctly international initiative, consisting of 27 authors representing 11 countries. The range of topics considered in this volume was broader than the title suggests, and included discussions on the nature of the colonized environments as well as the attributes and genetics of colonizing species. Chapters addressed a broad range of taxa, including plants, insects, birds, mammals, and microbes. The symposium and the book were defined by a strong evolutionary emphasis and participants included many of the prominent evolutionary biologists of the time, including Mayr, Stebbins, Dobzhansky, Wilson (E. O.), Carson, Lewontin, and Waddington. Significantly, authors considered colonists broadly, discussing colonizations that occurred during succession, colonizations of widespread species that had occurred without human assistance, colonizations of native weeds, as well as colonizations of species into new regions occurring as a direct result of human activity. Although the latter group did receive the most attention, participants recognized that common ecological and evolutionary processes underlay the different types of colonizations.

One noteworthy aspect of the book is the striking contrast between the language used by the authors and that commonly used by invasion ecologists today. A careful search of the book's text is necessary to find terms such as 'alien', 'exotic', 'invader', and 'invasion'. Most authors never used these words. A few, such as Wilson (E. O.), Mayr, and Wodzicki used them occasionally, but the articles and discussions were overwhelmingly guided by nouns, verbs, and adjectives, such as, 'colonizers', 'founding populations', 'introduced', 'non-native', 'new arrivals', 'migration', 'spread', 'geographically widespread'. Thus, participants did not adopt the effusive style used by Elton in his 1958 book, characterized by the frequent use of metaphor and analogy, often with explicit militaristic references. The one exception was Elton's colleague, John Harper, who usually did refer to the new species as 'aliens' and 'invaders' and whose chapter, *Establishment, Aggression, and Cohabitation in Weedy Species,* exhibited some of the same evocative language that Elton used in his book. The opening sentence of Harper's chapter is so reminiscent of the language used by Elton that the latter's influence on Harper can hardly be denied. Harper (1965) began, "The movements of man and his goods have resulted in a bombardment of areas of land and sea by alien species, both by chance and by the deliberate introduction of cultivated plants of the farm and garden."

In the preface to his book, Elton (1958) said his goal was to bring together three streams of thought — faunal history, ecology, particularly population ecology, and conservation. But the conservation theme dominated the book. He opened his book with graphic battlefield examples of invasions and ended the book with two chapters on the need for conservation. In contrast, the Asilomar participants did not consider conservation implications of species colonizations whatsoever (E. O. Wilson, personal communication). The clearly defined agenda of the 1964 Asilomar symposium was the search for generalizations regarding the evolutionary and ecological processes involved in species colonizations. In the 27 chapters and 562 pages of the symposium proceedings, Elton's 1958 invasion classic was cited only three times, once each by Birch, Mayr, and Wilson (E. O.).

As a publication focusing on introduced species, *The Genetics of Colonizing Species* stands starkly alone during this time period. Perhaps it was because the book focused more on evolutionary issues than ecological theory, but the symposium and companion volume elicited almost no response from the ecological community. Despite the publication of this volume and Elton's book seven years earlier, biological invasions did not attract the interest of many ecologists during the 1960s, at least from ecologists whose primary interests and objectives involved the development of theory and generalizations that transcend particular organisms and habitats. It is certainly true that with the development and dissemination of island biogeography theory (MacArthur and Wilson 1963, 1967; Simberloff and Wilson 1969), considerable attention was given to how arriving species would fit into resident communities. However, few ecologists at this time connected their research with the sort of invasion literature and issues summarized and presented by Elton (Simberloff, personal communication).

It must be pointed out that, during this time, some ecologists outside of North America were actively researching and publishing in the area of introduced species. Sukopp (1962) addressed both theory and terminology in his report on introduced species in the natural plant associations of central Europe. Holub and Jirásek (1967) and Schroeder (1969) presented classification schemes for non-native species (based on the mode of introduction, time of introduction, and degree of naturalization). Faliński (1966) wrote his dissertation on the distribution of introduced species in Poland's Bialowieza Primeval Forest) and Faliński (1968, 1969) and Kornas (1968a, 1968b) published a number of subsequent articles during the 1960s on the ecology of introduced species in natural and semi-natural communities in Poland. Hungarian ecologist Pal Juhasz-Nagy (1964) conducted field experiments in his studies of 'ecesis resistance' (invasibility) and used the term 'ecological homeostasis' to describe communities that were resistant to invasion. Hejný and Lhotská (1964) and Jehlík and Slavík (1968) described the ecology and distribution and spread in natural environments of several introduced species in the Czech Republic. Kohler and Sukopp (1964) studied the ecology of introduced species in cities, part of a long-standing interest among European plant ecologists. (One of the more memorable findings of Kohler and Sukopp was that *Robinia pseudoacacia* had spread spontaneously on the rubble of cities bombed during WWII.) Except for the abstracts, none of the above articles was published in English, and most English-speaking ecologists were probably not well aware of this research during the sixties.

Although biological invasions received little attention from ecologists during the 1960s, particularly in North America, this does not mean that research was not being conducted on introduced species and their impacts. In fact, considerable research on this topic was being conducted in Australia, New Zealand, South Africa, Europe, as well as North America. It was simply being conducted by biologists whose interests and priorities were more applied in nature, involving fisheries (e.g., Albrect 1964), wildlife (e.g., Bump and Bohl 1964, TCWF 1967, Warner 1968), forestry (e.g., Nichols 1961, Taylor 1969) and agriculture (e.g., Salisbury 1961, Metcalf *et al.* 1962, DeBach 1964). The applied and pest-control research conducted in during the 1960s did not represent any new initiative, and very little, if any, of it was inspired by Elton's book. Rather it was a continuation of extensive applied research in this area extending back to the beginning of the 20th century (e.g., Little 1916, Bailey 1924, Bryant 1927, Sim 1927, Wicht 1945), and even earlier (e.g., Merriam 1889, Palmer 1899).

From an historical perspective, a 1966 book by George Laycock was a noteworthy publication during this time. Titled, *The Alien Animals: The Story of Imported Wildlife*, the book was published by The American Museum of Natural History. Well researched, it contained nearly 200 references from the scientific literature. Like Elton's 1958 classic, this book was written from a conservation, and distinctly value-based, perspective, as well as for a larger public audience. On the opening page, Laycock (1966) refers to "man" as "the supreme meddler"; on the book's final page he warns of natural communities being "polluted" by alien species; and titles of his chapters included *Stangers in the Southwest*, *How the Gray Squirrels Invaded England*, and *The Conquering Mongoose*. In the book, Laycock presents twenty case studies of animal introductions around the world that had caused great economic and conservation harm. The book's language, tone, and series of case studies is very similar to Elton's 1958 book, although, rather inexplicably, Elton was not one of the nearly 200 references listed in the bibliography. However, unlike Elton, Laycock was not a scientist. He was a freelance writer, who, during his career, authored more than twenty books on animals and nature, many of them written for children and young adults. Nevertheless, Laycock's characterization of introduced species seems to have portended the perspective adopted in a large number of invasion ecology literature, both popular and scientific, later in the century, more so than has the Asilomar publication, *The Genetics of Colonizing Species*.

THE 1970S

On April 22, 1970, hundreds of thousands of students and others across the United States celebrated the first Earth Day, an event conceived and organized by Gaylord Nelson, then Senator of Wisconsin. Although Nelson, Rachel Carson, and others had been trying to raise environmental consciousness within the United States during the sixties, the 1970s is when the environmental movement finally captured the public's attention, in the US and throughout the world. SCOPE (Scientific Committee on Problems of the Environment, a committee established by the International Council of Scientific Unions) published the first of its SCOPE Series titles (now numbering more than sixty) in 1971. The 1970s is when conservation biology began to emerge as a discipline and ecologists worldwide began to study introduced species from a conservation perspective. In the early 1970s, the new journal *Biological Conservation* began publishing articles describing the threats posed by introduced species (e.g., Campbell and Ormond 1970, Schofield 1973).

Applied ecologists who had been studying introduced species and their ecological effects in the sixties and earlier, continued to do so in the seventies, publishing primarily in taxonomic-specific journals as they had previously. For example, Owre (1973) described the extent of introduced avifauna in southeastern Florida and raised the possibility of competition between the native and new species (in the conservation section of The Wilson Bulletin). Christie (1972) described the effects of species introductions on salmonid communities (Journal of the Fisheries Board of Canada), and Moyle (1973) reported on the effects on the native frogs of the introduced bullfrogs in California (Copeia). In a *Bioscience* review paper, Courtenay and Robins (1975) summarized the conservation problems (primarily in North America) created by the introduction of "exotic" animals by fish and game agencies and the pet trade. The European ecologists who had been studying introduced species and their ecological effects in the sixties also continued their research in the seventies, publishing hundreds of papers on this topic. For example, Faliński (1971) and Sukopp (1971) continued their work on the urban introduced and adventitious flora and Jehlík and Hejný (1974) continued their research in the Czech Republic, describing migration routes of introduced species and making what perhaps were the first predictions of when plant invasions would occur in particular locations (Hejný *et al.* 1973). Although this floristic approach to introduced species was not as common in the United States at this time, occasional studies of this type were undertaken, e.g., Muehlenbach's (1977) detailed account the adventive flora along railroads around Saint Louis, Missouri.

Occasional papers on biological invasions can be found in the ecological literature well before the 1970s (e.g., Egler 1942). However, the seventies was the first time since Elton that biological invasions began to appear in mainstream ecology literature with any frequency. In his review of the ecology of weeds, published in *Annual Review of Ecology and Systematics*, Baker (1974) explicitly extended the discussion of weeds beyond the disciplinary boundaries of horticulture and agriculture to include ecology. Other examples include Burdon and Chilver's 1977 *Oecologia* article on the impact of introduced pine species on Australia's eucalyptus forests, and Embree's (1979) chapter on the ecology of animal invaders in the book, *An Analysis of Ecological Systems* (Horn *et al.* 1979).

Although some ecologists were beginning to think about biological invasions during the seventies, there was still little coordination and communication among investigators. And, there was no formal infrastructure to support efforts to develop a theory of biological invasions. Ecology meetings at that time did not devote special symposia or paper sections to biological invasions, which would have served as important communication hubs for this topic. Thus, although scattered papers on biological invasions were appearing in the ecological literature, invasion ecology had not yet emerged as its own research specialty area.

1980-1984

The number of published studies on biological invasions increased dramatically during the 1980s, especially in particular taxa, such as plants (see review by Pyšek 1995). The emphasis on floristics that characterized much of the work in previous decades began to give way to the increasing interest on species biology and ecological interactions (Pyšek 1995). Due to the proliferation of invasion publications during this time, the historical account of this decade has been split into two periods.

Many of the European ecologists who had been studying introduced and adventive flora in prior decades continued to do so in the eighties, particularly focusing on the ecology of urban flora (e.g., Sukopp and Werner 1983, Kowarik 1984). In the United States, Richard Mack published his first article on biological invasions in 1981, reporting on the ecological impact in western North America of *Bromus tectorum*. Mack took a conservation approach in his article, quoting Elton and emphasizing the decline of native grass species. He chose to submit his article to the more applied journal *Agroecosystems* because its scope encompassed the ecosystems that cheatgrass had invaded. In the same year, Dan Simberloff contributed a chapter to the book *Biotic Crises in Ecological and Evolutionary Time*

(Nitecki 1981). The title of the book suggests that Simberloff (1981) might also have emphasized the conservation aspects of biological invasions. In fact, he did not do this, but rather used biological invasions as a way to test ecological theory, e.g., island biogeography theory and models of limiting similarity. Thus, whereas Mack's (1981) article was more in the conservation tradition of Elton's 1958 book, Simberloff's chapter was more in keeping with the emphasis on ecological and evolutionary theory that distinguished *The Genetics of Colonizing Species*. In the same year, Brown and Marshall (1981) published an article even more reminiscent of the 1965 Asilomar proceedings, emphasizing the evolutionary changes accompanying the colonization of plants. Two years later, Moulton and Pimm (1983) examined the impacts of the introduced Hawaiian avifauna as a way to assess the importance of competition in community assembly. Like Simberloff and Brown and Marshall, they also emphasized theory, as evidenced by their choice of journal (American Naturalist). The tone and style used by Simberloff (1981), Brown and Marshall (1981), and Moulton and Pimm (1983) was consistent to that used by the contributors to *The Genetics of Colonizing Species* (other than Harper).

The connection between introduced species and conservation continued to grow during the 1981. *Immigrant Killers* (King 1984), described the impact of introduced predators on native birds in New Zealand. Although written for a popular audience, King's book was well researched and referenced, and it represented an important scientific contribution as well.

In 1981, William Jordan III and the University of Wisconsin Arboretum published the first issue of *Restoration and Management Notes* (now published under the name of *Ecological Restoration*), an event that signified the beginning of an organized interest in habitat restoration in North America. Land managers had been thinking about restoration in the 1970s. For example, NATO sponsored a 1976 conference in Reykjavik, Iceland that resulted in the publication of *The Breakdown and Restoration of Ecosystems* (Holdgate and Woodman 1978). Europeans constituted the majority of participants at the Reykjavik conference and the interest at the meeting was more in "rehabilitating" the landscape rather than trying to restore the habitat to some prior native condition, as illustrated by a statement by Bradshaw *et al.* (1978), "from the point of view of nature conservation, rehabilitation should not necessarily mean restoration of the original, often uninteresting, ecosystem". This perspective contrasted greatly with the emerging interest in ecological restoration in the United States. In his opening editorial of the first volume of *Restoration and Management Notes*, William Jordan III (1981) stated that the new journal "will deal only with the development and management of communities that are native or at least ecologically appropriate to their site." The stated focus of the journal was to be the restoration and management of ecological communities for scientific and aesthetic purposes, e.g., prairies, wetlands, and forests, but not for range or timber management, and not for "reclamation efforts aimed solely at land stabilization." The subsequent development of the field of restoration ecology in future years was to have a major impact on invasion ecology, particularly in North America, through its emphasis on native environments and native species.

By the mid-1980s, more and more ecologists were beginning to think about invasions from an ecological perspective. However, what was still lacking was a synthesis of this thinking that would help define a field of invasion ecology. In fact, the groundwork for such an effort was already underway. In 1980, the Third International Conference on Mediterranean Ecosystems was held in Stellenbosch, South Africa, and interest in biological invasions at the meeting led to a proposal to the SCOPE general assembly in Ottawa in 1982. This proposal resulted in the creation in 1983 of a scientific advisory committee that was to encourage and facilitate focus on the impacts of biological invasions on natural ecosystems. The advisory committee articulated three questions to guide the SCOPE invasion program: 1) what factors determine whether a species will be an invader or not? 2) what are the characteristics of the environment that make it either vulnerable to or resistant to invasions? 3) how can the knowledge gained from answering the first two questions be used to develop effective management strategies? Developed with the intent of "building on the considerable knowledge base available on invaders of agricultural systems", the SCOPE program on biological invasions was clearly conceived in the conservation tradition of Elton. It is no coincidence that scientists from South Africa, Australia, New Zealand, and the United States contributed so significantly to the SCOPE initiative. The natural environments in these countries had been experiencing recent and substantial introductions of new species from other regions of the world and considerable conservation concerns were being raised in these countries regarding the impacts of these species.

1985-1989

South African scientists had been among the leaders in recognizing the conservation implications of introduced species (e.g., Wicht 1945, Taylor 1969) and they published the first SCOPE volume in 1984 (Macdonald and Jarman 1984). However, it was not until the latter half of the decade that most of the SCOPE regional workshops began publishing their proceedings. The North American contingent published *Ecology of Biological Invasions of North America and Hawaii* (Mooney and Drake 1986). The book was organized around various patterns of invasions, including associations with different taxonomic groups, attributes of invaders, and site characteristics promoting invasions. In addition a variety of case studies were provided. This publication was entirely a United States initiative, with both editors and all the contributors being from US universities. That biological invasions had finally captured the interest of prominent North American ecologists is evidenced by the contributors to the book, e.g., Simberloff, Ehrlich, Bazazz, Regal, Orians, Vitousek, Roughgarden, Ewel, Pimm, and Mooney.

Coincidentally, the symposium that led to this volume took place in Asilomar, California, the same site of the symposium on the genetics of colonizing species held twenty years earlier. The 1986 volume contains an interesting mixture of chapters. Most were clearly conceived and written with a conservation and environmental perspective, as would be expected given the nature of the SCOPE program. However, some chapters were written more in the spirit of the 1964 Asilomar symposium and focused more on ecological theory. Simberloff's (1986) chapter on the biogeography of insect introductions examined topics such as island-mainland patterns and issues of biotic resistance primarily from a theoretical perspective. Although he did cite Elton, Simberloff did not take a strongly conservation approach in his chapter, referring primarily to 'introduced' species and 'colonizers' rather than 'invaders' and 'exotics' and specifically recommended the "refinement of the approach of many of the papers in *The Genetics of Colonizing Species*. Bazzaz (1986) organized his chapter on life histories around the concept of 'colonizing species' and also made explicit references to the 1965 book. Moulton and Pimm (1986) expanded on their 1983 paper and showed how biological invasions can be used to test ecological theory.

Additional proceedings were published from SCOPE workshops held in South Africa, Australia, and Great Britain (e.g., Macdonald *et al.* 1986, Groves and Burdon 1986, Kornberg and Williamson 1987). Being SCOPE initiatives, these publications were also written from a conservation perspective. However, comments by the editors of the Australian proceedings indicate that they recognized an emerging tension that was developing in invasion ecology as some ecologists embraced the conservation and strong environmental emphasis articulated by Elton, while others expressed concern over the strong normative nature of this approach. In the Foreword to the 1986 publication (Groves and Burdon 1986), the editors stated that "the terminology associated with 'invasions' generally is unsatisfactory with its connotations of a military operation. .... Wherever possible we have tried to eliminate emotive terms such as 'weeds', 'pests', 'exotics', and 'aliens'." Instead, the editors strived to utilize the word 'introduced' as much as possible throughout the book.

Another important development that occurred during this time period was the founding in 1987 of the Society for Ecological Restoration (SER), now known as Society for Ecological Restoration International. SER emerged out of the rapidly developing interest in restoration by individuals and conservation groups interested in managing natural environments. Upon its founding, SER was primarily a North American initiative, and the restoration ecology movement that it represented and inspired sharply distinguished between species based on their geography of origin, with native species being desirable and non-natives being undesirable in natural environments. The stark dichotomy between native and non-native species particularly characterized conservation and restoration efforts in regions like North America, Hawaii, and New Zealand, which had experienced relatively few species introductions until the past few centuries. This distinction seemed to be less crucial among conservationists and restorationists in Europe where, as Usher (1988b) acknowledged, distinguishing between native and introduced species is often problematic due to the millennia of human movements and species introductions across the continent. Restoration ecology and invasion ecology emerged at about the same time, and, in North America, they developed as sister disciplines during the latter1980s, and continued to reinforce one another in subsequent years.

The growing relationship between invasion ecology and conservation ecology during this time was illustrated by the decision of the editors of the journal *Biological Conservation* to devote a special issue to this topic. Titled, *Biological Invasions of Nature Reserves*, this publication (Usher 1988a) represented another SCOPE product. Although it was becoming increasingly common for ecologists to adopt the more explicit conservation perspective emphasized by the SCOPE program, not all ecologists were doing so. For example, in his article on invasion windows, Johnstone (1986) examined habitat invasibility mostly from a theoretical perspective, and he did not link his article to conservation concerns regarding invasions.

The decade ended with the publication of *Biological Invasions: A Global Perspective* (Drake *et al.* 1989), the synthesis report of the various regional SCOPE symposia and workshops. This publication represented the 37th publication in the SCOPE series. Twenty-nine ecologists, representing seven countries (USA, UK, Australia, South Africa, France, Denmark, India), many of whom had contributed to the 1986 volume, authored chapters for this publication. Some of the prominent non-American contributors, many of whom have continued to focus much of their research on biological invasions up to the present, included Francesco di Castri, Peter Ashton, David Richardson, Ian Macdonald, Michael Usher, Mark Williamson, Richard Hobbs, Michael Crawley, Richard Groves and Marcel Rejmanek. (Rejmanek had moved from Czechoslovakia to the United States by the time the book was published.). Individually focusing on different taxa or regions of the world, contributors tried to answer the three questions that had been posed by the SCOPE advisory committee in 1983. In the book's final chapter, Mooney and Drake (1989) concluded that although knowledge permitted "generic guidelines" for assessing the likelihood of biological invasions, e.g., disturbances normally increased invasibility, they concurred with the conclusions reached by most of the authors that it was not possible at that time to make accurate predictions about individual cases.

Although some of the North American ecologists continued to use the terms 'introduction' and 'introduced species' rather than 'invasions' and 'invaders' in the 1989 volume (e.g., Simberloff and Pimm), possibly indicating some resistance to adopting the more value-based conservation approach, it is interesting that the frequency with which American ecologists cited Elton's 1958 book continued to increase. In 1965, only three of the eleven (27%) US contributors to *the Genetics of Colonizing Species* cited *The Ecology of Invasions by Animals and Plants*. In the North American 1986 SCOPE volume, which was authored entirely by US ecologists, 56% of the chapters cited Elton's book. And in the 1989 SCOPE synthesis volume, 8 of the 9 (89%) US authored chapters cited Elton's book. By contrast, only 2 of 11 (18%) of the chapters in the 1989 SCOPE volume authored entirely by non-US ecologists cited Elton. (Two of the chapters were authored by at least one US and one non-US author; one of these cited Elton and the other did not.) In the book's final chapter, Hal Mooney and James Drake (both US ecologists) specifically encouraged the readers to consult Elton's book: "Those wishing an in-depth treatment of the dynamics of biological invasions are referred to Elton's (1958) classic work, and to the publications which arose out of the program" [i.e., the SCOPE program].

The scattered ideas concerning biological invasions put forth by ecologists in the seventies and early eighties finally coalesced in the mid to late eighties, resulting in the publication of a number of important papers and edited volumes from 1986-1989. These publications reviewed knowledge and theory to that point, identified questions to be answered, and set much of the direction for future research of biological invasions. In short, invasion ecology emerged as a research specialty area during this time period as a direct result of these publications. Although, some investigators continued to pursue invasion ecology more in the tradition of the 1964 Asilomar symposium on colonization, i.e., focusing primarily on ecological and evolutionary theory, more and more ecologists, particularly North American ecologists, were presenting their writings in an explicitly environmental context, the path set forth by Elton in 1958.

1990-2005

Invasion ecology took the spotlight following the series of invasion books and articles published between 1986 and 1989. Biological invasions became a popular area of research and an increasingly controversial topic of debate during the 1990s resulting in a flood of publications, both scholarly and popular, that continues to this day. National and international ecology meetings, and some general science meetings (e.g., AAAS) began to regularly schedule symposia and paper sessions devoted to biological invasions and introduced species during the nineties. The first of seven international conferences on the ecology of alien plant invasions was held in Loughborough, UK in 1992, and the 7th conference was held in Fort Lauderdale, Florida, USA in 2003, and several of these conferences resulted in publications of the proceedings (e.g., Pyšek *et al.* 1995, Starfinger *et al.* 1998, Child *et al.* 2003). With the exception of the second and seventh conferences (held in Tempe, Arizona and Fort Lauderdale, Florida), the conferences took place in Europe, with European ecologists representing the majority of participants. Thus, this series provides an excellent overview of the research and ideas coming from Europe at this time, particularly from eastern European countries such as the Czech Republic, which, as stated above, have had a long tradition of studying invasions. Some of the prominent ecologists who played a significant role in these conferences and publications include Petr Pyšek and Karel Prach (Czech Republic), Ingo Kowarik and Uwe Starfinger (Germany), Mark Williamson, Max Wade, and Lois Child (UK), Giuseppi Brundu (Italy), and John Brock and Marcel Rejmanek (USA).

Unlike the time periods described above, the years 1990-2005 do not offer an obvious list of important publications, due to the sheer number of books and articles produced during this time. Thus, the publications briefly mentioned below constitute a very small representative subset of the large number of significant publications one could have included.

Lodge's 1993 TREE article provided a comprehensive overview and assessment of the field of invasion ecology to that point. Consistent with prior assessments, Lodge concluded that it was not yet possible to accurately predict the consequences of a single invasion event and that any hope of useful predictions could only emerge from focused studies on particular species and environments. Lodge's paper was distinctive because it contained well developed aspects of both the conceptual approaches to biological invasions that had developed during the preceding three decades. While Lodge began his paper emphasizing the environmental impacts of 'exotic' species, which have "serious consequences for both man and nature", he also stressed the scientific opportunity that invasions provide, "because the characteristics and ecological impact of exotic species may provide clues to longstanding issues in the study of community assembly". Consistent with this dual emphasis, Lodge cited both Elton (1958) and Baker and Stebbins (1965) in the same sentence, referring to them both as 'classics'. Williamson's 1996 book *Biological Invasions*, which reviewed and synthesized invasion research and theory conducted and developed during the prior several decades, made a very important contribution by providing investigators a comprehensive and up-to-date overview of the new and rapidly evolving field. Williamson's book primarily focused on the scientific aspects of species invasions, however he did set the book in a clear conservation context, emphasizing the management value of better scientific understanding of invasions. James Carlton's numerous publications on the impacts of invasions on marine ecosystems (e.g., Carlton 1996a, 1996b, 1999) prompted more marine ecologists to consider the importance of biological invasions. The publication of *Biological Invasions: Theory and Practice* by Japanese ecologists Shigesada and Kawasaki (1997), which reviewed mathematical models of species spread, illustrated the extent to which biological invasions had attracted the interests of ecologists from around the world.

The importance of understanding interactions with soil microbes (positive and negative) as part of the invasion process in plants was emphasized by Klironomos (2002) and Richardson *et al.* (2000) emphasized the role that mutualisms can play in the invasion process. The enemy-release hypothesis received considerable attention during this time, although studies produced conflicting results (Keane and Crawley 2002). Williamson (1993, 1996) proposed the since oft-cited 'tens rule' of biological invasions, which states that approximately 10% of imported species exist in some sense outside of captivity, 10% of these established self-sustaining populations, and 10% of these become pests. Reviving the emphasis on genetics of colonizing species that characterized the 1964 Asilomar symposium, many ecologists began focusing on the genetics of introduced species, recognizing the importance that new genotypes might play in species spread (Ellstrand and Schierenbeck 2000, Vila *et al.* 2000, Daehler and Carino 2001, Pysěk *et al.* 2003). Ewel *et al.* (1999) reviewed the benefits and risks associated with deliberate species introductions and outlined the research needed to provide the information required to increase the former and lower the latter. As invasion ecology continued to develop during this period, several scientists expressed concern that the field was not well connected with other disciplines and that invasion ecology was not being informed by, nor informing, disciplines such as biogeography, evolutionary biology, and ecological subdisciplines, such as succession ecology and weed ecology (Vermeij 1996, Davis *et al.* 2001, Davis *et al.* 2005). At the same time, some conscious efforts were being made to bridge these gaps (e.g., Booth *et al.* 2003).

During the nineties, ecologists continued the several decades-long effort to determine whether invaders possessed distinctive traits. Many researchers concluded that there was little difference between the traits of invasive and non-invasive species (e.g., Thompson *et al.*, 1995, Goodwin *et al.* 1999, Williamson 1999), although some taxonomically-based comparative studies did show distinct differences between resident and spreading introduced species (e.g., Rejmanek and Richardson 1996). Kolar and Lodge (2001) were more optimistic that traits could be used as predictors of invasions, concluding that quantitative approaches (e.g., Daehler and Carino 2000) were making progress and that these approaches should "allow us to predict patterns of invading species more successfully." Unlike Lodge's 1993 paper which emphasized both the conservation and scientific implications of biological invasions, the 2001 TREE paper by Kolar and Lodge was much more strongly organized around the conservation and environmental concerns about species invasions.

Factors that contribute to an environment's susceptibility to invasion (invasibility), one of the other organizing themes of the SCOPE initiatives in the 80s, also continued to be a central focus of research and discussion during the 90s and the early years of the new century. In his book *Biological Diversity*, Huston (1994) addressed invasibility and emphasized the combined importance of disturbance and productivity. Lonsdale (1999) presented his findings of a comprehensive review of global patterns of plant invasions and found a positive correlation between native and exotic species at a site, although he concluded that it was difficult to discriminate between the relative importance of invasibility and propagule pressure. Davis *et al.* (2000) argued for a theory of invasibility based on resource availability, emphasizing that invasibility is not a constant characteristic of an environment but is an attribute that varies over time as resource availability fluctuates. Davis and Pelsor (2001) tested this theory in a field experiment based on the introduction of native species (prairie forbs) into a plant community dominated by introduced species (North American old field), showing that it is the ecological processes that matter, not the geographic origin of the species involved.

Although much of the progress in invasion ecology during the eighties and nineties involved moving beyond purely floristic studies of native and introduced flora, more comprehensive efforts to describe and catalog flora were also undertaken during this time, providing valuable data bases for ongoing and future studies (e.g., Pysěk *et al.* 2003). The most comprehensive national or regional plant data bases at the current time are likely those of the United Kingdom (Clement and Foster 1994, Ryves *et al.* 1996), the Czech Republic (Pysěk *et al.* 2002), and Austria (Essl and Rabitsch 2002). Other excellent plant data bases include those in Germany (Klotz *et al.* 2003) and North America (Kartesz and Meacham 1999).

With the development of invasion ecology as a recognized research specialty area, two invasion related journals were established in the late nineties. *Diversity and Distributions* (Blackwell Publishing) was founded in 1998 (succeeding the journal *Biodiversity Letters*) with an emphasis on biological invasions and biodiversity. David Richardson has served as the journal's editor-in-chief from its outset. *Biological Invasions* (Kluwer) was founded in 1999 with James T. Carleton as the editor-in-chief, a position currently held by James A. Drake. The stated aims and scopes of both journals emphasize the science rather than the conservation aspects of biological invasions. However, it is not uncommon for authors to invoke conservation concerns in their articles, particularly in *Biological Invasions*.

Two developments in ecology during the 1990s particularly impacted invasion ecology — a renewed emphasis on the ecological effects of biodiversity and the increasing emphasis on global change. Research on the ecological impacts of biodiversity focused attention on Elton's (1958) hypothesis that species-rich communities should be more resistant to invasions than species-poor communities. A number of small-plot experimental studies involving constructed plant communities yielded results that supported Elton's diversity-invasibility hypothesis (Tilman 1997, Knops *et al.* 1999). However, these experiments were challenged on methodological grounds (Wardle 2001) and their relevancy was questioned since studies in natural communities often found that the most diverse environments were the most heavily invaded (Lonsdale 1999, Stohlgren *et al.* 1999). Efforts have been made to resolve this debate, e.g., emphasizing the role that spatial scale plays in interpreting the results and the fact that correlated environmental factors may affect both diversity and invasibity (Levine and D'Antonio 1999, Levine 2000, Shea and Chesson 2002). However, to date, these efforts have not quelled the controversy (Rejmanek 2003, Renne and Tracy 2003, Stohlgren 2003).

The interest in global change that developed in ecology in the nineties went beyond concerns involving increases in atmospheric CO_2 and climate change. Dukes and Mooney (1999) emphasized the global dimensions of biological invasions and researchers emphasized the cause and effect relationship that existed between biological invasions and other types of global change (Kowarik 2001). As a result of connections made between biological invasions and global change, the topic of biological invasion, historically a rather narrow and specialized area of research, has become part of a much larger and more prominent environmental agenda.

A number of books published during the 1990s and early 2000s, and, like Elton's 1958 book, were written from an explicit conservation and environmental perspective and aimed at the general public. Many of these books utilized similar evocative imagery to capture the public's attention. Examples of such books written or edited by ecologists, or science journalists, include *Life Out of Bounds: Bioinvasions in a Borderless World* (Bright and 1998), *Killer Algae* (Meinesz 1999), *Nature Out of Place: Biological Invasions in the Global Age* (Van Driesche and Van Driesche 2000), *Tinkering with Eden: A Natural History of Exotics in America* (Todd 2001), and *Plagues of Rats and Rubbervines* (Baskin 2002).

With their lamentation over the loss of native environments and decline of native species, the above books illustrate the increasingly strong synergy that had developed between restoration ecology and invasion ecology (particularly in North America) during the nineties, with the objectives of each reinforcing those of the other. Restoration ecology's emphasis on restoring environments with native species affirmed the importance of invasion ecology, and invasion ecology's emphasis on the harm caused by a small proportion of introduced species provided important justification for restoration ecology's preference for native species. However, by the end of the nineties and beginning of this century, more scientists were questioning the dichotomous and normative nature of a perspective that prefers some species over others on the basis of their geography of origin (Gould 1998, Slobodkin 2001), emphasizing the useful contributions that some introduced species might be able to make in restoration projects (D'Antonio and Meyerson 2002). Such objections were not new. In the SCOPE synthesis report (Drake *et al.* 1989), James Brown (1989) made the same arguments regarding the potentially positive value of some introduced species and expressed concern over "a kind of irrational xenophobia about invading animals and plants that resembles the inherent fear and intolerance of foreign races, cultures, and religions", an "attitude" he said he "detected" at the SCOPE conference. Actually, Egler (1942) raised these same concerns forty-seven years earlier. In his article in *Ecology*, he criticized the common resort to hyperbole and military imagery in discussions of alien species and warned against "anthropomorphic sentimentalism that breeds a spirit of defeat". During the nineties, writings of invasion biologists also attracted the attention of philosophers (Eser 1998, Sagoff 1999), who expressed similar concerns over the language and values they said permeated the writings of invasion ecology. Finally, although no ecologists questioned the fact that the earth's biota was becoming increasingly homogenized due to widespread species introductions, the common contention that species introductions lead to declines in regional and local biodiversity was challenged by several investigators (Rosenzweig 2001, Davis 2003, Sax and Gaines 2003), a point that had actually been made much earlier by Harper (1965).

By the end of the nineties, invasion biology was becoming increasingly institutionalized with the establishment of more national and international conservation and biodiversity initiatives relating to biological invasions. The United Nations sponsored conference on "invasive alien species" in Trondheim, Norway in 1996 (Sandlund *et al.* 1999) led to the formation in the following year of the Global Invasive Species Programme (GISP). GISP was developed explicitly as a conservation and environmental initiative, as evidenced by its mission statement: 'to conserve biodiversity and sustain human livelihoods by minimizing the spread and impact of invasive alien species'. To date, the GISP program has resulted in 27 publications.

In 1998, the European Environment Agency (EEA) highlighted the introduction of alien species as one of the principal threats to Europe's biodiversity in its year-end report on the status of the European environment (EEA 1998). In 1999, President Clinton signed Executive Order 13112, which authorized US federal agencies to prevent the introduction of alien invasive species, control the spread of alien invasive species in the United States, and restore native species and habitats that had been invaded. In 2000, IUCN (The World Conservation Union) passed 'Guidelines for the Prevention of Biodiversity Loss Caused by Alien Invasive Species' and published the guidelines on its website in 2001. In 2002, the Committee on the Scientific Basis for Predicting the Invasive Potential of Nonindigenous Plants and Plant Pests in the United States published its final report (National Research Council 2002). The final report of GISP's first phase was published in 2004 (Mooney *et al.*), and biological invasions was the theme of the 2004 annual meeting of AIBS (American Institute for Biological Sciences).

It is clear that forty-six years following the publication of Elton's 1958 classic, interest in biological invasions was stronger than ever both within and outside the scientific community, particularly with respect to the conservation and management aspects of introduced species.

LOOKING AHEAD: ANY LESSONS AND INSIGHTS FROM THE PAST?

Are there any lessons we can learn from this brief look over our shoulder? Can we gain any insights that might help us chart a productive path for the future? I think so. Of course, exactly what are those lessons and insights will likely be a matter of lively debate. Thus, the thoughts that follow are not intended to be prescriptive, but rather to focus discussion on how ecologists interested in biological invasions might best proceed from this point forward.

Two roads, one less traveled than the other

After the Asilomar conference in 1964, those interested in studying species introductions were presented with two distinct paths. They could take the strongly conservation and environmental approach advanced by Elton, or they could adopt the more value-neutral and more strictly scientific approach exhibited by *The Genetics of Colonizing Species*. The Eltonian path could be described as a top-down, deductive approach, in which an effort was made to apply general ecological theory and principles to biological invasions in order to help develop control management programs for specific invasions. Conversely, the Asilomar path could be considered more of a bottom-up, inductive approach, in which individual colonizations/invasions are examined in an effort to better inform general ecological theory and understanding of communities and populations. Another way to contrast the two paths might be to characterize the Eltonian path as a Platonic approach and the Asilomar path as an Aristotelian approach (*sensu* Sagoff 1993).

Although there have been some exceptions and occasional resistance by some ecologists over the years, invasion ecology has become increasingly more allied with the Eltonian approach, particularly in the United States, during past few decades. There are probably many explanations for this. Ecologists such as Aldo Leopold, Rachel Carson, and Paul Ehrlich were instrumental in raising the public consciousness on environmental issues which resulted in the emergence of the environmental movement during the 1970s. This movement attracted the attention not only of the general public, but also of many other ecologists. Whereas much of the focus in ecology in the 1950s and 60s had been on theoretical issues involving population dynamics and community assembly, desire, even pressure, to apply ecological knowledge to environmental problems began to be expressed, both from within the discipline and from society at large. Society in the US and elsewhere had long applied an ecological approach to introduced pests of fisheries, forests, and agricultural systems. The Eltonian path extended this approach to natural systems that were not necessarily providing direct economic benefits. Without question, the founding of the SCOPE initiative on invasions in the early 1980s played a major role in defining the conservation focus of invasion biology from that point onward. During the 1980s and 90s, ESA began to advocate a more active public role in general for ecologists, and for itself as an organization. In many ways, what had been a fairly clear distinction between "pure" ecology and "applied" ecology in the 50s, and 60s, dissolved during the latter decades of the twentieth century.

During the eighties and nineties in the United States there was increasing pressure for ecologists to justify their research in a larger social context. This was felt at all levels. To justify its budget to Congress, The National Science Foundation needed to more clearly articulate how the research it funded benefited the public good. In turn, in order to get funded by NSF, individual researchers needed to provide increasing evidence not only of the scientific value of their research but also of its broader impacts, including benefits to society. In addition, irrespective of these infrastructure pressures, some ecologists may have wanted to seek a more public forum for personal reasons, whether out of a sense of social responsibility or simply an interest in participating in policy-making and policy discussions. Probably all these factors in varying degrees played a role in the development of the strong environmental emphasis that now characterizes invasion ecology. Today, invasion ecology has become nearly an equal sibling (at least a close cousin) to the explicitly applied fields of conservation ecology and restoration ecology. One might consider the three fields as constituting a kind of triumvirate, leading the efforts to preserve the earth's biodiversity.

Most major invasion publications and conferences, whether targeted to the general public or to a more strictly scientific audience, are now explicitly framed from a conservation perspective, usually emphasizing current biodiversity threats and potential impending ecological crises, often using language and imagery similar to that used by Elton in 1958. Elton was not the father of invasion ecology, but he might be considered the founder of one path of invasion ecology. This path, the conservation and environmental path, has been well traveled and maintained during past several decades. However, the alternative path, embodied by the 1964 Asilomar conference, while certainly not wholly rejected, has not been as well traveled (Vermeij 1996). The difference has probably been most apparent during the past 20 years, and particularly in the United States. No doubt part of the explanation for this difference is that a large number of ecologists are employed by conservation groups and governmental agencies where they work primarily on applied problems.

Progress vs. effort

If one takes the time to go back and read the invasion and colonization literature from twenty, thirty, and forty years ago, one is struck by how little the questions, and, unfortunately, often the answers, seem to have changed over time. It is not easy to find either questions or answers (generalizations and theory) made in recent years that are not in the literature of thirty or forty years ago, or even earlier than that. The mission of the SCOPE invasion program was to try to answer three questions: what factors determine whether or not a species will be an invader, what properties determine whether or not a particular environment will prone to invasion, and, how can the knowledge gained from answering these questions help management efforts. The first question was actually the organizing question of the 1964 Asimolar symposium, and the second was articulated by Elton himself.

There is no question that our understanding of the process of biological invasions has increased considerably during the past forty-six years, particularly our knowledge of the details involved in particular biological invasions. However, the extent to which this understanding has enhanced our ability to develop effective management strategies (the 3rd SCOPE objective), is less clear and depends on whom you ask. Some ecologists maintain that good progress is being made and that, with the application of new analysis tools and approaches such as species screening (Daehler *et al.* 2004) and the use of climatic envelopes (Peterson 2003), there is good reason to be optimistic about invasion ecologists' ability to effectively inform land managers. Others are not so sanguine, arguing that the type of knowledge and understanding gained so far are seldom useful in specific management efforts (Moyle and Light 1996, Simberloff and Stiling 1996), meaning that efforts by invasion ecologists to effectively inform land managers continue to be frustrated (Williamson 1999, Hulme 2003).

CONCLUSION (WITH A FEW SUGGESTIONS TO CONSIDER)

Invasion ecology and conservation

All the evidence to date indicates that it is unlikely that ecologists will be able to develop a general theory of biological invasions that can be of significant practical value to land managers. Much research seems to have affirmed Asa Gray's observations of weeds more than a century ago. Trying to determine specific characteristics that make some plants weeds, Gray (1879) concluded that he "could discern nothing in the plant itself that would give it an advantage". Continuing, he wrote "the reasons for predominance may be almost as diverse as the weeds themselves".

With respect to its goal of assisting land managers in the control and management of introduced species, invasion ecology might consider emphasizing less the development of general theory and more the acquisition of system-specific knowledge. The fact that invasion ecology has consisted primarily as a series of case studies has generally been viewed as a weakness of the field (Williamson 1999). But, in reality, it may be very difficulty, perhaps even impossible, for the field to move much beyond this, at least at the level of providing useful management information. In this case, invasion ecologists might be better off viewing the "system specific" nature of the field's knowledge as a strength. Rather than lamenting the case-study nature of the field, ecologists might try to make increasing system-specific knowledge a primary objective (e.g., Rejmanek and Richardson 1996). While a land manager trying to manage the species composition of a grassland may not find invasion theory particularly helpful, the more system-specific knowledge that is available in his/her toolbox, the more successful the management efforts will be. To the extent that invasion ecology wants to inform conservation efforts, it may also make sense to for invasion ecologists to become more involved in developing formalized and comprehensive monitoring programs, which have the potential of greatly enhancing the effectiveness of management and control efforts (Lonsdale and Milton 2002).

Invasion ecology and science

The conservation and environmental emphasis in invasion ecology described above has been motivated by the conviction that ecological knowledge and theory can be used to better understand and predict biological invasions. The Asilomar approach was motivated by the opposite conviction — that biological colonizations/invasions can be viewed as natural experiments and used to inform more general ecological theory and understanding. Although there was some interest among ecologists in the early eighties to continue this emphasis (Brown and Marshall 1981, Simberloff 1981, Moulton and Pimm 1983), this approach has not been as well nurtured over the years by invasion ecology, compared to the

Eltonian conservation approach. The reinvigoration of a science of colonization and species spread as a way to study core concepts and theories of evolution and ecology would seem to be an exciting option for invasion ecology. Robert Frost's traveler could only choose one road, but there is no reason that invasion ecology couldn't embrace both a conservation and a more strictly scientific path. It has been forty years since the historic 1964 Asilomar symposium on colonization biology; perhaps it is time to organize another one.

Invasion ecology and language

High rates of species spread and introductions are here to stay. In some respects, one might question the sense of defining an ecological subdiscipline area around species introductions. Due to the abundance of introduced species in environments throughout the world, which is only going to increase in the future, it is becoming increasingly difficult to undertake any ecological study in which introduced species can be ignored. That biotic globalization is now the world in which we now live certainly does not mean we should cease efforts to control and manage the small number of introduced species that do cause great health, economic, and ecological harm (Ewel *et al.* 1999). However, it does raise questions as to the logic of dichotomously splitting species into natives and exotics, and even of using any terminology that singles out recently introduced species as a group (e.g., indigenous and non-indigenous) given that so few of these species are problematic. At the same time, particular research efforts will need to distinguish species that have been present in a region for a long time from those that have recently arrived through one means or another. Thus, from a practical perspective, categorizing species on their basis of geographic origin will be necessary in some cases (Webb 1985, Richardson *et al.* 2000, Colautti and MacIsaac 2004, Pyšek *et al.* 2004).

From a strictly scientific point of view, it would seem difficult to argue against the value of returning to the more value-neutral terminology used at the 1964 Asilomar symposium. If one is studying species introductions as a way illuminate basic ecological ideas and theory, then terms such as "introduced species", "colonizers", "new species", "new residents", and "geographic spread" seem more apt than the language often associated with the Eltonian path, e.g., "exotics", "aliens", and "invasions". Whether or not ecologists should modify their language when advocating environmental concerns will likely be a matter of disagreement within the field. Some will be concerned that, despite the best of intentions, the field's credibility and stature as a scientific discipline are being compromised when ecologists make their case to the public, and to one another, using value-laden terminology, an approach more typically associated with environmental-action groups. Those with this concern would probably be more comfortable if the field dropped the invasion/alien terminology altogether and replaced it with a terminology in which values were articulated more explicitly than implicitly. Using this approach, ecologists might refer to problem species such as zebra mussels and Japanese knotweed as 'harmful new species' or 'harmful introduced species', as opposed to using a terminology in which values are more subtly implicated, e.g., 'invasive alien species'. Others will argue that metaphoric, and even at times, hyperbolic language, is needed to capture the attention of policy makers and the general public and to persuade them of the need to act, just as Elton (1958) did when he characterized the global spread of species as a "convulsion in the world's flora and fauna", and framed the challenge confronting us as a "battle" to "determine the fate of the world", the latter a quote from Conan Doyle's book, *The Lost World.*

Looking ahead

The past twenty years have seen extraordinary development in the field of invasion ecology. With numerous national and international initiatives, the creation of two new journals, and the countless number of symposia, conferences, and workshops devoted to biological invasions, not to mention the thousands of invasion papers published, there has been little time for participants to step back and reflect on the way that the field has developed. Perhaps now would be a good time to do this before moving ahead. To what extent is invasion ecology becoming largely an applied discipline like conservation biology and restoration ecology? To what extent should the field try to revitalize efforts to study species introductions as a way to illuminate more general ecological theory, *a la* the 1964 Asilomar symposium approach? (The current book, and one similar to it (Sax *et al.* in press), appear to illustrate a revived interest in this approach.) Is it possible for invasion ecology, and individual ecologists, to embrace both a strongly value-based environmental path and a more value-neutral scientific approach to species introductions? Should an effort be made to keep the paths separate, in the way that the news pages and editorial pages are kept separate in a newspaper, or should the two paths be integrated? What language and imagery should ecologists utilize in their discourse, among themselves and with the public? These are some of the questions that the field of invasion ecology might want to address way before moving ahead.

ACKNOWLEDGMENTS

Writing a historical review of a scientific discipline is a challenge under any circumstances. Doing so while most of the participants are still alive and active in the field is a particularly daunting (some might say foolish) endeavor. I thank the many colleagues who read earlier drafts and provided thoughtful, and often detailed, comments and suggestions, particularly Curt Daehler, Dick Mack, Peter Pyšek, Dave Richardson, Mark Williamson, Marcel Rejmánek, and Dan Simberloff. The thoroughness of this chapter would not have been possible without their input. I also thank Ed Wilson for sharing his memories of the 1964 Asilomar symposium. There is no single correct characterization of the past, and my colleagues have not always fully agreed with the one I presented. Nevertheless they were always been generous and gracious in their advice during the writing of this chapter, for which I am indebted. Research and writing of this chapter was partially supported by the National Science Foundation under Grant No. 0208125.

REFERENCES

Albrect, A. B. 1964. Some observations on factors associated with survival of striped bass eggs and larvae. California Fish and Game **50**, 100-113.

Bailey, H. H. 1924. The armadillo in Florida and how it reached there. Journal of Mammology **5**, 264-265.

Baker, H. G. 1974. The evolution of weeds. Annual Review of Ecology and Systematics **5**, 1-24.

Baker, H. G. and G. L. Stebbins (eds). 1965. The Evolution of Colonizing Species. Academic Press, New York, NY.

Baskin, Y. 2002. A Plague of Rats and Rubbervines: The Growing Threat of Species Invasions Shearwater Books/Island Press, Washington D. C.

Bazzaz, F. A. 1986. Life history of colonizing plants: some demographic, genetic, and physiological features. Pages 96-110 in H. A. Mooney and J. A. Drake, editors. Ecology of Biological Invasions of North America and Hawaii. Springer-Verlag, New York, NY.

Binggeli, P. 1994. Misuse of terminology and anthropomorphic concepts in the description of introduced species. Bulletin of the British Ecological Society **25**, 10-13.

Birch, L. C. 1965. Evolutionary opportunity for insects and mammals in Australia. Pages 197-214 in H. G. Baker and G. L. Stebbins, editors. 1965. The Evolution of Colonizing Species. Academic Press, New York, NY.

Booth, B. D., S. D. Murphy and C. J. Swanson. 2003. Weed Ecology in Natural and Agricultural Systems. CABI Publishing, Wallingord, Oxon, UK.

Bradshaw, A. D., R. N. Humphries, M. S. Johnson and R. D. Roberts. 1978. The restoration of vegetation on derelict land produced by industrial activity. Pages 249-278 in M. W. Holdgate and M. J. Woodman, editors. The Breakdown and Restoration of Ecosystems. Proceedings, Conference, Rehabilitation of Severely Damaged Land and Freshwater Ecosystems in Temperate Zones, Reykjavik, Iceland, July 1976. Plenum Press, New York; USA.

Bright. C. 1998. Life Out of Bounds: Bioinvasion in a Borderless World. New York: W. W.

Brown, A. H. D. and D. R. Marshall. 1981. Evolutionary changes accompanying colonization in plants. Pages 351-363 in G. G. E. Scudder and J. L. Reveal, editors. Evolution Today. Carnegie-Mellon University Press, Pittsburgh, PA.

Bryant, H. C. 1927. The opossum reaches San Diego County. California Fish and Game **13**, 146.

Bump, G. and W. H. Bohl 1964. Summary of foreign game bird propagation and liberations 1960-1963. Special Scientific Report, Wildlife No. 80. Bureau of Sport Fisheries and Wildlife.

Burdon, J. J. and G. A. Chilvers. 1977. Preliminary studies on a native Australian eucalypt forest invaded by exotic pines. Oecologia **31**, 1-12.

Campbell, A. C. and R. F. G. Ormond. 1970. The threat of the "Crown-of-thorns" starfish (*Acanthaster planci*) to coral reefs in the Indo-Pacific area: Observations on a normal population in the Red Sea, Biological Conservation **2**, 246-251.

Carlton, J. T. 1996. Marine bioinvasions: the alteration of marine ecosystems by nonindigenous species. Oceanography **9**, 36-43.

Carlton, J. T. 1996. Biological invasions and cryptogenic species. Ecology **77**, 1653-1655.

Carlton, J. T. 1999. Molluscan invasions in marine and estuarine communities. Malacologia **41**, 439-454.

Child, L., J. H. Brock, G. Brundu, K. Prach, P. Pyšek, P. M. Wade and M. Williamson. (editors). 2003. Plant Invasions: Ecological Threats and Management Solutions. Backuys Publishers, Leiden, The Netherlands.

Christie. W. J. 1972. Lake Ontario: effects of exploitation, introductions and eutrophication on the salmonid community. Journal of the Fisheries Research Board of Canada **29**, 913-929.

Clement, E. J. and M. C. Foster. 1994. Alien Plants of the British Isles. A Provisional Catalogue of Vascular Plants (excluding grasses). Botanical Society of the British Isles, London.

Colautti, R. I. and H. J. MacIsaac. 2004. A neutral terminology to define 'invasive' species. Diversity and Distributions **10**, 135-141.

Courtenay, W. R., Jr. and C. R. Robins, 1975. Exotic organisms: An unsolved, complex problem. Bioscience **25**, 306-313.

Crawley, M. J. 1986. The population biology of invaders. Philosophical Transactions of the Royal Society B **314**, 711-731.

Crawley, M. J. 1987. What makes a community invasible? Symposia of the British Ecological Society **26**, 429-53.

Daehler C. C. and D. Carino D. 2001. Hybridization between native and alien plants and its consequences. Pages 81-102 in J. L. Lockwood J. L. and M. McKinney, editors. Biotic Homogenization. KluwerAcademic/Plenum Publishers, New York.

Daehler, C. C. and D. A. Carino. 2000. Predicting invasive plants: Prospects for a general screening system based on current regional models. Biological Invasions **2**, 92-103.

Daehler, C. C., J. S. Denslow, S. Ansari and H. Kuo. 2004. A risk-assessment system for screening out invasive pest plants from Hawaii and other Pacific Islands. Conservation Biology **18**, 1-9.

Daugherty, H. E. 1972. The impact of man on the zoogeography of El Salvador. Biological Conservation **4**, 273-278.

D'Antonio, C. and L. A. Meyerson. 2002. Exotic plant species as problems and solutions in ecological restoration: a synthesis. Restoration Ecology **10**, 703-713.

Davis, M. A. 2003. Biotic globalization: does competition from introduced species threaten biodiversity? Bioscience **53**, 481-489.

Davis, M. A., J. P. Grime and K. Thompson. 2000. Fluctuating resources in plant communities: a general theory of invasibility. Journal of Ecology **88**, 528-536.

Davis, M. A., J. P. Grime and K. Thompson. 2001. Charles S. Elton and the dissociation of invasion ecology from the rest of ecology. Diversity and Distributions **7**, 97-102.

Davis, M. A. and M. Pelsor 2001. Experimental support for a resource-based mechanistic model of invasiblity. Ecology Letters **4**, 421-428.

Davis, M. A., *et al.* (15 coauthors) 2005. Vegetation change: a reunifying concept in plant ecology. Perspectives in Plant Ecology, Evolution, and Systematics **7**, 69-76.

DeBach, P. (ed). 1964. Biological Control of Insect Pests and Weeds. Reinhold, New York, NY.

Drake, J. A., H. A. Mooney, F. di Castri, R. H. Groves, F. J. Kruger, M. Rejmanek and M. Williamson. 1989. Biological Invasions: A Global Perspective. SCOPE 37, John Wiley & Sons, Chichester, UK.

Dukes, J. S. and H. A. Mooney. 1999. Does global change increase the success of biological invaders. Trends in Ecology and Evolution **14**, 135-139.

EEA (European Environmental Agency). 1998. Environment in the European Union at the turn of the century.

Egler, F. E. 1942. Indigene versus alien in the development of arid Hawaiian vegetation. Ecology **23**, 14-23.

Ellstrand N. C. and K. A. Schierenbeck. 2000. Hybridization as a stimulus for the evolution of invasiveness in plants. Proceedings of the National Academy of Sciences **97**, 7043-7050.

Elton, C. S. 1958. The Ecology of Invasions by Animals and Plants. Methuen, London.

Embree, D. G. 1979. The ecology of colonizing species, with special emphasis on animal invaders. Pages 51-65 in D. J. Horn, G. R. Stairs and R. D. Mitchell, editors. Analysis of Ecological Systems. Ohio State University Press, Columbus, OH.

Eser, U. 1998. Assessment of plant invasions: theoretical and philosophical fundamentals. Pages 95-107 in U. Starfinger, K. Edwards, I. Kowarik and M. Williamson. Plant Invasions: Ecological Mechanisms and Human Responses. Backhuys Publishers, Leiden, The Netherlands.

Essl, F. and W. Rabitsch (editors) 2002. Neobiota in Österreich. Umweltbundesamt GmbH, Wien.

Ewel, J. J., *et al.* 1994. Deliberate introductions of species: research needs. Bioscience **49**, 619-630.

Faliński, J. B. 1966. Antropogeniczna roslinnosc Pusczy Bialowieskiej jako wynik synantropizacji naturalnego kompleksu lesnego. Vegetation anthropogene de la Grande Foret de Bialowieza comme un resultat de la synanthropisation du terrotoire silvestre naturel. Disseratationes Universitatis Varsoviensis **13**, 1-256.

Faliński, J. B. 1968. Stadia neofityzmu I stosunek neofitow do innych komponentow zbiorowiska. Stages of neophytism and the reaction of neophytes to other components of the community. In J. B. Faliński, editor. Synantropizacja szaty roslinnej. I. Neofityzm i apofityzm w szacie roslinnej Polski. Materialy Sympozjum w Nowogrodzie. Mater. Zakl. Fitosoc. Stos. UW **25**, 15-31.

Faliński, J. B. 1969. Neofity i neofityzm. Dyskusje fitosocjologiczne (5). Neophytes et neophytisme. Discussion phytosociologiques (5). Ekol. pol. B. **15**, 337-355.

Faliński J. B. (ed.) (1971): Synanthropisation of plant cover. II. Synanthropic flora and vegetation of towns connected with their natural conditions, history and function.

— Mater. Zakl. Fitosoc. Stos. Univ. Warsz. **27**, 1-317.

Goodwin, B. J., A. J. McAllister and L. Fahrig, L. 1999. Predicting invasiveness of plant species based on biological information. Conservation Biology **13**, 422-426.

Gould, S. J. 1998. An evolutionary perspective on strengths, fallacies, and confusions in the concept of native plants: Arnoldia. **58**, 2-10.

Gray, A. 1879. The pertinacity and predominance of weeds. The American Journal of Science and Arts, Vol XVIII.

Groves, R. H. and J. J. Burdon. (editors) Ecology of Biological Invasions: An Australian Perspective. Australian Academy of Science, Canberra, pp. 137-49.

Harper, J. L. 1965. Establishment, aggression, and cohabitation in weedy species. Pages 245-265 in H. G. Baker and G. L. Stebbins, editors. 1965. The Evolution of Colonizing Species. Academic Press, New York, NY.

Hejný, S. V. and M. Lhotská 1964.

Hejný S., V. Jehlík, K. Kopecký, Z. Kropáč Z. and M. Lhotská. . 1973. Karanténní plevele Československa. — Studie Čs. Akad. Věd., ser. math. -natur., **8**, 1-156.

Holdgate, M. W. and M. J. Woodman (editors). 1978. The Breakdown and Restoration of Ecosystems. Proceedings, Conference, Rehabilitation of Severely Damaged Land and Freshwater Ecosystems in Temperate Zones, Reykjavik, Iceland, July 1976. Plenum Press, New York; USA.

Holub J. and V. Jirásek. 1967. Zur Vereinheitlichung der Terminologie in der Phytogeographie. Folia Geobot. Phytotax. **2**, 69-113.

Horn, D. J., G. R. Stairs and R. D. Mitchell (editors). 1979. Analysis of Ecological Systems. Ohio State University Press, Columbus, OH.

Hulme, P. E. 2003. Biological invasions: winning the science battles but losing the conservation war? Oryx **37**, 178-193.

Huston, M. A. 1994. Biological Diversity: The Coexistence of Species on Changing Landscapes. Cambridge Univ. Press. Cambridge, UK.

Jehlik, V. and B. Slavik. 1968. Beitrag zum Erkennen des Verbreitungscharakters der Art *Bunias orientalis* L. in der Tschechoslowakei. Preslia 40: 274-293.

Jehlík V. and S. Hejný. 1974. Main migration routes of adventitious plants in Czechoslovakia. Folia Geobot. Phytotax. **9**, 241-248.

Johnstone, I. M 1986. Plant invasion windows: a time based classification of invasion potential. Biological Reviews **61**, 369-394.

Jordan, William R. III. 1981. Restoration and Management Notes: a beginning. Restoration and Management Notes **1**, 2.

Juhasz-Nagy, P. 1964. Investigations concerning ecological homeostasis. Abstract, Proceedings of the Tenth International Botanical Congress, Edinburgh.

Kartesz, J. T. and C. A. Meacham 1999. Synthesis of the North American Flora. Version 1. 0. North Carolina Botanical Garden, Chapel Hill, North Carolina.

Keane R. M. and M. J. Crawley. 2002. Exotic plant invasions and the enemy release hypothesis. Trends in Ecology and Evolution. **17**, 164-170,

King, C. 1984. Immigrant Killers. Oxford University Press, Auckland, NZ.

Klironomos, J. N. 2002. Feedback with soil biota contributes to plant rarity and invasiveness in communities. Nature **417**, 67-70.

Klotz, S., I. Kühn and W. Durka. 2003. BIOLFLOR — Eine Datenbank zu biologisch-öko-logischen Merkmalen der Gefäßpflanzen in Deutschland. Bundesamt für Naturschutz, Bonn. Klironomoz.

Knops, J. M. H., D. Tilman, N. M. Haddad, S. Naeem, C. E. Mitchell, J. Haarstad, M. E. Ritchie, K. M. Howe, P. B. Reich, E. Siemann and J. Groth. 1999. Effects of plant species richness on invasion dynamics, disease outbreaks, insect abundances and diversity. Ecology Letters **2**, 286-293.

Kohler, A. and H. Sukopp. 1964. Über die Gehölzentwicklung auf Berliner Trümmerstandorten. Zugleich ein Beitrag zum Studium neophytischer Holzarten. Berichte der Deutschen Botanischen Gesellschaft **76**, 389-407.

Kolar, C. and D. M. Lodge. 2001. Progress in invasion biology: predicting invaders. Trends in Ecology and Evolution **16**, 199-204.

Kornás, J. 1968a. Stadia neofityzmu i stosunek neofitow do innych komponentow zbiorowiska. Stages of neophytism and the relation of neophytes to other components of the community. In J. B. Faliński, editor. Synantropizacja szaty roslinnej. I. Neofityzm i apofityzm w szacie roslinnej Polski. Materialy Sympozjum w Nowogrodzie. Mater. Zakl. Fitosoc. Stos. UW **25**, 15-31.

Kornás, J. 1968b. Geograficzno-historyczna klasyfikacja roslin synantropijnych. A geographical-historical classification of synanthropic plants. In J. B. Faliński, editor. Synantropizacja szaty roslinnej. I. Neofityzm i apofityzm w szacie roslinnej Polski. Materialy Sympozjum w Nowogrodzie. Mater. Zakl. Fitosoc. Stos. UW **25**, 33-41.

Kornberg, H. and M. H. Williamson (eds). 1987. Quantitative Aspects of the Ecology of Biological Invasions. The Royal Society, London.

Kowarik, I. (1984). *Platanus hybrida* Brot. und andere adventive Gehölze auf städtischen Standorten in Berlin (West). Göttinger Floristische Rundbriefe **18**, 7-17.

Kowarik, I. 2001: Biological Invasions as result and vector of global change. Pages 80-88 in Contributions to Global Change Research. A Report by the German National Committee on Global Change Research, Bonn 2001.

Laycock, G. 1966. The Alien Animals. The Natural History Press, Garden City, NJ.

Levine, J. M. 2000. Species diversity and biological invasions: relating local process to community pattern. Science **288**, 852-854.

Levine, J. M. and C. M. D'Antonio. 1999. Elton revisited: A review of evidence linking diversity and invasibility. Oikos **87**, 15-26.

Little, E. V. 1916. The opossum in Los Angeles County. California Fish and Game **2**, 46-47.

Lodge, D. M. 1993 Biological invasions: lessons from ecology. Trends in Ecology and Evolution **8**, 133-136.

Lonsdale, W. M. 1999. Global patterns of plant invasions and the concept of invasibility. Ecology **80**, 1522-1536.

Lonsdale, W. M. and S. Milton. 2002 People and plant invasions of the rangelands. Pages 101-115 in A. C. Grice and K. C. Hodgkinson, editors. Global Rangelands: Progress and Prospects. CAB International, Wallingford UK.

Macdonald, I. A. W. and M. L. Jarnum (editors). 1984. Invasive Alien Organisms in the Terrestrial Ecosystems of the Fynbos Biome, South Africa. South African National Scientific Programmes Report No. 85, CSIRO, Pretoria.

Macdonald, I. A. W., F. J. Kruger and A. A. Ferrar. 1986. The Ecology and Management of Biological Invasions in Southern Africa. Oxford University Press, Cape Town.

Mack, R. N. 1981. Invasion of *Bromus tectorum* into western North America : an ecological chronicle. Agroecosystems. **7**,145-65.

MacArthur, R. H. and E. O. Wilson. 1963. An equilibrium theory of insular zoogeography. Evolution **17**, 373-387.

MacArthur, R. H. and E. O. Wilson. 1967. The Theory of Island Biogeography. Princeton University Press, Princeton.

Mayr, E. 1965. The nature of colonizations in birds. Pages 29-43 in H. G. Baker and G. L. Stebbins, editors. 1965. The Evolution of Colonizing Species. Academic Press, New York, NY.

Meinez, A. (translated by D. Simberloff) 1999. Killer Algae. University of Chicago Press, Chicago, Illinois.

Metcalf, C. L., W. P. Flint and R. L. Metcalf. 1962. Destructive and Useful Insects, 4th edition. McGraw-Hill, New York, NY.

Merriam, C. H. 1889. The English sparrow in North America. US Department of Agriculture Bulletin 1.

Mooney, H. A. and J. A. Drake (editors). 1986. Ecology of Biological Invasions of North America and Hawaii. Springer-Verlag, New York, NY.

Mooney, H. A. and J. A. Drake. 1989. Biological invasions: a SCOPE program overview. Pages 491-506 in J. A. Drake, H. A. Mooney, F. di Castri, R. H. Groves, F. J. Kruger, M. Rejmanek and M. Williamson. Biological Invasions: A Global Perspective. SCOPE 37, John Wiley & Sons, Chichester, UK.

Mooney, H. A., R. N. Mack, J. A. McNeely, L. E. Neville, P. J. Schei and J. K. Waage (editors). 2004. Invasive Alien Species: A New Synthesis. Island Press, Washington, D. C.

Moulton, M. P. and S. L. Pimm. 1983. The introduced Hawaiian avifauna: biogeographic evidence for competition. American Naturalist **121**, 669-690.

Moulton, M. P. and S. L. Pimm. 1986. Species introductions to Hawaii. Pages 231-249 in H. A. Mooney and J. A. Drake, editors. Ecology of Biological Invasions of North America and Hawaii. Springer-Verlag, New York, NY.

Moyle, P. B. 1973. Effects of introduced bullfrogs, Rana catesbeiana, on the native frogs of the San Joaquin Valley, California. Copeia **1973**, 18-22.

Moyle, P. B. and T. Light. 1996. Biological invasions of freshwater: empirical rules and assembly theory. Biological Invasions **78**, 149-161.

Muhlenbach, V. 1979. Contributions to the synanthropic (adventive) flora of the railroads in St. Louis, Missouri, U. S. A. Annals of the Missouri Botanical Garden **66**, 1-108.

National Research Council. 2002. Predicting Invasions of Nonindigenous Plants and Plant Pests. National Academy of Sciences, Washington D. C., 198 pp.

Nichols, J. O. 1961. The gypsy moth in Pennsylvania — its history and eradication. Pennsylvania Department of Agriculture Miscellaneous Bulletin No. 4404.

Nitecki, H. (editor). 1981. Biotic Crises in Ecological and Evolutionary Time. Academic Press, New York, NY.

Owre, O. T. 1973. A consideration of the exotic avifauna of southeastern Florida. Wilson Bulletin **85**, 495.

Palmer, T. S. 1899. The danger of introducing noxious animals and birds. US Department of Agriculture Yearbook for 1898, 87-110.

Peterson. A. T. 2003. Predicting the geography of species' invasions via ecological niche modeling. Quarterly Review of Biology **78**, 419-434.

Pimm, S. L. 1989. Theories of predicting success and impact of introduced species. Pages 351-368 in J. A. Drake, H. A. Mooney, F. di Castri, R. H. Groves, F. J. Kruger, M. Rejmanek and M. Williamson, editors. Biological Invasions: A Global Perspective. SCOPE 37, John Wiley & Sons, Chichester, UK.

Pyšek P. 1995. On the terminology used in plant invasion studies. Pages 71-81 in P. Pyšek, K. Prach, M. Rejmánek and M. Wade, editors. Plant Invasions: General Aspects and Special Problems. SPB Academic Publishing, Amsterdam.

Pyšek P., J. H. Brock, K. Bímová, B. Mandák, V. Jarošík, I. Koukolíková, J. Pergl and J. Štěpánek. 2003. Vegetative regeneration in invasive *Reynoutria* (Polygonaceae) taxa: the determinant of invasibility at the genotype level. American Journal of Botany. **90**, 1487-1495.

Pyšek P., D. M. Richardson, M. Rejmánek, G. Webster, M. Williamson and J. Kirschner 2004. Alien plants in checklists and floras: towards better communication between taxonomists and ecologists. Taxon **53**, 131-143.

Rejmánek, M. 2003. The rich get richer – responses. Frontiers in Ecology and the Environment **1**, 122-123.

Rejmánek, M. and D. M. Richardson. 1996. What attributes make some plant species more invasive? Ecology **77**, 1655-61.

Richardson, D. M., N. Allsopp, C. M. D'Antonio, S. J. Milton and M. Rejmánek. 2000. Plant invasions — the role of mutualisms. Biological Reviews **75**, 65-93.

Richardson, D. M., P. Pyšek, M. Rejmánek, M. G. Barbour, F. D. Panetta and C. J. West. 2000. Naturalization and invasion of alien plants: concepts and definitions. Diversity & Distributions **6**, 93-107.

Renne, I. J. and B. F. Tracy. 2003. The rich get richer – responses. Frontiers in Ecology and the Environment **1**, 122.

Rosenzweig, M. L. 2001 The four questions: What does the introduction of exotic species do to diversity. Evolutionary Ecology Research **3**, 361-367.

Sagoff, M. 1993. Biodiversity and the Culture of Ecology. Bulletin of the Ecological Society of America **74**, 374-381.

Sagoff, M. 1999. What's wrong with alien species? Report of the Institute for Philosophy and Public policy, University of Maryland, College Park, Maryland.

Sandlund, O. T., P. J. Schei and A. Viken (editors). 1999. Proceedings of the Norway/UN Conference on Alien Species. Directorate for Nature Management and Norwegian Institute for Nature Research. Trondheim, Norway.

Salisbury, E. 1961. Weeds and Aliens. Collins Publishing, London.

Sax, D. F. and S. D. Gaines. 2003. Species diversity: from global decreases to local increases. Trends in Ecology and Evolution **18**, 561-566

Sax, D. F., J. J. Stachowicz, and S. D. Gaines (editors) 2005. *Species Invasions: Insights into Ecology, Evolution, and Biogeography*. Sinauer Associates. Sunderland, MA.

Schroeder, F. -G. 1969. Zur Klassifizierung der Anthropochoren. Vegetatio **16**, 225-238.

Schofield, E. K. 1973. Galapagos flora: the threat of introduced plants. Biological Conservation: 48-51.

Shea K. and P. Chesson. 2002. Community ecology theory as a framework for biological invasions. Trends in Ecology and Evolution **17**, 170-176.

Shigesada, N and K. Kawasaki. 1997. Biological Invasions: theory and practice. Oxford Series in Ecology and Evolution. Oxford University Press, Oxford, UK.

Sim, T. R. 1927. Some effects of man's influence on the South African flora. South African Journal of Science **23**, 492-507.

Simberloff, D. 1981. Community effects of introduced species. Pages 53-81 in H. Nitecki, editor. Biotic Crises in Ecological and Evolutionary Time. Academic Press, New York, NY.

Simberloff, D. 1986. Introduced insects: a biogeographic and systematic perspective. Pages 3-26 in H. A. Mooney and J. A. Drake, editors. Ecology of Biological Invasions of North America and Hawaii. Springer-Verlag, New York, NY.

Simberloff, D. 1989. Which insect introductions succeed and which fail? Pages 61-75 in J. A. Drake, H. A. Mooney, F. di Castri, R. H. Groves, F. J. Kruger, M. Rejmanek and M. Williamson. Biological Invasions: A Global Perspective. SCOPE 37, John Wiley & Sons, Chichester, UK.

Simberloff, D. and P. Stiling. 1996. Risks of species introduced for biological control. Biological Conservation **78**, 185-192.

Simberloff, D. S. and E. O. Wilson. 1969. Experimental zoogeography of islands: the colonization of empty islands. Ecology **50**, 278-296.

Slobodkin, L. B. 2001. The good, the bad and the reified. Evolutionary Ecology Research **3**, 1–13.

Starfinger, U., K. Edwards, I. Kowarik and M. Williamson. 1998. Plant Invasions: Ecological Mechanisms and Human Responses. Backhuys Publishers, Leiden, The Netherlands.

Stohlgren, T. J., D. Binkley, G. W. Chong, M. A. Kalkhan, L. D. Schell, K. A. Bull, Y. Otsuki, G. Newman, M. Bashkin and Y. Son. 1999. Exotic plant species invade hot spots of native plant diversity. Ecological Monographs **69**, 25-46.

Stohlgren, T. J., D. T. Barnett and J. T. Kartesz. 2003. The rich get richer:patterns of plant invasions in the United States. Frontiers in Ecology and the Environment **1**, 11-14.

Sukopp, H. 1962. Neophyten in naturlichen Pflanzengesellschaften Mitteleuropas. Ber. Deutsch. Bot. Ges. **75**, 193-205.

Sukopp, H. 1971. Beiträge zur Ökologie von *Chenopodium botrys* L. I. Verbreitung und Geschichte. Verhandlungen des Botanischen Vereins der Provinz Brandenburg: **108**, 3-25.

Sukopp, H. and P. Werner. 1983. Urban environments and vegetation. In W. Holzner, M. J. A. Werger and I. Ikusima, editors. Man's Impact on Vegetation. Geobotany **5**, 247-260. Junk, Den Haag, Boston, London.

Taylor, H. C. 1969. Pest plants and nature conservation in the winter rainfall region. Journal of the Botanical Society of South Africa **55**, 32-38.

TCWF (Texas Chapter of the Wildlife Society). 1967. Introduction of Exotic Animals: Ecological and Sociological Considerations. Caesar Kleberg Research Program in Wildlife Ecology, Texas A & M University, College Station, Texas.

Thompson, K., J. G. Hodgson and T. C. G. Rich. 1995. Native and alien invasive plants: more of the same? Ecography **18**, 390-402.

Tilman, D. 1997. Community invasibility, recruitment limitation and grassland biodiversity. Ecology **78**, 81-92.

Todd, K. 2001. Tinkering with Eden: A Natural History of Exotics in America. W. W. Norton & Company, New York, NY.

Usher, M. B. 1988a. Biological invasions of nature reserves. *Biological Conservation* **44**, 1-135.

Usher, M. B. 1988b. Biological invasions of nature reserves: a search for generalisations. Biological Conservation **44**, 119-135.

Van Driesche, J. and R. Van Driesche. 2000. Nature Out of Place: Biological Invasions in the Global Age. Island Press, Washington, D. C.

Vermeij, G. J. 1996. An agenda for invasion biology. Biological Conservation **78**, 3-9.

Vila, M. and C. D'Antonio. 1998. Hybrid vigor for clonal growth in Carpobrotus (Aizoaceae) in coastal California. Ecological Applications **8**, 1196-1205.

Waddington, C. H. 1965. Introduction to the symposium. Pages 1-6 in H. G. Baker and G. L. Stebbins, editors. 1965. The Evolution of Colonizing Species. Academic Press, New York, NY.

Wardle, D. A. 2001. Experimental demonstration that plant diversity reduces invasibility — evidence of a biological mechanism or a consequence of sampling effect. Oikos **95**, 161-170.

Warner, R. E. 1968. The role of introduced diseases in the extinction of the endemic Hawaiian avifauna. Condor **70**, 101-120.

Webb, D. A. 1985 What are the criteria for presuming native status? Watsonia **15**, 231-236.

Wilson, E. O. 1965. The challenge from related :jb. Pages 7-24 in H. G. Baker and G. L. Stebbins, editors. 1965. The Evolution of Colonizing Species. Academic Press, New York, NY.

Wicht, C. L. 1945. Preservation of the vegetation of the South-western Cape. Special Publication of the Royal Society of South Africa, Cape Town.

Williamson, M. 1993. Invaders, weeds and the risk from genetically modified organisms. Ecperientia **49**, 219-224.

Williamson, M. 1996. Biological Invasions. Chapman & Hall, London.

Williamson, M. 1999. Invasions. Ecography **22**, 5-12.

Wodzicki, K. 1965. The status of some exotic vertebrates in the ecology of New Zealand. Pages 425-458 in H. G. Baker and G. L. Stebbins, editors. 1965. The Evolution of Colonizing Species. Academic Press, New York, NY.

Chapter four

Invasiveness in exotic plants: immigration and naturalization in an ecological continuum

H. T. Murphy, J. VanDerWal,
L. Lovett-Doust and J. Lovett-Doust

INTRODUCTION

A major goal of invasions ecology is to explain why some species are more successful as invaders than others, and ultimately to be able to predict traits predisposing a species to success in adventive environments. Unfortunately, to date few clear generalities exist and biologists are far from being able to predict future invasions. Recently several groups have shown how the lack of consensus on basic terminology has seriously limited progress in and application of invasions ecology, as well as its capacity to draw systematically on theory from ecological disciplines (Davis and Thompson 2000; Richardson *et al.* 2000a; Daehler 2001a; Davis and Thompson 2001, 2002; Rejmánek *et al.* 2002; Colautti and MacIsaac 2004).

Within the literature, in regard to very similar ecological situations a particular species may be regarded as alien, exotic, invasive, colonizing, weedy, introduced, non-indigenous, imported or non-native. Moreover a weedy species may

M.W. Cadotte et al. (eds.), Conceptual ecology and invasion biology, 65–105.

also be referred to as a naturalized-, transformer-, sleeper-, established-, casual-, persistent-, noxious-, native-, environmental-, or agricultural weed. Many of these terms connote only that a species is novel to a particular area. Others carry implications about actual or potential impact in novel habitats (see Davis and Thompson [2000] and Colautti and MacIsaac [2004] for a treatment of other terms). Here we are concerned primarily with species that are termed 'invasive' or 'invaders'. The scope of this notion alone covers much ground and in recent years debate over the criteria for regarding a species as invasive has received much attention (e.g., Davis and Thompson 2000; Richardson *et al.* 2000a; Davis and Thompson 2002; Rejmánek *et al.* 2002; Pyšek *et al.* 2004) and the term continues to be used inconsistently. The debate on terminology is more than simple semantics because the use of inconsistent terms impairs understanding of the processes they are meant to describe (Colautti and MacIsaac 2004).

Comparative analysis of alien species in regional floras has been used to characterize both invasive species and invaded ecosystems (e.g., Daehler, 1998; Pyšek, 1998; Lonsdale, 1999). However, such comparisons are flawed if the alien species in these floras have not been classified in the same ways. Researchers are frequently unable to distinguish clearly the particular characteristics of alien invaders (Thompson *et al.* 1995); discrepancies in terminology may contribute to this problem and confound meta-analyses.

To be most useful, it is logical that the terminology used to describe species at various stages of invasion should closely reflect the process by which invasions occur, and the various characteristics of those stages. The first part of this chapter is a review of the stages in the process of invasion. Later we test whether two recent classification schemes proposed by Richardson *et al.* (2000a, and see Pyšek *et al.* 2004) and Davis and Thompson (2000) are operational given a relatively rich species dataset and whether they succeed in identifying the same sets of species. We classify a sample of the exotic flora of Ontario using the two schemes. We identify limits to information available in this species dataset and show how the two classification systems can lead to different conclusions.

It is necessary that, for the purposes of clarity, we provide preliminary definitions of certain terms we will use in order to review the invasion process, before looking in more depth at the different approaches to classifying species at different stages of the process. Here we present some of the most common terms to describe generally the different stages of invasion, though we realize that these terms are often used interchangeably or as synonyms for other terms. Terms include "introduction" (introduced species are those that appear outside their native range), "establishment" (established species persist over relatively long periods outside their native range), "naturalization", (naturalized species reproduce and sustain populations over relatively long periods) and "invasion" (invasive species are widespread and abundant outside their native range).

In the final section of the chapter we describe a way to visualize and model the invasions process based on the status of populations of non-native species at a specific spatial and temporal scale. We draw on the 'blinking lights' metaphor of metapopulation ecology and show how this may be useful in describing the invasions process. ("Blinking lights" describes the colonization (light on) and extinction (light off) of populations in patches of habitat over time across a regional landscape.) Thus when population turnover is high (high colonization and high extinction) populations are rapidly blinking on and off, whereas where turnover is slower (high colonization, low extinction), populations blink on and remain on for extended periods. It is possible to envision this blinking on of populations during the invasion process, first as the non-native species is introduced, then establishes, naturalizes and finally as it becomes invasive. The probability of the population blinking off would decrease with time and with progress through the process. We attempt to show how this process may usefully be modeled through the use of population viability analysis (PVA), a tool from conservation biology for tracking populations of imperiled species. Indeed, populations of non-native species are characterized by the same large-scale processes (immigration, emigration, colonization and extinction) as native populations and so it makes sense to utilize the terminology and well-tested tools available to understand regional dynamics of native populations. We propose a modification to the Richardson *et al.* (2000a) framework for the invasion process, promoting a focus on demographic elements of casual, naturalized, spreading and invasive *populations*, rather than characterizing the species themselves with this terminology.

THE INVASION PROCESS: INTRODUCTION, ESTABLISHMENT AND NATURALIZATION

'Invasion' is considered to be the least likely outcome of a multistage process that begins when organisms are transported outside their native range (Mack *et al.* 2000). Moreover, it is commonly noted that many if not most introductions fail, and that even when a species becomes invasive in a new location there may have been numerous previous unsuccessful introductions (e.g., Kowarik 1995). This is well illustrated in birds by the classic example of the European starling (*Sturnus vulgaris*) which successfully colonized New York's Central Park in 1891, and then most of North America, but only after many failed introductions (Krebs 1985). Anderson *et al.* (2005) argue that most of the important infectious diseases of plants similarly depend upon multiple introductions for their success. Williamson and Fitter's (1996) 'tens rule' holds that just one in ten of those individuals transported to a new location will appear in the wild (i.e., become introduced); and only one in ten of those introduced will become established; and one in ten of those established will spread and become invasive.

Those individuals that do survive the initial transit are unlikely to persist for long in the new environment, due to a range of physical and biotic elements associated with the invading species itself, the recipient community and the rest of the environment. For example, Lodge (1993) described 'ecological resistance' as composed of demographic, biotic and environmental factors limiting the success of invaders. Other often-cited explanations for the success or failure of introductions are the 'enemy release' hypothesis, genetic factors and demographic and environmental stochasticity (see below).

'Demographic resistance' involves particular life-history parameters of the invading species influencing (typically) propagule pressure to the system (Von Holle *et al.* 2003). Successful introduction and establishment is often reported to be strongly related to propagule pressure (Kolar and Lodge 2001). The effectiveness of different barriers limiting the invasion process may depend on the *number* of propagules: the greater the number of propagules the greater the chance of a barrier being overcome and invasion proceeding (Rouget and Richardson 2003). Thus factors influencing the number and dispersal potential of propagules are likely to be critical in regulating the establishment and future spread of introduced species. For example, Rejmánek and Richardson (1996) found that invasive pine species had smaller seed mass, a shorter juvenile period, and shorter intervals between seed crops compared with introduced but non-invasive pines.

In both natural and human-facilitated introductions, varying inoculum numbers may bias observed patterns of invasion and lead to erroneous conclusions about the importance of other demographic or biotic parameters in invasion success. For example, a species could have a high rate of introduction success either because it was inherently a good invader, or because it was consistently introduced to locations that were easy to invade, or because it was introduced more frequently at difficult locations. Propagule pressure has been implicated in the introduction success of plants (Foxcroft *et al.* 2004, Rouget and Richardson, 2003), birds (Blackburn and Duncan 2001) and mammals (Forsyth and Duncan 2001). Studies of plants have shown that models incorporating propagule pressure are superior to those invoking only environmental parameters in explaining distribution and abundance of non-native species (Foxcroft *et al.* 2004, Rouget and Richardson, 2003). In an analysis of 1378 introduction events for 426 bird species across the world, Blackburn and Duncan (2001) found the broad pattern of introduction reflected patterns of European settlement, the proximity of species to introduction locations and the desirability of species in certain taxonomic families.

'Biotic resistance' refers to features shaping the invaded community and affecting its 'invasibility', which is often said to decrease with increasing species richness of the invaded community (Elton 1958; Case 1990; Tilman 1997). The idea that a community's resident biotic richness is inversely related to invasibility assumes that competition is a major force controlling community composition, and that diverse communities should use resources more fully, leaving little niche space for potential invaders (Bruno *et al.* 2003; but see Stohlgren *et al.* 1999). Besides competition, factors considered to be the major components of biotic resistance to exotics include predation, herbivory, and disease/parasite pressures (Levine *et al.* 2004).

Recently attention has been given to the role of facilitative interactions in promoting community coexistence of invasive species with native species, as well as with other invasive species (Richardson *et al.* 2000b). There is some evidence suggesting that invasions may depend on, or be enhanced by, the establishment of mutualisms or facilitative interactions between the invasive plant and other organisms (either introduced or native) (Richardson *et al.* 2000b). There are many examples of invasive weed species that are obligate outcrossers and require a pollinator mutualist to reproduce (e.g., purple loosestrife, *Lythrum salicaria* [Mal *et al.* 1992], and *Centaurea diffusa* and *C. maculosa* [Harrod and Taylor, 1995]). Carr (1993) reported that the invasive flora of Victoria, Australia is pollinated by a variety of both native and introduced insects (mainly honeybees, *Apis mellifera*).

Introduced plant species which alter soil nutrient availability can facilitate the invasion of other non-indigenous species. For example, the Atlantic nitrogen-fixing shrub *Myrica faya* has invaded nitrogen-poor volcanic soils in Hawaii and probably facilitates further plant invasions (Vitousek and Walker 1989, and see Simberloff and Von Holle 1999 for other examples). The interaction between plants and their soil community can result in dynamic feedback. Empirical evidence suggests a net positive feedback between plants and their soil community in communities outside their native range. Reinhart *et al.* (2003) found that the invasion of black cherry (*Prunus serotina*) into north western Europe was facilitated by features of the soil community. In its native range the soil community that develops around black cherry inhibits the growth of nearby conspecifics, whereas the opposite occurs in the non-native range. *Centaurea diffusa* has much stronger negative effects on grass species in its non-native North American range than in its native European range due to differences in the effects of its root exudates and how these affect competition for resources (Callaway and Aschehoug 2000).

Environmental resistance may involve abiotic resource availability (suitable climatic factors or light intensities, for example) and may preclude a species altogether, or may induce a significant time lag between arrival and establishment of a species (Von Holle *et al.* 2003). Blackburn and Duncan (2001) found, using a global dataset, that the success of introduced birds depends on the suitability of the abiotic environment for the species at the introduction site. For example, introduction success was significantly greater when the difference between a species' latitude of origin and the latitude of introduction was small.

The 'enemy release' hypothesis has been used to explain the success of some species in communities outside their native range, and states that outside of their native range plant species should experience a decrease in regulation by herbivores and other natural enemies, resulting in an increase in abundance and distribution. The success of biological control has been used as support for the enemy release hypothesis (Keane and Crawley 2002). Empirical evidence in support of the hypothesis is equivocal (Keane and Crawley 2002, Maron and Vilá 2001), however, in an analysis of 473 plant species naturalized in the U.S., Mitchell and Power (2003) found species in their non-native ranges had on average 84% fewer fungi and 24% fewer virus species than in the respective native ranges.

Demographic and environmental stochasticities probably promote local extinction in many small populations of species beginning to establish in new locations. Sax and Brown (2000) proposed two further phenomena that may account for the large fraction of failed introductions. They noted that species which are abundant and widespread tend to be rare or absent from most locations within their geographic range, due to spatial environmental variation. This suggests that *most* local introductions should fail because propagules would tend, by chance, to be introduced to marginal or unsuitable sites. Secondly, many established populations are sustained by the positive feedback effects of dispersal events on local population dynamics. Single, small, and newly established populations all lack opportunities for rescue by other populations and thus are more susceptible to extinction (Sax and Brown 2000).

There are also genetic obstacles to the persistence of populations of introduced organisms. Genetic diversity and phenotypic plasticity that can bring the lineage into accord with its new physical and biological environment may be difficult to achieve in far-flung populations (Levin 2003), because they are subject to genetic drift and founder effects (Eckert *et al.* 1996). Genetic variation in founder populations is likely to be lacking if the number of individuals introduced is not large or population growth rate is not substantial (Novak and Mack 1993). Kolbe *et al.* (2004) recently demonstrated that invasion success was enhanced when multiple introductions of the Caribbean brown anole, *Anolis sagrei*, to Florida came from different source populations in the lizard's geographic range. Populations in the adventive range contained more genetic variation than native populations.

INVASIVENESS

Among naturalized species persisting at new locations (despite the series of severe constraints described above), a few may go on to become invaders (Williamson and Fitter 1996; Mack *et al.* 2000). Darwin (1859, p. 60) was the first to propose a theory for invasion success in plant species. Using observations by Alphonse de Candolle, Darwin noted that "floras gain by naturalization, proportionally with the number of native genera and species, far more in new genera than in new species". Darwin gives the example that "In the last edition of Dr. Asa Gray's 'Manual of the flora of the United States', 260 naturalized species are enumerated, and these belong to 162 genera ... out of the 162 genera, no less than 100 genera are not there indigenous." Darwin reasoned that because differences in structure and functioning are greater between genera than between species, competition is likely to be greater between congeners. Results both support (e.g., Mack 1996; Rejmánek 1996) and argue against (Daehler 2001b) Darwin's hypothesis. Daehler's contradictory result for the Hawaiian flora indeed supports Darwin's (1859) original line of thinking: "It might have been expected that the plants which would succeed in becoming naturalized in any land would generally have been closely allied to indigenes; for these are commonly looked at as specially adapted for their own country".

Numerous, often disparate life-history factors have been suggested to explain the invasion success of particular plant species. Baker (1965, 1974) discussed several traits associated with invasive plant species and proposed that species having more of these traits would be more 'weedy' than species with fewer of these traits. Thus the 'ideal' weed would have the ability to reproduce sexually and grow clonally, show rapid growth, phenotypic plasticity and high tolerance of environmental heterogeneity. Cadotte and Lovett-Doust (2001) explicitly tested Baker's notion of a weed in the exotic flora of Ontario and found general concordance with the traits Baker described. Other traits have also been shown to correlate with invasion success. Rejmánek (1996) found that invasiveness of herb species was best predicted by native latitudinal range. Similarly, Cadotte *et al.* (2004) found that successful invaders in Ontario were significantly over-represented by species originating in Europe and Eurasia. Life history attributes particularly associated with invasion success in the Ontario flora were the presence of a clonal organ and a longer flowering season.

HOW TO CLASSIFY INVASIVES?

An ideal classification scheme should, to have utility, be capable of general application. Richardson *et al.* (2000a) offered operational definitions for terms such as 'alien', 'casual alien', 'naturalized', 'invasive' and 'transformer' with respect to plant species. In their scheme, invasion of a new region by an introduced species was envisioned as surmounting a series of 'barriers' (see Fig. 1). Thus 'casual alien species' have overcome geographic and local environmental barriers enabling them to survive in a new location, but have not overcome barriers to successful reproduction and population growth or expansion in the new locale. Naturalization (in Richardson *et al.*'s scheme) begins when various barriers to regular reproduction are overcome and population growth and survival (persistence) is maintained over the long term (functionally, long-term has been regarded as *c.* 25 years [e.g., Tutin *et al.* 1964] or 10 years [Pyšek *et al.* 2004]). 'Invasion' is considered to have occurred when the species has overcome the barriers to dispersal within the new region, and new populations are being founded away from the immediate area of introduction.

Davis and Thompson (2000) proposed a classification scheme modeled after Rabinowitz's (1981) tripartite scheme for rarity, to define 'invaders' as distinct from 'successional colonizers' and 'noninvasive colonizers'. Davis and Thompson used a taxonomy of invasion that raises three dichotomous questions: 1) is the species a long-distance or short-distance disperser; 2) is the species common, or novel to the region; and 3) does it have a small or large 'impact' on the new environment. This scheme generates eight potential conditions. The major operational difference between the system of Davis and Thompson and of Richardson *et al.* (2000a) lies in the emphasis in Davis and Thompson upon 'impact' in the new environment, whereas Richardson *et al.* rely on measures of natural

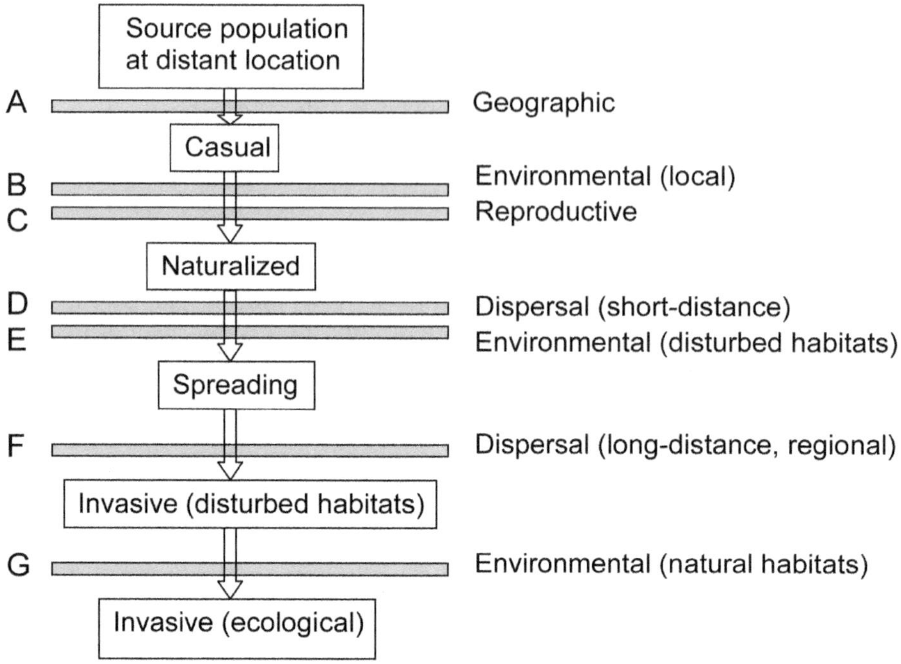

Fig. 1 A schematic representation of major barriers (A–G) limiting the spread of introduced plants. (Modified from Richardson *et al.* [2000a]).

reproduction and rates of spread in the new environment as major criteria determining invasiveness. Daehler (2001a) has argued against the utility of Davis and Thompson's scheme insofar as it hinged on a subjective criterion, 'impact', whereas the 'spread' criterion (of Richardson *et al.*) can be measured relatively more objectively. Daehler noted that both schemes probably embrace similar sets of species, since the concordance between invaders defined by the spread criterion, and those that have a great impact is likely close to a hundred percent.

METHODS

Over the past three decades, the Canadian Journal of Plant Science has published accounts of the biology of plants that are weedy in Canada. Each account follows a prescribed layout, and provides a detailed description of a single species of weed (or closely related group of species), the history of its introduction to North America and Canada, details of its economic importance, a Canadian distribution map and details of its distribution elsewhere in the world. Additionally, information on habitats occupied, means of seed and vegetative reproduction and dispersal, and details of growth and development are provided as well as response to weed control measures. Articles in the series bring together published and unpublished information on the biology, ecology and control of these weeds. Thus the information is relatively comprehensive and consistently presented. Independently, an exotic species list for Ontario was generated from the Ontario Ministry of Natural Resources, Natural Heritage Information Centre's database of vascular plants (NHIC http://www.mnr.gov.on.ca/MNR/nhic/nhic.cfm). For the present analysis we used the species occurring on the NHIC list of Ontario's exotic flora, for which there was a Biology of Canadian Weeds article (N = 95) (see Appendix 1 for list and references).

We attempted first to extract specific data required to meet the criteria for each category in the classifications of Richardson *et al.* (2000a) and Davis and Thompson (2000). This we term the 'ideal criteria'. That is, if available, the ideal criteria would best satisfy the definition of each category in the classification, and would divide species into the various categories described by the two schemes. Then using the information given in the Biology of Canadian Weeds reviews and the regional abundance ranking values determined by the NHIC, we developed 'realized' criteria for each category in the two classifications. Realized criteria are the best fit to the ideal criteria, using information that is actually available. Table 1 shows ideal and realized criteria for each category in the two schemes. In order to address each criterion, data were tabulated for each of the 95 weed species (see Table 2).

Certain limitations of both schemes became evident early. Davis and Thompson call for an assessment of 'dispersal distance'. In many cases an initial long-distance dispersal event by one agent may lead to an introduction in a new region, while another kind of agent is responsible for subsequent short-, medium- or long-distance dispersal in the new location (i.e., a form of diplochory, *sensu* Vander Wall and Longland [2004]). It seems unlikely that an initial long-distance dispersal event from a widely distant environment would have much effect in the new environment, or relevance to subsequent spread, if the species did not then successfully disperse from its initial point of establishment. Therefore in our analysis, the dispersal criterion is based on how the species is moved *following* its initial introduction to the adventive region. This is not always straightforward, since even approximate knowledge about dispersal distance, particularly long-distance dispersal, is typically scarce for even the best-studied species. We determined the principal modes of dispersal and assigned them to either a short- or long-distance category, bearing in mind that the overall scale of the analysis was the entire province of Ontario (total area 1,068,580 km^2 [Statistics Canada, www.statcan.ca]). Thus any species that did not have a known human-facilitated mode of dispersal was considered a short-distance disperser. In addition, some categories of human-facilitated dispersal that generally result in limited dispersal (such as within a crop field or farm property) were also considered short-distance (e.g., those associated with mowing, cultivation, farm machinery, domestic animals).

Table 1 Criteria used in classifying species according to the schemes of Davis and Thompson (2000) and Richardson *et al.* (2000a) 'Ideal' criteria are those that, if known, would best satisfy the criteria proposed by the authors of the two schemes. 'Realized' criteria are the best fit to the ideal criteria using information that is actually available.

Parameter		Ideal criteria	Realized Criteria
Davis and Thompson (2000)			
Invaders	1. Short or long distance dispersers	1. Dispersal has occurred gradually to nearby or adjacent environments without significant human assistance, OR Dispersal has occurred over larger distances, between widely distant environments, and may be facilitated by humans.	1. Dispersal is by vegetative or seed dispersal and/or is often facilitated by humans.
	2. Novel in the new environment	2. The species has expanded its range to occur in the region	2. Ontario is not within the native geographic range of the species.
	3. Great impact in the new environment	3. Community impacts — the species is noted to have a substantial impact on native communities, AND/OR Ecosystem impacts — the species is noted to have a substantial impact on ecosystem processes, AND/OR Economic impacts — the species is noted to have a substantial economic impact	3. The species is noted to regularly occur in natural or semi-natural (natural disturbed) communities, AND/OR The species is noted to substantially impact ecosystem processes, AND/OR The species is noted to have substantial economic impact.
Successional colonizers	1. Short or long distance dispersers	1. Dispersal has occurred gradually to nearby or adjacent environments without significant human assistance, OR Dispersal has occurred over larger distances, between widely distant environments, and may be facilitated by humans	1. Dispersal is by vegetative or seed dispersal and/or is often facilitated by humans

Successional colonizers	2. Common in the new environment	2. The species is native to the region — its occurrence in the region involves no range expansion	2. Ontario is within the native geographic range of the species.
	3. Small or great impact in the new environment	3. Community impacts — the species is not noted to have a substantial impact on native communities; AND Ecosystem impacts — the species is not noted to have a substantial impact on ecosystem processes (the species does not play a keystone role in the new community or ecosystem, AND Economic impacts — the species is not noted to have a substantial economic impact, OR Community impacts — the species is noted to have a substantial impact on native communities, AND/OR Ecosystem impacts — the species is noted to have a substantial impact on ecosystem processes, AND/OR Economic impacts — the species is noted to have a substantial economic impact.	3. The species does not generally occur in natural or semi-natural communities. It primarily invades disturbed 'unnatural' areas, AND The species is not noted to substantially impact ecosystem process, AND economic impacts associated with the species are limited in nature or scale, OR The species is noted regularly to occur in natural or semi-natural (natural disturbed) communities, AND/OR The species is noted to substantially impact ecosystem processes, AND/OR The species is noted to have substantial economic impact.

Table 1 Continued.

Parameter		Ideal criteria	Realized Criteria
Davis and Thompson (2000)			
Novel, non-invasive colonizers	1. Short or long distance dispersers	1. Dispersal has occurred gradually to nearby or adjacent environments without significant human assistance, OR Dispersal has occurred over larger distances, between widely distant environments, and may be facilitated by humans.	1. Dispersal is by vegetative means or seed, and/or is often facilitated by humans.
	2. Novel in the new environment	2. The species has expanded its range to occur in the region.	2. Ontario is not within the native geographic range of the species.
	3. Small impact in the new environment	3. Community impacts — the species is not noted to have a substantial impact on native communities, AND Ecosystem impacts — the species is not noted to have a substantial impact on ecosystem processes (the species does not play a keystone role in the new community or ecosystem, AND Economic impacts — the species is not noted to have a substantial economic impact.	3. The species does not generally occur in natural or semi-natural communities. It primarily invades disturbed 'unnatural' areas, AND The species is not noted to substantially impact ecosystem process, AND economic impacts associated with the species are limited in nature or scale.

Richardson *et al.* (2000a)		
Alien plants	1. The species occurs in an area, >100km from the nearest 'natural' population, due to introduction as a result of human activity.	1. All SEs not native to Ontario.
Casual alien plants	1. An alien plant.	1. SE Ranking in Ontario.
	2. Relies on repeated introductions for its persistence.	2. Extremely rare to very rare, high extinction probability (SE1-SE2).
	3. These species do not spread naturally past their point of introduction.	3. Human aided dispersal common AND occurrence only in human disturbed areas.
Naturalized plants	1. An alien plant.	1. SE Ranking in Ontario.
	2. Reproduces consistently and sustains populations over many lifecyles.	2. Uncommon to common, extinction probability medium to low (SE3-SE5).
	3. Has not spread into new environments beyond the point of introduction.	3. Occurrence only in human disturbed areas (have not dispersed to natural or semi-natural communities)[1].

Table 1 Continued.

Parameter	Ideal criteria	Realized Criteria
Richardson *et al*. (2000a)		
Invasive plants	1. An alien plant.	1.SE Ranking in Ontario.
	2. Produces reproductive offspring in large numbers at considerable distance from parent plant (for species primarily dispersed by seed >100 m in <50 years; for species spreading primarily vegetatively >6m/3years), or any species primarily spread by humans.	2.Common and widespread, low extinction probability (SE3-SE5).
	3. Has spread to areas distant to the point of introduction.	3. Occur in human disturbed areas and have spread to occur in disturbed natural and semi-natural communities.
Transformers	1. An invasive plant.	
	2. Changes the character, condition, form or nature of ecosystems over a substantial area.	

[1] Operationally, our definition of naturalized thus limits species to occurrence in mainly human-modified communities; this is not necessarily the intent of Richardson *et al.*'s definition (see text on this).

Long-distance dispersal vectors included weed-contaminated crop seed, soil or other agricultural products, boats, cars and other vehicles, waterfowl, fish, and irrigation water.

Rates of spread (as required by Richardson *et al.*'s scheme) prove to be only rarely available; only seven of the 95 articles we surveyed reported any rate of spread. Five of these were for species that spread primarily by vegetative means, and none had been measured or estimated in the adventive Ontario environment. Where dispersal of propagules is promoted by anthropogenic activities or other non-standard means, rates of spread cannot reasonably be estimated by seed or vegetative propagule characteristics (Higgins *et al.* 2003). Although in theory spread could be measured relatively objectively, using population growth and distance from the source, unless substantial historical information is available real rates of spread are quite difficult to determine. Furthermore for many of the species in our dataset, which are primarily agricultural weeds, long-distance dispersal may occur relatively regularly through transport in contaminated seed stocks, or infested hay. These methods have been considered a major contributor to spread for many species (e.g., *Silene alba*, McNeill 1977; *Melilotus* spp., Turkington *et al.* 1978; *Sorghum halepense*, Warwick and Black 1983).

The NHIC database includes estimates of abundance of exotic plants using a provincial ranking system (SE-rank [Sub-national exotic]). We used the provincial SE ranking as a surrogate for the amount of 'spread' that had been achieved. The SE ranking is essentially a distribution and abundance indicator, and so carries implications of extinction risk. Thus SE1- and SE2-ranked species are considered vulnerable to extinction, SE3 species have a medium risk of extinction and SE4 and SE5 have, generally, a low risk of extinction (see Master 1991). Thus,

- **SE1** **Extremely rare** in Ontario; usually 5 or fewer occurrences in the province or very few remaining individuals; often especially vulnerable to extirpation.
- **SE2** **Very rare** in Ontario; usually between 5 and 20 occurrences in the province or with many individuals in fewer occurrences; often susceptible to extirpation.
- **SE3** **Rare to uncommon** in Ontario; usually between 20 and 100 occurrences in the province; may have fewer occurrences, but with a large number of individuals in some populations; may be susceptible to large-scale disturbances. Most species with an SE3 rank are assigned to the watch list, unless they have a relatively high global rank.
- **SE4** **Common** and apparently secure in Ontario; usually with more than 100 occurrences in the province.
- **SE5** **Very common** and demonstrably secure in Ontario.

Community, ecosystem and economic 'impacts', as per Davis and Thompson's scheme, were quantified as follows (and see Table 2). For community impact, species occurrence in only community types 1 and/or 3 (both generally highly human-modified communities) was scored as 1 point, occurrence in any other

Table 2 Data extracted from the Biology of Canadian Weeds series and NHIC database, to address criteria outlined in Table 1.

Attribute	Description	Variable
Abundance Rank	SE Ranking	**SE1 — Extremely rare** in Ontario; usually 5 or fewer occurrences in the province or very few remaining individuals; often especially vulnerable to extirpation. **SE2 — Very rare** in Ontario; usually between 5 and 20 occurrences in the province or with many individuals in fewer occurrences; often susceptible to extirpation. **SE3 — Rare to uncommon** in Ontario; usually between 20 and 100 occurrences in the province; may have fewer occurrences, but with a large number of individuals in some populations; may be susceptible to large-scale disturbances. **SE4 — Common** and apparently secure in Ontario; usually with more than 100 occurrences in the province. **SE5 — Very common** and demonstrably secure in Ontario.
Date of first documented occurrence in Ontario	From herbarium specimen or, if no date given, date is noted as 1952 if the species is listed in Gleason (1952).	Date
Nativity to Ontario	Is Ontario within the native range of the species?	(1) yes; (2) no
Primary means of recruitment	Vegetative or seed	(1) vegetative; (2) seed

Table 2 Continued.

Attribute	Description	Variable
Primary dispersal mode	Primary mode of natural dispersal usually based on adaptations of the seed or fruit	(1) wind; (2) water; (3) bird; (4) animal; (5) gravity or ballistic
Secondary dispersal mode	Other documented, usually human-facilitated, means of dispersal	(1) contaminated crop seeds or infested hay; (2) farm machinery or vehicles; (3) in soil or manure; (4) livestock or domestic animals; (5) boats; (6) irrigation water; (7) human clothing or footwear
Communities in which species occurs	Documented occurrence in various community types	(1) roadsides, gardens, wastelands, other disturbed urban areas; (2) grazed pasture; (3) cultivated fields, crops; (4) disturbed natural communities; (5) open woodland; (6) forest margins or clearings; (7) stream banks, shorelines; (8) aquatic (lake, stream, pond, river etc.)
Economic impacts	Documented economic impacts	(1) competes with crop or pasture species; (2) is toxic or irritant to farm animals or humans; (3) destroys appearance of lawns or gardens; (4) harbours insects or disease organisms that attack other plant species; (5) interferes with water infrastructure; (6) impacts recreational areas (beaches, lakes etc.); (7) contaminates seed stocks, infests hay etc.[1]
Ecosystem impacts	Documented ecosystem impacts	(1) forms monocultures; (2) dominates aquatic ecosystems with major effects on flow, oxygen, light etc.; (3) significant nitrogen fixer.

[1] Economic impacts other than those listed here were occasionally reported (e.g., makes harvesting difficult; chemically inhibits crop species), however these impacts are closely related to (1) and always occurred with (1) and therefore were considered redundant and not included as separate impacts.

(more 'natural') community type(s) brought an additional point. (Thus, if the species occurred in community types 1, 2, 3 and 4 (see Table 2) the total score would be 3 points). The total number of points scored was then multiplied by the SE rank value. (For example, if the above species was ranked as SE5 the Community Impact Score (CIS) would be 15.) A similar system was applied to ecosystem impacts (Ecosystem Impact Score = EIS). For economic impact, a species scored 2 points if it was noted to have impacts related to descriptors 1, 2 and 4 (see Table 2), which in general involve both crop production losses and control costs, and 1 point if it was noted to have impact in any other category (which primarily involve control costs only). That total score was then multiplied by the SE rank value as above (Economic Impact Score = EconIS).

Criteria for determining whether the score corresponded to a 'great' or 'small' impact are shown in Table 3 ('D&T'). If a species was determined to have a great impact in any of the three impact categories, the overall impact was determined as 'great' (in the language of D&T). It is difficult to remove subjectivity from the scoring of impact. We tried several alternatives to the above schemes and present one of them here. We applied an alternative method to calculating CIS; if the species occurred only in medium to highly human-modified communities (i.e., 1, 2 or 3) a score of 1 was applied, and a score of 2 was applied to a species that occurred in any natural or semi-natural community. The score was then multiplied by the SE rank value. We also modified the criteria for determining whether the impact was small or great. These two modifications combined result in the alternative Davis and Thompson classification (D&T[1] in Table 3 and in Table 6 in results).

Table 3 Criteria to be satisfied for determination of impact as 'small' or 'great' in two alternative scoring schemes applied to Davis and Thompson's classification scheme.

Impact	D&T		D&T[1]	
	Small	Great	Small	Great
Community (CIS)	≤ 9	> 9	≤ 7	> 7
Ecosystem (EIS)	0	> 0	0	> 0
Economic (EconIS)	≤ 9	> 9	≤ 12	> 12

In our application of Richardson *et al.*'s scheme the SE rank value is the primary factor driving the distinction between invasive and other alien species. SE rank is also a key factor in our application of Davis and Thompson's scheme in the division between great and small impacts. A close association between SE rank and the raw impact factor (i.e., impact scores prior to multiplication by SE rank value), should result in a greater overlap between the two schemes. We used correlation analysis to determine how the SE rank value was related to the raw impact scores for community and economic impacts (only seven species scored above 0 in the ecosystem impact category so EIS was not tested for association). To examine differences between great and small impact species in life-history traits we used chi-squared contingency analysis with Yates correction. If two-way contingency tables showed non-independence ($P < 0.05$), the Freeman-Tukey deviate was used to determine if any individual cell was either under- or over-represented compared to the expected value. Life-history variables for which data were collected are shown in Table 4.

Table 4 Life-history traits recorded for each species.

Life-history trait	Parameters
Seed dormancy	Long (>3 years), Short (<3 years), None known
Primary pollination mode	Insect, wind, self, various, unknown
Clonal organ presence	Present, absent
Flowering season	Early, mid, late, all season
Flowering duration	Number of months
Soil moisture preference	Moist, dry, variable
Sex habit	Hermaphrodite, monoecious, dioecious, polygamous
Life cycle	Annual (and biennial), perennial, variable
Dominant recruitment mode	Vegetative, seed, variable

As with most macroecological datasets (Cadotte *et al.* 2005), ours has its limitations. The term 'weed' in the Biology of Canadian Weeds series refers to any vascular plant that perpetuates itself in habitats where it is "not wanted" (Cavers and Warwick, 2000); a meaning different from that commonly used for 'weeds' in the ecological literature and which usually includes that the species has detectable economic or environmental impacts (e.g., Pyšek *et al.* 2004; Richardson *et al.* 2000a). Thus, these species should not be considered to have been already classified as 'weeds' in the traditional ecological sense, rather, they still validly represent species in the range of categories defined by the two schemes. The species in our list are biased towards species occurring in agricultural areas since the original purpose of the series was to provide a basis for effective control methods and to meet the needs of regulatory agencies in Canada (Cavers and Warwick, 2000). However the dataset does include a range of species occurring in both natural and agricultural habitats and includes both abundant (SE4 and SE5) and rare (SE1-3) species. It is important to note that we are not attempting to test the *accuracy* of either scheme in classifying species into categories, since there is no 'right' answer. Rather we aim to determine if either scheme is more *functional*, or operational, based on the data available in a typical species dataset. Thus any conclusions we make as to which classification was 'better' refer only to which model could be applied more usefully given the data available to evaluate each criterion for categorization.

RESULTS

A total of 29 families was represented in the dataset of 95 species (Table 5). The Asteraceae was the only family represented in all SE ranks, with a total of 27 species (28%). Poaceae was next with a total of 10 species (11%) and only SE1 species (extremely rare) were not represented. Fabaceae was next with nine species (all ranked SE5) (10%). There were eight Brassicaceae (in SE2, SE3 and SE5 ranks) (9%). Twenty-two of the 28 families had only 1 or 2 species represented in the dataset. Asteraceae, Fabaceae and Brassicaceae are all over-represented in our dataset compared with their representation in the exotic flora of Ontario, and among native Ontario species (see Table 5).

Application of the Richardson *et al.* scheme classified 47% of the total species as invasive, as opposed to Davis and Thompson, where approximately 79% of species were classified as invaders using our first criteria (D&T) and 72% using the alternative criteria ($D\&T^1$) (Table 6). All but one of the species classified as invasive by Richardson *et al.* were also classified as invasive by both Davis and Thompson schemes. The exception was *Helianthus tuberosus* which was classified as a successional colonizer in both D&T and $D\&T^1$ due to its actually being considered native to Ontario. The distinction depends upon conflicting conclusions by experts. (Although considered non-native to Ontario by several authors (Morton and Venn [1990], Newmaster *et al.* [1998], and Scoggan [1979]), most regional floras and floras from nearby areas consider the species native (e.g., Gleason and Cronquist 1991). It is likely that at some time in the future the S-rank of this species will be changed from SE5 to S5 (indicating a native status rather than exotic in the province) (NHIC 2004)). All species ranked as SE1 were classified by Richardson *et al.* as casual and by D&T and $D\&T^1$ as novel non-invasive colonizers (Table 6). In our application of both classifications, it was not possible for an SE1 species to be regarded as invasive. Of the SE5 species, Richardson *et al.*'s scheme classified 64% as invaders, compared with 97% (D&T) and 91% ($D\&T^1$) (Table 6).

SE rank value was significantly correlated with raw community impact scores (D&T, $r = 0.381$, $p < 0.001$, $n = 95$; $D\&T^1$, $r = 0.457$, $p < 0.001$, $n = 95$), however there was no correlation between SE rank value and economic impact score (D&T, $r = -0.011$, $p = 0.920$, $n = 95$). Thus species with greater abundance tended to occur in more of the defined community types, including natural and semi-natural communities, than less abundant species but did not necessarily fall into more economic impact categories. In D&T all SE5 species had an overall impact of 'great' whereas in $D\&T^1$, three SE5 species had only 'small' impact. Great- and small-impact species differed only in their proportions in the seed dormancy life-history parameter (long, short or none) (D&T, $\chi^2 = 15.110$, $p = 0.002$;

Table 5 Family representation in SE rank and total number of family members in the database, compared with proportion of family in all species listed as 'exotic' in Ontario and in all native species of Ontario.

Family	SE Rank					Total	% of total dataset	% of family in all exotics of Ontario (n = 1086)	% of family in natives of Ontario (n = 2310)
	1	2	3	4	5				
Amaranthaceae					2	2	2.1	0.8	0.04
Apiaceae				1	1	2	2.1	2.2	1.3
Asteraceae	4	2	1	1	19	27	28.4	14.6	9.7
Boraginaceae					2	2	2.1	1.8	0.7
Brassicaceae		1	1		6	8	8.4	6.4	2.2
Cannabaceae	1					1	1.1	0.2	0.04
Caryophyllaceae			1		3	4	4.2	4.1	1.4
Chenopodiaceae					1	1	1.1	2.1	1.1
Clusiaceae					1	1	1.1	0.1	0.6
Convolvulaceae					1	1	1.1	0.6	0.1
Dipsacaceae					1	1	1.1	0.6	0
Euphorbiaceae					2	2	2.1	1.8	0.3
Fabaceae					9	9	9.5	5.9	2.6
Haloragaceae					1	1	1.1	0.1	0.3
Lamiaceae					1	1	1.1	4.1	1.8
Lythraceae					1	1	1.1	0.3	0.2
Malvaceae					1	1	1.1	1.4	0.1
Oxalidaceae	1					1	1.1	0.1	0.1
Plantaginaceae					2	2	2.1	0.6	0.2
Poaceae		2	2	1	5	10	10.5	10.2	8.5
Polygonaceae	1				1	2	2.1	2.3	1.4
Portulacaceae					1	1	1.1	0.2	0.1
Potamogetonaceae					1	1	1.1	0.1	1.6
Ranunculaceae					1	1	1.1	0.1	2.3
Rosaceae					2	2	2.1	5.8	6.2
Rubiaceae	1				1	2	2.1	0.7	0.9
Scrophulariaceae			1		2	3	3.2	3.9	2.3
Solanaceae	2		1		1	4	4.2	2.2	0.2
Violaceae				1		1	1.1	0.4	1.5
Total	10	5	7	4	69	95	100		

Table 6 Summary of results of classification schemes, for species in each SE rank (regional distribution and abundance) (D&T = Davis and Thompson).

	SE1	SE2	SE3	SE4	SE5	Total
Total in SE category	10	5	7	4	69	95
Richardson *et al.*						
• Casual	10	3				13
• Naturalized		2	7	3	25	37
• Invasive				1	44	45
D&T						
• Successional colonizer			1		2	3
• Novel-non-invasive colonizer	10	1	4	2		17
• Invader		4	2	2	67	75
D&T[1]						
• Successional colonizer			1		2	3
• Novel-non-invasive colonizer	10	3	4	3	4	24
• Invader		2	2	1	63	68
D&T = great impact		4	3	2	69	78
D&T = small impact	10	1	4	2		17
D&T[1] = great impact		2	3	1	66	72
D&T[1] = small impact	10	3	4	3	3	23

D&T[1], $\chi^2 = 8.851$, $p = 0.031$). Results of the Freeman-Tukey deviate test indicated that species with a small impact were significantly under-represented by the presence of a long dormancy period and species with a great impact were under-represented by presence of short and/or no dormancy period. Additionally, great- and small-impact species differed in their geographic origin (D&T, $\chi^2 = 17.068$, $p = 0.002$; D&T[1], $\chi^2 = 10.952$, $p = 0.027$). Species having a small impact were under-represented in originating from Eurasia, and were over-represented in originating from Asia proper and the Mediterranean area of Europe. Species with a great impact were over-represented in originating from Eurasia.

DISCUSSION

Davis and Thompson's classification scheme was clearly less discriminating in determining invasive species in this dataset. This could be ameliorated somewhat by increasing the score range for 'small' impacts. However, the original score range for EconIS (≤9) takes into account the fact that an SE5 species with an impact in one economic category (such as the most common category – 1) would score a 10. It makes little sense to call this a 'small' impact. However, even with the expanded range for small economic impacts (D&T[1]) there are only seven species (10%) that would be reassigned from the 'invasive' category to the 'non-invasive colonizer' category. It is noteworthy that SE rank and the economic impact score (before adjusting for abundance) were not correlated; SE1 species were just as likely to have a number of different types of economic impacts as SE5 species. However, more abundant species were more likely to have spread to natural or semi-natural habitats than less abundant species.

On a case-by-case basis, Richardson *et al.*'s scheme would seem to be more practically useful, since it is in theory less subjective. Yet when attempting to classify even a relatively detailed and consistent dataset as we have done, the 'spread' criterion presents almost as many problems and is perhaps just as subjective as Davis and Thompson's impact criterion. We are fortunate with the Ontario dataset to have the distribution and abundance ranking information that can be used as a surrogate for the 'spread' criterion. However the SE ranking system has several limitations in its use as a spread surrogate. First, it is based on current abundance in the province. SE5-ranked species are at the limit of the abundance ranking, but some may have exhausted the potential range of suitable habitats and be no longer capable of further spread, while others may not. SE1-3 species, although not currently abundant, may be spreading at a more rapid rate than SE4 and SE5 species, and this remaining potential for rapid spread is not captured in the SE ranking. Thus the spread criterion is complicated by the temporal 'snapshot' nature of the assessment.

A related problem is that the SE ranking gives no indication as to where the populations are located relative to each other. In fact, virtually all of the SE1 species in our dataset have been recorded from populations widely distant from one another (i.e., >500 km) (determined through a visual inspection of the distribution maps). If we were to base the rate-of-spread criterion on the time-since-introduction of the species, and the average distance between persistent populations in Ontario, we would most likely place all the species on our list (Appendix 1) in Richardson *et al.*'s 'invasive' category. The spread criterion for invasives (i.e., for seed-dispersed species >100m from parental plants in <50 years) begins to make little sense on regional and landscape scales, especially when most species have modes of dispersal with a potential to carry propagules great distances from parental individuals.

Richardson *et al.* (2000a) estimated that 50-80% of 'invaders', as defined by their classification, would have 'harmful' (detectable environmental or economic) effects and that the rest are 'benign invaders' whose environmental or economic impacts are beyond any practical detection limits. By our analysis, 100% of the species classified as invasive by Richardson *et al.*'s scheme have large environmental or economic impacts, and moreover between 60 and 78% of species defined as naturalized also have harmful effects. This is perhaps not surprising since our criteria for naturalization (in Richardson *et al.*'s scheme) included a requirement that the species had not yet spread to natural or semi-natural habitats. Thus while community impacts were minimized (occurrence in fewer community types), economic impacts for these species may still have been large enough to generate an overall great impact. For example, SE5 species that only occur in cultivated fields (i.e., Richardson *et al.*'s 'naturalized' species in our analysis) would still have scored a great impact if they compete with agricultural species in those fields, and therefore would have been classified as invaders in D & T's scheme. It should be noted that the distinction between naturalized and invasive species we used here may not strictly represent the intention of Richardson *et al.* Since Richardson *et al.* suggest naturalized plants have not overcome barriers to dispersal, we limited naturalized species to occurrence only in human-modified communities. Richardson *et al.*'s definition is unclear in this respect; they noted "Naturalized plants.....do not necessarily invade natural, semi-natural or human-made ecosystems."

Cadotte *et al.* (2004) have studied the relationship between SE ranks and life-history attributes in Ontario's exotic flora (n = 846). Their results showed that common exotics were more likely to have a clonal organ, a longer flowering period and a Eurasian origin. We found that, in terms of life-history traits, large-impact species differed from small-impact ones in the likelihood of having a lengthy seed dormancy period. Cadotte *et al.* found the variable most strongly associated with invasiveness (and higher SE-rank) was origin in Eurasia (and Europe). Similarly in our analysis, species having great impact came disproportionately often from Eurasia. This supports Cadotte *et al.* in that species native elsewhere in North America were not more likely to have a great impact in Ontario (however our sample size in this category [4 species] was small). Cadotte *et al.* suggest several reasons why Eurasian species may be more successful in Ontario. Species from Eurasia may be less phylogenetically related to species from North America and may be more likely to have evolved ecological novelties (in line with Darwin's naturalization hypothesis [Darwin 1859]). Alternatively, the enemy release hypothesis may be responsible, or the matching of provenance (i.e., coming from a comparable climatic zone) could be another possible factor.

We conclude that neither framework is especially useful in their present formulations for classifying regional (or larger) exotic floras, i.e., *lists* of known non-native species. The search for operational definitions of terminology in invasions ecology remains elusive. Relevant information is simply not available and the use of surrogates for criteria such as 'spread' and 'impact' are likely to introduce just as much subjectivity, inconsistency and confusion in the literature as already exists. Failure to operationalise definitions used in classifying exotic flora datasets will likely continue to lead to ambiguous generalizations and predictions. Notwithstanding this ambiguity, comparative studies of invasive floras have provided new insights in the understanding of general patterns of plant invasion and the value of these studies is not to be diminished. Rather, when classifying lists of flora the criteria used to assign each species to a category should be explicitly stated so that researchers can compare species based on like criteria rather than the category itself, which may have been derived from unlike data. As Pyšek *et al.* (2004) conclude in their recent paper on the treatment of invasions terminology in regional floras, statements preceded by 'probably' or 'possibly' may be temporarily the most honest way of classifying some taxa.

Colautti and MacIsaac (2004) recently proposed a supplementary lexicon to current terminology in invasions ecology. Their framework attempted to eliminate the need for universal definitions of current terms by using operational terms with no *a priori* meaning (i.e., as stages) in a process-based model. The framework is based on current models that envision processes in immigration and naturalization as a series of consecutive stages. The goal of the framework is to supplement ambiguous terms with the stage-based terminology. Such a framework is logical and intuitively appealing and would likely lead to some consistency in individual, population-based studies of invasive species (as would that of Richardson *et al.* [2000a]). However, it is also unlikely to be useful for classifying lists of alien species for comparative purposes, since the same data limitations would apply as described above for Richardson *et al.* and Davis and Thompson's schemes.

A PROPOSED MODIFICATION TO CURRENT CONCEPTUAL FRAMEWORKS FOR THE INVASIONS PROCESS

Colautti and MacIsaac (2004) noted that terminology used to describe introduced species are misnomers, in the sense that introduced, naturalized or invasive species etc. are of course introduced, naturalized or invasive *populations*. A focus on invasions at a population level rather than a species level has important implications for the development of invasions ecology. Indeed it makes little sense to assign categorical properties to a 'species'. Rather, demographic attributes of the adventive population and of any metapopulation or other regionally connected group (see Murphy and Lovett-Doust 2004) need to be characterized, as the attributes of longevity, fecundity and mortality rates are the key determinants of an invader's status within the ecological continuum representing the invasion process, and extending across the arc from immigration to naturalization. Furthermore when it does happen, classification should occur at a defined temporal or spatial scale, as populations of species will appear to be in different places on the continuum depending on the scale of the assessment.

The 'blinking lights' metaphor of metapopulation extinction and colonizations may be useful in the invasions process too. A metapopulation describes a 'population' consisting of a number of local populations, in the same sense in which a local population is a population consisting of many individuals. According to the Levins (1969) classical metapopulation concept, all local populations have a substantial probability of extinction and therefore long term persistence of a species is regulated at the regional, or metapopulation level (Hanski, 1999). Metapopulation theory proposes that the regional population persists as the result of a balance between local population extinction in patches and patch migrations leading to colonization. The key to population processes lies in understanding the shifting mosaic of patch occupancy as opposed to details of within-patch events (Kareiva and Wennergren, 1995). Freckleton and Watkinson (2002) describe other regional ensembles of plant population dynamics at large scales. In the following, we visualize Richardson *et al.*'s (2000) categories of alien species (plus an additional category, 'spreading' populations between naturalized and invasive stages) and the barriers they surmount through the invasions process, in a population-based context. The four categories of populations can be described in terms of the metapopulation parameters of extinction and colonization as shown in Table 7.

Table 7 Metapopulation conditions for population through the invasion process (e = extinction; c = colonization).

Category/Stage	Metapopulation dynamics
Casual	$e > c$
Naturalized	$e = c$
Spreading	$e < c$
Invasive	$e << c$

Casual

Adventive individual/s appear and may be introduced on multiple occasions before a population 'takes' or establishes. The recurrent but ephemeral introductions generate the status of 'casual'; i.e., the introduced species is at the early stage of being an intermittent member of the new flora. The population may 'blink' off and on again as it fails to sustain itself over longer periods but occasionally persists or becomes re-established because of repeated introductions from source populations outside the adventive area. These populations establish because some individuals of the species have surmounted a major geographical barrier (long distance barrier A of Fig. 1).

Naturalized

Eventually a casual population may blink on and remain on for an extended period. The first local population meets criteria set for steady-state maintenance (a sustainable population). This corresponds to the concept in Population Viability Analysis of an MVP (minimum viable population, the number of individuals that ensures a population's persistence [Shaffer 1981]). In the vocabulary of invasions, this can be termed 'naturalization'. At that point the founder population has become resilient to environmental and demographic stochasticities, and the probability of chance extinction has decreased. Demographic attributes of this population include a significant breeding population; for plants, an established seed bank is a benefit so that year-to-year variations in fecundity are not directly evident in recruitment. In addition, individuals likely develop facilitative associations such as pollination, seed dispersal, and perhaps mycorrhizal associations. Individuals in these populations have overcome barriers to survival in environmental conditions (barrier B in Fig. 1) at the site of introduction (soil and climate for example); as well as barriers to the prevention of consistent and long-term generation of recruits to the population (mating system, compatibility, fertility, fecundity and other reproductive barriers; barrier C in Fig. 1).

Spreading (metapopulation, or other regional entity)

'Spreading' populations can be visualized as including a population with a light permanently on (the naturalized population), surrounded by adjacent populations that may blink on and off (like casual populations, but the naturalized population is the source, rather than the source being from beyond the adventive area). This stage involves the relatively short-distance dispersal of propagules to adjacent suitable sites (thereby overcoming barrier D in Fig. 1). Spread from the initial naturalized population to other suitable sites requires sufficient dispersal by propagules, and/or sufficient landscape connectivity between suitable sites such that local extinctions may be replaced by recolonizations from nearby naturalized populations, if not from their own seed bank. (The discovery of contiguous populations of differing size around an older one is well documented in the herbarium collections used to develop the distribution maps of the species studied in the Biology of Canadian Weeds series.) This occurs when initial long-distance dispersal (e.g., into a new part of the province, or watershed) is followed by subsequent outbreaks of populations in the general area as suitable habitat becomes occupied. This corresponds to the stage we term landscape 'spreading' (Fig. 1). These populations have overcome barriers to relatively short-distance dispersal in the immediate area (barrier D in Fig. 1), and environmental conditions at adjacent (human-modified) areas (barrier E in Fig. 1) assuming availability of suitable disturbed sites. Spreading populations are self-sustaining, without input from sources outside the adventive region but may be recolonized from nearby sites if local extinction occurs.

Invasive (metapopulation or other regional entity)

The image of the spreading populations (at least one permanently 'on' light surrounded by numerous blinking lights) is repeated in many parts of the adventive range. Spreading populations become themselves the primary source for new casual, naturalized and spreading populations (though some may also continue to become established through separate dispersal events from the original source region). These populations have overcome barriers to long-distance dispersal in the adventive environment.

Some researchers also consider that abiotic and biotic barriers to establishment in *natural* habitats needs to be overcome at this stage of the invasions process (others consider that this barrier may be overcome by naturalized populations (see Richardson *et al.* [2000a]). From an ecological standpoint, the more important development occurs where invasive populations spread to natural environments. However from a management and economic point of view, invasive populations that only occur in, for example, agricultural settings, are equally as important. Thus metapopulations of the introduced species should not *have* to occur in natural habitat, to be regarded as invasive. We do think it is useful in our framework, however, to distinguish ecologically invasive populations from invasive populations in disturbed habitats. In terms of the major ecological impacts of invasive species on biodiversity, it may be important to understand how species that are capable of invading natural habitats differ from those that remain in disturbed or agricultural communities and have primarily economic impacts.

POPULATION VIABILITY ANALYSIS (PVA) TO TRACK POPULATIONS OF NON-NATIVE SPECIES

Population viability analysis has been used traditionally to estimate the risk, or time to extinction for an endangered species, and could equally be applied to predict the potential for establishment, growth or persistence in non-native species. PVA evaluates the likelihood that a given species will persist for a given time into the future. The goal of PVA is often conservation or management of rare or endangered species; this includes identifying threats to species persistence, and applying theoretical concepts of population and metapopulation ecology to improve survivorship (see e.g., Brook *et al.* 2000, Akçakaya and Sjögren-Gulve 2000).

Morris and Doak (2003) summarized three general levels of complexity available for PVA that incorporate multi-site data. All rely on use of matrices to characterize population status over time. The simplest approach is to track the number of populations (*sensu* metapopulation and other regional entities), that is, to estimate the presence or absence of populations in local sites of suitable habitat. It is relatively easy to enumerate sites where an introduced species is reported. A second, count-based (census) approach is a more useful form of PVA, but estimates of migration rates, and mean, variance and covariance in population growth rates are necessary to develop such models. Most useful of all, but also more data-demanding, is a PVA model including demographic analysis at each site, based on (for example) knowledge of birth and death rates, and rates of individual transitions through size- or age classes that govern fecundity rates. Such information would allow construction of a multi-site matrix model, which could more accurately predict the fate of the introduced species.

While count-based and demography-based PVA provide more robust results, our analysis here has shown that this level of species-specific information is often unavailable, and may be insufficient to estimate the parameters of the model. However, tracking presence or absence of a non-native species in suitable habitat is sufficient to estimate the parameters of the simplest PVA models — Patch Occupancy Models (POM) (Akçakaya and Sjögren-Gulve 2000). Such models, as with all 'patch' models of metapopulation dynamics (Hanski 1991), assume that local dynamics occur rapidly in comparison to metapopulation dynamics — i.e., changes in population sizes of local populations take place more rapidly than colonizations of currently empty habitat patches. With this assumption, it is reasonable to focus entirely on the presence or absence of the species of interest in the available habitat patches. Given this, the persistence of the species can be modeled by examining the probability of extinction of an occupied patch and the probability of colonization of an empty patch.

Simple POMs are based on classical metapopulation models *sensu* Levins (1969) and thus treat habitat patches as if they were identical, and movement between all pairs of patches is equally likely (Hanski and Gyllenberg 1993, Hanski 1994). Recently, more data-intensive and spatially explicit models have been created (see e.g., Hanski 1994, Moilanen 2000). We constructed a simple POM-based software module that demonstrates the usefulness of PVA for modeling contrasting levels of invasiveness in exotic species. The model uses two inputs: the probability that an occupied patch will become vacant due to local extinction; and the probability of an unoccupied patch becoming occupied by migrants from an occupied patch in each time step or iteration. Such values may be estimated from a series of 'snapshots' of the presence or absence of a species through time (and with particular life history characteristics influencing colonization ability). Gotelli and Taylor (1999a, b) estimated such probabilities for fish assemblages using the probability of colonization as the number of unoccupied sites in year (t) that were occupied in year (t+1) divided by the number of sites that were censused in year (t). Similarly, the probability of extinction can be estimated as the number of sites occupied in year (t) that were unoccupied in year (t + 1) divided by the number of sites occupied in year (t).

Using a combination of 3 probability levels for extinction (high, medium and low; P_{ext} values of 0.09, 0.0495 and 0.009, respectively) and three for the probability of colonization (high, medium and low; P_{col} values of 0.1, 0.055 and 0.01, respectively), clear, consistent outcomes emerged. Each was plotted for 200 random iterations to generate plots that correspond to the four categories of non-native populations of species. The values used to parameterize this model were based on a range of extinction and colonization estimates given for a variety of organisms (see e.g., Wahlberg *et al.* 1996, Moilanen *et al.* 1998, Moilanen 1999, Gotelli and Taylor 1999a, b). Simulations were run in a random landscape, with 100 available patches, initially with a single patch occupied (representing an introduction). In each case there were 100 iterations, i.e., 100 separate introductions.

Table 8 summarizes the simulation output and Fig. 2 depicts the change in proportion of patches occupied over time, for each scenario. The models provide an effective framework for generalization; we can state that where $P_{col} \leq P_{ext}$, the species is casual to naturalized. Where $P_{col} > P_{ext}$, the species is spreading; it is only where $P_{col} >> P_{ext}$ that the species is likely to become invasive. This generalization suggests that most introductions should result in casual or naturalized populations of species (67% of the outcomes), with only 23% being spreading and 10% invasive.

Independent of P_{col}, if P_{ext} is high then the species is merely casual. Even if the species has low P_{col} and a medium P_{ext}, it will occur as a casual. Populations may blink off and on again but fail to sustain themselves over longer periods — only occasionally persisting. Simulations showed that on average, with low colonization rate and medium extinction rate all populations blinked off after 12-25 iterations; however, two outlier populations persisted for the full 200 runs. Naturalized species were those with an equivalent (equally weak or strong) P_{col} and P_{ext}; that is having low P_{col} and P_{ext} (low-low), or medium P_{col} and P_{ext} (medium-medium). While most of the introductions ended in extinction, 38% (low-low) and 9% (medium-medium) persisted for all 200 iterations.

Spreading species would have a proportionately greater probability of P_{col} and P_{ext} (high-medium or medium-low). Invasive species are distinguished by having a high P_{col} and a low P_{ext}. We note these exemplars represent an ecological continuum from casual to invasive. It is interesting that the curves have distinctive shapes. For example, although spreading species may reach comparable levels of occupancy as invasives (see Figs. 2b and f), spreading species show a time lag associated with prolonged low site occupancy, resulting in a greater risk of "invasion failure".

Any introduction had the possibility of failure. Even the invasive species in our simulation failed to establish six percent of the time following introduction. Indeed, as noted earlier, the literature suggests that most introductions fail to establish and even those that successfully establish and persist may have had many previous failed attempts. It is important to note that even when populations have become invasive, the process is not static. Rather, populations may still be vulnerable to demographic and environmental factors acting to increase extinction and decrease colonization rates. We have observed dramatic changes over the past fifteen years in southwestern Ontario populations of purple loosestrife (*Lythrum salicaria*) (see too Mal *et al.* 1992). Following a surge of colonizations over the past 2-3 decades, and abundant large local populations, many of these have gone extinct or become extremely diminished in recent years.

Table 8 Results of population viability analysis for populations with differing extinction and colonization probabilities.

	casual				naturalized		spreading		invasive
Reference Fig. 2	a	d	g	h	e	i	b	f	c
Colonization probability	high	med	low	low	med	low	high	med	high
Extinction probability	high	high	high	med	med	low	med	low	low
Mean (± s.e.) patch occupancy (%)	0.14 (0.121)	0 (0)	0 (0)	0 (0)	0.66 (0.249)	1.16 (0.220)	25.32 (2.392)	64.3 (3.339)	85.1 (2.171)
Proportion (%) of introductions ending in extinction	98	100	100	100	91	62	45	21	6
Mean (± s.e.) time to patches going extinct (iterations)	24.24 (3.827)	13.26 (1.633)	12.02 (1.327)	24.77 (2.402)	30.34 (4.778)	63.77 (6.531)	18.95 (3.368)	23.1 (3.418)	18.5 (7.074)

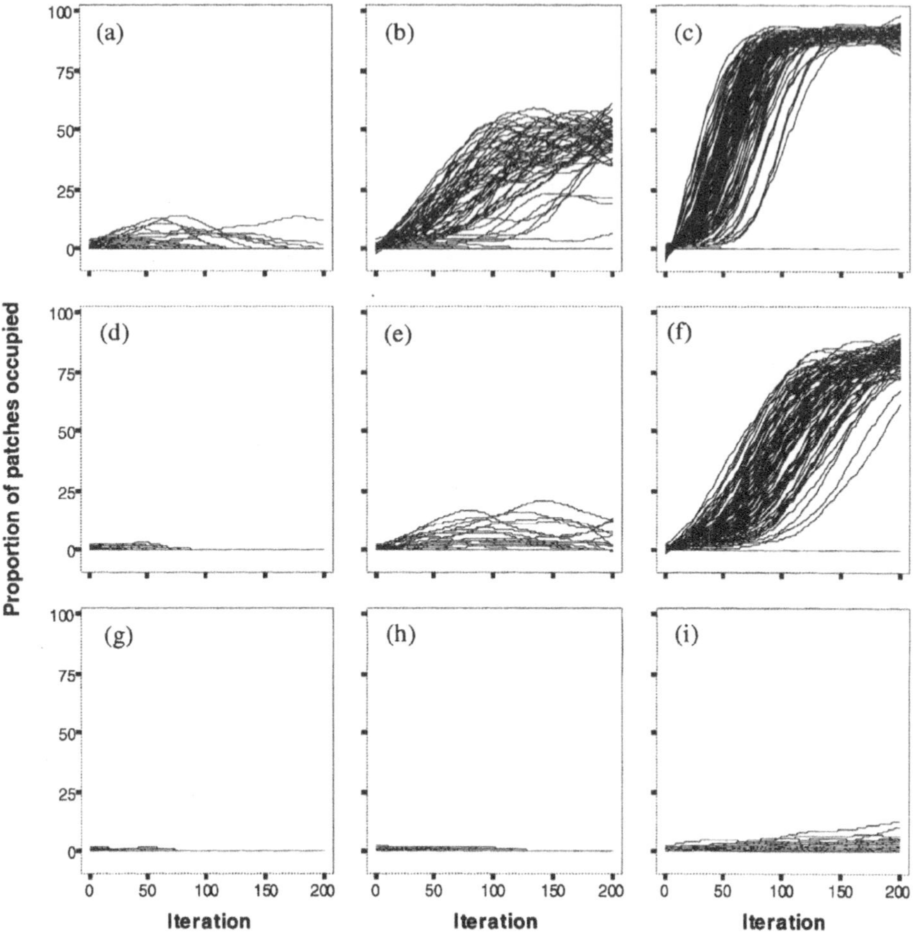

Fig. 2 Results of Population Viability Analysis for populations with differing extinction and colonization rates (see Table 8). (a) casual, (b) spreading, (c) invasive, (d) casual, (e) naturalized, (f) spreading, (g) casual, (h) casual, (i) naturalized.

The population-based framework we have described is informed by the modified scheme in Table 1, where the environmental and life-history parameters that are case dependent are not the focus (as in Richardson *et al.* [2000a]), but rather the filter regulating the changing status of populations of an introduced species. We acknowledge it is difficult to build operational definitions for regional lists of alien flora with the limited life-history data on record, and suggest that it is time for research in invasions ecology to move beyond comparative analyses of regional

Appendix 1 Species included in the dataset with their associated Biology of Canadian Weeds (BCW) series number and reference.

BCW series #	Family	Species	Reference
44	Amaranthaceae	*Amaranthus powellii*	Mulligan 1984
44	Amaranthaceae	*Amaranthus retroflexus*	Mulligan 1984
111	Apiaceae	*Anthriscus sylvestris*	Darbyshire *et al.* 1999
5	Apiaceae	*Daucus carota*	Mulligan 1979
52	Asteraceae	*Achillea millefolium* ssp. *millefolium*	Mulligan 1984
43	Asteraceae	*Acroptilon repens*	Mulligan 1984
38	Asteraceae	*Arctium lappa*	Mulligan 1984
38	Asteraceae	*Arctium minus*	Mulligan 1984
66	Asteraceae	*Artemisia absinthium*	Cavers 1995
118	Asteraceae	*Artemisia vulgaris*	Barney and DiTommaso 2003
89	Asteraceae	*Carduus acanthoides*	Cavers 2000
89	Asteraceae	*Carduus nutans*	Cavers 2000
6	Asteraceae	*Centaurea diffusa*	Mulligan 1979
6	Asteraceae	*Centaurea maculosa*	Mulligan 1979
13	Asteraceae	*Cirsium arvense*	Mulligan 1979
54	Asteraceae	*Crepis tectorum*	Mulligan 1984
58	Asteraceae	*Galinsoga parviflora*	Mulligan 1984
58	Asteraceae	*Galinsoga quadriradiata*	Mulligan 1984
101	Asteraceae	*Helianthus tuberosus*	Cavers 2000
50	Asteraceae	*Hypochaeris radicata*	Mulligan 1984
122	Asteraceae	*Lactuca serriola*	Weaver and Downs 2003
99	Asteraceae	*Matricaria perforata*	Cavers 2000
96	Asteraceae	*Senecio jacobaea*	Cavers 2000
123	Asteraceae	*Senecio vulgaris*	Robinson *et al.* 2003
94	Asteraceae	*Sonchus arvensis*	Cavers 2000
63	Asteraceae	*Sonchus asper*	Cavers 1995
63	Asteraceae	*Sonchus oleraceus*	Cavers 1995
117	Asteraceae	*Taraxacum officinale*	Stewart-Wade *et al.* 2002
110	Asteraceae	*Tragopogon dubius*	Clements *et al.* 1999
110	Asteraceae	*Tragopogon porrifolius*	Clements *et al.* 1999
110	Asteraceae	*Tragopogon pratensis*	Clements *et al.* 1999
87	Boraginaceae	*Cynoglossum officinale*	Cavers 2000
116	Boraginaceae	*Echium vulgare*	Klemow *et al.* 2002
35	Brassicaceae	*Alliaria petiolata*	Mulligan 1984
97	Brassicaceae	*Barbarea vulgaris*	Cavers 2000
3	Brassicaceae	*Cardaria draba*	Mulligan 1979

Appendix 1 Continued.

BCW series #	Family	Species	Reference
22	Brassicaceae	*Descurainia sophia*	Mulligan 1979
108	Brassicaceae	*Erucastrum gallicum*	Warwick and Wall 1998
120	Brassicaceae	*Neslia paniculata*	Francis and Warwick 2003
8	Brassicaceae	*Sinapis arvensis*	Mulligan 1979; Warwick *et al.* 2000
9	Brassicaceae	*Thlaspi arvense*	Mulligan 1979; Warwick *et al.* 2002
119	Cannabaceae	*Cannabis sativa*	Small and Cavers 2002
14	Caryophyllaceae	*Gypsophila paniculata*	Mulligan 1979
25	Caryophyllaceae	*Silene latifolia* ssp. *alba*	Mulligan 1979
46	Caryophyllaceae	*Silene noctiflora*	Mulligan 1984
42	Caryophyllaceae	*Stellaria media*	Mulligan 1984
32	Chenopodiaceae	*Chenopodium album*	Mulligan 1979
83	Clusiaceae	*Hypericum perforatum*	Cavers 1995
53	Convolvulaceae	*Convolvulus arvensis*	Mulligan 1984
12	Dipsacaceae	*Dipsacus fullonum* ssp. *sylvestris*	Mulligan 1979
85	Euphorbiaceae	*Euphorbia cyparissias*	Cavers 2000
39	Euphorbiaceae	*Euphorbia esula*	Mulligan 1984
41	Fabaceae	*Lotus corniculatus*	Mulligan 1984
33	Fabaceae	*Medicago lupulina*	Mulligan 1984
29	Fabaceae	*Melilotus alba*	Mulligan 1979
29	Fabaceae	*Melilotus officinalis*	Mulligan 1979
57	Fabaceae	*Trifolium repens*	Mulligan 1984
76	Fabaceae	*Vicia cracca*	Cavers 1995
76	Fabaceae	*Vicia sativa*	Cavers 1995
76	Fabaceae	*Vicia tetrasperma*	Cavers 1995
76	Fabaceae	*Vicia villosa*	Cavers 1995
34	Haloragaceae	*Myriophyllum spicatum*	Mulligan 1984
80	Lamiaceae	*Galeopsis tetrahit*	Cavers 1995
100	Lythraceae	*Lythrum salicaria*	Cavers 2000
90	Malvaceae	*Abutilon theophrasti*	Cavers 2000
71	Oxalidaceae	*Oxalis corniculata*	Cavers 1995
47	Plantaginaceae	*Plantago lanceolata*	Mulligan 1984
4	Plantaginaceae	*Plantago major*	Mulligan 1979
72	Poaceae	*Apera spica-venti*	Cavers 1995
27	Poaceae	*Avena fatua*	Mulligan 1979
75	Poaceae	*Bromus tectorum*	Cavers 1995
77	Poaceae	*Echinochloa crusgalli*	Cavers 1995

82	Poaceae	*Holcus lanatus*	Cavers 1995
31	Poaceae	*Hordeum jubatum*	Mulligan 1979
37	Poaceae	*Poa annua*	Mulligan 1984
59	Poaceae	*Setaria verticillata*	Mulligan 1984
70	Poaceae	*Setaria viridis*	Cavers 1995
61	Poaceae	*Sorghum halepense*	Mulligan 1984
74	Polygonaceae	*Fagopyrum tataricum*	Cavers 1995
60	Polygonaceae	*Polygonum convolvulus*	Mulligan 1984
40	Portulacaceae	*Portulaca oleracea*	Mulligan 1984
69	Potamogetonaceae	*Potamogeton crispus*	Cavers 1995
95	Ranunculaceae	*Ranunculus repens*	Cavers 2000
18	Rosaceae	*Potentilla argentea*	Mulligan 1979
18	Rosaceae	*Potentilla recta*	Mulligan 1979
121	Rubiaceae	*Galium mollugo*	Mersereau and DiTommaso 2001
86	Rubiaceae	*Galium spurium*	Cavers 2000
106	Scrophulariaceae	*Linaria dalmatica*	Vujnovic and Wein 1997
105	Scrophulariaceae	*Linaria vulgaris*	Saner *et al.* 1995
28	Scrophulariaceae	*Verbascum thapsus*	Mulligan 1979
64	Solanaceae	*Datura stramonium*	Cavers 1995
78	Solanaceae	*Solanum carolinense*	Cavers 1995
67	Solanaceae	*Solanum nigrum*	Cavers 1995
78	Solanaceae	*Solanum rostratum*	Cavers 1995
99a	Violaceae	*Viola arvensis*	Cavers 2000

flora datasets in order to develop general principles that will allow carrying out more accurate risk assessment of casual, introduced populations of species before they naturalize and spread to become serious and costly invasive populations.

ACKNOWLEDGMENTS

HTM was supported by SHARCNET, JVDW, JLD and LLD are supported by Natural Sciences and Engineering Research Council of Canada. We thank Rob Colautti, an anonymous reviewer and the editors for comments that greatly improved the manuscript.

REFERENCES

Akçakaya, H. R. and P. Sjögren-Gulve. 2000. Population viability analysis in conservation planning: An overview. Ecological Bulletin **48**, 9-21.

Anderson, P. K., A. A. Cunningham, N. G. Patel, F. J. Morales, P. R. Epstein and P. Daszak. 2004. Emerging infectious diseases of plants: pathogen pollution, climate change and agrotechnology drivers. Trends in Ecology and Evolution **19**, 535-544.

Baker, H. G. 1965. Characteristics and mode of origin of weeds. pp. 147-168 in H. G. Baker and G. L. Stebbins, editors. The genetics of colonizing species. Academic Press, New York, NY.

Baker, H. G. 1974. The evolution of weeds. Annual Review of Ecology and Systematics **5**, 1-24.

Barney, J. N. and A. DiTommaso. 2003. The Biology of Canadian Weeds. 118. *Artemisia vulgaris* L. Canadian Journal of Plant Science **83**, 205-215.

Blackburn, T. M. and R. P. Duncan. 2001a. Determinants of establishment success in introduced birds. Nature **414**, 195-197.

Blackburn, T. M. and R. P. Duncan. 2001b. Establishment patterns of exotic birds are constrained by non-random patterns in introduction. Journal of Biogeography **28**, 927-939.

Brook B. W., J. J. O'Grady, A. P. Chapman, M. A. Burgman, H. R. Akçakaya and R. Frankham. 2000. Predictive accuracy of population viability analysis in conservation biology. Nature **404**, 385-387.

Bruno, J. F., J. J. Stachowicz and M. D. Bertness. 2003. Inclusion of facilitation into ecological theory. Trends in Ecology and Evolution **18**, 119-125.

Cadotte, M. W. and J. Lovett-Doust. 2001. Ecological and taxonomic differences between native and introduced plants of southwestern Ontario. Ecoscience **8**, 230-238.

Cadotte, M. W., B. R. Murray and J. Lovett-Doust. Evolutionary and ecological factors influencing Ontario plant invader success. Ecoscience (in press).

Cadotte, M. W., B. R. Murray and J. Lovett-Doust. 2005. Ecological patterns and biological invasions: Using regional species inventories in macroecology. Biological Invasions **8**, 1-13.

Callaway, R. M. and E. T. Aschehoug. 2000. Invasive plants versus their new and old neighbors: A mechanism for exotic invasion. Science **290**, 521-523.

Carr, G. W. 1993. Exotic flora of Victoria and its impact on indigenous biota. pp. 256-297 in D. B. Foreman and N. G. Walsh, editors. Flora of Victoria. Inkata Press, Melbourne, Australia.

Case, T. J. 1990. Invasion resistance arises in strongly interacting species-rich model competition communities. Proceedings of the National Academy of Science **87**, 9610-9614.

Cavers, P. B., editor. 1995. The Biology of Canadian Weeds, Vol. 3. Contributions 62-83. The Agricultural Institute of Canada, Ottawa, ON.

Cavers, P. B., editor. 2000. The Biology of Canadian Weeds, Vol. 4. Contributions 84-102. The Agricultural Institute of Canada, Ottawa, ON.

Cavers, B. P. and S. Warwick. 2000. The Biology of Canadian Weeds. Revised format and instructions for preparation of updated accounts. Canadian Journal of Plant Science **80**, 237-240.

Clements, D. R., M. K. Upadhyaya and S. J. Bos. 1999. The Biology of Canadian Weeds. 110. *Tragopogon dubius* Scop., *Tragopogon pratensis* L., and *Tragopogon porrifolius* L. Canadian Journal of Plant Science **79**, 153-163.

Colautti, R. I. and H. J. MacIsaac. 2004. A neutral terminology to define 'invasive' species. Diversity and Distributions **10**, 135-141.

Daehler, C. C. 1998. The taxonomic distribution of invasive angiosperm plants: Ecological insights and comparison to agricultural weeds. Biological Conservation **84**, 167-180.

Daehler, C. C. 2001a. Two ways to be an invader, but one is more suitable for ecology. ESA Bulletin **82**, 101-102.

Daehler, C. C. 2001b. Darwin's naturalization hypothesis revisited. American Naturalist **158**, 324-329.

Darbyshire, S. J., R. Hoeg and J. Haverkort. 1999. The Biology of Canadian Weeds. 111. *Anthriscus sylvestris* (L.) Hoffm. Canadian Journal of Plant Science **79**, 671-682.

Darwin, C. 1859. On the origin of species. Murray, London.

Davis, M. A. and K. Thompson. 2000. Eight ways to be a colonizer; two ways to be an invader: A proposed nomenclature scheme for invasion ecology. ESA Bulletin 81, 226-230.

Davis, M. A. and K. Thompson. 2001. Invasion terminology: should ecologists define their terms differently than others? No, not if we want to be of any help! ESA Bulletin **82**, 206.

Davis, M. A. and K. Thompson. 2002. "Newcomers" invade the field of invasion ecology: question the field's future. ESA Bulletin **83**, 196-197.

Eckert, G. C., D. Manicacci and S. C. H. Barrett. 1996. Genetic drift and founder effect in native versus introduced populations of an invading plant, *Lythrum salicaria* (Lythraceae). Evolution **50**, 1512-1519.

Elton, C. S. 1958. The ecology of invasions by animals and plants. Methuen, London.

Forsyth, D. M. and R. P. Duncan. 2001. Propagule size and the relative success of exotic ungulate and bird introductions to New Zealand. The American Naturalist **157**, 583-595.

Foxcroft, L. C., M. Rouget, D. M. Richardson and S. Mac Fayden. 2004. Reconstructing 50 years of Opuntia stricta invasion in the Kruger National Park, South Africa: environmental determinants and propagule pressure. Diversity and Distributions **10**, 427-437.

Francis, A. and S. I. Warwick. 2003. The Biology of Canadian Weeds. 120. *Neslia paniculata* (L.) Desv. Canadian Journal of Plant Science **83**, 441-451.

Freckleton, R. P. and Watkinson, A. R. 2002. Large-scale spatial dynamics of plants: metapopulations, regional ensembles and patchy populations. Journal of Ecology **90**, 419-434.

Gleason, H. A. 1952. The new Britton and Brown illustrated flora of the Northeastern United States and adjacent Canada. Hafner Press, New York.

Gleason, H. A. and A. Cronquist. 1991. Manual of vascular plants of northeastern United States and adjacent Canada. Second Edition. New York Botanical Garden, New York, NY.

Gotelli, N. and C. M. Taylor. 1999a. Testing metapopulation models with stream-fish assemblages. Evolutionary Ecology Research **1**, 835-845.

Gotelli, N. and C. M. Taylor. 1999b. Testing macroecology models with stream-fish assemblages. Evolutionary Ecology Research **1**, 847-858.

Hanski, I. 1991. Single-species metapopulation dynamics — concepts, models and observations. Biological Journal of the Linnean Society **42**, 17-38.

Hanski, I. 1994. A practical model of metapopulation dynamics. Journal of Animal Ecology **63**, 151-162.

Harrod, R. J. and R. J. Taylor. 1995. Reproduction and pollination biology of Centaurea and Acroptilon species, with emphasis on *C. diffusa*. Northwest Science **69**, 97-105.

Higgins, S. I., R. Nathan and M. L. Cain. 2003. Are long-distance dispersal events in plants usually caused by nonstandard means of dispersal? Ecology **84**, 1945-1956.

Kareiva, P. and Wennergren, U. 1995. Connecting landscape patterns to ecosystem and population processes. Nature **373**, 299-303.

Keane, R. M. and M. J. Crawley. 2002. Exotic plant invasions and the enemy release hypothesis. Trends in Ecology and Evolution **17**, 164-170.

Klemow, K. M., D. R. Clements, P. F. Threadgill and P. B. Cavers. 2002. The Biology of Canadian Weeds. 116. Echium vulgare L. Canadian Journal of Plant Science **82**, 235-248.

Klironomos, J. N. 2002. Feedback with soil biota contributes to plant rarity and invasiveness in communities. Nature **417**, 67-70.

Kolar, S. C. and D. M. Lodge. 2001. Progress in invasion biology: predicting invaders. Trends in Ecology and Evolution **16**, 199-204.

Kolbe, J. J., E. E. Glor, L. R. Schettino, A. C. Lara, A. Larson and J. B. Losos. 2004. Genetic variation increases during biological invasion by a Cuban lizard. Nature **431**, 177-181.

Kowarik I. 1995. Clonal growth in Ailanthus altissima on a natural site in West Virginia. Journal of Vegetation Science **6**, 853-856.

Krebs, C. J. 1985. Ecology: The experimental analysis of distribution and abundance. Harper & Row, New York, NY.

Levin, D. A. 2003. Ecological speciation: lessons from invasive species. Systematic Botany **28**, 643-650.

Levine, J. M., P. B. Adler and S. G. Yelenik. 2004. A meta-analysis of biotic resistance to exotic plant invasions. Ecology Letters **7**, 975-989.

Levins, R. 1969. Some demographic and genetic consequences of environmental heterogeneity for biological control. Bulletin of the Entomological Society of America **15**, 237-240.

Lodge, D. M. 1993. Biological invasions: lessons for ecology. Trends in Ecology and Evolution **8**, 133-137.

Lonsdale, M. W. 1999. Global patterns of plant invasions and the concept of invasibility. Ecology **80**, 1522-1536.

Mack, R. N. 1996. Predicting the identity and the fate of plant invaders: emergent and emerging approaches. Biological Conservation **78**, 107-121.

Mack, R. N., D. Simberloff, W. M. Lonsdale, H. Evans, M. Clout and F. A. Bazzaz. 2000. Biotic invasions: causes, epidemiology, global consequences, and control. Ecological Applications **10**, 689-710.

Mal, T. K., Lovett-Doust, J., L. Lovett-Doust and G. A. Mulligan. 1992. The Biology of Canadian Weeds. 100. Lythrum salicaria. Canadian Journal of Plant Science **72**, 1305-1330.

Maron, J. L. and Vilá, M. 2001. When do herbivores affect plant invasion? Evidence for the natural enemies and biotic resistance hypotheses. Oikos **95**, 361-373.

Master, L. L. 1991. Assessing threats and setting priorities for conservation. Conservation Biology **5**, 559-563.

McNeill, J. 1977. The Biology of Canadian Weeds. 25. Silene alba (Miller) E. H. L. Krause. Canadian Journal of Plant Science **57**, 1103-1114.

Mersereau, D. and A. DiTommaso. 2001. The Biology of Canadian Weeds. 121. *Galium mollugo* L. Canadian Journal of Plant Science **83**, 453-465.

Mitchell, C. E. and A. G. Power. 2003. Release of invasive plants from fungal and viral pathogens. Nature **421**, 625-627.

Morris, W. F. and D. F. Doak. 2003. Quantitative conservation biology: theory and practice of population viability analysis. Sinauer Associates, Sunderland, MA.

Moilanen, A. 1999. Patch occupancy models of metapopulation dynamics: Efficient parameter estimation using implicit statistical inference. Ecology **80**, 1031-1043.

Moilanen, A. 2000. The equilibrium assumption in estimating the parameters of metapopulation models. Journal of Animal Ecology **69**, 143-153.

Moilanen, A., A. T. Smith and I. Hanski. 1998. Long-term dynamics in a metapopulation of the American pika. American Naturalist **152**, 530-542.

Morton, J. K. and J. M. Venn. 1990. A checklist of the flora of Ontario – Vascular plants. University of Waterloo Biology Series No. 34, Waterloo, ON.

Mulligan, G. A., editor. 1979. The Biology of Canadian Weeds. Contributions 1-32. Agriculture Canada Publ. 1693, Ottawa, ON.

Mulligan, G. A., editor. 1984. The Biology of Canadian Weeds. Contributions 33-61. Agriculture Canada Publ. 1765, Ottawa, ON.

Murphy, H. T. and J. Lovett-Doust. 2004. Context and connectivity in plant metapopulations and landscape mosaics: does the matrix matter? Oikos **105**, 3-14.

Natural Heritage Information Centre. 2004. Database of vascular plants. http://www.mnr.gov.on.ca/MNR/nhic/nhic.cfm

Newmaster, S. G., A. Lehela, M. J. Oldham, P. W. C. Uhlig and S. McMurray. 1998. Ontario plant list. Forest Research Information Paper No. 123; Ontario Forest Research Institute, Sault Ste. Marie, ON.

Novak, S. J. and R. N. Mack. 1993. Genetic variation in *Bromus tectorum* (Poaceae): comparison between native and introduced populations. Heredity **71**, 167-176.

Pyšek, P. 1998. Is there a taxonomic pattern to plant invasions? Oikos **82**, 282-294.

Pyšek, P., D. M. Richardson, M. Rejmánek, G. L. Webster, M. Williamson and J. Kirschner. 2004. Alien plants in checklists and floras: towards better communication between taxonomists and ecologists. Taxon **53**, 131-143.

Rabinowitz, D. 1981. Seven forms of rarity. pp. 205-218 in H. Synge, editor. The biological aspects of rare plant conservation. Wiley, New York, NY.

Reinhart, K. O., A. Packer, W. H. Van der Putten and K. Clay. 2003. Plant-soil biota interactions and spatial distribution of black cherry in its native and invasive ranges. Ecology Letters **6**, 1046-1050.

Rejmánek, M. 1996. A theory of seed plant invasiveness: the first sketch. Biological Conservation **78**, 171-181.

Rejmánek, M. and D. M. Richardson. 1996. What attributes make some plant species more invasive? Ecology **77**, 1655-1661.

Rejmánek, M., D. M. Richardson, M. G. Barbour, M. J. Crawley, G. F. Hrusa, P. B, Moyle, J. M. Randal, D. Simberloff and M. Williamson. 2002. Biological invasions: politics and the discontinuity of the ecological terminology. ESA Bulletin **83**, 131-133.

Richardson, D. M., P. Pyšek, M. Rejmánek, M. G. Barbour, F. D. Panetta and C. J. West. 2000a. Naturalization and invasion of alien plants: concepts and definitions. Diversity and Distributions **6**, 93-107.

Richardson, D. M., N. Allsopp, C. M. D'Antonio, S. J. Milton and M. Rejmánek. 2000b. Plant invasions — the role of mutualisms. Biological Reviews **75**, 65-93.

Robinson, D. E., J. T. O'Donovan, M. P. Sharma, D. J. Doohan and R. Figueroa. 2003. The Biology of Canadian Weeds. 123. *Senecio vulgaris* L. Canadian Journal of Plant Science **83**, 629-644.

Rouget, M. and D. M. Richardson. 2003. Inferring process from pattern in plant invasions: a semi-mechanistic model incorporating propagule pressure and environmental factors. American Naturalist **162**, 713-724.

Saner, M. A., D. R. Clements, M. R. Hall, D. J. Doohan and C. W. Crompton. 1995. The Biology of Canadian Weeds. 105. *Linaria vulgaris* Mill. Canadian Journal of Plant Science **75**, 525-537.

Sax, D. F. and J. H. Brown. 2000. The paradox of invasion. Global Ecology and Biogeography **9**, 363-371.

Shaffer, M. L. 1981. Minimum population sizes for species conservation. BioScience **31**, 131-134.

Scoggan, H. J. 1979. The flora of Canada. National Museums of Canada. Ottawa, ON.

Simberloff, D. and B. von Holle. 1999. Positive interactions of non-indigenous species: invasional meltdown? Biological Invasions **1**, 21-32.

Small, E. and P. B. Cavers. 2002. The Biology of Canadian Weeds. 119. *Cannabis sativa* L. Canadian Journal of Plant Science **83**, 217-237.

Stewart-Wade, S. M., S. Neumann, L. L. Collins and G. J. Boland. 2002. The Biology of Canadian Weeds. 117. Taraxacum officinale G. H. Weber ex Wiggers. Canadian Journal of Plant Science **82**, 825-853.

Stohlgren T. J., D. Binkley, G. W. Chong, M. A. Kalkhan, L. D. Schell, K. A. Bull, Y. Otsuki, G. Newman, M. Bashkin and Y. Son. 1999. Exotic plant species invade hot spots of native plant diversity. Ecological Monographs **69**, 25-46.

Thompson, K., J. G. Hodgson and T. C. G. Rich. 1995. Native and alien invasive plants: more of the same? Ecography **18**, 390-402.

Tilman, D. 1997. Community invasibility, recruitment limitation, and grassland biodiversity. Ecology **78**, 81-92.

Tutin, T. G., V. H. Heywood, N. A. Burgess, D. H. Valentine, S. M. Walters and D. A. Webb. 1964. Flora Europaea, Vol 1. Lycopodiaceae to Platanaceae. Cambridge University Press, Cambridge, UK.

Turkington, R. A., P. B. Cavers and E. Rempel. 1978. The Biology of Canadian Weeds. 29. *Melilotus alba Desr.* and *M. officinalis* (L.) Lam. Canadian Journal of Plant Science **58**, 523-537.

Vander Wall, S. B. and W. S. Longland. 2004. Diplochory: are two seed dispersers better than one? Trends in Ecology and Evolution **19**, 155-161.

Vitousek, P. M. and L. R. Walker. 1989. Biological invasion by *Myrica faya* in Hawai'i: Plant demography, nitrogen fixation, ecosystem effects. Ecological Monographs **59**, 247-265.

Von Holle, B., H. R. Delcourt and D. Simberloff. 2003. The importance of biological inertia in plant community resistance to invasion. Journal of Vegetation Science **14**, 425-432.

Vujnovic, K. and R. W. Wein. 1997. The Biology of Canadian Weeds. 106. *Linaria dalmatica* (L.) Mill. Canadian Journal of Plant Science **77**, 483-491.

Wahlberg, N., A Moilanen and I. Hanski. 1996. Predicting the occurrence of endangered species in fragmented landscapes. Science **273**, 1536-1538.

Warwick, S. I. and L. D. Black. 1983. The Biology of Canadian Weeds. 61. *Sorghum halapense* (L.) Pers. Canadian Journal of Plant Science **63**, 997-1014.

Warwick, S. I., H. J. Beckie, G. A. Thomas and T. McDonald. 2000. The Biology of Canadian Weeds. 8. *Sinapsis arvensis* L. (updated). Canadian Journal of Plant Science **80**, 939-961.

Warwick, S. I. and D. A. Wall. 1998. The Biology of Canadian Weeds. 108. *Erucastrum gallicum* (Willd.) O. E. Schulz. Canadian Journal of Plant Science **78**, 155-165.

Warwick, S. I., A. Francis and D. J. Susko. 2002. The Biology of Canadian Weeds. **9**. *Thlaspi arvense* L. (updated). Canadian Journal of Plant Science **82**, 830-823.

Weaver, S. E. and M. P. Downs. 2003. The Biology of Canadian Weeds. 122. *Lactuca serriola* L. Canadian Journal of Plant Science **83**, 619-628.

Williamson, M. and A. Fitter. 1996. The varying success of invaders. Ecology **77**, 1661-1666.

Populations at play

Chapter five

Density dependence in invasive plants: demography, herbivory, spread and evolution

Y. M. Buckley and J. Metcalf

INTRODUCTION

Because of their environmental and economic costs (Pimentel *et al.* 2000), some invasive plant species are well studied over relatively long time scales (Buckley *et al.* 2003a) and in different habitats (Paynter *et al.* 2003). These data are valuable for the development or testing of general theoretical models of population dynamics and species interactions. Furthermore, knowing the kind of dynamics exhibited by invasives matters if we are attempting to predict their impact on an ecosystem, and to anticipate how they will respond to novel environments over ecological and evolutionary time-scales. General recent reviews of density dependence can be found in: Turchin (1999), Krebs (2002), Sibly and Hone (2002); here, we concentrate on how recent advances in our understanding of how density dependent processes contribute to our understanding of the demography, management and evolution of invasive plant species.

Despite considerable controversy, stretching back to debates between Nicholson and Andrewartha over 50 years ago, and continuing to some extent today,

M.W. Cadotte et al. (eds.), Conceptual ecology and invasion biology, 109–123.

a combination of endogenous and exogenous forces is generally accepted to regulate population size (Turchin 1999). Endogenous mechanisms are usually referred to as "density dependence", which is a shorthand term encompassing the range of mechanisms that modify vital rates according to the density of conspecifics. These are theoretically of great importance as a necessary condition for population regulation, but are often difficult to detect consistently under natural conditions (see examples in Krebs (2002)). By contrast, exogenous mechanisms incorporate external factors such as resource availability, herbivory, or harvesting. Exogenous and endogenous factors are intimately linked. For example, the strength of exogenous factors, such as resource availability, can determine the equilibrium of a simple density dependent model. The well known Hassell model (Hassell 1975) relates population size in year $t+1$ (N_{t+1}) to population size in the previous year (N_t) modified by a density dependent function incorporating maximum fecundity at low densities (λ), a measure of habitat carrying capacity (a) and the strength and form of competition (b). All three parameters are positive.

$$N_{t+1} = N_t \frac{\lambda}{(1 + aN_t)^b}$$

The equilibrium population size depends on all three parameters,

$$N^* = \frac{\lambda^{1/b} - 1}{a}$$

but stability only depends on the values of λ and b, stability only being possible providing

$$b\,(1 - \lambda^{-1/b}) < 2$$

Changing the carrying capacity of the environment (changing a in the Hassell model) will therefore lead to changes in the equilibrium density of the population (Fig. 1). This is particularly relevant for invasive species taken from their native range to a novel environment, often without a suite of specialist herbivores and pathogens. The enemy release hypothesis (ERH, or natural enemies hypothesis) (Keane and Crawley 2002), posits that invaders are particularly successful in their exotic habitat because they are released from regulation by specialist herbivores, pathogens or parasites present in their native habitat. Most predictions stemming from the ERH relate to increased vigour, be it phenotypic or genotypic, exhibited by individuals. However, if resources are not limiting in either exotic or native ranges and enemy release is operating we might expect populations of invasives in the exotic range to have a higher carrying capacity and exhibit higher equilibrium densities. This appears to be the case for scotch broom (*Cytisus scoparius*)

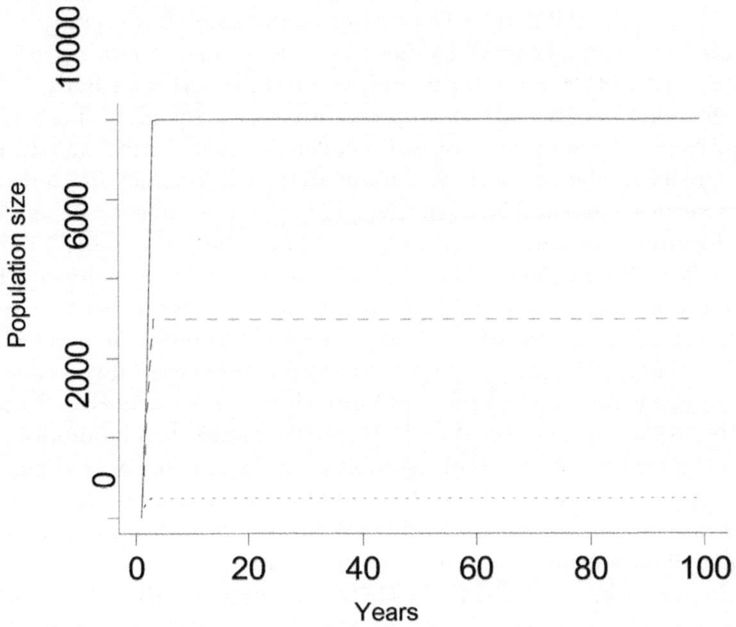

Fig. 1 Hassell model at different carrying capacities, equilibrium population densities achieved over time (years on the x-axis) with a = 0.5, solid line, a = 1, dashed line and a = 10, dotted line.

which was found to have significantly higher densities in two countries in its exotic range (Australia and New Zealand) compared with its native range (Europe) (Paynter *et al.* 2003). The absence of specialist herbivores or pathogens may also affect the competition parameter *b*, which at its extreme values represents ideal scramble and ideal contest competition. Competition for enemy-free space may have a different form from competition for soil resources, and different values of *b* in different environments would change not just the equilibrium population densities but could potentially lead to populations with different stability properties. This very simple model illustrates how exogenous changes to conditions can lead to different density dependent dynamics. To our knowledge, a detailed study of the population dynamics of an invasive plant in native and exotic habitats has not been carried out in order to investigate how density dependent conditions differ between habitats. With the use of observations, experiments and modelling, an invasive plant system would be ideal for investigating the influence of different environments on density dependent processes.

The importance of exogenous factors for density dependent population regulation is particularly well illustrated by a recent analysis of the dynamics of ragwort, *Senecio jacobaea* L., and its specialist herbivore the cinnabar moth, *Tyria jacobaeae* L. (Bonsall *et al.* 2003). Long term (≥20 years) time series for both species exist for two sites, one in Silwood, UK and the other in Meijendel, The Netherlands. In Silwood ragwort populations, only direct density dependence (i.e., the influence of plant density in the previous year) was detected; whereas in Meijendel both direct (1 year previous) and delayed (two years previous) density dependence was detected, leading to qualitatively different dynamics at the two sites. Populations are relatively stable in Silwood but exhibit oscillations in Meijendel. The driver for the differences in dynamics between the sites may be differences in recruitment. Recruitment is micro-site limited at Silwood but not at Meijendel, implying strong density dependence operating at the seedling stage in Silwood populations. *Senecio jacobaea* is native to Europe and introduced to Australasia, the Americas and South Africa (Harper and Wood 1957). This study is a landmark for ecologists working on invasives, as the variability in the qualitative dynamics of this species within its native range alone should lead us to ask questions of attempts to extrapolate model or experimental results from native to exotic ecosystems in the absence of data from the exotic habitat. It should also encourage us to move on from documenting trait differences between species in their native and exotic ranges (Siemann and Rogers 2001, Buckley *et al.* 2003b, Siemann and Rogers 2003) to investigating the population dynamic consequences of those trait differences and their interaction with habitat differences. In addition, context dependency of this sort might result in density dependence being overlooked if the system studied includes habitat heterogeneity. For example, in a study of reef fish Shima and Osenberg (2003) found that the strength of density dependent mortality depended on site quality, and a failure to take heterogeneity in site quality into account could have lead to the conclusion that density dependence was unimportant. Shima & Osenberg refer to the patterns observed as "cryptic density dependence".

Within a population, differences between individuals or stages will also affect density dependence and resulting dynamics. For example, population density and asymmetric competition can increase the differential in size between levels in hierarchies of plant populations (Weiner 1985), and this can create differences between individual responses to different densities of different stages. Simple models like the Hassell model shown above do not address this. However, we can incorporate such differences by using models with age or stage structure. Benton *et al.* (2004) and Cameron and Benton (2004) use experimentally manipulated soil mite populations and models to explore the interaction of density dependence and environmental variability on the outcome of harvesting different stages. They found that harvesting different stages (reducing survival of eggs, juveniles or adults) had different impacts on the mean population size and its variance in each stage. In a variable environment they found that removal of 15% of eggs actually resulted in an increase in mean population size, whereas removal of corresponding proportions of juveniles and adults led to a decrease in mean population size. This would appear to be a result of density dependence acting differently in different stages. They liken removal of eggs to thinning seedlings, this releases the survivors from density dependent suppression and if density dependence is strong enough can lead to an increase in overall population size. The effects of egg removal may have been more pronounced in the variable environment because of the alternating periods of strong and weak competition for resources. The same counter-intuitive result (harvesting leading to increased population size) can occur in a constant environment where density dependence is very strong or over-compensating (e.g., Buckley *et al.* 2001). Of relevance to the management of invasives is Benton *et al.*'s (2004) conclusion that density independent models failed to predict the negative elasticity of survival at the egg stage, therefore management models which do not include density dependence, where it is known to be important, run the risk of giving the wrong answer.

Incorporating the added structure from individual variation to models is often important; with this added complexity however, greater care should be taken to choose the appropriate model. Fox and Kendall (2002) illustrate that failure to model individual variability correctly as systematic variation among individuals, rather than random variation, may lead to inaccurate predictions. They conclude "almost all population viability analyses (PVAs) over-estimate the importance of demographic stochasticity and, therefore, the risk of extinction". Murrell (Chapter 7) presents a modelling framework that takes account of local interactions, the "plant's eye view" of its neighbourhood density, which even without spatial or temporal heterogeneity leads to invasion dynamics quite different from those predicted by mean field approximations, which take overall density into account. New tools which allow us to take account of variation in resources at different spatio-temporal scales, and how plants respond to variable resources and clustering, allow both increasingly accurate predictions of dynamics, and a better understanding of the scale and mechanisms involved in density dependent responses.

So far we have only addressed negative effects of density dependence on fitness or population growth; however, density dependence can also operate positively. Where any component of fitness increases as density or numbers increase, this is known as an Allee effect, this may translate into demographic Allee effects where population growth rate declines with decreasing density, depending on the strength of negative density dependence in the component of fitness in question (Stephens *et al.* 1999). An example of a component Allee effect is that there might be a greater probability of finding a mate in a larger population. For insect pollinated plants, seed set may depend on there being nearby con-specifics; Parker (1997) and Parker and Haubensak (2001) give examples of pollinator limitation in the invasive Scotch broom (*Cytisus scoparius*). An important consequence of the Allee effect for invasions is that isolated plants or isolated populations can experience a slower population growth rate relative to larger populations. This may be particularly important during the establishment phase of invasive species, or determine what level of control will lead to eradication or slower population growth. Many invasive species experience a "lag time" between initial colonization and rapid population growth. This might simply be due to the lag phase in an exponential growth curve, but it could also indicate the action of strong or weak Allee effects. The distinction between strong and weak Allee effects is an important one and is outlined in Wang and Kot (2001). Demographic Allee effects occur when the population growth rate at some set of low densities is slower than population growth rate at higher densities. Strong demographic Allee effects occur when population growth rate at low density is negative. If strong Allee effects were operating during the establishment phase of an invasive species, the population would need to reach a threshold size before being able to spread rapidly. This could be achieved through one large initial release or through repeated invasions equivalent to a rescue effect in meta-population models (Amarasekare 1998). Equally, that population could be driven to extinction by reducing it below the threshold. Weak demographic Allee effects result in a population that grows more slowly at low densities than it does at high densities but that growth is always positive so a threshold does not exist. A consequence of the operation of weak Allee effects would be that the lag phase of an invasion would be longer than predicted by an exponential growth curve.

Allee effects can be important during invasive species expansion. Davis *et al.* (2004) demonstrate this for *Spartina alterniflora*, which "sets very little viable seed at the leading edges of an invasion in Willapa Bay, Washington, USA where it was introduced *c.* 100 years ago". *Spartina alterniflora* has two dispersal mechanisms, short distance spread from rhizomatous growth in existing plants and long distance spread via seeds. Long distance spread from the leading edge of the population is limited however by a lack of viable seed until the population has filled in through rhizomatous growth or germination of other seeds dispersed from existing dense populations. Weak Allee effects such those found in the invasive *S. alterniflora* not only affect population growth rates at low density, equilibrium densities and stability (Fowler and Ruxton 2002) but their inclusion in models of spread can alter spread velocities. The impact of weak Allee effects on equilibria, stability and spread velocities has only been addressed relatively recently, previously only strong Allee effects had been considered in any detail. For integrodifference equation (IDE) models of spread, Kot *et al.* (1996) showed that strong Allee effects can turn accelerating invasions into constant speed invasions and Wang *et al.* (2002) suggest, using numerical solutions, that including weak Allee effects can also slow down invasions from accelerating to constant speeds. Wang *et al.* (2002) also show that Allee effects determine the direction of a travelling wave solution of IDE models, where the direction of the travelling wave solution determines whether an invasion will succeed or not. Traditional reaction-diffusion models of spread have also been investigated with regard to the inclusion of Allee effects, again showing that Allee effects slow down spread (Cruickshank *et al.* 1999) and even that an invasion can only succeed once a certain critical area is occupied (Lewis and Kareiva 1993).

Whether density dependence acts negatively or through Allee effects, it will have implications for population stability. This is true even in the simplest models (see the Hassell model above for negative density dependence and Scheuring (1999) who shows that Allee effects promote stability in simple models). Interest in density dependence often stems from the search for mechanisms either promoting stability in plant populations or destabilising populations. Most plant populations are believed to exhibit stable dynamics due to stabilizing features of their life-histories such as soil seed banks (Watkinson 1980, Pacala and Silander 1990, Jarry *et al.* 1995), phenotypic plasticity and asymmetric competition (Palmblad 1968, Weiner 1985, Pacala and Weiner 1991, Silvertown 1991), small size thresholds for reproduction (Rees and Crawley 1989) and self-thinning (Watkinson 1980, Watkinson and Davy 1985). Destabilizing processes include over-compensating density dependent fecundity (Thrall *et al.* 1989), time lags caused by maternal effects (Crone 1997), large size thresholds for reproduction (Thrall *et al.* 1989) and delayed flowering at high density (Buckley *et al.* 2001). Cyclic or chaotic dynamics can result from an over-compensating density dependent response combined with a high finite rate of population growth from low densities. Over-compensating density dependence occurs "when the proportional decline in parameter value with increasing density exceeds the change in density" (Freckleton and Watkinson 2002). Mechanisms giving rise to over-compensating density dependence are rare in plant systems, and therefore unstable dynamics such as cycles or chaos are equally rare. Chaotic dynamics in annual plant systems are reviewed by Freckleton & Watkinson (2002) and there is little evidence for chaotic dynamics in other plant life-histories (Rees and Crawley 1989, 1991). However, Rees *et al.* (2002) have shown that flowering dynamics close to chaotic occur in a mast-seeding tussock grass, *Chionochloa pallens*, possibly as an evolutionary response to attack by seed and flower feeders.

Stability is an important issue for populations of invasives for a number of reasons; knowing whether a population is likely to be stable or exhibit cycles may allow prediction of levels of impact likely to occur, stability may also affect the success of biocontrol programs. For plant species with long-lived seed banks it is important to maintain populations of a biocontrol agent in the system in order to control populations arising from the seed bank at a later date, or continuous re-introduction of the biocontrol agent is necessary. When choosing a biocontrol agent therefore it may be better to choose an agent which maintains the plant population at a low stable level rather than one which drives the above ground population to extinction, causing its own extinction and the subsequent resurgence of the invasive plant. Even if the herbivore does not cause plant extinction directly, unstable cycles can result in a higher probability of stochastic extinction of the plant population. Buckley *et al.* (2005) use an experimentally parameterised plant-herbivore model to show that density dependence in both plant and herbivore populations contribute to stability in the system but that the form of density dependence in the herbivore, in combination with high attack rates, can destabilise the dynamics. Studying the form of density dependence (e.g. scramble or contest) in potential agents may therefore give us some idea of which agents would be most appropriate to introduce; in systems like this a biocontrol agent combining strong scramble competition with a high attack rate may not be a good choice.

The exact way in which density dependent processes affect stability can be subtle. Density dependence may occur at several life history stages within a species, and which stage it acts on can be of critical importance to its effect for population stability. For example, Buckley and colleagues (2001) found evidence for density dependent survival, flowering and fecundity in the invasive annual *Tripleurospermum perforatum*. These different occurrences had different implications for population stability. Density dependence in survival and fecundity was stabilising, whereas density dependence in flowering led to instability in facultative biennial systems (see Fig. 2 for density dependent flowering probability and Fig. 3 for model stability). However, density dependent flowering also means that in this facultative biennial, delayed flowering is favoured at high densities. This leads to asymmetric competition between larger over-wintered rosettes and new seedlings in the following year, which is generally a stabilising force as the larger rosettes are more competitive with seedlings. So density dependence not only affects the dynamics in this predominantly annual system, but also plays a part in determining the life-history strategy adopted by the plant, and the timing of density dependence is important for both of these outcomes.

This example also illustrates how the timing of density dependence can be critical for decisions relating to management of invasive species. Management strategies generally result in exogenous mortality, through mowing, for example. Due to over-compensating density dependence, in *T. perforatum*, the timing of exogenous mortality affects the equilibrium seed density dramatically. If exogenous mortality occurs early in the life-cycle, freeing the few surviving plants from density dependent depression of seed production, seed output can actually increase, exacerbating the weed problem in the long-term (Buckley *et al.* 2001).

Generally, if density dependence is important for population regulation, the removal of seeds, seedlings or plants before density dependence occurs is unlikely to lead to successful control. This is often relevant when considering biocontrol agents. As pointed out by Alexander and Schrag (2003) seed predators or seedling pathogens can have a strong effect on seedling numbers (Edwards and Crawley 1999, Alexander and Mihail 2000, Cummings and Alexander 2002) but not necessarily impact on final reproductive output of a patch due to subsequent density dependent processes. Where seed input is reduced, low density seedling populations are produced which may have lower rates of intraspecific competition and thus higher per capita seed production, compensating fully or to some extent for the initial seed or seedling loss. The case of *Acacia nilotica* illustrates this point. *Acacia nilotica*, an invasive shrub or small tree is dispersed by cattle, which eat seeds and deposit them in cow pats. The seedlings are therefore aggregated in small patches and self-thinning ensures that no more than approximately one seedling survives to adulthood. Kriticos *et al.* (1999) and Kriticos *et al.* (2003) estimate that the introduction of a seed-eating biocontrol agent is therefore

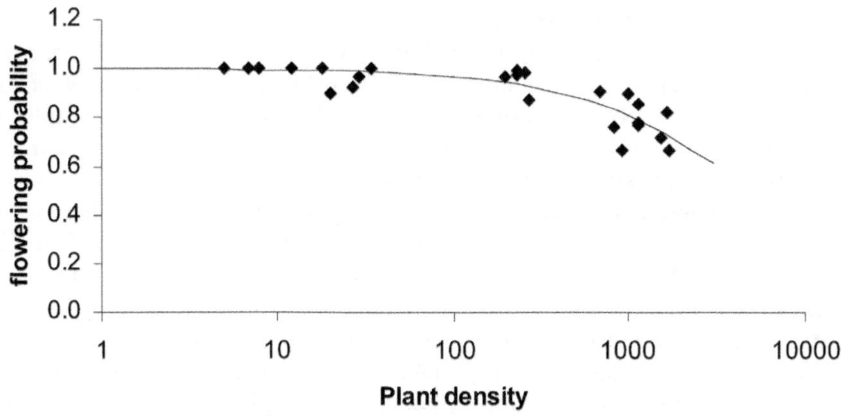

Fig. 2 Density dependent function for flowering probability in *Tripleurospermum perforatum* (with permission — Ecology Letters).

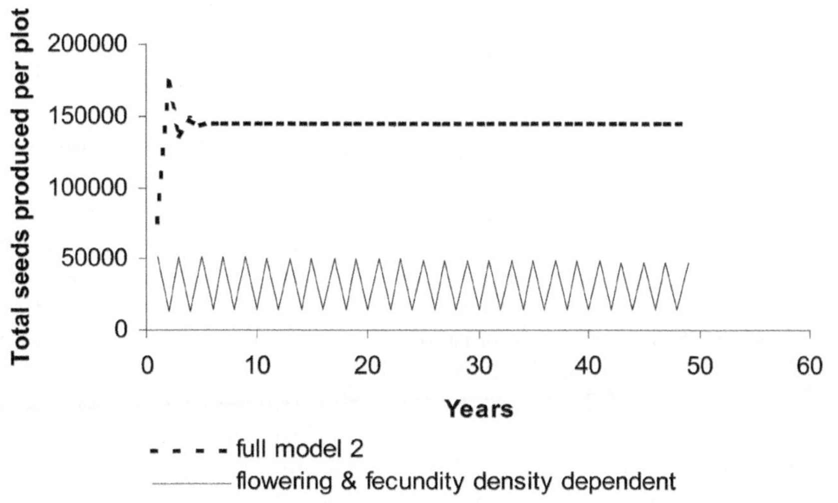

Fig. 3 A comparison of the dynamics predicted by the full model incorporating density dependent survival, flowering probability and fecundity, with a model incorporating density dependent flowering probability and fecundity only. A model using density dependent fecundity alone produces damped oscillations and an equilibrium level similar to that observed in the full model. (with permission — Ecology Letters).

unlikely to affect adult population sizes unless the number of seeds per cow pat is reduced dramatically, to two or less.

Several authors have called for a thorough understanding of the occurrence and effects of density dependence at different stages (Mortimer *et al.* 1989, Watkinson *et al.* 1989, Gillman *et al.* 1993). As shown by the examples given above, this is especially relevant for weed species where density is manipulated at one or more specific stages for control purposes.

In view of these results it is obvious that models seeking to predict the effects of control strategies at different stages should incorporate density dependence. In a sustainable harvesting study Freckleton *et al.* (2003) point out that the use of density independent matrix models, where the transition probabilities are calculated when the population is close to equilibrium (λ close to 1), could give misleading results when used to predict population increase from low density. They call for explicit incorporation of density dependence in harvesting models, and the same call should be made for control models of invasives where density dependence is known or suspected to be important.

A thorough understanding of the occurrence and effects of density dependence would also greatly contribute to investigations into selection pressures and evolution. Invasive plants can be excellent model systems for the development of models of life-history evolution. Plants in the exotic range may be under different selection pressures due to enemy release or competitive release. Contrasting models predicting evolutionarily stable strategies in each environment could be developed. This approach would complement observational and experimental studies aimed at detecting evolutionary change between invasives in their native and exotic habitats (Blossey and Nötzold 1995, Willis *et al.* 1999, 2000, Siemann and Rogers 2001).

Historically, when exploring life history strategies, investigators ignored density dependence, and simply used *r*, from the Euler-Lotka equation, to define fitness (Stearns 1992). However, it became apparent that density dependence could not be ignored, because in a density dependent situation, the fitness of one strategy will depend on the actions of other strategies (Grant and Benton 2000). Because of this, the evolutionary stable strategy (ESS) is best determined by using the invasion exponent ϑ, or the rate of increase of a rare mutant into an "environment" set by a resident strategy, and characterised by the resident's density. If $\vartheta > 0$ then the invader is successful. The ESS is simply the strategy that cannot be invaded by any other. Using this framework, Mylius and Diekmann (1995) showed that in a constant environment, the ESS may sometimes be the strategy at which a relatively simple function is maximised. However, which function is appropriate is determined by *when* density dependence acts in the life cycle. If density dependence acts so that only the expected lifetime offspring production is reduced, then an appropriate fitness measure is lifetime reproductive success, or R_0. If density dependence affects only the probability per unit time of dying, then population rate of increase, λ, is appropriate. The two may provide different predictions, so it is important to use the appropriate measure.

This framework has been used to predict the ESS flowering strategy in monocarpic plants (Rees *et al.* 2002, Rees and Rose 2002), a category to which many nuisance weeds and invasive species belong. Data from a range of species shows that neighbour density or size has no effect on growth, survival, or reproduction in monocarpic plants (Metcalf *et al.* 2003). However, recruitment is generally uncorrelated with seed production, suggesting that seed establishment is limited by suitable microsite availability. Together, these observations imply the operation of density dependent processes at the seedling establishment level only. We can deduce that the ESS strategy corresponds to maximising R_0 (assuming the environment is constant). A matrix model, or its continuous analogue, the Integral Projection Model (Easterling *et al.* 2000), can be used to describe the population, and calculate R_0 for a range of flowering strategies. This approach has been used to predict ESS flowering strategy values that are very close to the observed (Rees and Rose 2002), and contribute to our understanding of selection pressures on moncarpic plants.

CONCLUSIONS

Density dependent processes can determine equilibrium densities achievable by an invasive population, stability of that population, population growth rates at the front of an invasion, population recovery after control measures have been implemented and the evolutionarily stable strategies acquired by populations in different environments. In this chapter we have outlined how theoretical ecology has contributed both to our understanding of how invasive populations are regulated by density dependence, and how work on invasives has stimulated theoretical work on important ecological problems like Allee effects and spatial spread of populations. As we develop models for an increasing number of the complex processes that govern population dynamics we can start to combine models and investigate the interaction of processes like Allee effects and the shape of dispersal kernels (Wang *et al.* 2002), on spread rates or density dependence at different stages in the life cycle (Buckley *et al.* 2001). Ultimately these models will enable us to apply targeted and effective management strategies to invasive plant populations in order to reduce their impact on biodiversity loss and ecosystem function.

ACKNOWLEDGEMENTS

We thank Darren Kriticos, Karen Rose, Caz Taylor, Dave Murrell and an anonymous reviewer for helpful additions. JCM was supported by a NERC studentship.

REFERENCES

Alexander, H. M. and J. D. Mihail. 2000. Seedling disease in an annual legume: consequences for seedling mortality, plant size, and population seed production. Oecologia **122**, 346-353.

Alexander, H. M. and A. M. Schrag. 2003. Role of soil seed banks and newly dispersed seeds in population dynamics of the annual sunflower, *Helianthus annuus*. Journal of Ecology **91**, 987-998.

Amarasekare, P. 1998. Interactions between local dynamics and dispersal: Insights from single species models. Theoretical Population Biology **53**, 44-59.

Benton, T. G., T. C. Cameron and A. Grant. 2004. Population responses to perturbations: predictions and responses from laboratory mite populations. Journal of Animal Ecology **73**, 983-995

Blossey, B. and R. Nötzold. 1995. Evolution of increased competitive ability in invasive nonindigenous plants: a hypothesis. Journal of Ecology **83**, 887-889.

Bonsall, M. B., E. van der Meijden and M. J. Crawley. 2003. Contrasting dynamics in the same plant-herbivore interaction. Proceedings of the National Academy of Sciences **100**, 14932-14936.

Buckley, Y. M., D. T. Briese and M. Rees. 2003a. Demography and management of the invasive plant species *Hypericum perforatum*. I. Using multi-level mixed-effects models for characterizing growth, survival and fecundity in a long-term data set. Journal of Applied Ecology **40**, 481-493.

Buckley, Y. M., P. Downey, S. V. Fowler, R. L. Hill, J. Memmott, H. Norambuena, M. Pitcairn, R. G. Shaw, A. W. Sheppard, C. Winks, R. Wittenberg and M. Rees. 2003b. Are invasives bigger? A global study of seed size variation in two invasive shrubs. Ecology **84**, 1434-1440.

Buckley, Y. M., H. L. Hinz, D. Matthies and M. Rees. 2001. Interactions between density-dependent processes, population dynamics and control of an invasive plant species, *Tripleurospermum perforatum* (scentless chamomile). Ecology Letters **4**, 551-558.

Buckley, Y. M., M. Rees, A. W. Sheppard and M. J. Smyth. 2005. Stable coexistence of an invasive plant and biocontrol agent: a parameterised coupled plant-herbivore model. Journal of Applied Ecology **42**, 70-79.

Cameron, T. C. and T. G. Benton. 2004. Stage-structured harvesting and its effects: an empirical investigation using soil mites. Journal of Animal Ecology **73**, 996-1006.

Caswell, H. 2001. Matrix population models: construction, analysis and interpretation, 2nd edition. Sinauer Associates, Inc., Sunderland, Massachusetts.

Crone, E. E. 1997. Parental environmental effects and cyclical dynamics in plant populations. The American Naturalist **150**, 708-729.

Cruickshank, I., W. S. C. Gurney and A. R. Veitch. 1999. The characteristics of epidemics and invasions with thresholds. Theoretical Population Biology **56**, 279-292.

Cummings, C. L. and H. M. Alexander. 2002. Population ecology of wild sunflowers: effects of seed density and post-dispersal vertebrate seed predation. Oecologia **130**, 274-280.

Davis, H. G., C. M. Taylor, J. C. Civille and D. R. Strong. 2004. An Allee effect at the front of a plant invasion: *Spartina* in a Pacific estuary. Journal of Ecology **92**, 321-327.

Easterling, M. R., S. P. Ellner and P. M. Dixon. 2000. Size-specific sensitivity: Applying a new structured population model. Ecology **81**, 694-708.

Edwards, G. R. and M. J. Crawley. 1999. Rodent seed predation and seedling recruitment in mesic grassland. Oecologia **118**, 288-296.

Fowler, S. V. and G. D. Ruxton. 2002. Population dynamic consequences of Allee effects. Journal of Theoretical Biology **215**, 39-46.

Fox, G. A. and B. E. Kendall. 2002. Demographic stochasticity and the variance reduction effect. Ecology **83**, 1928-1934.

Freckleton, R. P., D. M. Silva Matos, M. L. A. Bovi and A. R. Watkinson. 2003. Predicting the impacts of harvesting using structured population models: the importance of density-dependence and timing of harvest for a tropical palm tree. Journal of Applied Ecology **40**, 846-858.

Freckleton, R. P. and A. R. Watkinson. 2002. Are weed population dynamics chaotic? Journal of Applied Ecology **39**, 699-707.

Gillman, M., J. M. Bullock, J. Silvertown and B. Clear Hill. 1993. A density-dependent model of *Cirsium vulgare* population dynamics using field-estimated parameter values. Oecologia **96**, 282-289.

Grant, A. and T. G. Benton. 2000. Elasticity analysis for density dependent populations in stochastic environments. Ecology **81**, 680-693.

Harper, J. L. and W. A. Wood. 1957. Biological Flora of the British Isles: *Senecio jacobaea* L. Journal of Ecology **45**, 617-637.

Hassell, M. P. 1975. Density-dependence in single-species populations. Journal of Animal Ecology **44**, 283-295.

Jarry, M., M. Khaladi, M. Hossaert-McKey and D. McKey. 1995. Modeling the population dynamics of annual plants with seed bank and density dependent effects. Acta Biotheoretica **43**, 53-65.

Keane, R. M. and M. J. Crawley. 2002. Exotic plant invasions and the enemy release hypothesis. Trends in Ecology & Evolution **17**, 164-170.

Kot, M., M. A. Lewis and P. van den Driessche. 1996. Dispersal data and the spread of invading organisms. Ecology **77**, 2027-2042.

Krebs, C. J. 2002. Two complementary paradigms for analysing population dynamics. Philosophical Transactions of the Royal Society of London Series B-Biological Sciences **357**, 1211-1219.

Kriticos, D., J. Brown, I. Radford and M. Nicholas. 1999. Plant population ecology and biological control: *Acacia nilotica* as a case study. Biological Control **16**, 230-239.

Kriticos, D. J., J. R. Brown, G. F. Maywald, I. D. Radford, D. M. Nicholas, R. W. Sutherst and S. W. Adkins. 2003. SPAnDX: a process-based population dynamics model to explore management and climate change impacts on an invasive alien plant, *Acacia nilotica*. Ecological modelling **163**, 187-208.

Lewis, M. A. and P. Kareiva. 1993. Allee dynamics and the spread of invading organisms. Theoretical Population Biology **43**, 141-158.

Metcalf, J. C., K. E. Rose and M. Rees. 2003. Evolutionary demography of monocarpic perennials. Trends in Ecology & Evolution **18**, 471-480.

Mortimer, A. M., J. J. Sutton and P. Gould. 1989. On robust weed population models. Weed Research **29**, 229-238.

Mylius, S. D. and O. Diekmann. 1995. On evolutionarily stable life histories and the need to be specific about density dependence. Oikos **74**, 218-224.

Pacala, S. W. and J. A. Silander. 1990. Field tests of neighborhood population dynamic models of two annual weed species. Ecological Monographs **60**, 113-134.

Pacala, S. W. and J. Weiner. 1991. Effects of competitive asymmetry on a local density model of plant interference. Journal of Theoretical Biology **149**, 165-179.

Palmblad, I. G. 1968. Competition in experimental studies on populations of weeds with particular emphasis on the regulation of population size. Ecology **49**, 26-34.

Parker, I. M. 1997. Pollinator limitation of *Cytisus scoparius* (Scotch broom), an invasive exotic shrub. Ecology **78**, 1457-1470.

Parker, I. M. and K. A. Haubensak. 2001. Comparative pollinator limitation of two non-native shrubs: do mutualisms influence invasions? Oecologia.

Paynter, Q., P. O. Downey and A. W. Sheppard. 2003. Age structure and growth of the woody legume weed *Cytisus scoparius* in native and exotic habitats: implications for control. Journal of Applied Ecology **40**, 470-480.

Pimentel, D., L. Lach, R. Zuniga and D. Morrison. 2000. Environmental and economic costs of nonindigenous species in the United States. Bioscience **50**, 53-65.

Rees, M. and M. J. Crawley. 1989. Growth, reproduction and population dynamics. Functional Ecology **3**, 645-653.

Rees, M. and M. J. Crawley. 1991. Do plant populations cycle? Functional Ecology **5**, 577-582.

Rees, M., D. Kelly and O. N. Bjornstad. 2002. Snow tussocks, chaos and the evolution of mast seeding. The American Naturalist **160**, 44-59.

Rees, M. and K. E. Rose. 2002. Evolution of flowering strategies in *Oenothera glazioviana*: an integral projection model approach. Proceedings of the Royal Society London: Series B **269**, 1509-1515.

Scheuring, I. 1999. Allee effect increases the dynamical stability of populations. Journal of Theoretical Biology **199**, 407-414.

Shima, J. S. and C. W. Osenberg. 2003. Cryptic density dependence: effects of covariation between density and site quality in reef fish. Ecology **84**, 46-52.

Sibly, R. M. and J. Hone. 2002. Population growth rate and its determinants: an overview. Philosophical Transactions of the Royal Society of London Series B-Biological Sciences **357**, 1153-1170.

Siemann, E. and W. E. Rogers. 2001. Genetic differences in growth of an invasive tree species. Ecology Letters **4**, 514-518.

Siemann, E. and W. E. Rogers. 2003. Reduced resistance of invasive varieties of the alien tree *Sapium sebiferum* to a generalist herbivore. Oecologia **135**, 451-457.

Silvertown, J. 1991. Modularity, reproductive thresholds and plant population dynamics. Functional Ecology **5**, 577-582.

Stearns, S. C. 1992. The evolution of life histories. Oxford University Press, Oxford.

Stephens, P. A., W. J. Sutherland and R. Freckleton. 1999. What is the Allee effect? Oikos **87**, 185-190.

Thrall, P. H., S. W. Pacala and J. A. Silander. 1989. Oscillatory dynamics in populations of an annual weed species *Abutilon theophrasti*. Journal of Ecology **77**, 1135-1149.

Turchin, P. 1999. Population regulation: a synthetic view. Oikos **84**, 153-159.

Wang, M. H., M. Kot and M. G. Neubert. 2002. Integrodifference equations, Allee effects and invasions. Journal of Mathematical Biology **44**, 150-168.

Watkinson, A. R. 1980. Density-dependence in single-species populations of plants. Journal of Theoretical Biology **83**, 345-357.

Watkinson, A. R. and A. J. Davy. 1985. Population biology of salt-marsh and sand dune annuals. Vegetatio **62**, 487-497.

Watkinson, A. R., W. M. Lonsdale and M. H. Andrew. 1989. Modelling the population dynamics of an annual plant *Sorghum intrans* in the wet-dry tropics. Journal of Ecology **77**, 162-181.

Weiner, J. 1985. Size hierarchies in experimental populations of annual plants. Ecology **66**, 743-752.

Willis, A. J., J. Memmott and R. I. Forrester. 2000. Is there evidence for the post-invasion evolution of vigour among invasive plant species? Ecology Letters **3**, 275-283.

Willis, A. J., M. B. Thomas and J. H. Lawton. 1999. Is the increased vigour of invasive weeds explained by a trade- off between growth and herbivore resistance? Oecologia **120**, 632-640.

Chapter six

Stochasticity, nonlinearity and instability in biological invasions

R. P. Freckleton, P. M. Dowling and N. K. Dulvy

INTRODUCTION

Compared with other sciences ecology has few general theories that can be applied to predict the dynamics of its systems from first principles. This is perhaps not surprising given the huge variability of ecological systems. A number of broad insights have emerged, however, and one of the most important of these is that predicting the dynamics of ecological systems requires that we understand the interplay of two kinds of processes, stochastic and deterministic (Lewontin and Cohen 1968; May 1973; Lande *et al.* 2003). This conclusion is essentially a consensus resulting from a great deal of debate in the ecological literature, dating back to the 1950s (e.g., Andrewartha and Birch 1954, Nicholson 1954; Hassell 1986; White 2001).

Predicting biological invasions is an extremely important applied ecological problem because invasive species are economically enormously significant in agricultural and conservation terms, and methods for predicting the effects of control measures are urgently required (Byers *et al.* 2002). One of the most important approaches to studying biological invasions is through population modelling. Population models may take a range of forms including simple

M.W. Cadotte et al. (eds.), Conceptual ecology and invasion biology, 125–146.

mathematical models (Hassell 1975; May 1973, May *et al.* 1974; Maynard Smith 1978; Watkinson 1980), statistical models (Buckley *et al.* 2003*a*, *b*) and complex simulations (Pacala *et al.* 1996). These models can be used in a range of ways, from making highly specific predictions about the dynamics of species in given areas, through to general analyses aimed at developing a general understanding of the processes determining invasion success.

Unfortunately population models are frequently equated with an equilibrium or deterministic outlook on population dynamics, which is frequently criticised. However, even simple models may frequently yield rather complex outcomes (May and Oster 1976). For instance, a simple logistic model of population growth might be expected to yield a straightforward outcome, namely the smooth growth of a population to its carrying capacity. When implemented as a discrete time model, however, the outcome may be far more complex and chaotic dynamics may be possible. From the point of view of predicting population dynamics this raised the possibility that completely deterministic systems may yield unpredictable changes in population numbers. The early 1970s also saw an increasing recognition that various forms of interactions and intervention could yield unstable population dynamics (May 1977). For example hunting, harvesting or predation could potentially destabilise population dynamics to the extent that otherwise stable populations might become extinct, given a perturbation to the system.

A final element that is important in predicting population change is the stochastic component of population change. Stochasticity results from random variations in the environment, or from random variation between individuals. The theory for stochastic population dynamics has a long history (Cohen 1966; Lewontin and Cohen 1969; Tuljapurkar 1990), and there have been numerous recent developments of this (Lande *et al.* 2003). The major conclusion from this work has been that the effects of stochasticity on population dynamics may be complex. Predicting the impacts of stochasticity is not a straightforward as making some parameters of models random variables, or of calculating an average of some form. Rather, stochasticity may impact in a profound and a qualitative manner on population dynamics.

The problem of predicting and managing populations of invasive species requires that we are able to understand the impacts of the factors detailed above on several phases of the invasion process. The elements of population dynamics of invasion can be conceptually viewed as consisting of three phases, arrival, establishment and spread (Dobson and May 1986; Williamson 1996). The processes described above might be expected to play different roles depending on which of this phases the invading population is in. In the arrival and establishment phases, populations are growing from low densities and expanding into new habitat. In this phase stochastic factors may arguably be important, and limiting factors less so. In contrast, following successful establishment and during the spread phase populations may be at much higher densities and factors such as density-dependence may dominate population changes. Correspondingly, these differences in dynamics may influence control strategies (Taylor and Hastings

2004). Moreover, when factors such as Allee effects operate, the deterministic component of dynamics may be significant at low densities and influence population persistence. The key task for population modelling is to work out how these various factors interact with each other.

This chapter is organised into three sections. In the first section we review the principles of simple stochastic models and how they may be applied to invasions. We then examine the deterministic counterparts of these models, and the problems these present. Finally we briefly review two studies that exhibit unstable dynamics and speculate on the generality of these.

STOCHASTIC POPULATION DYNAMICS AND THE STABILITY OF POPULATIONS

It is useful to recognise two sources of stochasticity in population dynamics (Engen *et al.* 1999) since the source of stochasticity is important in understanding its effects on population dynamics. Environmental stochasticity results from random variations in demographic rates as a consequence of variation in environmental conditions, for example resulting from the effects of weather. A second source of stochasticity is demographic stochasticity. Under demographic stochasticity the expected contribution of each individual to the next generation is also independent of the state of the population in the previous generation, however the variance about this expectation does depend on the state of the population (as distinct from an Allee effect in which the expectation depends on density; Stephens *et al.* 1999).

Simple stochastic population growth

The simplest model for population dynamics is one in which the expected rates of death and fecundity are assumed to be constant, and vary randomly about these means. If the net rate of population change in year t is $\lambda(t)$, and $N(t)$ and $N(t + 1)$ are population sizes in successive years, then the dynamics of such a population can be modelled as:

$$N(t + 1) = \lambda(t)N(t) \tag{1}$$

For a species to be capable of invasion, the expected value of $N(t + T) / N(t)$ must be greater than one as T becomes large. In a constant environment this requires that the value of λ must be greater than unity. In an environment in which there is stochasticity the condition on λ is not as straightforward, but depends on the distribution of λ about its mean. According to equation (1), population size at time $t + T$ is a non-linear function of λ_{t+1}, $\lambda_{t+2}...\lambda_{t+T}$. This means that the distribution of λ has to be understood in order to predict long-term dynamics. More complex effects such as the autocorrelations between population growth rates in successive years may also come into play and further complicate predictions of population persistence (Heino *et al.* 2000). Although the possibilities are complex, two simple examples illustrate some of the main features of stochastic models. The main conclusion at the end of this section is that the outcome of stochastic models can be more complex than one might first imagine.

Environmental stochasticity

Firstly, assume that λ is log-normally distributed, as may be the case in populations subject to environmental stochasticity. In that case, equation (1) may be written as, where $n \equiv \log N$:

$$n(t + 1) = n(t) + \log [\lambda(t)] \tag{2}$$

This is linear in n, with the consequence that the arithmetic mean of log λ predicts the long term population trajectory. It is easy to show that the expected value of $n(t + T)$ is normally distributed with mean $n(0)+ T\,\overline{\log(\lambda)}$ and variance T var $[\log(\lambda)]$ if the lambdas are independent and normally distributed. Fig. 1*a* shows examples of a series of populations growing in this way, and Fig. 1*c* shows the distribution of population sizes after a period of population growth.

As noted above, predicting the outcome of stochasticity requires that the effects of stochasticity on population growth are understood in detail. For instance, if the distribution of population growth rates is not log-normal, or if successive population growth rates are not independent, then the results in the previous paragraph do not hold exactly. In such cases stochastic calculus has to be used in order to determine the appropriate method of transforming population densities in order to predict persistence (e.g., Lande 1998).

Demographic stochasticity

The second example is a population subject to demographic stochasticity (May 1981, Lande 1998; Lande *et al.* 2003). Demographic stochasticity results in variability in which the variance declines with the number of individuals in the population:

$$\mathrm{var}\,(\lambda) = \frac{\sigma^2_D}{N} \tag{3}$$

or as a hyperbolic function of density if N is the average number of individuals per unit area, and individuals are assumed to be randomly distributed (Freckleton unpub.):

$$\mathrm{var}\,(\lambda) = \frac{\sigma^2_D}{1 + dN} \tag{4}$$

When combined with environmental variability (which has variance σ^2_E), the variance in population growth rate in this model is:

$$\text{var}(\lambda) = \sigma^2_E \quad \frac{\sigma^2_D}{N} \tag{5}$$

Examples of populations growing according to this model are shown in Fig. 1*b*. The most important feature of this model is that if density falls below a critical threshold

$$(N^* = \frac{1}{4}\sigma^2_D \; / \; \left(\lambda - 1 - \frac{1}{2}\sigma^2_E\right))$$

then extinction is highly likely (Lande *et al.* 2003). The result of this is a bimodal distribution of population sizes, as shown in Fig. 1*d*, compared with the normal distribution in Fig. 1*c*.

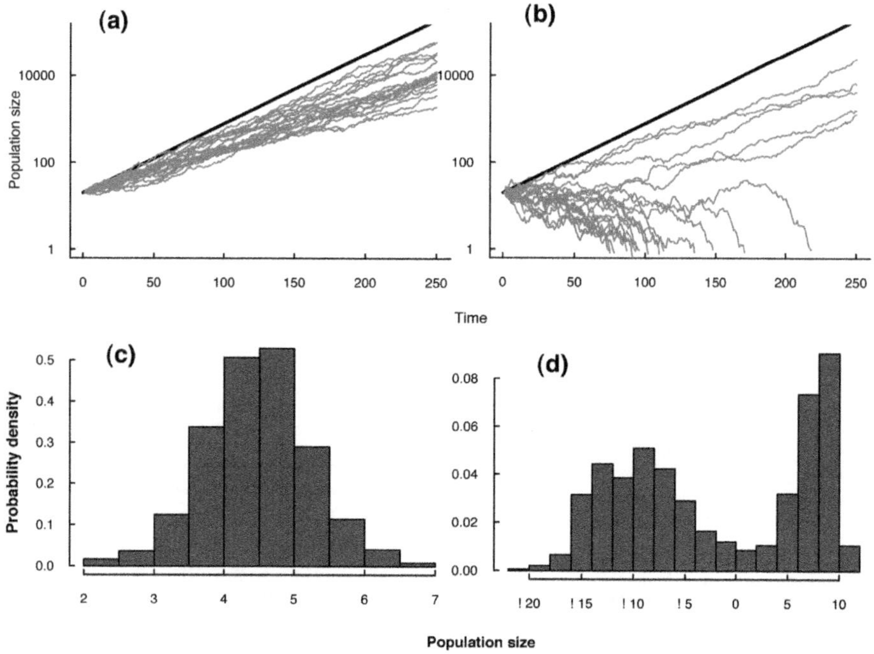

Fig. 1 Sources of stochasticity and effects on population growth (a & c) and population size (b & d). Population growth was density-independent with stochasticity either environmental (a & c) or environmental and demographic (b & d).

In the model with environmental stochasticity only, populations are able to invade if the population growth rate exceeds a critical value. The greater the difference between the long-term average and this critical value, the more likely a species is to invade. On the other hand, if a population is subject to demographic stochasticity, populations must exceed a critical density before invasion can occur. Therefore the two forms of stochasticity have rather different implications for population dynamics and predicting invasion requires that both of these sources of stochasticity are understood in detail. In non-invasive species, estimates of demographic and environmental stochasticity from bird populations suggest that both sources may be extremely significant, and hence that the two forms need to be measured separately (Saether *et al.* 2004, 2005).

Implications of stochastic models for modelling invasions

The main conclusion from the example models outlined above is that in a stochastic environment the long-term population growth is not a linear function of the average population growth rates (Cohen 1966; Lewontin and Cohen 1969; Tuljapurkar 1990). Not only is it important to estimate the amount of variability in population growth rates (Freckleton and Watkinson 1998), but also to disentangle the relative roles of demographic and environmental variation. This requires that temporally replicated measures of population growth, or the components of population growth, are available. Without such information it is not possible to predict which species are capable of invading and which are not, or conditions under which a given species will become invasive or not, or to derive realistic models.

Unfortunately such information is rarely available for many invasive species, often because invasive species have recently been introduced to novel environments. Consequently long-term datasets tend to be rare. Under such circumstances there are two choices for the modeller, either to give up, or to try to make do with whatever information is available. As an example of the latter Rees and Paynter (1997) developed a model for the dynamics and control of invasive populations of Scotch Broom. This model was parameterised using a several data sources from both the native and exotic range of the species, as well as an understanding of its basic autecology. It seems likely, however, that many existing models are compromised to some extent in their predictive ability through a lack of data on stochastic variation in key parameters.

Recent work has suggested that the demographic and environmental components of population growth may be related to life history traits in birds (Saether *et al.* 2004, 2005). For example, demographic stochasticity tends to be larger in species with greater adult survival, age at maturity and generation times (Saether *et al.* 2004). Similarly the expected time to extinction of populations increased with decreasing clutch size. These results are important because they show that it is possible to generalize about how sensitive populations are to different forms of variability, based on life-history traits. To date there have been no comparable analyses of the demographics of population during the initial stages of invasion, however such analysis could be enormously revealing.

DETERMINISTIC POPULATION DYNAMICS

If left unchecked, the system defined by equation (1) will ultimately become extinct (Lewontin & Cohen 1969). If the value of λ is greater than one, the expected mean population size as T becomes large is infinite (although, counterintuitively, the probability of extinction remains equal to one, albeit with a very long expected time to extinction). This behaviour does not sit well with ecological intuition. Common observation appears to indicate that the dynamics of many species are relatively invariant. Some invasive species do exhibit intrinsically complex population dynamic behaviour, particularly the viruses that cause diseases such as measles and flu (Earn *et al.* 1998). Cyclic dynamics are reasonably common in a range of taxa (Kendall *et al.* 1999; Ginzburg & Colyvan 2004), however these tend to be extrinsically driven or the consequence of species interactions such as predator-prey dynamics, rather than intrinsic properties of populations. A great deal of effort has been expended on understanding, firstly why populations do not simply boom or bust in accord with the most simple density-independent models, and secondly, why different systems exhibit alternative kinds of population dynamics.

Density-dependence

Density-dependence occurs when any vital rate changes systematically with density. Strictly speaking this could include a range of population level phenomena (Royama 1992), but typically this definition is taken to imply that some *per capita* rate, such as mortality or fecundity varies systematically with density. Density-dependence of this sort may result from a range of ecological processes. The most commonly recognised form is negative density-dependence which results from direct competition for resources (Nicholson 1951), such as food, shelter or mates. More indirectly, density-dependence may arise from the impact of other species, for instance predators or parasites which may increase mortality rates at high prey or host densities respectively.

As noted above density-independent models suggest that density-dependence should be highly prevalent, because in the absence of density-dependence, populations are likely to either expand without limit or become extinct. Despite this, many models for population dynamics ignore density-dependence, and this can severely compromise model predictions (e.g., see Freckleton *et al.* 2003). This problem may be especially acute for invasive species which might be expected to have a high average value of λ, and hence be capable of rapidly growing to high densities. In the initial stages of an invasion the stochastic factors that determine λ seem likely to play a key role in determining invasion success. However, as shown below, following successful invasion, the ultimate size reached by the invader population will be in large part determined by the form, nature and strength of density-dependence.

Models for density-dependent population growth

A suite of models have been employed to model the dynamics of species growing in a density-dependent fashion. The best known of these is the logistic model, here the discrete time version:

$$N(t+1) = N(t) + rN(t)(1 - N/K) \tag{6}$$

In this form the logistic model is a poor model for describing ecological dynamics. There are a number of problems with this model, including the equilibrium is independent of the finite rate of increase, density-dependence is linear, and stability behaviour is determined solely by the maximum rate of increase, r (see below). Despite this, the logistic model is widely employed in theoretical modelling, probably because it is analytically very convenient to work with.

More realistic models have been suggested, including modifications to the logistic in order to make it more realistic (e.g., the theta-logistic of Lande *et al.* 2003). One particularly useful model is that of Hassell (1975):

$$N(t+1) = \lambda N(t)\,[1 + aN(t)]^{-b} \tag{7}$$

This model has two parameters that model different aspects of the density-dependent response. a determines the density at which the *per capita* effects of density-dependence become significant, whilst b varies the rate of change in population growth rate as density becomes high and may be related to the nature of competition (contest versus scramble; Hassell 1975) or the efficiency with which resources are converted into population growth (Firbank and Watkinson 1985; Freckleton and Watkinson 2001). The equilibrium of this model is given by:

$$\hat{N} = (\lambda^{1/b} - 1)\,/\,a \tag{8}$$

The important advance over the logistic model is that the equilibrium is a compound of three model parameters. Increasing the finite rate of increase, or decreasing a, leads to increases in population density. These effects are then modulated by b, which also determines the stability properties of the model.

It seems likely that the three parameters may be related to each other at a number of scales. Within populations, temporal variability in the environment may lead to variation in any of the parameters. As pointed out by Chesson and Huntley (1988, 1989), understanding how the effects of stochasticity interact with the strength of competition is the key to predicting the outcome of competition in stochastic environments. In the model above, the problem is to understand how the parameters a and λ covary with each other. If λ and a are independent, then population growth rates are termed additive (Fig. 2*a*) with the consequence that stochastic environmental effects impact directly on population sizes (Fig. 2*c*). In contrast, sub-additive population growth rates occur when λ and

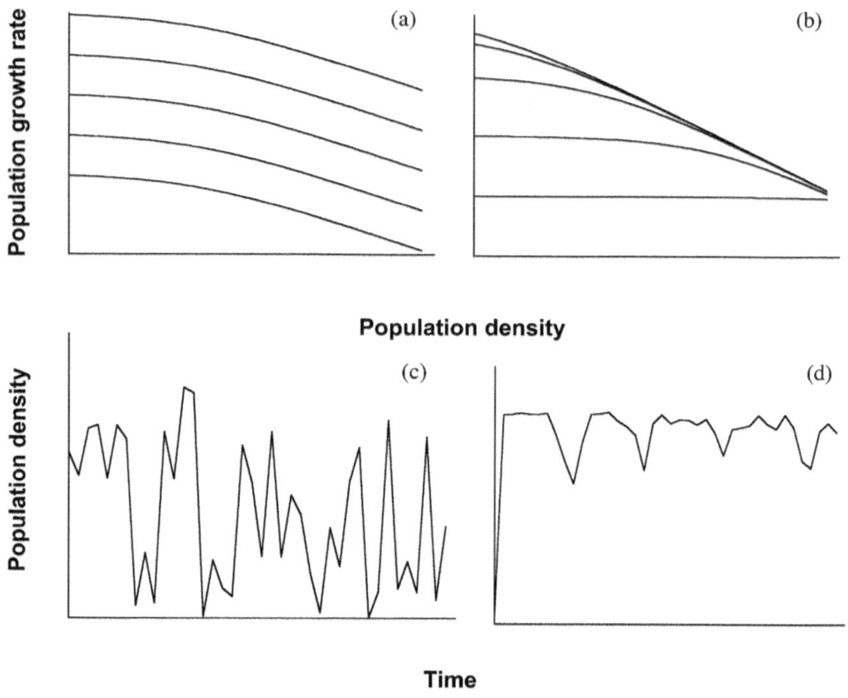

Fig. 2 Interactions between density-dependence and stochasticity. Lines represent population growth under different environmental conditions. (a) Additive population growth rates; (b) sub-additive population growth rates; (c) a population subject to stochasticity and additive population growth; (d) a population subject to stochasticity and sub-additive population growth.

a are correlated with each other (Fig. 2*a*). In such populations, the strength of competition changes so as to buffer population growth from stochastic environmental changes. The consequence is that populations subject to sub-additive population growth are buffered against environmental change: Fig. 2*d* shows a population subject to perturbations identical to those in Fig. 2*c*, however the sub-additive population growth rates clearly have an important buffering effect.

From the point of view of predicting the dynamics of invasive species, the patterns shown in Figs. 2*a* and *b* have two important consequences. First, the form of density-dependence and additivity influences the effects of stochasticity. Second, the form of additivity can influence the effects of changes to management. For instance if population growth rates are sub-additive then reductions in λ

resulting from attempts at control or eradication will be compensated for by reduction in the strength of density-dependence, yielding no net effect on population numbers. Thus, it is not sufficient simply to document the effects of stochasticity or density-dependence in isolation, but the interaction between these also needs to be understood. In practice this is very difficult as this requires that the form of density-dependence is measured under a range of environmental conditions, which is highly data intensive.

Non-linear dynamics and stability

As noted above, in the mid-1970s it was found that simple ecological models could yield unexpectedly complicated patterns of population dynamics (May 1973, 1974; May and Oster 1976; Fig. 3*a*, *b*). In short, in models for density-dependent population growth for organisms with discrete generations, density-dependence at high density may overcompensate for changes in density, with the result that increases in population size above the deterministic equilibrium tend to be followed by disproportionately large decreases. The consequence of this is cyclic dynamics, or more complex behaviour such as limit cycles and chaos. Under such circumstances populations exhibit sustained fluctuations in population size that are apparently random and superficially similar to the fluctuations in population sizes that are seen in real populations.

Extensive attempts have been made to find examples of population exhibiting chaotic dynamics. However few examples exist. In fact in some cases there are good reasons to expect that chaotic dynamics are rather unlikely. In plants, for example, the absence of reproductive thresholds, the presence of a seedbank, and asymmetric competition leading to compensating density responses, all lead to generally stable population dynamics (Rees and Crawley 1989, 1991; Freckleton and Watkinson 2003). Importantly, this multiplicity of stability-generating mechanisms means that even when one condition for instability holds, others tend to cancel this out (e.g., Buckley *et al.* 2001). Moreover general reviews of the stability properties of natural populations appear to show that most single species populations show stable dynamics (Hassell *et al.* 1976; Lonsdale 1996; Freckleton and Watkinson 2003). As noted above, in the majority of documented cases cyclic dynamics tend to be driven by exogenous factors, or interactions with other species (e.g., see Kendall *et al.* 1999).

Positive density-dependence and instability

The models described above emphasise the negative effects of density-dependence on population growth rates. However recent reviews have pointed out that not all density-dependent effects on population growth are of this form (Stephens and Sutherland 1999; Stephens *et al.* 1999). Positive density-dependence (rather confusingly termed "inverse" density-dependence by some authors) is most well known through Allee effects. Allee effects arise most commonly in social species,

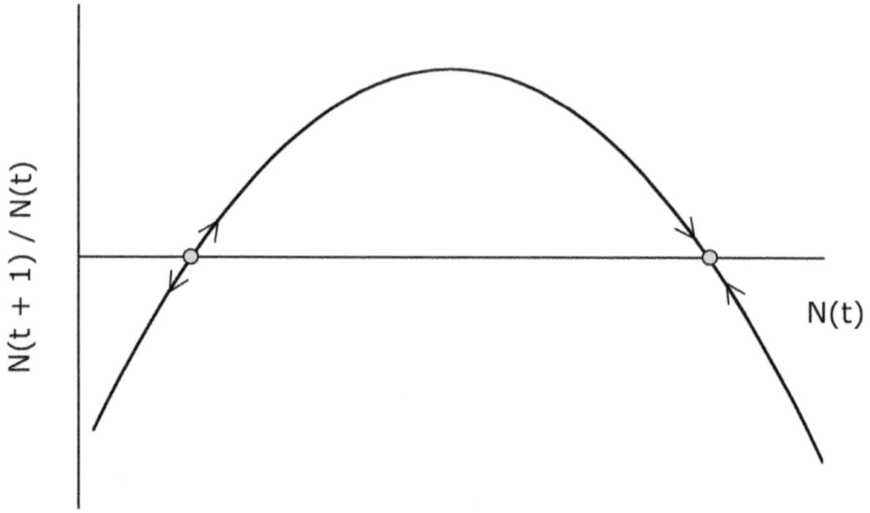

Fig. 3 Stable and unstable equilibria resulting from Allee effects. The relationship between population growth rate and density is humped. Where the curve cuts the line of zero population growth rate there is an equilibrium. As indicated by the arrows, the upper equilibrium is stable, whereas the lower one is unstable.

where cooperation between individuals breaks down at low densities, leading to a decline in some aspects of fitness with decreasing densities. In the extreme such *component Allee effects* (which affect one component of fitness) may be manifest at the level of population growth leading to demographic Allee effects (i.e., are so important that they impinge on rates of population change; Stephens and Sutherland 1999). The importance of Allee effects in population growth is that these can lead to a threshold density below which populations cannot persist (see Fig. 3). This is of course highly significant for the dynamics of invasive species since invasions require that species are capable of arriving in new environments at low densities and then subsequently increasing. This can have important implications for rates of spatial spread (Lewis and Kareiva 1993), as well as responses of populations to stochastic variability.

Liebhold and Bascompte (2003) explored the consequences of Allee effects and stochasticity for the control of invasive species. The most important outcome of Allee effects in their models was that Allee effects can influence the outcomes of eradication programmes. Frequently it is found that close to 100% eradication is required for control programmes to be effective. Because such high levels of control are generally difficult to achieve, this may prohibit attempts at control. However, Allee effects may lead to extinction thresholds at low densities, with the result that lower rates of control may yield eradication. Liebhold and Bascompte (2003) studied populations of the Gypsy Moth (*Lymantia disper*) in which they found that eradication could be achieved with c. 80% mortality in a model for the population which included an Allee effect, whereas in the absence of the Allee effect close to 100% mortality would be required. In the case study below, we describe in more detail a system in which Allee effects are extremely important in explaining invasions.

Multispecies interactions

The main focus on this review, so far, has been on single species dynamics. However interactions between species can affect the stability of population dynamics. Indeed most species do not live in isolation, so this possibility may be quite common, and we review two examples below.

In an influential paper May (1977) suggested that instability may be a common property of many forms of ecological systems. He showed that the same modelling framework could be applied to predator-prey, harvested and or indeed any consumer-resource system. This effect arises as a consequence of the function response of the consumer. When consumers are at low densities the rate of consumption of resources increases with increasing resource density, with the consequence that the rate of resource loss is positively related to resource density. At low densities consumers can potentially eradicate their resources at low densities. By contrast, at high resource densities, resources become saturating, and effectively the rate of resource loss to consumers is negatively related to resource density at high densities of resources. This dual behaviour leads to the potential for two equilibria, one at low resource densities, the other at high resource densities. If it exists the equilibrium at low densities is typically unstable. Recently Gascoigne and Lipcius (2004) have extended this framework to consider the aggregative responses of resources and have emphasised that such behaviour has been largely ignored in the applied and conservation literature, and show that there are a growing number of systems that implicate such a mechanism.

Competitive interactions may also yield instabilities, in two somewhat different ways. First, simple models for the dynamics of two competing species show that three distinct outcomes may occur: (i) one species or another excludes the other; (ii) the competing species settle at a stable joint equilibrium; (iii) there is a joint equilibrium, but which is unstable, so that perturbations from the joint equilibrium lead to the system becoming dominated by one species or another. This latter unstable condition may arise if the per capita effects of inter-specific competition are greater than the per capita effects of intra-specific competition. In the case-studies below, we detail such an example.

Second, competitive interactions may yield unstable dynamics when multi-species systems in interactions are strong are subjected to perturbations (Tilman *et al.* 1994). The reason for this is that in such a mixture the dynamics of any single species are a complex function of the densities of all the species. Consequently, the response of the system to perturbation may be complex, and perturbation may lead to complex intermediate or transient dynamics. The significance of this is that following a change, such as the invasion of a new species, the entire system may take many generations to return to equilibrium.

CASE STUDIES

The previous sections have emphasised that predicting the dynamics of invasive species may require that a suite of processes are dis-entangled. These include the effects of stochasticity, as well as the strengths of intra- and inter-specific interactions in the widest sense. For invasive species the key questions are how these processes affect the invasion of populations into new environments, as well as how they determine population size and the outcome of control strategies.

In this section we review in detail two case studies of invasive species which have been studies using simple modelling approaches. Necessarily it is impossible to consider all of the processes described above. The first example illustrates the key role that extreme stochasticity may play in determining invasion success, and how this may be modulated by the effects of competition. The second example shows the effect of perturbation on a predator-prey system with an instability.

Vulpia bromoides *in Australian pastures — the interplay of stochasticity and instability*

This example shows how instability and stochasticty simultaneously play roles in determining whether a species invades a system. *V. bromoides* is a winter annual which originates from the Mediterranean, and has become a problem in annual pasture systems in Australia. In contrast in Spain and Portugal, where it occurs naturally in pastures, *V. bromoides* is not regarded as a problem species. *V. bromoides* was probably introduced to Australia as a contaminant of seed or forage. Several species of the genus *Vulpia* were introduced to Australia. We studied the invasion dynamics of annual pastures in NSW, in which *V.bromoides* is the commonest of the species and occurs in c. 75% of the pastures where the genus is present (Dowling 1996). Given that the *Vulpia* genus is well established in Australia, the current problem is to determine what factors enable the persistence and spread of the species.

V. bromoides is a problem because, although in the early stages of the growing season it may provide some useful forage for live-stock, the forage produced late in the season is of very low nutritional value, and because the seeds produced by the weed become tangled in the wool of sheep and may even cause physical injuries to live-stock. Long-term management of *Vulpia* is difficult in pastures that are dominated by annuals, i.e., in which the perennial component is low. The problem of controlling *Vulpia* is compounded by the high seed production of the species which allows fast population recovery in the years following herbicide application, as well as high rates of compensatory growth of survivors following herbicidal control (Dowling *et al.* 1995).

The main questions that are important in understanding the dynamics of *V. bromoides*, as well as in predicting the effects of management, are: (i) how do interactions within and between species affect long-term dynamics? (ii) How are population dynamics and long-term persistence affected by stochasticity? Freckleton *et al.* (2000) developed a model to predict the dynamics of this species in a multi-species pasture mixture in order to address these questions.

The population dynamics of *V. bromoides* were studied in a 5-year field experiment, conducted between 1989 and 1994. Artificial pastures were established, comprising three species, *V. bromoides*, together with two 'desirable' species, *Trifolium subterraneum* (a legume) and *Lolium rigidum* (another grass). Pastures were set up originally with low, medium and high densities of *Trifolium*, low or zero densities of *Lolium* and low or high densities of *Vulpia*. These artificial pastures were allowed to grow until spring 1992, when half of the plots were sprayed with herbicide in order to create a further density differential of *Vulpia*. The year 1993-1994 was a drought year, and we estimated that droughts of this severity occur every 10 years or so (Freckleton *et al.* 2000).

We used the data from this experiment to parameterise a model predicting the numbers numbers of mature plants (N) and seeds in the seedbank (S) of each of the three species, i.e.,

$$N_{i=1...3}(t+1) = \lambda_i N_i(t)\left[1+\sum_{j=1}^{3}\alpha_{ij}N_j(t)\right]^{-1} + \nu_i S_i(t)$$

$$S_{i=1...3}(t+1) = \lambda'_i N_i(t)\left[1+\sum_{j=1}^{3}\alpha_{ij}N_j(t)\right]^{-1} + \nu'_i S_i(t) \qquad (9)$$

λ and λ' measure the per capita rate of change in numbers of mature plants and seeds, respectively, owing to seed production at the vegetative stage; υ and υ', respectively, measure the recruitment of plants from and persistence of seeds within the seed bank; competition between any pair of species i and j is measured by a parameter α_{ij}, which is the *per capita* reduction in population growth. Note that when i and j are different, α_{ij} refers to inter-specific competition (i.e., competition between species), but when $i = j$, α measures intra-specific competition (i.e., competition within the species). For the three species mixture, therefore, there are nine (i.e., 3×3) competition coefficients.

Analysis of the fitted model indicated a clear division within this simple community. First, competition between *Trifolium* and the two grasses was very weak indeed, and there was very little influence of the *Trifolium* on the dynamics of either of the grasses, or of the grasses on the *Trifolium*. This is perhaps unsurprising as legumes and grasses represent rather different functional groups. On the other hand, competition between the two grasses was very strong, and they influenced each others' dynamics profoundly. Since neither of the grasses possess very persistent seedbanks (in contrast to the *Trifolium*), the dynamics of these two species can be more simply modelled by a simple 2-species model:

$$N_L(t+1) = \lambda_L N_L(t)\,[1 + \alpha_{LL} N_L(t) + \alpha_{LV} N_V(t)]^{-1}$$
$$N_V(t+1) = \lambda_V N_V(t)\,[1 + \alpha_{VV} N_V(t) + \alpha_{VL} N_L(t)]^{-1} \qquad (10)$$

The dynamics of this system may be analysed using a simple phase plane analysis (Fig. 4). The fitted competition coefficients for equations (10) had the property that in both models the inter-specific coefficients were larger than the intra-specific coefficients. As a consequence the equilibrium for the two species is predicted to be unstable (Fig. 4*a*). The unstable equilibrium has an important consequence for the invasion of *Vulpia*. *Vulpia* is only able to invade communities when the density of *Lolium* is low. Examples of the predicted dynamics are shown in Fig. 4*b* and 4*c*. *Vulpia* is unable to invade a high density of *Lolium* (Fig. 4*b*) it is able to invade a low density (Fig. 4*c*).

At first sight it appears that the invasion of *Vulpia* into pastures may be difficult because it needs to exceed a critical density in order to invade. If the environment were constant, then this would be the case. In reality, however, the environment is not constant. As noted above, periodically severe droughts occur. These greatly facilitate the invasion of *Vulpia*, because these droughts reduce the densities of all species to low levels. Following drought the finite rate of increase of *Vulpia* is much greater than that of *Lolium* ($\lambda_V = 49$, whilst $\lambda_L = 9.7$). Consequently *Vulpia* populations respond much faster following droughts than do *Lolium* populations, and hence the invasion of *Vulpia* is facilitated. We estimated that severe droughts occurring with a frequency of every 1 in 10 to 1in 5 years would greatly facilitate the invasion of *Vulpia*. Suggestively, anecdotal evidence suggests that *Vulpia* became noticed as a problem following severe droughts in the early 1980s, although we have no quantitative data with which to support this contention.

In summary, understanding the invasion of *Vulpia* into pastures, and its persistence, requires that we not only understand the details of competition with other species, but also the effects of long-term stochasticity. Stochasticity plays a key role because the unstable equilibrium would make invasion of *Vulpia* unlikely from low densities into established pastures. However the effects of competition become modulated by periodic droughts making invasion far more likely.

Instability and outbreaks: the crown of thorns starfish

The second example illustrates how Allee effects resulting from predator-prey interactions can lead to unstable population dynamics, and consequently determine whether invasions are possible or not. There has been a great deal of recent

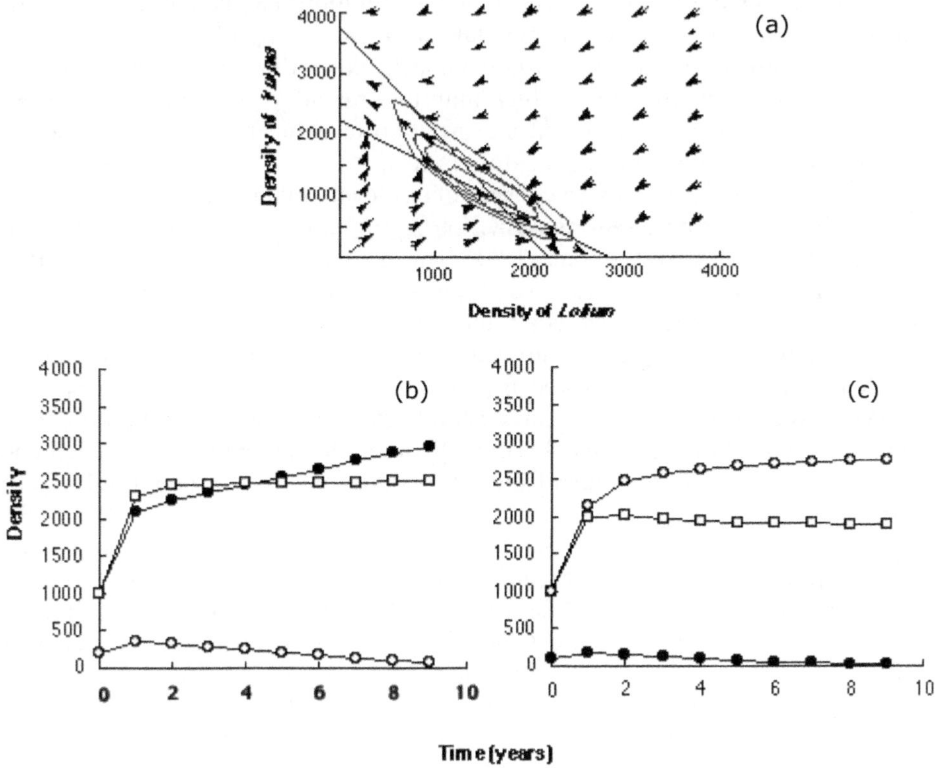

Fig. 4 (a) Phase-plane diagram showing unstable interactions in mixtures of *Lolium* and *Vulpia*. (b) & (c) Frequency dependent invasions of communities by either *Lolium* (filled circles) or *Vulpia* (open circles): the invader depends on the relative initial densities of the two species. The dynamics of the third species, *Trifolium* (squares) are essentially neutral with respect to the other two species.

interest in such systems (reviewed by Gascoigne and Lipcius 2004), inspired mainly by the results of simple models. In brief, predators can generate Allee effects in the population dynamics of their prey as a consequence of their functional and numerical responses. This happens in systems where the dynamics of the prey population are driven mainly by predation, and when prey have no refuges (spatial or temporal) from the effects of predation. Allee effects arise in such systems when predators hone in on prey at low densities, so that low density populations of prey tend to have negative population growth rates.

Such predator induced Allee effects are significant for invasive species because the potential exists for predators to eradicate unwanted invasive prey by reducing them to low densities. Alternatively if prey populations can be reduced by control measures, the potential exits for predators to lead to the ultimate extinction of the prey. Conversely, however, the possibility exists that prey populations may be held in check by predators, which maintain negative prey population growth rates from low densities, but that perturbation to either population may lead to the prey population escaping predator control.

The system we worked with consists of a 3-level food web involving predatory fishes, the invasive coral-eating crown-of-thorns starfish (*Acanthaster planci*) and reef-building corals (Dulvy *et al.* 2004). The crown-of-thorns starfish feeds upon live corals causing the largest known pest-related disturbances on Indo-Pacific coral reefs and it is regarded as a major management problem. Several mechanisms, including hydrography, hurricane disturbance, nutrient inputs and predator removal, have been suggested as potential causes of starfish invasions.

We studied a series of 13 oceanic islands varying in size and human population. For each island an index of human fishing intensity was estimated. At each of the sites densities of predatory fishes, starfish and barrier reef communities were surveyed by divers. Surveying was conducted in successive years. We wished to test the hypothesis that starfish populations outbreak and invade reefs when perturbations to their predators remove the predator control and, moreover, that predators are responsible for generating an Allee effect in starfish populations.

We did not have sufficiently detailed data to be able to parameterise a full population model. However we were still able to use a modeling framework to analyse the dynamics of this system. For the starfish we calculated per-capita rates of population change, $N(t)^{-1}\,(N(t+1) - N(t))$, for each island. To test the hypothesis that predators were responsible for generating Allee effects in starfish populations we plotted the per capita rate of population change against starfish density and predator density, as well as to look at how predator densities are affected by human fishing pressure.

First, we found that predatory fishes were 61% less abundant at the most intensively fished sites as the least fished sites. This indicates that there is an enormous range of variation in the predatory fishing pressure in this system. Given this, the data on rates of population growth of the starfish (Fig. 5) suggest that predator removal by subsistence exploitation may be sufficient to allow outbreaks of the crown-of-thorns. As shown in Fig. 5*a*, the relationship between per-capita rate of population change and starfish density is humped, as would be expected if populations were subject to an Allee effect. The data appear to indicate that there are two equilibria, one at low densities, the other at high densities. The lower equilibrium will be unstable. In a variable environment low density populations will either outbreak, or will become extinct.

That the Allee effect shown in Fig. 5*a* actually results from the effects of predators is confirmed in Fig. 5*b*. As shown there is a negative relationship between the rate of change in starfish density and predator density, confirming that starfish population dynamics are driven by predation, thus fulfilling the conditions required for predator-induced instability.

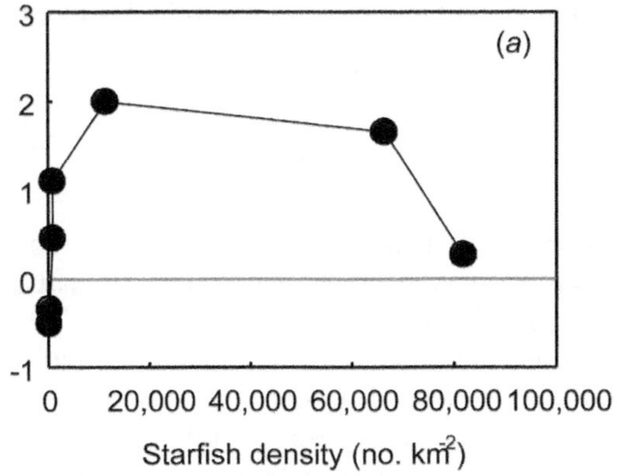

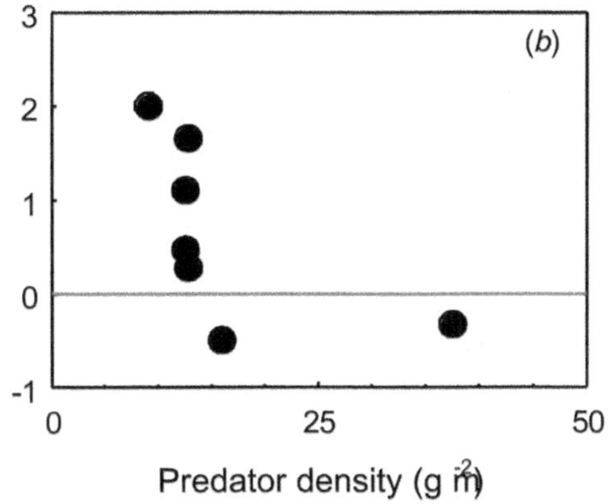

Fig. 5 The relationship between average starfish density and per capita population growth of starfish and (a) average starfish density, (b) average density of predatory fishes. Data are taken from 13 islands, presented in Dulvy *et al.* (2004).

In summary this system suggests that invasion of the crown of thorns starfish may be governed by underlying unstable dynamics and how this interacts with variations in the densities of predators. In this case-study variation in predator density was spatial and the consequence of differences in fishing intensity among islands. In other systems variation could be the result of other factors, such as habitat loss or climatic effects. What this example illustrates, however, is that predicting such effects will require a detailed understanding of the nature of population stability.

CONCLUDING REMARKS

Except in the unlikely event that population dynamics are completely density-independent, the long-term dynamics of populations will be determined by the effects of various forms of intra- and inter-specific interactions. In addition to this the effects of density-independent stochasticity may be complex, particularly when populations are forced to low densities. This means that forecasting the outcome of invasion may be complex and requires a detailed understanding of the structure of the system in question. Simple population models can be an extremely valuable guide in analysing such problems, both in theory and in practice.

REFERENCES

Andrewartha, H. G. and Birch, L. C. (1954) The distribution and abundance of animals University of Chicago Press, Chicago.

Buckley, Y. M., Briese, D. T. and Rees, M. (2003a) Demography and management of the invasive plant species *Hypericum perforatum*. I. Using multi-level mixed-effects models for characterizing growth, survival and fecundity in a long-term data set. Journal of Applied Ecology, **40**, 481-493.

Buckley, Y. M., Briese, D. T. and Rees, M. (2003b) Demography and management of the invasive plant species *Hypericum perforatum*. II. Construction and use of an individual-based model to predict population dynamics and the effects of management strategies. Journal of Applied Ecology, **40**, 494-507.

Byers, J. E., Reichard, S., Randall, J. M., Parker, I. M., Smith, C. S., Lonsdale, W. M., Atkinson, I. A. E., Seastedt, T. R., Williamson, M., Chornesky, E. and Hayes, D. (2002) Directing research to reduce the impact of non-indigenous species. Conservation Biology, **16**, 630-640.

Chesson, P. L. and Huntly, N. (1988) Community consequences of life-history traits in a variable environment. *Ann. Zool. Fennici*, **25**, 5-16.

Chesson, P. L. and Huntly, N. (1989) Short-term instabilities and long-term community dynamics. Trends in Ecology and Evolution, **4**, 293-297.

Cohen, D. (1966) Optimizing reproduction in randomly varying environments. Journal of Theoretical Biology, **12**, 119-129.

Dobson, A. P. and R. M. May. 1986. Patterns of Invasions by Pathogens and Parasites. The Ecology of Biological Invasions of North America and Hawaii. 1986. Springer-Verlag, New York.

Dowling, P. M. (1996) The ecology of Vulpia. Plant Protection Quartely, Supplement, **11**, 204-206.

Dowling, P. M., Verbeek, B., Lemerle, D. and Leys, A. R. (1995) Rapid regeneration of *Vulpia* (*Vulpia bromoides*) in pastures. In Proceedings 15th Asian-Pacific Weed Society Conference, pp. 497-500, Tosukuba, Japan.

Duffy, J. E. (2003) Biodiversity loss, trophic skew and ecosystem functioning. Ecology Letters, **6**, 680-687.

Dulvy, N. K., Freckleton, R. P. and Polunin, N. V. C. (2004) Coral reef cascades and the indirect effects of predator removal by exploitation. Ecology Letters, **7**, 410-416.

Earn, D. J. D., Rohani, P. and Grenfell, B. T. (1998) Persistence, chaos and synchrony in ecology and epidemiology. Proceedings of the Royal Society, Series B, **265**, 7-10.

Engen, S., Ø. Bakke and A. Islam. 1998. Demographic and environmental stochasticity–concepts and definitions. Biometrics **54**, 840–846.

Firbank, L. G. and Watkinson, A. R. (1985) A model of interference within plant populations. Journal of Theoretical Biology, **116**, 291-311.

Freckleton, R. P. and Watkinson, A. R. (1998) How does temporal variability affect predictions of weed population numbers? Journal of Applied Ecology, **35**, 340-344.

Freckleton, R. P. and Watkinson, A. R. (2001) Asymmetric competition between plant species. Funcional Ecology, **15**, 615-623.

Freckleton, R. P. and Watkinson, A. R. (2002) Are weed populations chaotic? Journal of Applied Ecology, **39**, 699-707.

Freckleton, R. P., Watkinson, A. R., Dowling, P. M. and Leys, A. R. (2000) Determinants of the abundance of invasive annual weeds: community structure and non-equilibrium dynamics. Proceedings of the Royal Society Series B, **267**, 1153-1161.

Freckleton, R. P., Silva Matos, D. M., Bovi, M. L. A. and Watkinson, A. R. (2003) Problems with predicting the impacts of harvesting using structured population models: an example based on tropical trees. Journal of Applied Ecology, **40**, 846-858.

Gascoigne, J. C. and Lipcius, R. N. (2004) Allee effects driven by predation. Journal of Applied Ecology, **41**, 801-810.

Ginzburg, L. R. and Colyvan, M. (2003) Ecological orbits: how planets move and populations grow. Oxford University Press, Oxford.

Hassell, M. P. (1975) Density-dependence in single species populations. Journal of Animal Ecology, **44**, 283-295.

Hassell, M. P. (1986) Detecting density dependence. Trends in Ecology and Evolution, **1**, 90-93.

Hassell, M. P., Lawton, J. H. and May, R. M. (1976) Patterns of dynamical behaviour in single-species populations. Journal of Animal Ecology, **45**, 471-486.

Heino, M., Ripa, J. and Kaitala, V. 2000 Extinction risk under coloured environmental noise. Ecography **23**, 177-184.

Kendall, B. E., *et al.* Why do populations cycle? A synthesis of statistical and mechanistic modelling approches. Ecology, **80**, 1789-1800.

Lande, R. (1998) Demographic stochasticity and Allee effect on a scale with isotropic noise. Oikos, **83**, 353-358.

Lande, R., Engen, S. and Saether, B.-E. (2003) Stochastic population dynamics in ecology and conservation Oxford Univsersity Press, Oxford.

Lewis, M. A. and Kareiva, P. (1993) Allee dynamics and the spread of invading organisms, Theoretical Population Biology **43**, 141-158.

Lewontin, R. C. and Cohen, D. (1969) On population growth in a randomly varying environment. Proceedings of the National Academy of Sciences (USA), **62**, 1056-1060.

Liebhold, A and Bascompte, J. (2003) The Allee effect, stochastic dynamics and the eradication of invasive species, Ecology Letters, 2003, **6**, 133-140.

Lonsdale, W. M. (1996). The balance of weed populations. In Frontiers of Population Ecology (eds. R. B. Floyd and A. W. Sheppard). CSIRO Publishing, Melbourne.

May, R. M. (1973) Stability and complexity in model ecosystems Princeton University Press, Princeton, New Jersey.

May, R. M. (1977) Thresholds and breakpoints in ecosystems with a multiplicity of stable states. Nature, **269**, 471-477.

May, R. M. (1981). Models for two interacting populations. In Theoretical Ecology (ed. R. M. May), pp. 78-104. Blackwell Scientific Publications, Oxford.

May, R. M., Conway, G. R., Hassell, M. P. and Southwood, T. R. E. (1974) Time delays, density-dependence and single species oscillations. Journal of Animal Ecology, **43**, 747-770.

May, R. M. and Oster, G. F. (1976) Bifurcations and dynamic comlexity in simple ecological models. American Naturalist, **110**, 573-599.

Maynard Smith, J. (1973) Models in ecology Cambridge University Press, Cambridge.

Nicholson, A. J. (1954) An outline of the dynamics of animal populations. Australian Journal of Zoology, **2**, 9-65.

Pacala, S. W., Canham, C. D., Saponara, J., Silander, J. A., Kobe, R. K. and Ribbens, E. (1996) Forest models defined by field measurements: estimation, error analysis and dynamics. Ecological Monographs, **66**, 1-43.

Rees, M. and Crawley, M. J. (1989) Growth, reproduction and population dynamics. Functional Ecology, **3**, 645-653.

Rees, M. and Crawley, M. J. (1991) Do plant populations cycle? Functional Ecology, **5**, 580-582.

Rees, M. and Paynter, Q. (1997) Biological Control of Scotch Broom: modelling the determinants of abundance and the potential impact of introduced insect herbivores. Journal of Applied Ecology, **34**, 1203-1221.

Royama, T. (1992) Analytical Population Dynamics Chapman & Hall, London.

Scheffer, M. and Carpenter, S. (2003) Catastrophic regime shifts in ecosystems: linking theory to observation. Trends in Ecology and Evolution, **18**, 648-656.

Saether, B.-E., *et al.* (2004) Life-history variation predicts the effects of demographic stohcasticity on avian population dynamics. American Naturalist, **164**, 793-802.

Saether, B.-E., *et al.* (2005) Time to extinction of bird populations. Ecology in press.

Stephens, P. A. and Sutherland, W. J. (1999) Consequences of the Allee effect for behaviour, ecology and conservation. Trends in Ecology and Evolution, **14**, 401-405.

Stephens, P. A., Sutherland, W. J. and Freckleton, R. P. (1999) What is the Allee effect? Oikos, **87**, 185-190.

Taylor, C. M. and Hastings, A. (2004) Finding optimal control strategies for invasive species: a density-structured model for *Spartina alterniflora*. Journal of Applied Ecology, **41**, 1049-1057.

Tilman, D., May, R. M., Lehman, C. L. and Nowak, M. A. (1994) Habitat destruction and the extinction debt. Nature, **371**, 65-66.

Tuljapurkar, S. (1990) Population dynamics in variable environments Springer, New York.

Watkinson, A. R. (1980) Density-dependence in single-species populations of plants. Journal of Theoretical Biology, **83**, 345-357.

White, T. C. R. (2001) Opposing paradigms: regulation or limitation of populations? Oikos, **93**, 148-152.

Williamson, M. (1996) Biological Invasions. Chapman & Hall, London, UK.

Chapter seven

Local interactions and invasion dynamics: population growth in space and time

D. J. Murrell

INTRODUCTION

The study of biological invasions is essentially the study of population growth in both space and time. Not only are we interested in the population size of an invader, but also its rate of spread across a landscape or region. Hence theoretical models of invasions very often incorporate some form of spatial heterogeneity (Skellam 1951, Boerlijst and Hogeweg 1991, Hassell *et al.* 1994, Kot *et al.* 1996, Keeling *et al.* 1997, Lewis 2000, Lewis and Pacala 2000, Rietkerk *et al.* 2002).

Broadly speaking, there are two classes of spatial heterogeneity; the so-called endogenous or self-induced spatial heterogeneity (type i), and exogenous heterogeneity (type ii). Type (i) heterogeneity is caused by interactions between individuals (or populations) and leads to what is sometimes called self-organised spatial patterns (Boerlijst and Hogeweg 1991, Hassell *et al.* 1994, Keeling *et al.* 1997, Rietkerk *et al.* 2002). For example, in monocultures of pines, intense competition between close neighbours leads to an even spacing of adult trees (Kenkel 1988).

M.W. Cadotte et al. (eds.), Conceptual ecology and invasion biology, 147–168.

Type (ii) heterogeneity is caused by processes beyond individual interactions, such as soil depth, temperature gradients etc. This leads to the aggregation of individuals caused by species being associated with certain habitat types more than others. For example, some tropical rainforest trees have been found to be associated with certain soil types, and it has been argued that this is an important factor in determining their aggregated spatial structure (Palmiotto *et al.* 2004). Undoubtedly both classes of heterogeneity are important during invasions; interactions with other organisms may determine whether an invasion succeeds (Fagan and Bishop 2000, Owen and Lewis 2001), as well as the availability and spatial patterning of suitable habitat (Richardson *et al.* 2000, Hill *et al.* 2001, Holway *et al.* 2002, With 2002). However, this chapter will focus on only endogenous spatial heterogeneity in an attempt to show how important local spatial structure can be in determining the rate of spread of a population starting from a low density (see Bolker 2003 for results for a similar model in heterogeneous space).

Ecological interactions for most organisms tend to occur between individuals that are nearby in space. Coupled to this, most organisms larger than 1mm are argued to be in some way dispersal limited (Finlay and Clarke 1999, Finlay *et al.* 1999, Finlay 2002). (Many microbial organisms are so small that they can be carried large distances by wind and together with long-lived resting stages this means that many microbial species may enjoy a near global distribution.) Early in its history ecological theory neglected both these aspects and it was assumed that individuals randomly bump into one another; the so-called mean-field assumption. More recently, non-uniform dispersal has been incorporated and has been used with much success in describing the rates of spread (invasion) of numerous species (Kot *et al.* 1996, Veit and Lewis 1996). However, most of these models have not included the discrete nature of individuals and without it, the tight small-scale clustering that is often evident in natural populations cannot be included.

As has been often observed, there is a crucial feedback between the spatial pattern of individuals across the landscape and the dynamics of the population (Bolker and Pacala 1997, Durrett and Levin 1994). Local dispersal, even without any density dependent effects on death rates or fecundity, results in a clustered or patchy distribution of individuals across the landscape (Young *et al.* 2001). In such a scenario the spatial pattern is not important for the population dynamics other than in determining the rate of spread of an invader across a landscape. However, as shall be expanded upon below, once local density dependent processes are incorporated, the spatial pattern of individuals may become very important to many features of the invasion dynamics and may ultimately determine whether an exotic can invade or not. With local spatial structure, what an individual experiences in its local neighbourhood may be very different to the overall density of individuals across the landscape. This local spatial structure then helps to determine the levels of competition that an individual experiences and hence indirectly determines the level of competition (and local spatial structure) in the next time interval.

The rest of this chapter will focus on a general model for the spatio-temporal population dynamics of an invading species. We will consider individuals similar to perennial plants that reproduce, dispersing their offspring locally, and die at rates that are dependent on their local (neighbourhood) density. Two models will be presented; one individual-based model, and its deterministic approximation. The link between these models and the classical (non-spatial) logistic model for population growth is made explicit and will act as a reference point for the results that are obtained when interactions and dispersal are localised in space. Both spatial models will be used to investigate the scales of dispersal and neighbourhood interactions on the transient dynamics (success and speed of invasion) and asymptotic state of the population (the final population density). It is also observed that the expected reproductive output of an individual in the absence of competition is also important in determining the success and speed of an invasion as well as the final population size.

A STOCHASTIC INDIVIDUAL-BASED MODEL (IBM)

Invasions are inherently stochastic; they very often start with just a few individuals and the success of an invasion owes a great deal to chance; therefore it makes sense to start from a stochastic model that focuses on the individual-level (microscopic) processes of births and deaths. We will concentrate on a spatial extension to the familiar logistic model for population growth (Pearl and Read 1920); one where both interactions and dispersal are localised in space.

Consider a population of organisms of a single species invading a two-dimensional landscape such as a field. In the IBM space is continuous as opposed to a lattice or site-based model, and large enough for edge effects to be negligible. To reduce the edge effects the landscape is in the shape of a torus (formerly speaking the boundaries are periodic). An individual *i* located at coordinates $x_i = \{x_1, x_2\}$ is denoted by a Dirac delta function $\delta_{x_i}(x)$. (The Dirac delta function serves to pick out individuals; it is a peak of infinite height and infinitely small width centred on the location of each individual, giving an area underneath it that sums to one.) The state of the population at time *t* is given by the function $p(x, t)$ which is the sum of all the delta functions and therefore describes the number and locations of all individuals in the population.

There are three different types of event that may occur in the IBM, birth, death and movement. To keep in close association with the classical logistic model, only deaths are density dependent, although the model is flexible so that births and movements may also be dependent upon local densities (Bolker and Pacala 1997).

The probability per unit time, $D(x, x', p)$, that an individual at location x dies is given by

$$D(x, x', p) = d + d' \cdot \int w(x' - x) \cdot [p(x') - \delta_x(x')]\, dx'. \qquad (1)$$

The death term consists of two parts. The first is an intrinsic per capita death rate, d, and the second part modifies the death rate according to the presence of nearby neighbours. Neighbours have a per capita effect d' on the death rate of the focal individual. However, it is assumed that competitive effect of neighbours is dependent upon the distance between them. The term $w(x' - x)$, otherwise called the competition kernel, describes the weighting that neighbours at location x' have on the death rate at location x and is a function of the distance, $x' - x$, that separates the pair. This weighting is multiplied by the density, $p(x')$, of individuals at location x', and the integral takes into account the contributions to the death rate at x of all locations x' in the neighbourhood of x. The last term, $\delta_x(x')$ is only switched on when $x' = x$, thereby removing the spurious effect that an individual would otherwise have on its own death rate.

The probability per unit time, $B(x, x', p)$, that an individual located at x produces an offspring at x' is given by

$$B(x, x', p) = b \cdot m(x' - x), \tag{2}$$

where b is the intrinsic per capita birth rate and $m(x' - x)$ is the dispersal kernel which describes the probability that the newborn individual comes to rest at a displacement $x' = x$ from its parent.

The dispersal and competition kernels are very flexible and can take on practically any shape (see for variations see Bolker and Pacala 1997). It seems sensible to ensure that they are both monotonically decreasing functions and we choose to use bivariate Gaussian distributions:

$$m(x' - x) = \frac{1}{M}\exp\left[-\frac{|x' - x|^2}{2s^2_m}\right], \tag{3a}$$

$$w(x' - x) = \frac{1}{W}\exp\left[-\frac{|x' - x|^2}{2s^2_w}\right], \tag{3b}$$

each having one parameter, s_m, s_w, that measures the width of the kernel. A small value for s_m, means that offspring fall very close to their parent (short-scale dispersal). A small value for s_w means that competition is very intense between close neighbours but that it soon diminishes with distance (short-scale interactions). On the other hand large values for s_m, s_w, reduces the importance of space. For practical purposes we truncate the kernels at a distance of 3 times the parameter value; the effect this has on the dynamics is negligible. The kernels are normalised so that that their integrals over $x' - x$ are unity, M and W being the normalisation constants.

SPATIAL STATISTICS

In the IBM each individual and each event needs to be constantly tracked. In order to investigate the implications of incorporating local interactions and local dispersal into the logistic model for population growth, it helps to have some summary statistics that can be output from the IBM. These will later become state variables in the deterministic approximation so it is important that they are defined.

The first spatial moment is the familiar average density:

$$N(p, t) = \frac{1}{A} \cdot \int p(x, t)\, dx'. \tag{4}$$

This simply sums over all individuals in the population at time t and divides by the area A in which the population lives.

The first spatial moment does not give any information on how individuals are spread across the landscape, in order to get a feel for the spatial pattern, higher order spatial moments are required. There are a number of higher order spatial moments that could be used (Diggle 1983, Cressie 1991) but the one used here is

$$C(\xi, p, t) = \frac{1}{A} \cdot \int p(x, t) \cdot [p(x + \xi, t) - \delta_x(x + \xi)]\, dx'. \tag{5}$$

The second spatial moment is the average density of pairs separated by a distance ξ. It can be thought of as the individual's eye view since it describes the expected neighbour density at different distances, and hence describes the variation in the density of individuals across the landscape. Once again, the delta term removes the self-pair term that arises when $x + \xi = x$. This form of second moment has the benefit of being mathematically simple, a feature that is useful for the deterministic approximation that follows. In the results given below the second moments are normalised by dividing by $N^2(p, t)$. This means that when $C(\xi, p, t) > 1$, this indicates a more crowded neighbourhood than would be expected if the spatial pattern of individuals were random (aggregation); when $C(\xi, p, t) < 1$ this indicates a relatively competitor-free neighbourhood compared to a random pattern (spatial segregation) and when $C(\xi, p, t) = 1$ this indicates a random spatial association of individuals.

A DETERMINISTIC APPROXIMATION

There is much that can be gained from the IBM, but keeping track of all individuals and their dynamics is computationally very intensive. The IBM defined above is stochastic (any 2 independent realisations are expected to have slightly different patterns) and Markovian (the rate of change of the system depends only on the current state). These make it hard to discern the ecological signal from the background noise; many realisations are required to get a better picture of properties of the model (Law *et al.* 2003), and hence full analysis of the IBM is very difficult. For example, knowing when the population has reached its end state can be very difficult to determine.

Complementing the IBM with an approach that deals with the macroscopic properties (such as population densities) is therefore desirable. Scaling up from small-scale processes to large-scale properties is not a trivial task (Levin 1992), but recent advances in mathematical techniques have expanded the range of cases for which this is possible in spatial models (Bolker and Pacala 1997, Bolker 1999, Bolker and Pacala 1999, Dieckmann and Law 2000, Law and Dieckmann 2000, Murrell and Law 2000, Bolker 2003, Bolker *et al.* 2003, Law *et al.* 2003, Murrell and Law 2003, Murrell *et al.* 2004).

From the description of the IBM above it is possible to derive a deterministic approximation that describes the expected rate of change of the population (Dieckmann and Law 2000). The expected rate of change in population density is given by

$$\dot{N}(t) = bN(t) - dN(t) - d' \cdot \int w(\xi)\, C(\xi)\, d\xi \qquad (6)$$

The first and second terms deal with the contributions to population density of the intrinsic rates of birth and death. These are made explicit in the spatial model rather than substituting $r = b - d$, because, as will be shown below, births and deaths have very different effects on the spatial structure. The third term is where the spatial structure first enters the dynamical system. Competition is now dependent on the expected density of neighbours around an individual, $C(\xi, t)$, and is scaled by both the interaction kernel $w(\xi)$, which takes into account the distance between pairs, and by the per capita competitive effect d'.

The spatial structure summarised by $C(\xi, t)$ has dynamics of its own that also need to be made explicit. The expected rate of change of pair densities are described by

$$
\begin{aligned}
\dot{C}(\xi, t) = \quad & 2\, b \int m(\xi')\, C(\xi + \xi', t)\, d\xi' && \text{(7a)}\\
+ \quad & 2\, b\, m(\xi)\, N(t) && \text{(7b)}\\
- \quad & 2\, d\, C(\xi, t) && \text{(7c)}\\
- \quad & 2\, d' \int w(\xi')\, T(\xi, \xi', t)\, d\xi' && \text{(7d)}\\
- \quad & 2\, d'\, w(\xi)\, C(\xi, t). && \text{(7e)}
\end{aligned}
$$

Each term describes events that lead to the creation and deletion of pairs (or more formerly pair densities). The first term (a) describes how pairs are formed when a newborn disperses a distance ξ' from its parent. The second term (b) takes into account pairs formed when one of the members of the pair is the parent.

The other three terms deal with loss of pairs through deaths. Term (c) describes how pairs are lost through density independent deaths. Term (d) modifies this rate due to neighbours at a distance ξ' with the interaction kernel, $w(\xi')$ weighting the effect of the neighbours according to distance between locations. Since the pair is affected by a third individual, information on the expected density of triplets is required and this is denoted by T, the third spatial moment. This represents a non-trivial technical problem that is expanded upon in the Appendix. Finally, (e) accounts for density dependent deaths caused by competitive interactions within the pair. Note that each term is doubled because births or deaths can occur at either location in the pair.

Having defined both the stochastic and the deterministic models, we can now turn to investigate the effect of incorporating local spatial structure on the success and speed of invasion, and also on the final population density. As a reference point, the results are compared to the conclusions that are reached from the classical logistic model for population growth. Recall that in the classical logistic model population growth is reached at exactly half the final population size; the growth trajectory shows an S-shaped curve; and that from low initial densities, populations that have lower final densities reach their equilibrium the fastest. The reader should note that the classical logistic model is a special case of both models defined here, and can be recovered whenever interactions occur over very large spatial scales. This is true even when dispersal is short-range because, when interactions occur over sufficiently large spatial scales, any spatial pattern that is evident is irrelevant to the dynamics (the positions of individuals in space becomes unimportant).

RESULTS

Dispersal distance

The importance of dispersal distance for the invasion dynamics is shown in Fig. 1. Shown are 2 populations, one with relatively poor dispersal, one with relatively good dispersal abilities; the invaders are otherwise identical. Snapshots of the IBM show that the poorer disperser soon develops a clumped distribution and this inhibits population growth because individuals experience crowded neighbourhoods. On the other hand the better disperser seems to have faster population dynamics and reaches a higher population density. This is because more offspring can escape intense competition with their parents and hence their neighbourhood is relatively competitor-free. Inspection of the spatial statistics during the invasion confirms this; the better disperser has faster population dynamics, experiences a less crowded neighbourhood and reaches a higher final density (Fig. 2). Note however, that until about 20 time units have passed, even the better disperser shows some degree of aggregation (Fig. 2); but this then disappears as population grows and spreads, and eventually the individuals are spatially segregated.

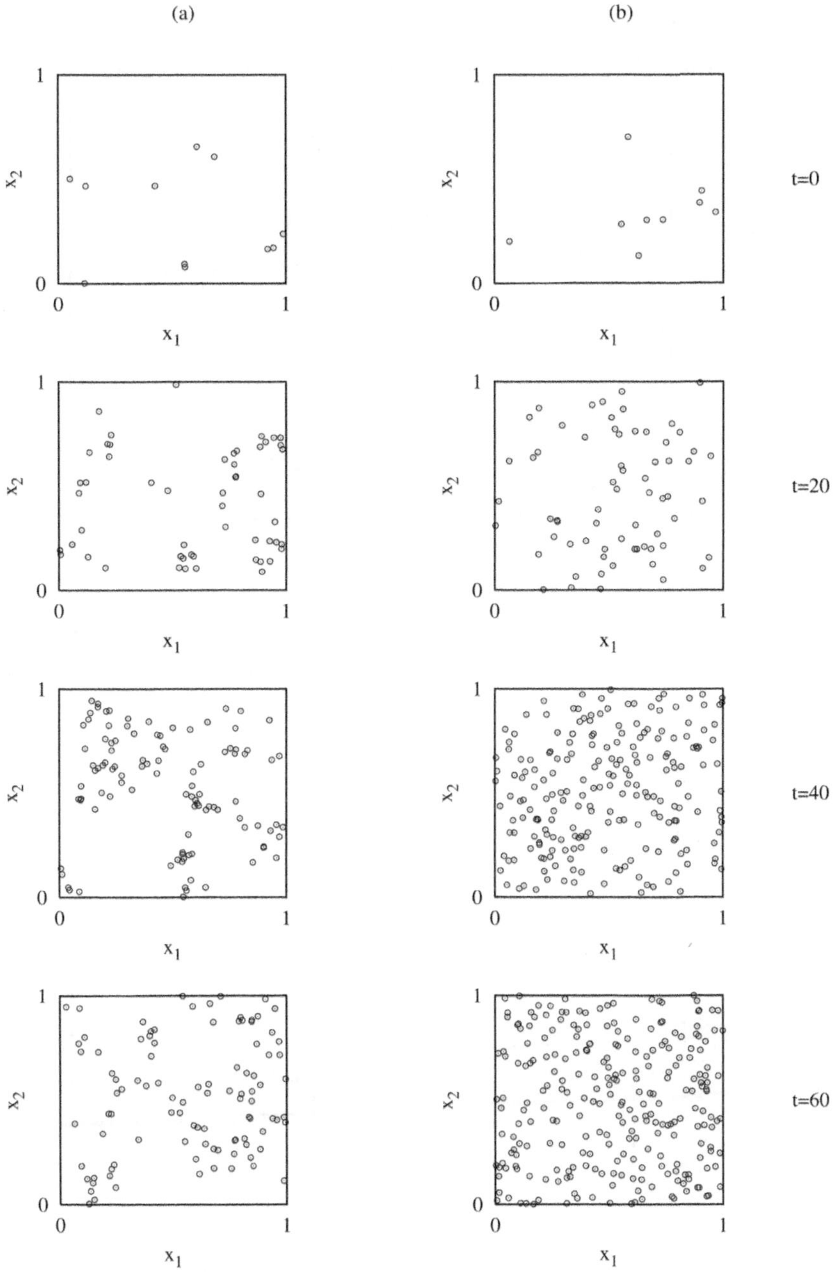

Fig. 1 Snapshots from the IBM showing the invasion of two different populations. In (a) the individuals have short-range dispersal (s_m = 0.04), whereas in (b) individuals have longer-range dispersal (s_m = 0.12). In both cases the starting population has 10 individuals scattered randomly across the landscape which is of unit area. Other parameters held constant are b = 0.4; d = 0.2; d' = 0.001; s_w = 0.04.

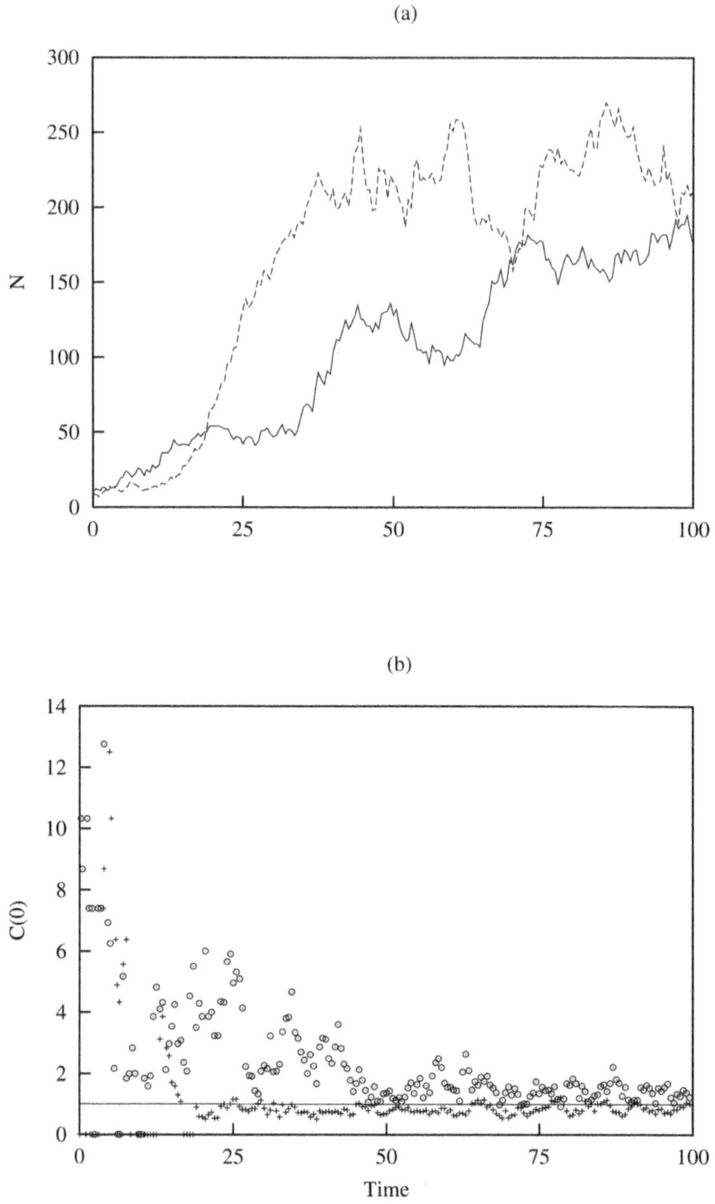

Fig. 2 Temporal dynamics for (a) the mean density (first spatial moments) and (b) second spatial moments at short distances for the invasions of the short-range disperser (solid line and circles) and the long-range disperser (broken line and crosses) shown in Fig. 1. In (b) values greater than 1 (above the solid line) indicate aggregation of individuals, whereas values less than 1 (below the solid line) indicate spatial segregation. Parameters and starting conditions are as described in Fig. 1.

We now turn to see how local interactions and local dispersal affect the invasion success, P_s, of a population. For a range of scales of dispersal, 50 populations of 10 individuals are initialised in the IBM to investigate how b_0, the critical birth rate required for invasion, changes with the dispersal of the invader. For each invading population the model is run for 50 time units and an invasion is deemed a success if it has expanded after this time i.e. if $N(t = 50) > N(t = 0)$. The probability of invasion success, P_s, is then calculated from the proportion of the 50 populations that show this increase in population density. When the invader is restricted to only short-range dispersal it takes a relatively large b_0 for the population to be able to invade with any degree of certainty (Fig. 3). For example for $s_m = 0.03$ invasion success does not exceed 50% until $b = 0.33$. In fact for many values for intrinsic birth rates, b, that would result in almost guaranteed invasion success in the stochastic version of the non-spatial logistic model none of the 50 initial populations are able to invade.

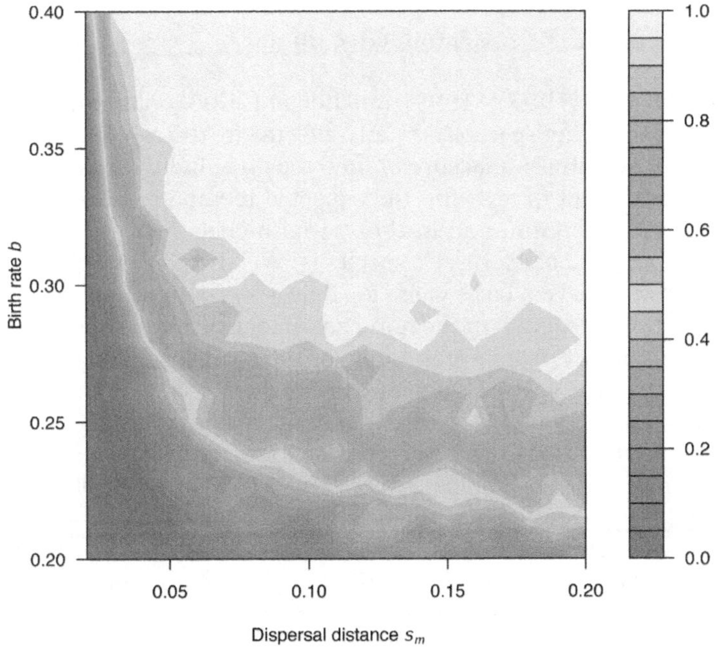

Fig. 3 The effect of intrinsic birth rate (b) and dispersal distance (expressed by the parameter s_m that determines the width of the dispersal kernel) on the invasion success (P_s) of a single species. The landscape (of unit area) is seeded with 10 randomly located individuals and each square represents the proportion of 50 realisations that show population growth after 50 time units have elapsed. Other parameters held constant throughout are $d = 0.2$, $d' = 0.001$, $s_w = 0.02$.

There are numerous examples showing how the invasion rate is affected by the dispersal kernel. For example, Caswell *et al.* (2003) showed that the Pied Flycatcher (*Ficedula hypoleuca*) in Europe is a slower invader than the Starling (*Sturnus vulgaris*) in North America and that 70% of the difference between invasion rates could be explained by the differences in dispersal abilities, with the rest being explained by the differences in demography. Over longer periods of time the dispersal abilities of an invader might be expected to be under some strong selective pressure. Using a simulation model Travis and Dytham (2002) were able to show that early on during an invasion on an island, long-distance invasion is selected for, but that shorter distance subsequently evolves. Evidence for such patterns can be seen in the difference between wind-dispersed plants in mainland and small island subpopulations (Cody and Overton 1996). In new island populations the dispersal ability of the plants tends to be greater than the mainland populations. Over time selection is for a reduction in dispersal ability on the islands so that old island plant populations have a lower dispersal ability than the mainland populations (see also Darlington 1943, Brown and Lomolino 1998, Whittaker 1998, Filin and Ziv 2004).

Reproductive number

As Bolker and Pacala (1997), (1999), and Bolker (2003) point out the population dynamics of an invading species is greatly affected by the reproductive number R (where $R = b/d$) which is a measure of how sensitive individuals are to competition. R can be thought of as being the expected lifetime reproductive output in the absence of competition, and in the spatial models R becomes important in determining how good a coloniser a species is. Weedy species, (and hence invaders) are expected to have a large value for R, less weedy species will have a low R number. Note that varying R in the non-spatial logistic model has no outcome on population dynamics; R is important only for the spatial dynamics.

When R is small the species is a poor coloniser and the population growth rate is slow, or if R is small enough the invasion may fail altogether (Fig. 4). Moreover, note that low values for R are associated with long periods of near linear population growth; the classical sigmoidal growth curve characteristic of the non-spatial logistic model only occurs when R is high and/or when dispersal is long-range (see also Law *et al.* 2003). Another feature is that R greatly affects the final population size, with large values for R leading to higher population densities (Fig. 4). Coupled to this is the interesting observation that the larger populations take the shortest amount of time to reach their equilibrium; this is in stark contrast to the non-spatial logistic model where the smallest population reach their final size before those with larger population densities. Note also that the agreement between the IBM and the deterministic approximation (moment dynamics) is very good (Fig. 4). Other results confirm these findings (Murrell and Law 2000, Law *et al.* 2003, Murrell and Law 2003), and the rest of the chapter shall concentrate on the deterministic approximation.

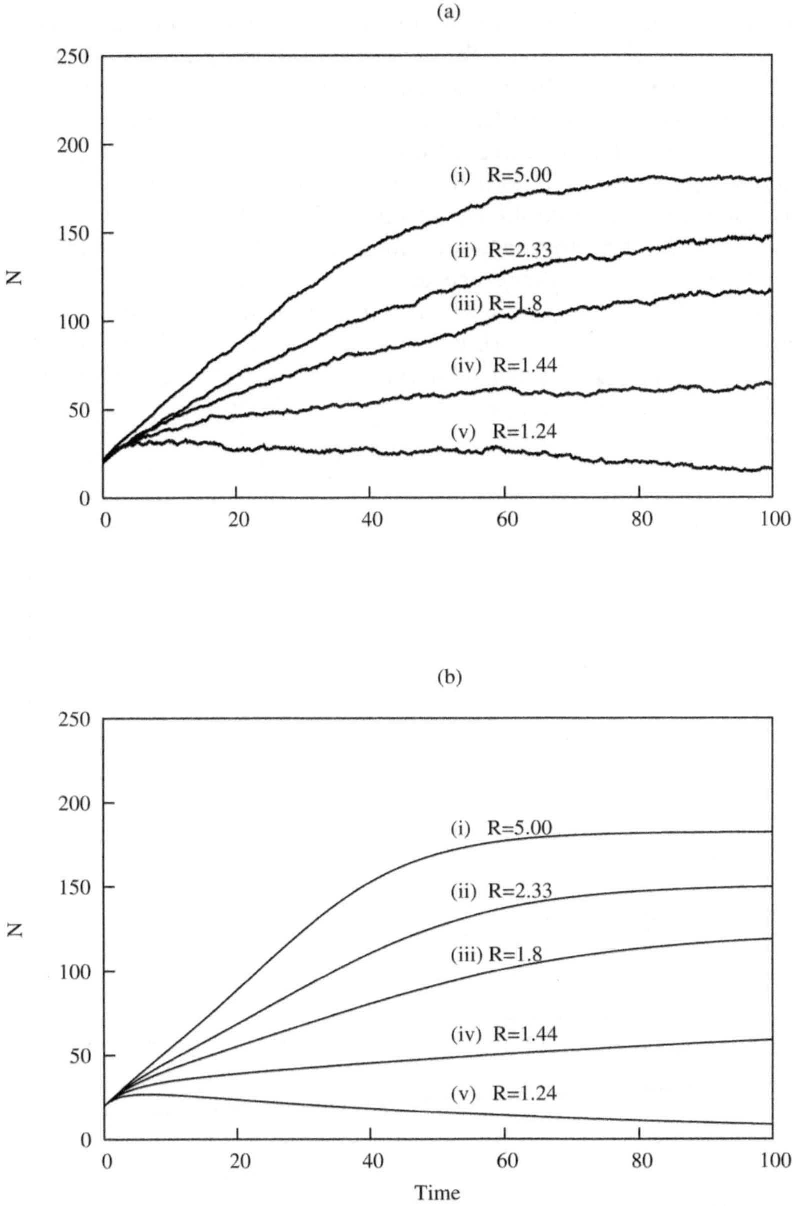

Fig. 4 The effect of the reproductive number R on population growth. Shown in (a) are mean trajectories for 100 realisations of the IBM and in (b) the deterministic approximation using the dynamical system (6) and (7) for 5 different values for R. In all cases the intrinsic growth rate r is held constant, $r = 0.2$. In (i) $b = 0.25$, $d = 0.05$; (ii) $b = 0.35$, $d = 0.15$; (iii) $b = 0.45$, $d = 0.25$; (iv) $b = 0.65$, $d = 0.45$; (v) $b = 1.05$, $d = 0.85$. Other parameters held constant throughout are $d' = 0.001$, $s_w = s_m = 0.04$.

Another example shows how important R can be to the invasion dynamics and how it can reverse some of the conclusions of the non-spatial logistic model (Fig. 5). Consider 2 imaginary species (labelled 1 and 2 respectively), if $r_2 > r_2$ (i.e., species 1 has a higher intrinsic rate of growth than species 2), but $R_2 < R_2$ and if interactions and dispersal are localised in space then it is possible for species 2 to be the better invader both in terms of speed of invasion and equilibrium density. Similar results could easily be produced for 2 species with different scales of dispersal, and this sends the message that knowing r may not be sufficient in determining whether a species can invade or not. These results all point to the fact that a high reproductive number, R, is advantageous for an invader in homogeneous space, but other results (Bolker 2003) show that low values for R can be beneficial if there is spatial variation in the external environment (type (ii) spatial heterogeneity).

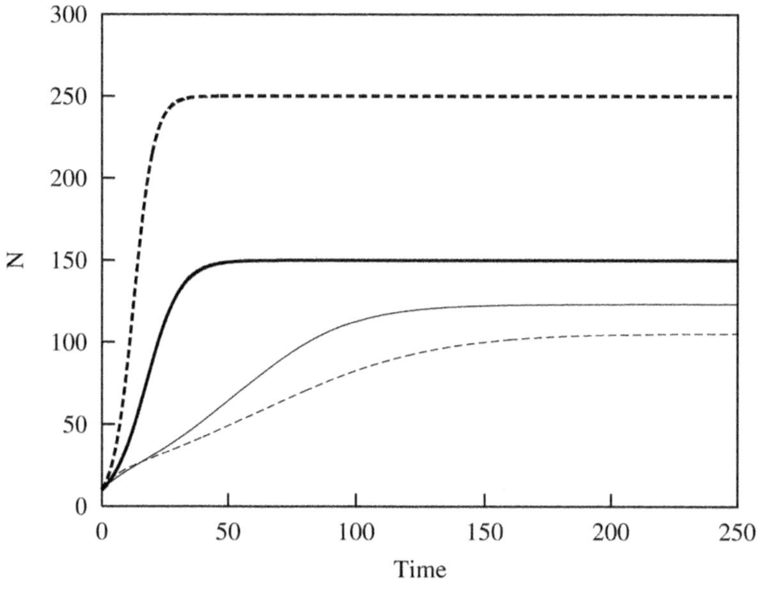

Fig. 5 An example showing how R may reverse a main conclusion of the non-spatial logistic model. Species 1 (broken lines) has a high value for r but a low R number. Species 2 (solid lines) has a lower value for r, but a higher R number. In the non-spatial model (thick lines) species 1 is the better invader and has a higher density at equilibrium. When interactions and dispersal are localised in space (thin lines) this result is reversed; now the species 2 has the faster invasion rate and higher density at equilibrium. For species 1, $b = 0.85$, $d = 0.6$ giving $r = 0.25$, $R = 1.4167$; and for species 2, $b = 0.2$, $d = 0.05$ giving $r = 0.15$, $R = 4.00$. Other parameters are identical for both species and are $d' = 0.001$, $s_w = s_m = 0.04$.

Final population size

So far we have concentrated on the transient dynamics, but neighbourhood interactions and local dispersal can also greatly affect the final population size (Law *et al.* 2003, Etheridge 2004). If neighbourhood interactions are sufficiently large-scale then the position of individuals across the landscape is of little importance in determining the competitive interactions. In such cases the results of the mean-field model are asymptotically approached (Fig. 6). This is in contrast to the case where dispersal is large-scale. If interactions are localised in space (small value for s_w) then many offspring can escape from competing with their parent. This leads to a spatially segregated population where most individuals experience a relatively competitor-free neighbourhood and hence the final population size is much larger than that under mean-field conditions. On the other hand if both dispersal and interactions are short-range (small values for s_m and s_w) then the competitive neighbourhood is perceived as being crowded with the result that the final density is much lower than the mean-field expectation. In extreme cases, where both interactions and dispersal occur over short spatial scales then self-driven extinction occurs (Fig. 6, Figs. 1 and 3: Law *et al.* 2003).

Self-driven extinction might at first thought seem only an interesting mathematical artefact and of no concern to invasion ecology; after all if an organisms dispersal is so short-range how could it persist at another location and send out emmigrants? However, it is entirely possible for a species' dispersal kernel to differ between its natural and exotic habitats. Many plants rely on secondary dispersal of seeds by animals (Andresen 1999, Bohning-Gaese *et al.* 1999, Hoshizaki *et al.* 1999, Milton and Dean 2001, Gordon and van der Valk 2003, Guitian *et al.* 2003) and, if in a new environment the secondary dispersers are not present the dispersal kernel is likely to be much shorter, and hence self-driven extinction is much more likely. Another feature of the presence of spatial aggregation of conspecifics is that when looking at the invasion dynamics of a species into a community it is no longer possible to ignore the effect of intraspecific competition as has traditionally been the case (Morton *et al.* 1996, Morton and Law 1997).

→

Fig. 6 Equilibrium values for (a) N and (b) $C(0)$ as a function of the spatial scale of the interaction kernel and dispersal kernel. The results are obtained by numerically integrating the deterministic approximation (6) and (7) until the change in N over an integration step 0.1 falls below 0.0001. Other parameters are held constant are $b = 0.4$; $d = 0.2$; $d' = 0.001$. For comparison this set of parameters non-spatial logistic model gives an equilibrium density of 200.

(a)

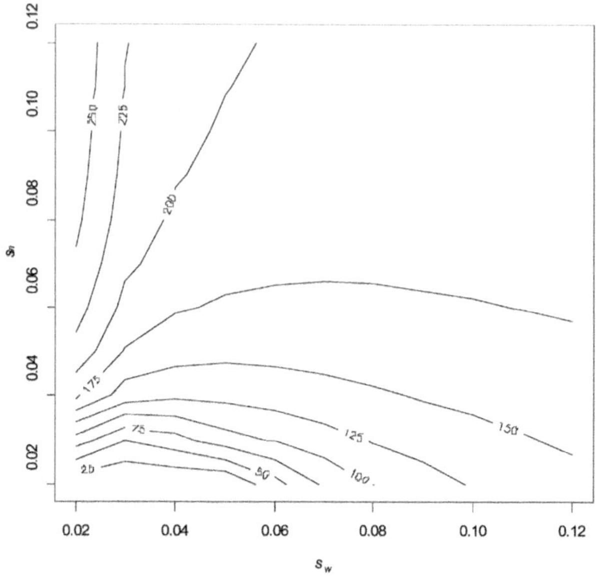
250
225
200
175
150
125
100
75
50
25
0.02
0.04
0.06
0.08
0.10
0.12
s_n
s_w

(b)

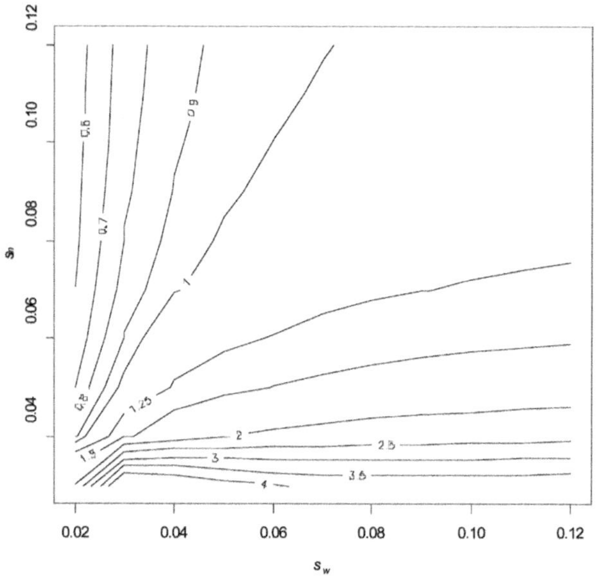
0.6
0.7
0.9
1
0.8
1.25
2
2.5
1.5
3
3.5
4
0.02
0.04
0.06
0.08
0.10
0.12
s_n
s_w

This is because even though globally the invading species is at a low density, locally the invader still 'sees' a high density of conspecifics thanks to the local nature of dispersal and competition (Bolker and Pacala 1999, Murrell *et al.* 2001, Murrell and Law 2003).

These results all point to the importance of the relative scales of dispersal and neighbourhood interactions to both the transient dynamics and final population density (Bolker and Pacala 1997, Ellner 2001, Law *et al.* 2003). For many animal species the interaction kernel is likely to be over a much shorter distance than the dispersal kernel. However for plants and other sedentary organisms this may not be the case, and it is entirely possible that the effects of density may be felt over relatively large spatial scales (Peters 2003). Ecologists have put much effort into measuring dispersal rates and distances (Clobert *et al.* 2000, Bullock *et al.* 2002, Levine and Murrell 2003), but rather less is known about the shapes and sizes of interaction kernels (Purves and Law 2002). Clearly models of this type show that the interaction kernels are potentially very important in determining invasion dynamics.

LINKS TO OTHER MODELS

As has been suggested at various points in this chapter, the models described and analysed here are closely related to a number of other spatial models. Young *et al.* (2001) showed that a population of discrete individuals inhabiting a landscape and with local dispersal and post-birth movement will quickly aggregate even without any density dependent processes. The models above have shown that incorporating local density-dependent processes can lead to any of the 3 main classes of spatial structure: aggregation; segregation or random patterning. Further the dynamical system (6) and (7) collapses to the model of (Young *et al.* 2001) when $d' = 0.0$, and the spatial pattern is now described by only three pairs of terms (7(a)-(c)), two of which are concerned with births and only one of which is concerned with the loss of pairs. It can be easily seen that in the first instance the birth terms will outweigh the death term and so aggregations of individuals will develop. However, it is not clear that the aggregations will ever stop, although the rate of change will slow over time.

The theta-logistic model (Gilpin and Ayayla 1973) has been used with some success to model the population dynamics of a number of species (Saether and Engen 2002, Saether *et al.* 2002a, Saether *et al.* 2002b). The main basis for the theta-logistic model is that the maximum rate of population growth may be achieved at densities other than half the carrying capacity (Turchin 2003). The spatial version of the logistic model also shows this feature (Law *et al.* 2003), and may therefore act as a mechanistic foundation for the theta-logistic models. When interactions are localised in space, how crowded the neighbourhood is determines when the maximum growth rate is achieved. When both dispersal and interactions occur over short-scales, the neighbourhood quickly becomes crowded, even if the overall density is low, and this leads to the maximum growth rate occurring at less than half the final density. On the other hand, if dispersal is long-range but interactions are short-scale, the individuals tend not to experience many neighbours until the overall density is much higher, so the maximum population growth rate tends to occur at densities that are higher than the half the final population density (Law *et al.* 2003).

This chapter has focused on a model for competition investigating the invasion dynamics of a single species. Many economically important invasions are by parasites/pathogens and theoretical models have been used to investigate the effects of local spatial structure on invasion dynamics of the pest/pathogen (Bolker 1999, Keeling 1999). In a susceptible-infected-recovered (SIR) host-pathogen model with local dispersal of the pathogen (Keeling 1999) shows how the reproductive ratio required for invasion in a network with local and finite connections is always larger than the mean-field case. Generally speaking the aggregation of hosts is good at slowing the rate of spread of a pest/pathogen if its dispersal is fairly short-range, and this has important implications for evolution of virulence (Boots and Sasaki 1999, Boots *et al.* 2004).

CONCLUSIONS

Invasions are essentially the study of population growth in space and time. Both dispersal and competitive interactions tend to be localised in space, and acting together they may produce strong spatial structures that feed back onto population dynamics and so greatly affect the invasion dynamics. This chapter has investigated the effects of local interactions and local dispersal on population growth using spatial extensions of the classical logistic equation. From a stochastic individual-based model a deterministic approximation has been derived which greatly aids the analysis of the model. A number of qualitative and quantitative patterns emerge: (1) population growth may be either faster or slower than the classical model; (2) the ratio of density independent births to density independent deaths (R) can greatly affect the probability and speed of invasion as well as the final population size; (3) population growth may be linear rather than the familiar sigmoid shape; (4) deterministic extinction may occur purely due to parent-offspring competition meaning that intraspecific competition cannot be ignored in the analysis of invasions; and that (5) counter-intuitively, populations with the lowest equilibrium density may take the longest to reach this state. All of these properties are dependent not only on the absolute spatial scales of neighbourhood interactions and local dispersal but also their relative scales. It has long been known that dispersal is important for the invasive ability of a species but these results also show that R may be just as important as the intrinsic growth rate, r, in determining population growth.

ACKNOWLEDGEMENTS

The work presented here owes a great deal to Richard Law and Ulf Dieckmann who solved many of the difficult initial problems to do with deriving the deterministic approximation and who helped and encouraged me to think long and hard about the problem of moment closures. I would also like to thank Ben Bolker, Calvin Dytham, Daniel Llambi, and Drew Purves for various helpful and informative discussions along the way, and to Marc Cadotte, Rene Salinas, Jessica Metcalf and Yvonne Buckley for comments on earlier drafts of this chapter. This work was partly supported by the Natural Environment Research Council and also the Central Science Laboratory (DEFRA). Imperial College Parallel Computing Centre (ICPC) kindly donated computer processor time that helped to generate some of the results reported in the figures.

APPENDIX

The presence of the third moment in the dynamics for the second moments means that the dynamical system is not yet closed. In principle it is possible to derive the dynamics of the third moments, but apart from causing a headache due to the number of terms involved they would themselves be functions of fourth moments (quad densities). How to deal with this moment hierarchy is the central problem to this technique. Classical theory effectively truncates at the first moment, by assuming that there is no spatial structure, but this only deals with a limiting case of the IBM (see below). A better solution is to close the hierarchy at the second moments by replacing the third moments by a term that is a function of the first and second moments. Recent work has suggested a form of second order moment closure that gives good approximations to the IBM for a wide range of spatial structures (Law *et al.* 2003, Murrell and Law 2003, Murrell *et al.* 2004) and it is used here:

$$T(\xi, \xi') = \frac{1}{5}\left[\frac{4\,C(\xi)\,C(\xi')}{N} + \frac{C(\xi)\,C(\xi'-\xi)}{N} + \frac{C(\xi')\,C(\xi'-\xi)}{N} - N^3 \right] \qquad \text{(A1)}$$

This closure does however mean that analytical solutions to the dynamical system are very hard to achieve, and so the analysis of the dynamical system proceeds by numerical methods.

REFERENCES

Andresen, E. 1999. Seed dispersal by monkeys and the fate of dispersed seeds in a Peruvian rain forest. Biotropica **31**, 145-158.

Boerlijst, M. C. and P. Hogeweg. 1991. Spiral Wave Structure in Pre-Biotic Evolution – Hypercycles Stable against Parasites. Physica D **48**, 17-28.

Bohning-Gaese, K., B. H. Gaese and S. B. Rabemanantsoa. 1999. Importance of primary and secondary seed dispersal in the Malagasy tree *Commiphora guillaumini*. Ecology **80**, 821-832.

Bolker, B. and S. W. Pacala. 1997. Using moment equations to understand stochastically driven spatial pattern formation in ecological systems. Theoretical Population Biology **52**, 179-197.

Bolker, B. M. 1999. Analytic models for the patchy spread of plant disease. Bulletin of Mathematical Biology **61**, 849-874.

Bolker, B. M. 2003. Combining endogenous and exogenous spatial variability in analytical population models. Theoretical Population Biology **64**, 255-270.

Bolker, B. M. and S. W. Pacala. 1999. Spatial moment equations for plant competition: Understanding spatial strategies and the advantages of short dispersal. American Naturalist **153**, 575-602.

Bolker, B. M., S. W. Pacala and C. Neuhauser. 2003. Spatial dynamics in model plant communities: What do we really know? American Naturalist **162**, 135-148.

Boots, M., P. J. Hudson and A. Sasaki. 2004. Large shifts in pathogen virulence relate to host population structure. Science **303**, 842-844.

Boots, M. and A. Sasaki. 1999. 'Small worlds' and the evolution of virulence: infection occurs locally and at a distance. Proceedings of the Royal Society of London Series B-Biological Sciences **266**, 1933-1938.

Brown, J. H. and Lomolino, M. V. 1998. Biogeography. Sinauer Associates, Sunderland, MA.

Bullock, J. M., R. E. Kenward and R. S. Hails, editors. 2002. Dispersal ecology. Blackwell, Oxford.

Caswell, H., R. Lensink and M. G. Neubert. 2003. Demography and dispersal: Life table response experiments for invasion speed. Ecology **84**, 1968-1978.

Clobert, J., E. Danchin, A. A. Dhondt and J. D. Nichols, editors. 2000. Dispersal. Oxford University Press, Oxford.

Cody, M. L. and J. M. Overton. 1996. Short-term evolution of reduced dispersal in island plant populations. Journal of Ecology **84**, 53-61.

Cressie, N. A. C. 1991. Statistics for spatial data. Wiley, New York.

Darlington, P. J., Jr. 1943. Carabidae of mountains and islands: data on the evolution of isolated faunas, and on atrophy of wings. Ecological Monographs **13**, 37-61.

Dieckmann, U. and R. Law. 2000. Relaxation projections and the method of moments. Pages 412-455 *in* U. Dieckmann, R. Law and J. A. J. Metz, editors. The Geometry of ecological interactions: simplifying spatial complexity. Cambridge University Press, Cambridge.

Diggle, P. J. 1983. Statistical analysis of spatial point patterns. Academic Press, London.

Ellner, S. P. 2001. Pair approximation for lattice models with multiple interaction scales. Journal of Theoretical Biology **210**, 435-447.

Etheridge, A. M. 2004. Survival and extinction in a locally regulated population. Annals of Applied Probability **14**, 188-214.

Fagan, W. F. and J. G. Bishop. 2000. Trophic interactions during primary succession: Herbivores slow a plant reinvasion at Mount St. Helens. American Naturalist **155**, 238-251.

Finlay, B. J. 2002. Global dispersal of free-living microbial eukaryote species. Science **296**, 1061-1063.

Finlay, B. J. and K. J. Clarke. 1999. Ubiquitous dispersal of microbial species. Nature **400**, 828-828.

Finlay, B. J., G. F. Esteban, J. L. Olmo and P. A. Tyler. 1999. Global distribution of free-living microbial species. Ecography **22**, 138-144.

Filin, I. and Ziv, Y. 2004. New theory of insular evolution: unifying the loss of disersability and body-mass change. Evolutionary Ecology Research **6**, 115-124.

Gilpin, M. E. and F. J. Ayayla. 1973. Global models of growth and competition. Proceedings of the National Academy of Sciences of the United States of America **70**, 3590-3593.

Gordon, E. and A. G. van der Valk. 2003. Secondary seed dispersal in *Montrichardia arborescens* (L.) schott dominated wetlands in laguna grande, Venezuela. Plant Ecology **168**, 177-190.

Guitian, P., M. Medrano and J. Guitian. 2003. Seed dispersal in *Erythronium dens-canis* L. (Liliaceae): variation among habitats in a myrmecochorous plant. Plant Ecology **169**, 171-177.

Hassell, M. P., H. N. Comins and R. M. May. 1994. Species Coexistence and Self-Organizing Spatial Dynamics. Nature **370**, 290-292.

Hill, J. K., Y. C. Collingham, C. D. Thomas, D. S. Blakeley, R. Fox, D. Moss and B. Huntley. 2001. Impacts of landscape structure on butterfly range expansion. Ecology Letters **4**, 313-321.

Holway, D. A., A. V. Suarez and T. J. Case. 2002. Role of abiotic factors in governing susceptibility to invasion: A test with argentine ants. Ecology **83**, 1610-1619.

Hoshizaki, K., W. Suzuki and T. Nakashizuka. 1999. Evaluation of secondary dispersal in a large-seeded tree Aesculus turbinata: a test of directed dispersal. Plant Ecology **144**, 167-176.

Keeling, M. J. 1999. The effects of local spatial structure on epidemiological invasions. Proceedings of the Royal Society of London Series B-Biological Sciences **266**, 859-867.

Keeling, M. J., D. A. Rand and A. J. Morris. 1997. Correlation models for childhood epidemics. Proceedings of the Royal Society of London Series B-Biological Sciences **264**, 1149-1156.

Kenkel, N. C. 1988. Pattern of self-thinning in jack pine: testing the random mortality hypothesis. Ecology **69**, 1017-1024.

Kot, M., M. A. Lewis and P. vandenDriessche. 1996. Dispersal data and the spread of invading organisms. Ecology **77**, 2027-2042.

Law, R. and U. Dieckmann. 2000. A dynamical system for neighborhoods in plant communities. Ecology **81**, 2137-2148.

Law, R., D. J. Murrell and U. Dieckmann. 2003. Population growth in space and time: Spatial logistic equations. Ecology **84**, 252-262.

Levin, S. A. 1992. The Problem of Pattern and Scale in Ecology. Ecology **73**, 1943-1967.

Levine, J. M. and D. J. Murrell. 2003. Community-level consequences of seed dispersal patterns. Annual Review of Ecology, Evolution and Systematics **34**, 549-574.

Lewis, M. A. 2000. Spread rate for a nonlinear stochastic invasion. Journal of Mathematical Biology **41**, 430-454.

Lewis, M. A. and S. Pacala. 2000. Modeling and analysis of stochastic invasion processes. Journal of Mathematical Biology **41**, 387-429.

Milton, S. J. and W. R. J. Dean. 2001. Seeds dispersed in dung of insectivores and herbivores in semi-arid southern Africa. Journal of Arid Environments **47**, 465-483.

Morton, R. D. and R. Law. 1997. Regional species pools and the assembly of local ecological communities. Journal of Theoretical Biology **187**, 321-331.

Morton, R. D., R. Law, S. L. Pimm and J. A. Drake. 1996. On models for assembling ecological communities. Oikos **75**, 493-499.

Murrell, D., U. Dieckmann and R. Law. 2004. On moment closures for population dynamics in continuous space. Journal of Theoretical Biology **229**, 421-432.

Murrell, D. J. and R. Law. 2000. Beetles in fragmented woodlands: a formal framework for dynamics of movement in ecological landscapes. Journal of Animal Ecology **69**, 471-483.

Murrell, D. J. and R. Law. 2003. Heteromyopia and the spatial coexistence of similar competitors. Ecology Letters **6**, 48-59.

Murrell, D. J., D. W. Purves and R. Law. 2001. Uniting pattern and process in plant ecology. Trends in Ecology & Evolution **16**, 529-530.

Owen, M. R. and M. A. Lewis. 2001. How predation can slow, stop or reverse a prey invasion. Bulletin of Mathematical Biology **63**, 655-684.

Palmiotto, P. A., Davies, S. J., Vogt, K. A., Ashton, M. S., Vogt, D. J. and Ashton, P. S. 2004. Soil-related habitat specialization in dipterocarp ran forest tree species in Borneo. Journal of Ecology **92**, 609-623.

Pearl, R. and L. J. Read. 1920. On the rate of growth of the population of the United States since 1790, and its mathematical representation. Proceedings of the National Academy of Sciences of the United States of America **6**, 275-288.

Peters, H. A. 2003. Neighbour-regulated mortality: the influence of positive and negative density dependence on tree populations in species-rich tropical forests. Ecology Letters **6**, 757-765.

Purves, D. W. and R. Law. 2002. Experimental derivation of functions relating growth of *Arabidopsis thaliana* to neighbour size and distance. Journal of Ecology **90**, 882-894.

Richardson, D. M., N. Allsopp, C. M. D'Antonio, S. J. Milton and M. Rejmanek. 2000. Plant invasions – the role of mutualisms. Biological Reviews **75**, 65-93.

Rietkerk, M., M. C. Boerlijst, F. van Langevelde, R. HilleRisLambers, J. van de Koppel, L. Kumar, H. H. T. Prins and A. M. de Roos. 2002. Self-organization of vegetation in arid ecosystems. American Naturalist **160**, 524-530.

Saether, B. E. and S. Engen. 2002. Pattern of variation in avian population growth rates. Philosophical Transactions of the Royal Society of London Series B-Biological Sciences **357**, 1185-1195.

Saether, B. E., S. Engen, F. Filli, R. Aanes, W. Schroder and R. Andersen. 2002a. Stochastic population dynamics of an introduced Swiss population of the ibex. Ecology **83**, 3457-3465.

Saether, B. E., S. Engen, R. Lande, C. Both and M. E. Visser. 2002b. Density dependence and stochastic variation in a newly established population of a small songbird. Oikos **99**, 331-337.

Skellam, J. G. 1951. Random dispersal in theoretical populations. Biometrika **38**, 196-218.

Travis, J. M. J. and C. Dytham. 2002. Dispersal evolution during invasions. Evolutionary Ecology Research **4**, 1119-1129.

Turchin, P. 2003. Complex population dynamics. Princeton University Press, Princeton.

Veit, R. R. and M. A. Lewis. 1996. Dispersal, population growth, and the Allee effect: Dynamics of the house finch invasion of eastern North America. American Naturalist **148**, 255-274.

With, K. A. 2002. The landscape ecology of invasive spread. Conservation Biology **16**, 1192-1203.

Whittaker, R. J. 1998. Island Biogeography: Ecology, Evolution and Conservation. Oxford University Press, Oxford.

Young, W. R., A. J. Roberts and G. Stuhne. 2001. Reproductive pair correlations and the clustering of organisms. Nature 412, 328-331.

Chapter eight

A guide to calculating discrete-time invasion rates from data

M. A. Lewis, M. G. Neubert, H. Caswell, J. S. Clark and K. Shea

INTRODUCTION

One measure of biological invasiveness is the rate at which an established invader will spread spatially in its new environment. Slow spread signifies slow increase in ecological impact, whereas fast spread signifies the converse. If one can predict spread rates from life history attributes, such as growth rates and dispersal distances, then potential invasiveness can be assessed before the invasion occurs.

* This work was conducted as part of the Demography and Dispersal Working Group supported by the National Center for Ecological Analysis and Synthesis, a Center funded by NSF (DEB-0072909), the University of California, and the Santa Barbara campus. This research was also supported in part by a Canada Research Chair, the Natural Sciences and Engineering Research Council of Canada, and the National Science Foundation (DEB-0315860, DEB-9527400, DEB-9808501, DMS-9973312, DMS-0213698, DEB-0315860). Thanks to the Demography and Dispersal Working Group members for their help and support with the research.

M.W. Cadotte et al. (eds.), Conceptual ecology and invasion biology, 169–192.

A prediction of this sort requires models for population spread. As outlined below, such models have a long and distinguished history in quantitative ecology.

Whereas early mathematical models for population spread were primarily conceptual and qualitative in nature, a new generation of realistic models is emerging. These new models are tied directly to the demography and dispersal of individuals. However, there are new challenges in the linking of these models to the biological processes.

As we will illustrate in this chapter, spread rate predictions are very sensitive to assumptions about long-distance dispersal. Are there robust methods for estimating spread rates? This is one question we will address.

Furthermore almost all mathematical models assume that the spread occurs in one spatial dimension, along a line. This is not because mathematicians have not noticed that most population spread takes place in two dimensions (except cases like dispersal along a coastline or river). Rather, it is because the main qualitative features of invasions are apparent in one-dimensional models, although, as we will show, the quantitative results in specific cases can depend very much on the the dimensionality of the dispersal data and of the model used for analysis. In particular, we will show that the naive application of one dimensional models to two dimensional dispersal data will produce a systematic bias (sometimes positive, sometimes negative) in spread rate estimates. In this chapter we will outline new fitting methods for avoiding these biases.

MODELING BACKGROUND

The issue of spread rate for biological invaders was considered in detail by Skellam (1951) where Fisher's 1937 partial differential equation model was used to describe the rate of change of local population density with time

$$\frac{\partial n}{\partial t} = \rho n \left(1 - \frac{n}{\kappa}\right) + D \frac{\partial^2 n}{\partial x^2}. \tag{1}$$

Here $n(x, t)$ is the local population density, ρ is the intrinsic growth rate, κ is the carrying capacity, D is the discusion coeficient, x is the one-dimensional space coordinate and t is time. The rate of spread of a population obeying this equation asymptotically approaches $c = 2\sqrt{\rho D}$ for large times (Kolmogorov *et al.* 1937; Aronson and Weinberger 1975) (Fig. 1). Whereas the spread rate predictions are on a landscape scale, the parameters r and D can be measured on the individual level, using life table analysis, and mark-recapture. Furthermore, comparisons of historically observed invasive spread rates with the asymptotic spread rate formula have held up for a wide variety of species (Andow *et al.* 1990), making the interplay between invasion theory and spread data a modern-day success story in quantitative ecology (Shigesada and Kawasaki 1997).

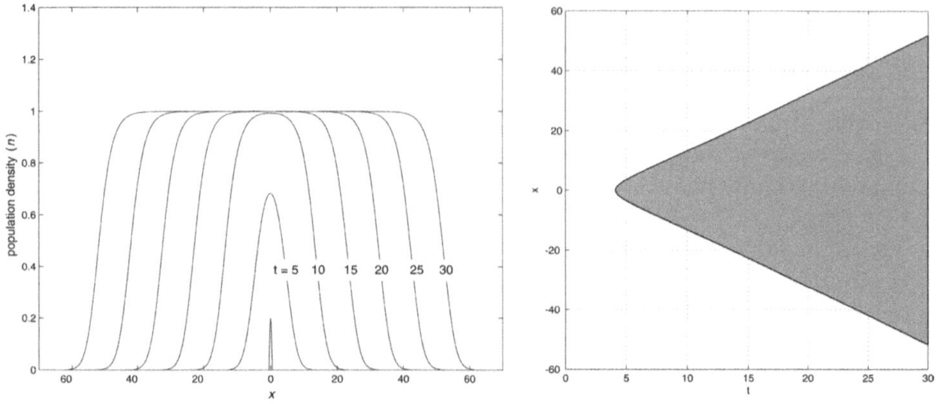

Fig. 1 Population spread for Fisher's equation. A typical solution of Fisher's equation (1) illustrates growth and spread in one-dimensional linear space. At left, the solution is plotted for equally-spaced time intervals. At right, the gray area denotes the region in space where the population is larger than a threshold level $n = 0.5$. The boundaries of this area have slopes equal to the asymptotic spread rate $c = 2\sqrt{\rho D}$. For this figure, $\rho = \kappa = D = 1$, and $n(x, 0) = 0.2 \cos(\pi x)$ for $|x| \leq 0.5$. Based on Neubert and Parker (2004).

A major problem with Fisher's 1937 model is that it imposes one particular form of dispersal. The assumption of diffusion in equation (1), which implies normally distributed dispersal propagules, is often violated when dispersal is measured for biological populations (Lewis 1997). While there is tremendous variability in such dispersal data, there is a strong tendency for the distribution of dispersal distances to be leptokurtic, with a larger number of distances near the center and in the tails than in a normal distribution with comparable variance. The effect of the long-distance dispersers, as described by the tails of the distribution of distances, is dramatic (Caswell *et al.* 2003). Predictions for invasion rates can speed up by an order of magnitude or more when the long-distance dispersers are included (Kot, Lewis, and van den Driessche 1996).

When the dispersal distribution is far from normal, the partial differential equation model (1) no longer succes. It is then necessary to define a dispersal kernel $k(x, y)$, in one spatial dimension, which describes the probability that a propagule that starts at y moves to the interval $(x, x + dx)$ by $k(x, y)\, dx$. The units for the dispersal kernel k are per unit length. The spatial region of interest is the interval $\Omega = (-\ell, \ell)$. When modelling invasions, we typically assume ℓ is arbitrarily large. In the absence of immigration from outside Ω, every disperser must originate at some other point in, so that

$$\int_{-\ell}^{\ell} k(x, y)\, dy = 1. \tag{2}$$

In a homogeneous habitat, dispersal between two locations will only depend upon upon the relative locations of the start and finish points. In this case, to which we will restrict our attention for the remainder of the chapter, we write $k(x, y) = k(x - y)$. (Shigesada, Kawasaki, and Teramoto (1986) examine invasions in heterogeneous environments.) To include population dynamics, population growth from one generation to the next can be described with a non-linear function

$$n_{t+1} = f(n_t) = n_t g(n_t), \tag{3}$$

where g defines the per capita growth rate as a function of local population density n_t. Non-overlapping generations are assumed, but, here and in the subsequent analysis, the assumption can be relaxed to include models with stage-structure (see Neubert and Caswell (2000) for details).

The nonspatial population model (3) is then modified to allow for dispersal between reproduction events. If we designate the population density at location x and time t by $n_t(x)$, these models take the form of a scalar *integrodifference equation*

$$n_t(x) = \int_{-\ell}^{\ell} f(n_t(y))\, k(x - y)\, dy\,. \tag{4}$$

Early analyses of population spread using integrodifference equations appeared in the mathematical literature primarily in the context of genetics (Weinberger 1982; Lui 1982a, b; Weinberger 1984; Lui 1985, 1986, 1989a, 1989 b). However, integrodifference equations, and generalizations of them, are now being used by a growing list of ecologists to investigate the spread rate of ecological populations (Kot 1992; Allen *et al.* 1996; Allen *et al.* 1996; Kot *et al.* 1996; Veit and Lewis 1996; Lewis 1997; Hart and Gardner 1997; Clark *et al.* 1998; Clark 1998; Higgins and Richardson 1999; Neubert and Caswell 2000; Neubert *et al.* 2000; Takasu *et al.* 2000; Woolcock and Cousens 2000; With 2002; Schofield 2002; Caswell *et al.* 2003; Clark *et al.* 2003; Marchant 2003; Neubert and Parker 2004; Powell and Zimmermann 2004). Among the results of these investigations are two key findings. First, integrodifference models produce a richer set of invasion dynamics than can be generated by the reaction-diffusion equation (1), including, for example, the possibility of accelerating spread (Kot *et al.* 1996). Second, the shape of the dispersal kernel — especially the shape of the tails which determine the probability of long-distance dispersal — plays a crucial role in determining the rate of spread.

The effect of long-distance dispersal on spread rates was highlighted by Kot *et al.* (1996), who fit dispersal kernels to a data set describing the displacement of of genetically marked *Drosophila* (Dobzhansky and Wright 1943), and predictions for the corresponding asymptotic spread rate were linked to the shapes of the kernels. These predicted spread rates varied over an order of magnitude, depending upon the fatness of the tails of the related dispersal kernels (see Fig. 2).

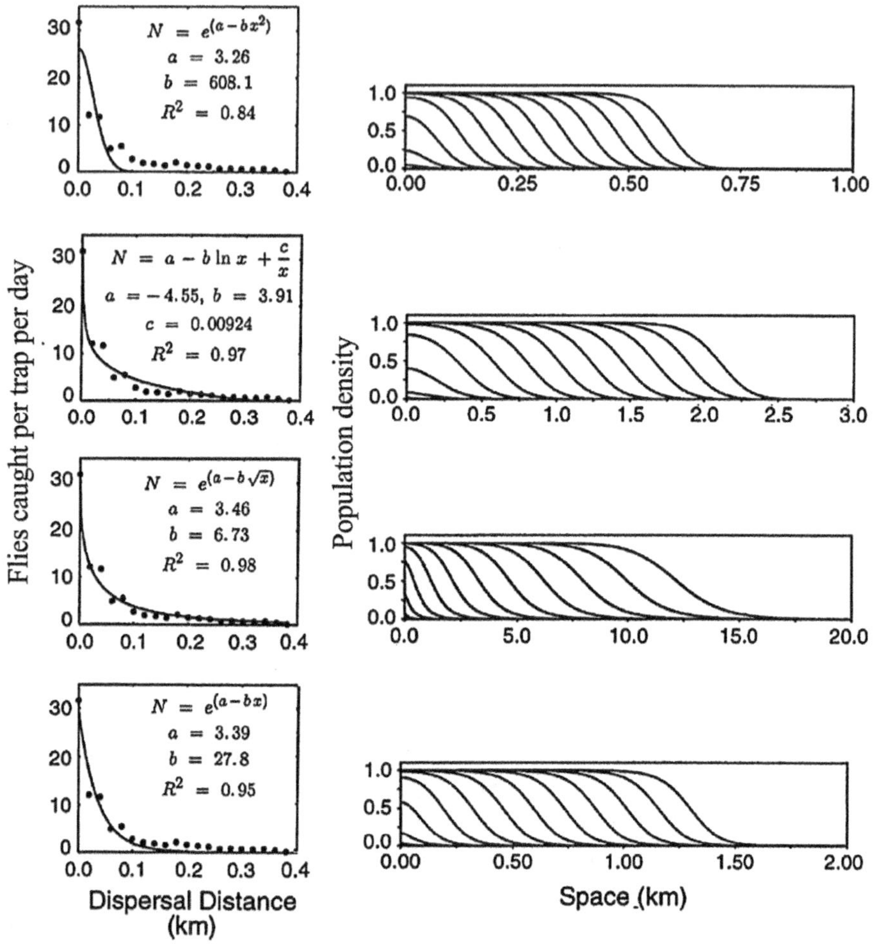

Fig. 2 Fitted functions to *D. pseudoobscura* dispersal data provide ingredients for an integrodifference model for insect spread 4. The left panels show average number of insects caught per trap per day in Dobzhansky and Wright's experiments. It was assumed that dispersal was equally likely in both directions, so the dispersal kernels were $k(x) = (g(x) + g(-x))/2$, where g is the fitted function. The right panel shows simulations of the integrodifference equations. Simulations assume Beverton-Holt population dynamics for $f(n)$, with a geometric growth rate of $\lambda = 10$. The carrying capacity was scaled to equal one. Each integrodifference was iterated for 12 generations. Based on Kot *et al.* (1996). Here is is assumed that all reproduction and dispersal occurs along a one-dimensional strip of suitable habitat. The spread rate can be calculated by dividing the total distance moved by the population front by the 12 generations taken to move the distance.

The issue of long-distance dispersal, however, goes beyond the choice of a parametric dispersal distribution to describe a set of data. It can also reflect real biological processes. For example, Neubert and Caswell (2000) computed the spread rate for the herbaceous plant *Dipsacus sylvestris* based on data from a seed trap experiment (Werner 1975). Teasel seeds are known to float, but dispersal by streams or rivers was obviously not measured by the seed trap experiment. Neubert and Caswell calculated the asymptotic spread rate resulting from hypothetical mixtures of the seed trap data and dispersal by water with a longer mean distance. They found that long-distance dispersal of even one seed in a million was enough to make the spread rate dependent on the water dispersal alone. Similarly, the seeds of the tropical plant *Calathea ovandensis* are dispersed by at least four species of ant, each with its own typical dispersal distance (Horvitz and Schemske 1986). Neubert and Caswell found that over 99% of the asymptotic spread rate was accounted for by the ant species with the longest dispersal distance, even though it dispersed only 7% of the seeds.

Classical models for population spread like (1) and (4) consider the case where the spatial domain is one-dimensional and linear, and a small beachhead of individuals is introduced locally. This is only directly applicable to cases such as population spread along a roadside, coastline (Lubina and Levin 1988) or a river (Speirs and Gurney 2001; Pachepsky *et al.* 2004).

Later, we will introduce models that describe dispersal in two spatial dimensions. These models produce asymptotic spread rate predictions that can differ in different directions if the dispersal kernel is not radially symmetric. In both one and two dimensions, the dispersal kernel plays a crucial role in determining the asymptotic rate of spread. There are many methods for estimating dispersal kernels, and for estimating the properties of those kernels that enter into the formulae for spread rate, from data (see, for example, Silverman (1986)). We will discuss some of these methods later in the chapter. First, we discuss some of the kinds of dispersal data that are typically collected and/or published.

DISPERSAL KERNELS IN ONE AND TWO DIMENSIONS

We first consider forms in which data are collected. These fall loosely into two kinds: dispersal data and density data. Dispersal data describe the location of dispersers relative to the parents. These data come from following individual dispersers (e.g., banding and recapture of birds, mark and recapture of seeds using coloring and/or radio-tagging) and is recorded as either displacements or displacement distances. In contrast, density data describe the density of dispersers (number per unit area) observed at a given point, typically as a function of distance from a natal site or source of dispersers. These data come from seed traps, pheremone traps for insects and so forth.

If we assume that the population lives along a one-dimensional strip of suitable habitat, along which all dispersal and reproduction occurs, the one-dimensional dispersal kernel (number of dispersers per unit length) is needed in equation (4). This kernel can be fitted either directly from the one-dimensional dispersal data or from the density data (number of dispersers per unit area) multiplied by the width of the strip. The constraint that the kernel must be scaled to integrate to 1 (equation (2)), means that the kernel k will actually be independent of the width of the strip — multiplication is only done formally to ensure the correct units for k.

If we assume that the population lives in a two dimensional habitat, it is first necessary to extend the definition of dispersal to two spatial dimensions. Here dispersal is between points $\mathbf{x} = [x_1, x_2]^T$ and $\mathbf{y} = [y_1, y_2]^T$ in two dimensional space. The the two-dimensional dispersal kernel $K(\mathbf{x}, \mathbf{y})$ describes the probability of a propagule which starts at $\mathbf{y}$ moving to the rectangle with corners $\mathbf{x}$ and $\mathbf{x} + \mathbf{dx}$ by $K(\mathbf{x}, \mathbf{y})\, dx_1\, dx_2$.[1] The spatial region of interest is given by Ω. In the case of invasions, this is typically assumed to be arbitrarily large. In the absence of immigration from outside Ω, every disperser must originate at some other point in Ω, so that

$$\int_{\Omega} K(\mathbf{x}, \mathbf{y})\, \mathbf{dy} = 1\,. \tag{5}$$

If the kernel $K(\mathbf{x}, \mathbf{y})$ depends only upon the relative locations of the start and finish points we write $K(\mathbf{x}, \mathbf{y}) = K(\mathbf{x} - \mathbf{y})$.

To understand the difference between one and two dimensional dispersal kernels we consider the case where the dispersal is isotropic (identical in all directions). In this case the two dimensional dispersal kernel K can be written as a function of the dispersal radius $r = |\mathbf{x} - \mathbf{y}|$. In a linear one-dimensional environment, the scaled distribution of densities (dispersers per unit length) and a distribution of distances that the dispersers travel from the parent are the same and are given by the kernel k. In a two-dimensional environment, the scaled distribution of densities (dispersers per unit area), given by $K(r)$, and the distribution of distances that dispersers travel from from the parent (dispersers per unit length), given by $\tilde{K} = 2\pi r K(r)$, are not the same, because there is more area available at distances further from the parent.

POPULATION SPREAD IN A ONE-DIMENSIONAL LINEAR ENVIRONMENT

The Dobzhansky and Wright insect data were collected from traps placed along linear transects radiating from a point source. Thus the traps give a relative measure of the density of dispersers (number per unit area), as a function of distance

---

[1] Throughout this chapter, we will use boldface Roman characters such as **u** and **v** to represent vectors. Here, $|\mathrm{u}| = \sqrt{u_1^2 + u_2^2}$ denotes the length of the vector $\mathbf{u}$, and $\mathbf{u} \cdot \mathbf{v} = u_1 v_1 + u_2 v_2 = |\mathrm{u}|\,|\mathrm{v}| \cos\theta$ denotes the 'dot product' or projection of one vector onto the other, where θ is the angle between $\mathbf{u}$ and $\mathbf{v}$.

from the release site. The most reasonable assumption is that the transect data describe the radial drop-off in settled insect density in a two-dimensional habitat (next section). However, for the sake of illustration, in this paper we first consider the assumption that the insects and transects are found along a one-dimensional strip of suitable habitat, along which all dispersal and reproduction occurs (next section). This is a standard assumption in population spread models and is the assumption made in the original analysis of the insect spread rates in Kot *et al.* (1996). As we will show, the different assumptions about the habitat and dispersal give rise to quite different spread rate estimates.

Theory

As the introduced beachhead of individuals grows and disperses we expect growth of the range boundary with time. Figure 1 shows a typical progression. A plot of range boundary versus time gives lines whose slope eventually become constant. The eventual slope of these lines — the asymptotic rate of spread of the invasion (henceforth referred to simply as spread rate) — can be predicted using mathematical theory which relates the slope to model parameters. In this section we outline the theory.

It is first necessary to make some assumptions about the growth dynamics. The simplest population dynamics exhibit no overcompensation or Allee effect. This assumption translates into a growth function f in (4) that is monotonically increasing with maximum per capita growth rates at lowest population levels.

The assumption that the maximum per capita growth rate λ occurs at the lowest possible density means $\lambda = g(0) \geq g(n)$ for $n > 0$. A growing population requires $\lambda > 1$. As described above, local introduction of individuals, coupled with a growth rate $\lambda > 1$ and a dispersal kernel $k(z)$, $z = x - y$, in (4), means the population spreads as it grows and disperses (Fig. 2). Weinberger (1982) showed that, under the above assumptions on growth dynamics, the population spreads to the right at a rate which approaches speed c as the time since introduction increases, where

$$c = \min_{s>0} \frac{1}{s} \ln [\lambda M(s)], \tag{6}$$

and $M(s)$ is the moment generating function for the dispersal kernel $k(z)$

$$M(s) = \int_{-\infty}^{\infty} k(z) \exp(sz)\, dz. \tag{7}$$

The parameter s can be understood as a measure of the steepness of the wave: $n \propto \exp(-sz)$ at the leading edge of the rightward-spreading population. To find the speed of the leftward spreading front, one should use the moment generating function for $k(-z)$ in equation (7).

Thus the population spread rate depends only upon two features: the geometric growth rate of the population λ, and the shape of the dispersal kernel k. The speed for Fisher's equation (1), $c = 2\sqrt{\rho D}$, can be regained from equations (6)-(7) by the choice of $k = N(0; 2D)$ and $\rho = \log(\lambda)$. Details are given in Kot *et al.* (1996).

In practice, calculation of the spread rate must be done numerically, either by using a standard minimization routine, or by solving the double root condition equations, $\exp(sc) = \lambda M(s)$ and $c \exp(sc) = \lambda M'(s)$, for the wave speed c and wave steepness s. For a simple example, written in Maple code see the Appendix.

Here it is assumed that the kernel k is exponentially bounded so that the moment generating function (7) can be calculated. When the kernel k has tails that are fatter than exponential, there is no asymptotic rate of spread — the spread rate continues to increase with increasing time (Kot *et al.* 1996) (Fig. 2, third panel from the top). In this situation, an alternative definition of spread rate, based on the change in location of the furthest forward individual in the population from generation to generation (furthest forward velocity) is appropriate (Clark, Lewis, and Horvath 2001) (see Discussion).

When the growth rate λ is known, but the dispersal kernel is unknown, estimates for population spread rates using equation (6) can vary widely, depending upon the parametric form of the kernel chosen, as with the simulations shown in Fig. 2.

When appropriate data are available, this problem can be addressed by means of a nonparametric estimator for the moment generating function (7) which makes no assumption about the form of the underlying kernel. In this case the moment generating function is estimated from raw one-dimensional linear dispersal displacement data, $z_1 \ldots z_N$.

$$M^E(s) = \frac{1}{N} \sum_{i=1}^{N} \exp(sz_i). \tag{8}$$

The superscript is used to indicate an empirical estimate of the moment generating function (Clark, Horvath, and Lewis 2001). Here the dispersal measurements arise from tracking a series of individuals. As a displacement, x_i gives the distance and direction that the *i-th* individual moves. By convention, leftward movements are assigned negative values. It is assumed that the tracking effort and tracking efficiency remain constant over the entire linear one-dimensional dispersal region (Fujiwara *et al.* (2004) consider the effects of changes in sampling effort or detection probability.) In the case where $z_1 \ldots z_N$ are (nonnegative) distances, rather than displacements, the assumption of a symmetric dispersal kernel (whereby individuals are as likely to move to the left as to the right) leads to

$$M^E(s) = \frac{1}{N} \sum_{i=1}^{N} \cosh(sz_i). \tag{9}$$

Substitution of $M^E(s)$ instead of $M(s)$ in (6) leads to an empirically estimated wave speed c^E. This empirically estimated wave speed has many nice properties (Clark *et al.* 2001). For example it is unbiased: c^E converges to to the true population spread rate c as $N \rightarrow \infty$. This is not generally the case when parametric kernels are fitted to the dispersal data. In that case, the true population dispersal kernel is not known, and different fitted kernels can give very different wave speed predictions.

When the number of data points N is finite, each empirically estimated wave speed will be different, as it will depend on the precise data set used. However, the distribution of the empirical wave speed c^E about the true wave speed c is approximately Gaussian and the variance of the Gaussian approaches zero as N approaches infinity (Clark *et al.* 2001). Although there is no closed form expression for the variance, it can be estimated using bootstrapping methods (Clark *et al.* 2001). When the data collection or recording methods do not provide dispersal data, but histogram (density) data are available, it may be necessary to use such a histogram as an estimator for the kernel

$$k^H(z) = \begin{cases} f_i & \text{if} \quad \xi_{i-1} \le z < \xi_i \\ 0 & \text{otherwise} \end{cases} \tag{10}$$

where f_i is the bin height for the histogram, $1 \le i \le L$, and

$$\sum_{i=1}^{L} (\xi_i - \xi_{i-1}) f_i = 1 \tag{11}$$

This yields

$$M^H(s) = \frac{1}{s} \sum_{i=1}^{L} f_i [\exp(s\xi_i) - \exp(s\xi_{i-1})], \tag{12}$$

and substitution of $M^H(s)$ instead of $M(s)$ in (6) leads to a 'histogram' estimator for the wave speed c^H. Due to the arbitrary nature of location of the histogram bins, it can be shown that the histogram estimator does not provide an unbiased estimator for the true speed c. However, in the absence of other data, this histogram estimator is a useful alternative to the empirical estimator given above and, in practice, gives very similar results. As the sizes of the bins $\xi_i - \xi_{i-1}$ approaches zero the two estimators are identical. In the case where the histogram data is for distances as opposed to displacements, the assumption of a symmetric redistribution kernel causes (12) to be modified to

$$M^H(s) = \frac{1}{s} \sum_{i=1}^{L} f_i [\sinh(s\xi_i) - \sinh(s\xi_{i-1})]. \tag{13}$$

We now apply the histogram spread rate estimate (13) and (6) to the Dobzhansky and Wright (1943) data, under the assumption that the insects and transects are found along a one-dimensional strip of suitable habitat, along which all dispersal and reproduction occurs (see discussion in Modeling Background). This estimate, which uses the data shown in Fig. 2, gives a spread rate of 0.258 km per year. This is higher than the spread rate prediction made by using the exponential and Gaussian kernels in Fig. 2, but is substantially lower than the prediction made by the fat tailed kernel. As with the empirical estimator, bootstrapping gives the distribution of wave speeds, from which confidence intervals can be calculated (Fig. 3).

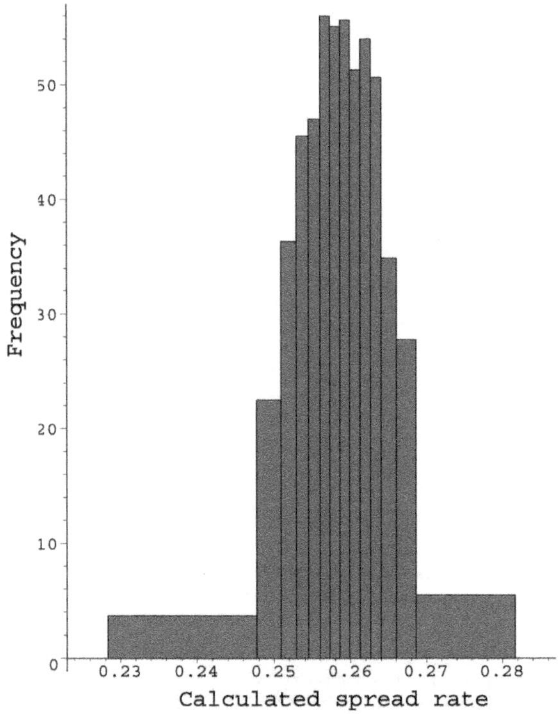

Fig. 3 Bootstrapping gives a range of possible values for the histogram wave speed. The histogram wave speed estimator (equations (13) and (6)) is applied to a histogram based on the total number of insects caught per trap, starting at distance zero, finishing at distance 0.38 km, with inter-trap spacing of 0.02 km were: 317, 121, 117, 50, 55, 27, 19, 18, 14, 21, 15, 13, 11, 7, 7, 6, 6, 7, 4, and 2. Bootstrapping was done by re-sampling from the 837 dispersal distances, with replacement, to produce 5000 new data sets. The wave speed was calculated for each of these new data sets, and the distribution of speeds is shown here. Ninety per cent of the speeds fell in the range (0.246 km/year, 0.270 km/year). Compare the estimate here with the spread rates simulated in Fig. 2.

POPULATION SPREAD IN TWO DIMENSIONS

Invasion in two spatial dimensions involves local introduction at a beachhead, followed by growth and spread in space. Here the process can be divided into three stages: the initial establishment, the early radial expansion until well established in space, and the later spread of the established population (Fig. 4). Although each of these three stages is of interest biologically, the focus of this chapter is on analyzing the later spread of an established population. At this stage we can approximate the invading front by a planar front, moving with a well-defined speed.

The calculation of spread in two dimensions requires a unit vector $\mathbf{u} = [u_1; u_2]^T$ describing the direction, perpendicular to the wave front, in which the spread is being considered (Fig. 4). The asymptotic spread rate in the direction $\mathbf{u}$ is given by

$$c_u = \min_{s>0} \frac{1}{s} \ln [\lambda\, M_u(s)], \tag{14}$$

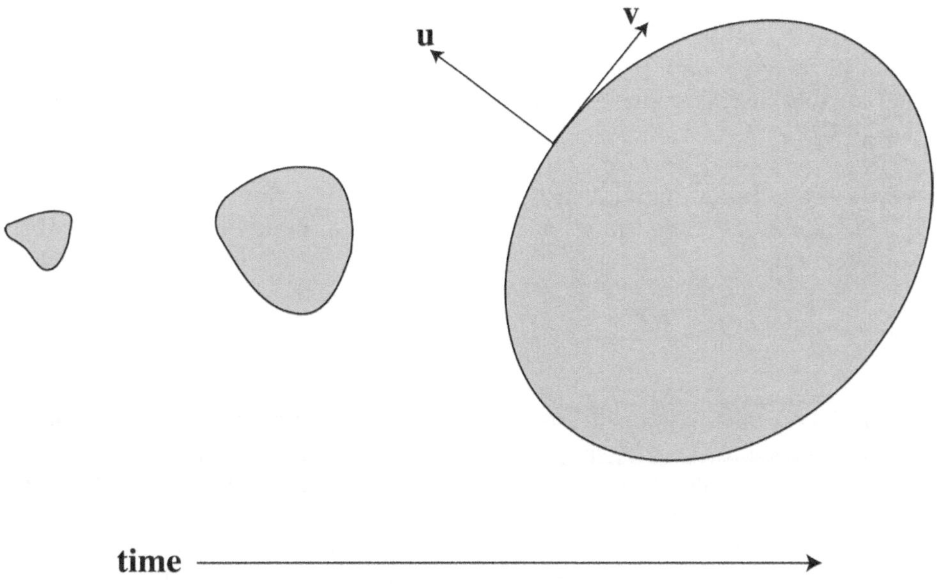

Fig. 4 A sketch of the three stages of population spread in a homogeneous environment. Shaded areas indicate invaded habitat. As time progresses, the initial "beachhead" grows and becomes more elliptically shaped. For long times, the invasion front is approximately planar in all directions. The speed in the direction **u**, perpendicular to the front, is found using the marginal dispersal kernel in that direction $k_u(u)$ (c.f. equation 28). The marginal kernel, in turn, is found by integrating the original 2-dimensional distribution over the direction **v**.

where $M_{\mathbf{u}}(s)$ is the 'directional' moment generating function, the moment generating function of K evaluated in the direction of **u** (Appendix). Two approaches for calculation of the directional moment generating function are given in the the Appendix. In general the planar spread rate cu will depend on the direction **u**. However, when the dispersal kernel is directionally isotropic (is identical in all directions) the planar spread rate will also be isotropic. We now consider the case with directional isotropy, so that $K(\mathbf{z})$, $\mathbf{z} = \mathbf{x} - \mathbf{y}$, can be rewritten as $K(r)$, $r = \sqrt{z_1^2 + z_2^2}$. Note that $K(r)$ is a two-dimensional density function which denotes the relative number of seeds *per unit area* falling at distance r from the source. The area under $K(r)$ is equal to one:

$$\int_{-\infty}^{2\pi} \int_{-\infty}^{\infty} K(r)\, r\, dr d\theta = 1. \tag{15}$$

A related kernel denotes the number of seeds *per unit* length falling a distance r from the source. This related kernel is found by multiplying $K(r)$ by the perimeter of a circle of radius r to account for the fact that, at larger radii, there is more available area for seeds to fall $\tilde{K}(r) = 2\pi r K(r)$. If the data have been collected in this form, a simple rescaling by $2\pi r$ will transform $\tilde{K}$ to K.

We now consider how to calculate the planar spread rate (see equation (14)) when the kernel is radially symmetric. We illustrate the two options that are given in the Appendix for calculating the directional moment generating function for this case. Here we consider a wave spreading in the x_1-direction so that $\mathbf{u} = [1, 0]^T$.

1. Evaluate the marginal distribution of K by integrating over the z_2-direction to yield a one dimensional dispersal kernel that describes dispersal in the z_1 direction,

$$K_{\mathbf{u}}(z_1) = \int_{-\infty}^{\infty} K(\sqrt{z_1^2 + z_2^2})\, dz_2, \tag{16}$$

   and then calculate the directional moment generating function $M_{\mathbf{u}}$ of the kernel $k_{\mathbf{u}}(z_1)$, and thus the speed $c_{\mathbf{u}}$ (14) (see equations (27) and (28) for details). This method effectively reduces the two dimensional spread problem to one spatial dimension, by first taking the marginal distribution of the dispersal kernel, and then proceeding as with the one dimensional case. This approach is conceptually straightforward, but many marginal distributions cannot be calculated analytically, even for simple kernels.
2. Evaluate the moment generating function in the z_1-direction (see equation (26)) directly as

$$M_{\mathbf{u}}(s) = \int_{-\infty}^{\infty} \int_{-\infty}^{\infty} K(r) \exp(sz_1)\, dz_1 dz_2 \tag{17}$$

$$= \int_0^{2\pi} \int_0^{\infty} K(r) \exp(sr \cos\theta)\, r\, dr d\theta \tag{18}$$

$$= 2\pi \int_0^{\infty} K(r)\, r\, I_0(sr) dr, \tag{19}$$

$$= \int_0^{\infty} \tilde{K}(r)\, I_0(sr) dr, \tag{20}$$

and then use this to calculate of the speed c_u (14). Here I_0 is the modified Bessel function of the first kind and zeroth order (Abramowiz and Stegun 1970).

When there are raw one-dimensional radial dispersal distance data, $r_1 \ldots r_N$, then the empirical moment generating for the planar wave speed calculation (20) becomes

$$M_u^E(s) = \frac{1}{N} \sum_{i=1}^{N} I_0(sr_i)\,, \tag{21}$$

and substitution into (14) gives the empirical estimator for the planar wave speed. Here the r_i measurements arise from tracking a series of individuals as they disperse. It is assumed that the tracking effort per unit area and tracking efficiency per unit area remains constant over the entire dispersal area. Given a histogram of radial density data

$$K^H(r) = \begin{cases} f_i & \text{if} \quad \rho_{i-1} \le r < \rho_i \\ 0 & \text{otherwise} \end{cases} \tag{22}$$

where $1 \le i \le L$, the constraint that the area under the histogram integrates to one (15) means

$$\pi \sum_{i=1}^{L} (\rho_i^2 - \rho_{i-1}^2) f_i = 1 \tag{23}$$

If the histogram is measured in terms of the relative *number* of dispersers that move a distance r rather than the relative *density*, it is necessary to rescale the kernel by $(2\pi r)^{-1}$, as above.

The directional moment generating function for the histogram comes from substituting (22) into (19) to obtain

$$M_u^H(s) = \frac{2\pi}{s} \sum_{i=1}^{L} f_i \left(\rho_i I_1(s\rho_i) - \rho_{i-1} I_1(s\rho_{i-1})\right), \tag{24}$$

(compare with (12)), and substitution into (14) gives the planar wave speed. Here I_1 is the modified Bessel function of the first kind and first order (Abramowiz and Stegun 1970).

For the sake of illustration, we earlier considered the case where the Dobzhansky and Wright data described insect movement along linear transects of available habitat. Accordingly, the histogram estimator for the spread rate for the situation shown in Fig. 2 was calculated, using equations (12) and (6) as 0.258 km per year.

To compare the difference between the results of this section and our earlier results, we revisit the Dobzhansky and Wright data, under the more reasonable assumption that the transect data describe the radial drop-off in settled insect density in a two-dimensional habitat. In this case the histogram estimator for the planar spread rate is calculated from equations (24) and (14) as 0.288 km per year, 16% higher than the previous estimate.

Repeating the comparison of spread rates for different λ's shows that the spread rate, assuming radial dispersal and growth in the two-dimensional habitat, is significantly higher than the spread rate, assuming growth and dispersal in a linear one-dimensional habitat, (see Table 1). This difference is most pronounced for a low growth rates λ. This ordering of spread rates (2D speed larger than 1D speed, Table 1) appears only for density data, not dispersal data. In fact, if the original data describe dispersal, then it can be shown that the ordering will be reversed, with the 1D speed (based on growth and dispersal in a linear one-dimensional habitat, e.g., equation 8 and 6) larger than 2D speed (based on radial dispersal an growth in a two-dimensional habitat, e.g., equation 21 and 14).

Table 1 Spread rates for varying growth rates λ. Dispersal data are as given by the left panels in Fig. 2. In column labeled 1D it is assumed that the growth and dispersal occur in a 1D linear habitat. Spread rates are calculated using equations (12) and (6). In column labeled 2D it is assumed that the dispersal is radial, and that growth and dispersal occur in a 2D habitat. Here the planar spread rate is calculated from equations (24) and (14). The last column shows the ratio of the second to the first columns entries. The difference between the spread rates is most pronounced for low growth rates.

λ	1D speed (km/year)	2D speed (km/year)	ratio of speeds
10	0.258	0.288	1.16
4	0.194	0.231	1.19
2	0.130	0.166	1.27
1.1	0.0466	0.0619	1.39

MONTE CARLO METHODS

The simplest estimator of the moment generating function is the empirical estimator calculated from one-dimensional displacement data (equation (8)). It is not only simple to implement numerically, but in many cases is appealingly nonparametric. It is sometimes useful to generate appropriate 1-D displacement data from some other form, using Monte Carlo methods. The resulting displacements can then be used as input to the empirical estimator. For example, Caswell *et al.* (2003) analyzed the invasion rates of several species of European birds, using dispersal data compiled by van den Bosch *et al.* (1992). The dispersal data were obtained in the form of histograms of displacement distances, not densities. Thus they were interpreted as giving a set of distances, in two-dimensional space, moved by a set of marked individuals. Obtaining the appropriate one-dimensional dispersal kernel requires the marginal distribution of the two-dimensional distribution of displacements. This was obtained by generating a large number of random displacement distances from the histogram (assuming a uniform distribution of displacement distance within each histogram bin). Assuming two-dimensional isotropy, each of these distances was assigned a direction uniformly distributed between 0 and 2π. This produced a set of artificial 2-dimensional displacement data, the distance component of which matches the reported histogram. The marginalized distribution was easily generated by taking the z_1 component of each of the points. The resulting set of distances was then input to the empirical moment-generating function to produce an estimate of wave speed.

This Monte Carlo method is appropriate because we know mathematically that, providing the dispersal kernel is exponentially bounded, the empirical estimator for the wave speed is unbiased (i.e., approaches the true wave speed as $N \rightarrow \infty$). Details on convergence can be found in Clark *et al.* (2001).

DISCUSSION

The focus of this paper is methods to reliably connect population spread rate theory to biological data. There are two features that we focus on: (i) sensitivity of spread rate estimates to model assumptions about long-distance dispersal, and (ii) model fitting issues that arise from fitting two-dimensional dispersal data to one-dimensional models. The empirical and histogram estimators provide a method to bypass assumptions about long distance dispersal. These are really 'what you see is what you get' estimators. Indeed, formally substituting the observed sum of point dispersal jumps $k(z) = \sum_{i=1}^{N} \delta(z - z_i)$ into equation (7) leads to the empirical moment generating function (8). Whereas the above sum of delta functions provides a poor estimator for the kernel, the process leads to the empirical moment generating function and a good estimate for the spread rate, which is the quantity of interest. The histogram estimator behaves in a manner which is similar to the empirical estimator, although the location of bins can produce small biases in the spread rate estimate.

We recommend that the empirical and histogram spread rate estimators should be a basic tool for any modeling exercise which involves spread rates where long-distance dispersal plays a role. Fitted parametric dispersal kernels may be preferable when there is reason to believe that extrapolation of the dispersal function beyond the furthest observed dispersal distance can be justified. However, even in this case, it is useful to compare spread rate results from the parametric fitted kernel with the empirical or histogram spread rates. The difference between the spread rates will highlight the implications of assumptions about long distance dispersal distances that go into the parametric kernel.

The new model fitting issue we focus on is the correct method to calculate planar population spread from radially symmetric dispersal data. Naively fitting the linear one-dimensional dispersal kernel to the radially symmetric dispersal data will give the wrong spread rate when the one-dimensional spread model is used, biased downward for density data, and upward for dispersal data.

In a linear one-dimensional environment, the scaled distribution of densities (dispersers per unit length) and a distribution of distances that the dispersers travel from the parent are the same and are given by the kernel k. In a two-dimensional environment, the scaled distribution of densities (dispersers per unit area), given by K, and the distribution of distances that dispersers travel from from the parent, given by $\tilde{K} = 2\pi r K(r)$, are not the same, because there is more area available at distances further from the parent. The correct method for calculating the spread in the two-dimensional environment involves a directional moment generating function of the two-dimensional dispersal kernel. This is found either by evaluating the moment generating function of the marginal distribution of the dispersal kernel $K(r)$, or by using $\tilde{K}(r)$ in a modified moment generating function calculation.

To see the difference between the linear one-dimensional histogram density kernel and the marginal distribution of it's radially symmetric analog, we can compute the marginal distribution of the radially symmetric histogram density kernel (22) directly

$$k_{\mathrm{u}}^{H}(z_1) = \begin{cases} 2\sum_{i=1}^{L} (f_i - f_{i+1})\sqrt{\rho_i^2 - z_1^2} & \text{if } \rho_{i-1} \leq |z_1| < \rho_i \\ 0 & \text{otherwise.} \end{cases} \tag{25}$$

The two kernels $k^H(z)$ (10) and $k^H_{\mathrm{u}}(z)$ (25) are shown in Fig. 5. As can be seen from the results in Table 1, the increase in number of long-distance dispersers in the marginal distribution speeds up the spread rate significantly.

The correct formulae for calculating spread rates are given in Tables 2 and 3. To the best of our knowledge, these formulae have not been widely developed or discussed elsewhere in the literature. It is our hope that these tables, along with the discussion in the chapter will form a user's guide to calculating spread rates

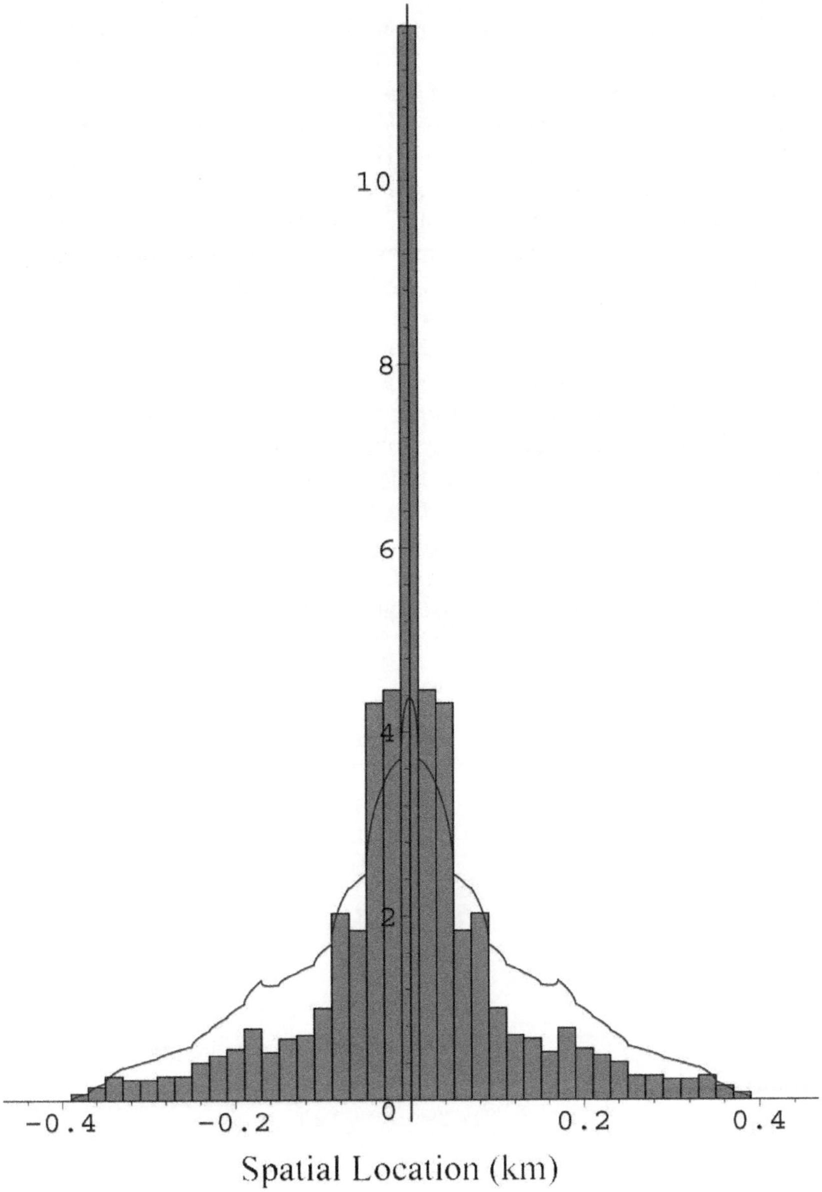

Fig. 5 Histogram dispersal kernel (equation (10)) forDobzhansky and Wright's Drosophila dispersal data under the assumption that dispersal occurs in a linear one-dimensional habitat (grey). Marginal distribution (equation (25)) of histogram dispersal kernel (equation (22)) under the assumption that dispersal occurs in a two-dimensional habitat and is radially symmetric.

for discrete-time models, and, in turn, will lead to a closer connection between ecological theory and data.

When the dispersal data come from a 'fat-tailed' kernel (kernel with no moment generating function), theory predicts a constantly accelerating wave with an asymptotically infinite speed. However, the spread rates based on the empirical moment generating function (e.g., equation 8) or the histogram moment generating function (e.g., equation 12) are finite. This is because the these spread rates are calculated from a finite number of dispersal observations or on densities measured over a finite region. Here additional sampling effort may find rare, long-distance dispersers or may measure nonzero densities of dispersers far from their natal source, and thus the spread rate estimate can increase with sampling effort. When there is good reason to believe that the dispersal kernel is fat-tailed or nearly fat-tailed, a more appropriate measure of spread rate is the 'furthest-forward' velocity (Clark, Lewis, and Horvath 2001) which remains bounded. A more general discussion of uncertainty in spread rates associated with long-distance dispersal is given in Clark *et al.* (2003).

Table 2 Methods to calculate the moment generating function needed for the spread rate calculation (equation (6)) when the population lives in a one-dimensional linear habitat.

Data	MGF [$M(s)$]	Equation
parametric kernel	$\int_{-\infty}^{\infty} k(z)\exp(sz)\,dz$	7
dispersal displacement data	$\frac{1}{N}\sum_{i=1}^{N}\exp(sz_i)$	8
displacement distance data	$\frac{1}{N}\sum_{i=1}^{N}\cosh(sz_i)$	9
displacement histogram	$\frac{1}{s}\sum_{i=1}^{L} f_i[\exp(s\xi_i)-\exp(s\xi_{i-1})]$	12
distance histogram	$\frac{1}{s}\sum_{i=1}^{L} f_i[\sinh(s\xi_i)-\sinh(s\xi_{i-1})],\ \xi_0=0$	13

Table 3 Methods to calculate the directional moment generating function needed for the planar spread rate calculation (equation (14)) when the population lives in a two dimensional habitat.

Data	MGF $[M_x(s)]$	Equation
parametric 2D kernel	$\int_{-\infty}^{\infty}\int_{-\infty}^{\infty} K(\mathbf{z}) \exp(sz_1)\, dz_1 dz_2$	26
parametric radial kernel	$2\pi \int_{0}^{\infty} K(r)\, r\, I_0(sr) dr,$	19
radial displacement data	$\frac{1}{N} \sum_{i=1}^{N} I_0(r_i s)$	21
radial histogram data	$\frac{2\pi}{s} \sum_{i=1}^{L} f_i (\rho_i I_1(s\rho_i) - \rho_{i-1} I_1(s\rho_{i-1}))$	24

I_0 and I_1 are modifed Bessel functions of zeroth and first order (cf. Abramowiz and Stegun (1970)).

APPENDIX

Calculating the spread rate with maple

The following is Maple code that can be used to calculate the spread rate for a population with geometric growth rate λ, and the composite Laplace dispersal kernel $k(z) = p\alpha_1 \exp(-\alpha_1|z|) + (1-p)\alpha_2 \exp(-\alpha_2|z|)$.

```
Digits:=20;
#
# Work out speed gam for the following parameter values
#
lambda:=1.2;
alpha_1:=1.0;
alpha_2:=1.0;
```

```
p:=0.99;
#
# define the moment generating function and exponential
#
v_1:=alpha_1^2/(alpha_1^2-s^2);
v_2:=alpha_2^2/(alpha_2^2-s^2);
fn1:=exp(s*gam);
fn2:=lambda*(p*v_1+(1-p)*v_2);
#
# Find the double root
#
eq1:=fn1=fn2;
eq2:=diff(fn1,s)=diff(fn2,s);
fsolve({eq1,eq2},{s,gam},s=0..alpha_2);
#
```

The directional moment generating function

The directional moment generating function is

$$M_{\mathbf{u}}(s) = \int_{-\infty}^{\infty}\int_{-\infty}^{\infty} K(\mathbf{z}) \exp(sz_1)\, dz_1 dz_2 \qquad (26)$$

Recall that the dispersal kernel $K(\mathbf{z})$ now describes the probability density for jumps whose directions and magnitude are described by the vector $[z_1, z_2]^T$. The term $\mathbf{u} \cdot \mathbf{z} = u_1 z_1 + u_2 z_2$ in equation (26) is the component of the dispersal jump $\mathbf{z}$ that lies in the $\mathbf{u}$-direction. If the kernel has no directional bias (is isotropic), depending only upon distance (and hence $K(\mathbf{z}) = K(\sqrt{z^2_1 + z^2_2})$), then the moment generating function $M_{\mathbf{u}}(s)$ and hence the speed cu is independent of the direction vector $\mathbf{u}$. We consider this case in the section Population Spread in Two Dimensions. The 'directional' moment generating function $M_{\mathbf{u}}$ can be interpreted as the moment generating function of the kernel $K(z)$ marginalized in the direction of the unit vector $\mathbf{v}$ which is perpendicular to $\mathbf{u}$

$$M_u(s) = \int_{-\infty}^{\infty} k_u(u) \exp(su)\, du. \qquad (27)$$

where

$$k_u(u) = \int_{-\infty}^{\infty} K(\mathbf{z})\, dv. \qquad (28)$$

and $u = \mathbf{u} \cdot \mathbf{z}$, $v = \mathbf{v} \cdot \mathbf{z}$.

REFERENCES

Abramowiz, M. and I. Stegun (1970). Handbook of Mathematical Functions. Dover Publications Inc.

Allen, E. J., L. J. S. Allen, and X. Gilliam (1996). Dispersal and competition models for plants. Journal of Mathematical Biology **34**, 455-481.

Allen, L. J. S., E. J. Allen, and S. Ponweera (1996). A mathematical model for weed dispersal and control. Bulletin of Mathematical of Biology **58** (5), 815-834.

Andow, D. A., P. M. Kareiva, S. A. Levin, and A. Okubo (1990). Spread of invading organisms. Landscape Ecology **4**, 177-188.

Aronson, D. G. and H. F. Weinberger (1975). Nonlinear dilusion in population genetics, combustion, and nerve pulse propagation. In J. A. Goldstein (Ed.), Lecture Notes in Mathematics, Volume 446, pp. 5-49. Berlin: Springer-Verlag.

Caswell, H., R. Lensink, and M. G. Neubert (2003). Demography and dispersal: Comparing invasion speeds using life table response experiments. Ecology **84**, 1968-1978.

Clark, J. S. (1998). Why trees migrate so fast: Confronting theory with dispersal biology and the paleorecord. The American Naturalist **152** (2), 204-224.

Clark, J. S., C. Fastie, G. Hurtt, S. T. Jackson, C. Johnson, G. A. King, M. Lewis, J. Lynch, S. Pacala, C. Prentice, E. W. Schupp, T. W. III, and P. Wycoff (1998). Reid's paradox of rapid plant migration. BioScience **48** (1), 13-24.

Clark, J. S., L. Horvath, and M. A. Lewis (2001). On the estimation of spread rate for a biological population. Statistics and Probability Letters **51**, 225-234.

Clark, J. S., M. A. Lewis, and L. Horvath (2001). Invasion by extremes: Population spread with variation in dispersal and reproduction. American Naturalist **157**, 537-554.

Clark, J. S., M. A. Lewis, J. S. McLachlan, and J. H. RisLambers (2003). Estimating population spread: What can we forecast and how well? Ecology **84**, 1979-1988.

Dobzhansky, T. and S. Wright (1943). Genetics of natural populations. x. Dispersion rates in *drosophila pseudoobscura*. Genetics **27**, 304-340.

Fisher, R. A. (1937). The wave of advance of advantageous genes. Ann. Eugen. London, **37**, 355-369.

Fujiwara, M., K. Anderson, H. Caswell, and M. G. Neubert (2004). On the estimation of dispersal kernels from individual capture-recapture data. Environmental and Ecological Statistics In press.

Hart, D. R. and R. H. Gardner (1997). A spatial model for the spread of invading organisms subject to competition. J. Math. Biol. **35**, 935-948.

Higgins, S. I. and D. M. Richardson (1999). Predicting plant migration rates in a changing world: the role of long-distance dispersal. Am. Nat. **153**, 464.

Horvitz, C. C. and D. W. Schemske (1986). Seed dispersal of a neotropical *myrmecochore*: variation in removal rates and dispersal distance. Biotropica **18**, 319-323.

Kolmogorov, A., Petrovsky, and N. I. Piscounov (1937). Étude de l'équation de la diffusion avec croissance de la quantité de matière et son application a un problème biologique. Bull. Moscow Univ. Math. Mech **1** (6), 1-26.

Kot, M. (1992). Discrete-time travelling waves: Ecological examples. J. Math. Biol. **30**, 413-436.

Kot, M., M. A. Lewis, and P. van den Driessche (1996). Dispersal data and the spread of invading organisms. Ecology **77** (7), 2027-2042.

Lewis, M. A. (1997). Variability, Patchiness, and Jump Dispersal in the Spread of an Invading Population, Chapter 3, pp. 46-69. Princeton, New Jersey: Princetin Univesity Press.

Lubina, J. A. and S. A. Levin (1988). The spread of a reinvading species: range expansion in the California Sea Otter. American Naturalist **151** (4), 526-543.

Lui, R. (1982a). A nonlinear integral operator arising from a model in population genetics, I. monotone initial data. SIAM J. Math. Anal. **13** (6), 913-937.

Lui, R. (1982b). A nonlinear integral operator arising from a model in population genetics, II Initial data with compact support. SIAM J. Math. Anal. **13** (6), 938-953.

Lui, R. (1985). A nonlinear integral operator arising from a model in population genetics, III. Initial data with compact support. SIAM J. Math. Anal. **16** (6), 1180-1206.

Lui, R. (1986). A nonlinear integral operator arising from a model in population genetics IV. Clines. SIAM J. Math. Anal. **17** (1), 152-168.

Lui, R. (1989a). Biological growth and spread modeled by systems of recursions. I Mathematical theory. Mathematical Biosciences **93**, 269-295.

Lui, R. (1989b). Biological growth and spread modeled by systems of recursions. II Biological theory. Mathematical Biosciences **93**, 297-312.

Marchant, S. (2003). Analysis of an integrodifference model for biological invasions with a quasi-local interaction. Master's thesis, University of British Columbia.

Neubert, M. G. and H. Caswell (2000). Demography and dispersal: Calculation and sensitivity analysis of invasion speed for stage-structured populations. Ecology **81**, 1613- 1628.

Neubert, M. G., M. Kot, and M. A. Lewis (2000). Invasion speed in fluctuating environments. Proc. Roy. Soc. Lond. B **267**, 1603-1610.

Neubert, M. G. and I. Parker (2004). Using integrodifference equations to project rates of spread for invasive species. Risk Analysis, in press.

Pachepsky, E., F. Lutscher, and M. A. Lewis (2004). Persistence, spread and the drift paradox. Theoretical Population Biology, in press.

Powell, J. and N. E. Zimmermann (2004). Analysis of active seed dispersal contributes to resolving Reid's paradox. Ecology **85**, 490-506.

Schofield, P. (2002). Spatially explicit models of Turelli-Hoffmann wolbachia invasive wave fronts. Journal of Theoretical Biology **215**, 121-131.

Shigesada, N. and K. Kawasaki (1997). Biological Invasions: Theory and Practice. Oxford: Oxford University Press.

Shigesada, N., K. Kawasaki, and E. Teramoto (1986). Traveling periodic waves in heterogeneous environments. Theor. Pop. Biol. **30** (1), 143-160.

Silverman, B. W. (1986). Density Estimation for Statistics and Data Analysis. Chapman and Hall.

Skellam, J. G. (1951). Random dispersal in theoretical populations. Biometrika **38**, 196-218.

Speirs, D. C. and W. S. C. Gurney (2001). Population persistence in rivers and estuaries. Ecology 82, 1219-1237.

Takasu, F., N. Yamamoto, K. Kawasaki, K. Togashi, and N. Shigesada (2000). Modeling the range expansion of an introduced tree disease. Biological Invasions **2**, 141-150.

van den Bosch, F., R. Hengeveld, and J. A. J. Metz (1992). Analysing the velocity of animal range expansion. Journal of Biogeography **19**, 135-150.

Veit, R. R. and M. A. Lewis (1996). Disperal, population growth and the allee effect: Dynamics of the house finch invasion of eastern north America. The American Naturalist **148** (2), 255-274.

Weinberger, H. F. (1982). Long-time behavior of a class of biological models. SIAM journal on mathematical analysis **13** (3), 353-396.

Weinberger, H. F. (1984). Long-time behaviour of a class of biological models. In W. E. F. III (Ed.), In: Research Notes in Mathematics No. 101, pp. 323-352. Marshfield, MA: Pitman Publishing Ltd.

Werner, P. A. (1975). A seed trap for determining patterns of seed deposition in terrestrial plants. Canadian Journal of Botany **53**, 810-813.

With, K. A. (2002). The landscape ecology of invasive spread. Conservation Biology **16**, 1192.

Woolcock, J. L. and R. Cousens (2000). A mathematical analysis of factors affecting the rate of spread of annual weeds in an arable field. Weed Science **48**, 141-150.

Chapter nine

The role of evolutionary genetics in studies of plant invasions

K. A. Schierenbeck and M. L. Aïnouche

INTRODUCTION

Invasive species have long raised the interest of evolutionists, as they provide ideal systems to understand the evolutionary processes that determine and accompany the expansion of a species (Baker and Stebbins 1965). Pragmatically, invasions of non-native species or genotypes pose a major biological threat to native biodiversity and ecosystem functioning. While the ecology of these invasions has received considerable international attention, understanding the evolutionary dimensions of this problem remains rudimentary. The observation can be made however, that invasions have two main evolutionary consequences; first, they can result in the rapid evolution of introduced populations in new environments, and secondly, they can promote reticulate gene flow with related taxa at the population, ecotype, or species level (Mooney and Cleland 2001). Both processes will result in new genotype-by-environment interactions and subsequent genetic diversification.

Despite much conjecture about the genetic characteristics of invasive species, data collection has resulted in ambiguous conclusions. Allozyme data indicate *Bromus tectorum* (cheatgrass) may have a "general purpose genotype" (Novak

M.W. Cadotte et al. (eds.), Conceptual ecology and invasion biology, 193–221.

and Mack 1993), quantitative genetic data indicate *Sapium sebiferum* (Chinese tallow) may undergo post-introductory adaptations (Siemann and Rogers 2001), and other species may require a novel combination of intra- or interspecific genetic variation from the home range to produce a successful invader (*e.g.*, *Tamarix* species, Gaskin and Schaal 2002).

It is becoming clear that invasions can result in rapid evolutionary events via a number of genetic mechanisms (Ellstrand and Schierenbeck 2000; Lee 2002). The recent development of genetic and genomic approaches has allowed the investigation of model or cultivated systems at the sequence level. Data have accumulated rapidly in the past few years to elucidate molecular level mechanisms that promote species formation and expansion (Rieseberg and Wendel 2004). Among the questions that can now be addressed are (1) the geographic and evolutionary origin of invasive species (2) the identification of the parental genotypes that are involved in the formation of new (recent) invasive species (3) the possible genetic exchanges between invasive and native species (4) genome evolution of invasive species compared to related non-invasive species (5) the molecular basis of adaptive traits that may promote invasiveness (6) importance of neutral versus adaptive processes that have accompanied genome evolution of invasive species.

Allendorf and Lundquist (2003) and Lee (2002) are among those that emphasize that genetic considerations should play a larger role in the development of invasive species management and control. Sadly, adequate genetic data rarely exist to set policy at a regional or global level. We hope to compel the reader of this chapter that the genetics of invasive species are interesting and important from a theoretical evolutionary perspective and merit fundamental consideration at the management level. Researchers just entering the study of evolutionary genetics can be assured of many decades of work with invasive species at the molecular, population, or species level. However unfortunate the circumstance, species invasions are ideal for the provision of syntheses between ecology and genetics.

ROLE OF MOLECULAR EVOLUTION IN INVASIVE ABILITY

Possible heritable factors that contribute to invasive ability have been conjectured since Darwin (1872). Although sophisticated techniques to explore the genetics of invasive ability are now available, the field still has far to go to find practical methodologies to elucidate the complex genetic and ecological interactions of invasive species. Whether invasive plant species are "born or made" (Ellstrand and Schierenbeck 2000), if a particular genetic characteristic is identified, we can now determine whether allelic variability for a trait exists and assess its evolutionary importance in synthesis pathways or gene expression.

As agricultural weed scientists realized over 60 years ago, weed control, genetics, and molecular biology are basically and intimately related via studies on chemical disruption of complex metabolic pathways (Gressel 2000). As a result of enzymes that are blocked by herbicides, the pathways in which they occur become elucidated and may subsequently become important in transgenic experimentation (Gressel 2000). Of particular note, agricultural scientists use basic evolutionary principles when faced with the often inevitable resistance problems that result from herbicide use. For example, the herbicide paraquat has been successful in a "pretest" for a genetic predisposition for transient drought tolerance because both drought and herbicide resistance cause oxidative stress (Gressel 2000). Tests for screening favorable genotypes in crop situations may be useful in screening unwanted non-native horticultural genotypes that may contain traits such as resistance to drought stress that are known to contribute to invasiveness (Bazzaz 1986). An important step in the development of a management policy of invasives is the integration of techniques that are flexible enough to respond to genetic traits that may change as a result of adaptation and control. The role of the ecologist and land manager should not be underestimated in the determination of the ecological importance of the genetic traits of invaders.

We considered five major aspects of the role of molecular evolution to invasive ability. First, the genetic variability necessary to respond to new environmental pressures exists within some taxa, *i.e.*, either "untapped" genetic variation or a "general purpose genotype" (Baker 1965). Secondly, more rapid evolution can result from adaptive radiation into uninhabited niches, similar to that found with adaptive radiation in island ecosystems. Depending on age and disturbance history, some of these niches may be more available than others and thus become receptacles for rapid evolutionary change. Our third topic considers an idea hypothesized upon for many years (Stebbins 1950), mainly that polyploidy may create an advantage in invasions due to molecular evolution of gene duplications and increased genetic variation. Fourth, the creation of new variation can result from hybridization at either the inter- or intraspecific level and this genetic variation can originate from repeated introductions and seed banks. Lastly, not all aspects of the evolution of invasive species result from genetic variation; insights have recently accumulated, that stress the importance of epigenetic mechanisms and the role of expression plasticity in shaping phenotypes.

Existing genetic variation

Molecular Genetics

Questions regarding the importance of molecular markers to measure response to selection are part of the larger debate over "nearly-neutral" versus selectionist theory (Ohta and Gillespie 1996). If the question is, 'is the heterozygosity of molecular markers necessary for a response to selection?', the answer is a qualified "no". Reed and Frankham (2001) have gone as far as to suggest that genetic diversity data have little place in the context of the invasive species question. Despite this view, a number of examples demonstrate that variable molecular markers can be useful in understanding plant invasions. Genotypes that are clearly more invasive than others have been identified in *Tamarix* (salt cedar) species

(Gaskin and Schaal 2002) and *Phragmites australis* (common reed) (Saltonstall 2002). In addition, we are now able to identify variation in chromosomal regions, if not for particular genes, that confer a fitness advantage in invasions. Clauss and Mitchell-Olds (2003) have demonstrated a direct relationship between trypsin inhibitor loci, important in plant defense in *Arabidopsis* (mouse-ear cress), and the evolution of life history traits. A quantitative trait locus in *Helianthus paradoxus* (pecos sunflower) suggests Ca-dependent salt tolerance and higher fitness in a range of environments in the hybrid species (Lexer *et al.* 2003). Although admittedly there is much to learn about the relationship between the molecular genetic variation of populations and selection, it remains a fruitful and seminal area of inquiry, particularly with plant invasions.

Genetic markers are often used in selection and other genetic manipulations of agricultural crops (Dekker 2003). Basic population genetics have been useful in the separation of two hybridizing *Amaranthus* (pigweed) species to prevent the introgression of transgenics (Wetzel *et al.* 1999). Molecular markers have been useful in *Oryza* (rice) species in the identification of varieties that may or may not be able to hybridize and subsequently whether these varieties are appropriate for a particular growing region (Cohen *et al.* 1999). The similar application of such markers to invasive species will lead to removal prioritization of problematic genotypes in particular geographic regions, assuming they can be visually differentiated in the field.

One of the problems with using existing genetic variation as an explanation for invasive ability is that it often requires widespread and intensive sampling from both the native and introduced ranges. The most definitive work in the comparison of native and introduced genotypes is with *Bromus tectorum* (Novak and Mack 1993, Novak *et al.* 1993; Bartlett *et al.* 2002). *Bromus tectorum* is a cleistogamous plant with various homozygous multilocus genotypes; repeated introductions of different Old World genotypes that have invaded North America have resulted in the redistribution of the genetic diversity available in the native area of the species. Although the *B. tectorum* work is very thorough, further data collection is needed to determine whether predictions can be made about invasive *B. tectorum* genotypes. Broad tolerances needed for phenology, seed set, drought, and freezing are found in *B. tectorum* and obviously these traits have some genetic basis. For species like *B. tectorum* that have reduced genetic variation, the ideal match between invader characteristics and new range environment may be simply idiosyncratic, *i.e.*, a genotype is "pre-adapted" for the new range. On the other hand, millenia of selective processes and inbreeding in *B. tectorum* may have worked together to result in the "ideal invader". These are the "general purpose genotypes" of Baker, genotypes that can grow in a variety of different environments, with no need to undergo further selection (Williams 1992).

There remains a lack of consensus regarding the relationship of genetic variability to the invasive ability of plants, *i.e.*, some invasive species maintain low levels of genetic variation (Baumel *et al.* 2001; Ye *et al.* 2003) whereas others are highly variable (Barrett and Richardson 1986). Despite conjecture on this matter since 1965 (Baker and Stebbins 1965), experimental evidence in this area has been slow to accumulate. A number of authors have found that invasive species with low levels of genetic diversity often share traits such as inbreeding or an annual life habit with a self-breeding system (Brown and Marshall 1981, Warwick, 1990, Novak and Mack 1993, Wang *et al.* 1995, Squirrell *et al.* 2001, Bartlett *et al.* 2002). These conclusions, as with many other conclusions about invasive species, have exceptions, for example *Viola riviniana* (common dog violet), is a primarily clonal, cleistogamous species with high levels of genetic variation (Auge *et al.* 2001).

Grouping species by life history traits and geographical range has been successful in detecting association between genetic diversity and life history characteristics for plant species in general (Hamrick *et al.* 1979; Loveless and Hamrick 1984; Hamrick *et al.* 1992). Thus, it maybe useful to first categorize invasive species by life history traits and then look for patterns in genetic diversity (Schierenbeck *et al.* 1995). Hamrick *et al.* (1979) found that weedy and early successional species are less variable than species of mid- and late-successional stages. A recent study with the herbaceous perennial, *Alternanthera philoxeriodes* (alligator weed, a native of South America) supports this pattern. Considered one of the world's worst weeds, *A. philoxeriodes* is a partially submerged aquatic with widespread clonal propagation and very low genetic diversity throughout southern China ($H_T = 0.0286$, Xu *et al.* 2003).

Among long-lived woody species for which genetic variation has been measured, correlations between life history traits and genetic variation are less clear and for most studies on invasive woody angiosperms, genetic variation is reported for the home range only. For example, there are home range studies for *Robinia pseudoacacia* (black locust) (Surles *et al.* 1980), *Prosopis glandulosa* (honey mesquite) (Paneida and Carstairs 1989), *Casuarina cunninghamiana* (river-she oak) (Moore and Moran 1989), *Acacia melanoxylon* (blackwood wattle) (Moran *et al.* 1989), *Acacia decurrens* (green wattle) (Moran *et al.* 1989), and *Eucalyptus obliqua* (messmate) (Brown *et al.* 1975) and all show higher levels of variation than expected for species with similar life history traits (Schierenbeck *et al.* 1992). All but one of these species (*Eucalyptus obliqua*) are nitrogen fixers and it should be noted that the trait for nitrogen fixation is well-associated with ability to invade new ranges rapidly (Vitousek 1986). *Lonicera japonica* (Japanese honeysuckle) an invasive woody vine native to Asia and invasive throughout warm temperate and tropical climates worldwide, has levels of variation within the range expected for species with similar life history traits (Schierenbeck *et al.* 1995). An invasive clonal vine of the southeastern U.S., *Pueraria lobata* (kudzu) has a high level of genetic variation in its introduced range that is suggestive of multiple introductions. Excess heterozygosity in introduced populations of *P. lob*ata also suggests that selection may be act in favor of heterozygous individuals (Pappert *et al.* 2000). *Rubus alceifolius* (giant bramble), a shrub native to southeast Asia, has greater genetic variation in its home range, reduced variation in its new range of Madagascar, and a single invasive genotype as measured by

AFLPs, in its new range in the Indian Ocean (Amsellem *et al.* 2000). In a more recent example, DeWalt (2003) found *Clidemia hirta* (soapbush), a shrub native to Central and South America and invasive in Hawai'i, had low levels of genetic variation in both the introduced and home range. Surprisingly, to our knowledge, there are no more recent reviews that examine the relationship between genetic variability, invasiveness, and life history traits.

Regardless of the expectations of genetic variability in native or introduced populations, effective studies of intraspecific genetic variation require the use of a number of genes. Coalescence analyses of gene genealogies allow a determination of the geographic origins of invasive genotypes (Gaskin and Schaal 2000; Schaal *et al.* 2003) and the expansion of genotypes during invasion can be retraced through star-like phylogenetic patterns that indicate recent and rapid population growth (Slatkin and Hudson 1991). Although haplotype phylogenies are useful in the reconstruction of historical and recent gene flow patterns, the analytical techniques for the determination of historical gene flow patterns have only recently become available (Templeton 2004). Complications can arise however, in hybrid lineages that form new, recombining, coaslescent complexes.

Quantitative Genetics

The long-term conjecture that certain traits are associated with invasiveness (Baker 1965, 1974; Bazzaz 1986) is now supported by experimental evidence for the genetic basis of phenology in *Capsella bursa-pastoris* (shepard's purse) (Neuffer and Hurka 1999), *Solidago altissima* (tall goldenrod), *S. gigantea* (late goldenrod) (Weber and Schmid 1998), high relative growth rates and early reproduction for *Pinus* (pine) spp. (Rejmanek and Richardson 1996; Grotkopp *et al.* 2002), and trade-offs between herbivore defense and fitness (Strauss *et al.* 2002). Parker *et al.* (2003) concluded based on measured growth rates, freezing tolerance, and growth habits, that the invasiveness of *Verbascum thapsus* (mullein) is more likely due to a general purpose genotype that is influenced more by environmental conditions than by heritable factors.

A small number of quantitative genetic studies of native and introduced genotypes have found post-colonization evolution of genetic traits. For example, in *Sapium sebiferum*, introduced populations had a greater seed set than native populations but less protection from herbivory (Siemann and Rogers 2001). Leger and Rice (2002) found selection within 150 years for genetically based growth and reproductive traits in genotypes of the California native *Eschscholzia californica* (California poppy), invasive in Chile.

Lee (2002) suggests invasive ability may be more of reflection of ability to respond to selection than to phenotypic plasticity; but this leads us back to the question of whether invasive species are born or made, and definitive data with which to answer this question simply do not exist.

Rapid evolution resulting from adaptive radiation

Adaptive radiation, in the classical sense, is the colonization and subsequent diversification of species from a common ancestor into new habitats. The process of adaptive radiation, in part, inspired Darwin's theory of natural selection and has been supported empirically for at least four decades at many spatial and temporal scales. The recent spread, radiation, and evolution of invasive species likely follows similar processes of adaptive radiation. Questions remain, however, about the rapidity and spatial scale with which this process can occur. Can fragmented, disturbed landscapes devoid or partially devoid of native biota be considered island situations and receptacles for adaptive radiation? We know virtually nothing about what evolutionary processes will occur in situations in which species from remote areas of the globe are brought together into a new habitat. The human mediated migration of propagules is not unlike non-human mediated dispersal, although on a different temporal scale. Reznick and Ghalambor (2001) reviewed 47 studies to conclude that the rapid evolution following colonization of new habitats is promoted by new ecological conditions. Novel ecological conditions in their study included new food resources, biotic or abiotic interactions, predators, and competitors. Species poor communities that subsequently became vessels for rapid evolutionary change often were a result of anthropogenic disturbance (Reznick and Ghalambor 2001). We know of no work which has examined the rapid evolutionary consequences of dispersal of a single plant species into a range of new and different habitats, although there are some animal examples (Huey *et al.* 2000, Losos *et al.* 1997). We primarily can draw from examples which illustrate the rapidity with which evolution can occur following adaptive radiation.

Classic examples of adaptive radiation in island habitats are not only examples of rapid evolutionary change but provide ideal opportunities to study the genetics and ecology of invasions. An example of very effective colonization into new, unoccupied habitats has been well-supported in the magnificent diversification of the Hawaiian silversword alliance over the last 6 million years (Barrier *et al.* 1999; Barrier *et al.* 2001). Comparisions of mutation rates between genes important in the regulation of floral and inflorescence development and non-regulatory genes in the Hawaiian silversword alliance provide evidence that adaptive radiation may be more correlated with variation in regulatory loci (Barrier *et al.* 2001).

Reticulate gene flow can facilitate adaptive radiation via new gene combinations (Seehausen 2004). If reticulate gene flow is important in the spread of colonizing species, any gene flow needs to occur prior to the spread, but there could be repeated opportunities for this to happen through repeated introductions. Support of hypotheses for rapid adaptive radiation requires variation at function loci and multiple "opportunities" for adaptive divergence with repeated introductions (Seehausen 2004). Thus a combination of repeated introductions, new gene combinations and unoccupied or partially filled niches result create a vulnerability to invasion from previously unseen genotypes. The human-mediated adaptive radiation of plant species into new habitats provides an ideal situation for the experimental study of human induced evolutionary change. However, high rates of human-facilitated plant dispersal may also prevent or slow radiation by promoting gene flow and panmixis.

Hybridization

The prevalence of reticulate evolution, that is, the merging of divergent genomes through interspecific gene flow is known as an important evolutionary force in plants (Anderson and Stebbins 1954). The use of molecular markers has greatly helped to document origins and occurrence of hybrid lineages and the genetic consequences of introgressive hybridization (Rieseberg 1997; Arnold 1997). Molecular phylogenetic approaches that combine multiple sequence data sets have allowed the detection of ancient introgression events and reveal that reticulation is even more frequent than previously thought (*e.g.*, Doyle *et al.* 2004; Cronn and Wendel 2004; Small *et al.* 2004).

Hybridization is related to invasion in two ways: First, introduced invasive plants may hybridize with native species, and give rise to new successful and rapidly expanding taxa (Abbott 1992, Abbott *et al.* 2003). Second, hybridization between non-invasive species can result in new aggressive hybrids that compete with the parents and invade new habitats (Rieseberg and Wendel 1993). Genetic introgression between invasive species and closely related natives may have critical and rapid evolutionary consequences (Huxel 1999). The saltmarsh species *Spartina alterniflora* was deliberately introduced from the Atlantic American coast to California where it hybridized with the native *S. foliosa* (Daehler and Strong 1997). Hybridization was shown to occur in both directions, although the introduced species has higher male fitness (Antilla *et al.* 1998). Rather than suffering from competition with *S. alterniflora*, *S. foliosa* is now threatened by introgressant hybrids that result from recurrent backrosses, and that threaten the genetic integrity of the native species (Ayres *et al.* 2000). Pollen swamping is also thought to have represented an important mechanism allowing hybridization and invasion in oaks (Petit *et al.* 2004).

Human activities have increased ecologically disturbed areas, bringing together previously isolated taxa and generating open arrays of niches that are better suited to hybrids than to their parents (Ellstrand and Schierenbeck 2000). Hybrids do well in disturbed habitat (Anderson 1949) and invasive species are empirically associated with disturbed ecosystems (Vitousek 1986). This is particularly well-illustrated in the two classical examples of introgressive hybridization and hybrid speciation in the Louisiana irises (Anderson 1949, Arnold 1997) and in *Helianthus* species (sunflowers) (Rieseberg *et al.* 2003). Hybridization and introgression between *Iris hexagona* and *Iris fulva* occur primarily in disturbed areas that allow sympatry between the parental species. Hybrid genotypes display various combinations of parental ecological traits, such as shade tolerance that confer different fitnesses across different environments (Arnold 2004 and references therein). Studies on the *Iris fulva* x *I. brevicaulis* complex have also demonstrated the importance of considering all life stages in experimentation to understand hybrid evolution (Johnston *et al.* 2003). *Helianthus* is a genus particularly affected by reticulate evolution involving introgressive hybridization and homoploid hybrid speciation, as illustrated by *H. annuus* and *H. debilis* (Kim and Rieseberg 1999). Additionally, hybridization between *H. annuus* and *H. petiolaris* gave rise to three homoploid hybrid species (*H. anomalus, H. deserticola, H. paradoxus*); these stable new lineages diplay novel ecological adaptations. Transgressive segregation has resulted in extreme phenotypes in these *Helianthus* species and is thought to be the key for their ability to invade novel habitats (Rieseberg *et al.* 1999), *e.g.*, as demonstrated with salt adaptation in the hybrid species *H. paradoxus* (Lexer *et al.* 2004).

There are now many examples in which hybrid genotypes are more fit than one or both of the parental genotypes (Burke and Arnold 2001). Ellstrand and Schierenbeck (2000) found 28 examples in which the occurrence of new invasive taxa ness was preceded by hybridization and for which there was strong molecular evidence. Gaskin and Schaal (2002) provide both nuclear and cpDNA evidence that the most common invasive haplotypes of the voracious *Tamarix* in the U.S. are post-introduction hybrid combinations between primarily *T. ramosissima* and *T. chinensis* with some additional gene flow from *T. parviflora* and *T. gallica.*

Birchler *et al.* (2003) suggest that regulatory gene allelic interaction in hybrid genotypes might account for the well-known heterosis effect, *i.e.*, hybrid heterozygosity results in greater vigor, biomass, speed of development, and fertility than in the parental genotypes. For instance, upregulation of housekeeping genes may cause gene expression that is different in hybrids than the midparent predictions. Future studies, linking phenotypic changes and investigations at the genome level should provide new insights into molecular mechanisms that are involved in the adaptive success of hybrid lineages.

There is now little doubt that hybridization is an important evolutionary mechanism in plants, and the concern with hybridization in invasive species is not whether it can happen, but the speed with which humans accelerate this evolutionary process. Homogenization is a process that is occurring not only at the community level but also at the genetic level within taxa (Olden *et al.* 2004). Locally adapted genotypes are becoming lost through homogenization and invasion of dominants.

Polyploidy

Polyploidy, resulting from whole genome duplication, is a widespread evolutionary phenomenon and a common speciation mechanism in plants (Stebbins 1950, Lewis 1980, Grant 1981). One of the most conspicuous contributions that has resulted from the development of recent genomic approaches is the awareness of the prevalence of polyploidy in most eukaryotic lineages (*e.g.*, Wolfe 2001; Blanc *et al.* 2003). This has contributed to a renewed interest in the evolutionary success and potential selective advantage of genome duplication (Otto and Whitton 2000).

Many polyploid species are well-adapted, successful weedy species, which reinforces the idea that polyploidy may have predisposed species to become invaders (Brown and Marshall 1981, Barrett and Richardson, 1986). Interestingly, newly formed polyploids are frequently invasive species, which suggests that polyploidy confers an immediate ecological aptitude to invade new habitats. Invasive species of recent origin are excellent model systems to investigate the early evolutionary mechanisms associated with invasiveness, and provide the unique opportunity to compare the new lineage to its parents that are generally identified and still extant. The allopolyploids *Spartina anglica* (cordgrass) (Aïnouche *et al.* 2004a), *Tragopogon mirus* and *T. miscellus* (goatsbeard) (Soltis *et al.* 2004), *Senecio cambrensis* (Welsh ragwort) and *S. eboracencis* (Abbott and Lowe 2004), and *Cardamine schulzii* (bittercress) (Urbanska *et al.* 1997) formed during the last 100-120 years, have well-documented origins, have spread rapidly, and display a larger ecological amplitude than their progenitors. These species are either perennials or annuals-biennials and display various breeding or pollination systems.

Genome duplication may have different impacts on fertility and modes of inheritance that is dependent on chromosome behavior and genetic segregation. It is generally predicted that duplication of the same genome within species (*i.e.*, strict autopolyploidy) will result in random pairing (polysomic inheritance at duplicated loci), irregular meiosis, and thus limited fertility, whereas the duplication of more differentiated (homoeologous) genomes (*i.e.*, allopolyploidy) will result in preferential chromosome pairing (bivalents), regular meiosis, high fertility and disomic inheritance (Da Silva and Sobral 1996). In fact, autopolyploids and allopolyploids occur on a continuum in nature, as the ongoing evolutionary process results in more or less divergent parental populations of the polyploid (Stebbins 1971; Wendel and Doyle 2004). Moreover, chromosome pairing may be affected by various genetic and genomic factors and vary with the age of the polyploid; it is then recommended to distinguish between the mode of formation of a polyploid species and its mode of chromosomal segregation (Otto and Whitton 2000).

Recent research has resulted in a particularly dynamic vision of polyploid genomes over both a short- and long-term evolutionary time scale (Wendel 2000, Soltis and Soltis 2000). The development of molecular markers and particularly, the combined use of maternally-inherited cytoplasmic markers with biparentally-inherited nuclear markers has allowed the detection of multiple and recurrent origins of polyploid species (Soltis and Soltis 1999). As multiple-sequence datasets are now available for phylogenetic analyses, it is possible to detect recurrent and bi-directional reticulate evolution even in old polyploid lineages (*e.g.*, *Glycine* Doyle *et al.* 2004) where they otherwise would have been undetected.

The process of recurrent polyploid formation with reticulate gene flow may involve various parental genotypes and increases the level of genetic diversity available to newly formed species (Schierenbeck *et al.* 1992). For instance, at least 21 lineages of separate origins for the allotetraploid *Tragopogon miscellus* and 11 for *T. mirus* have been documented in the Palouse according to various morphological and molecular (allozymes, chloroplast and nuclear DNA) lines of evidence (reviewed in Soltis *et al.* 2004); these recurrent origins involve either only one diploid species as the maternal parent (*e.g.*, *T. porrifolius* for the allotetraploid *T. mirus*) or alternatively both the parental species *T. porrifolius* and *T. dubius* in reciprocal crosses (*e.g.*, for the allotetraploid *T. miscellus*). This has resulted in various genotypes and dramatic floral differences in the allopolyploid populations that are progressively replacing diploids as prevalent weeds (Soltis *et al.* 2004). Similarly, two separate origins have been documented in North Wales and Scotland for the ruderal allohexaploid *Senecio cambrensis* that originated in Britain during the past 85 years (reviewed in Abbott and Lowe 2004). One of the most complex examples of multiple allopolyploid origins is represented by the polyploid agamic complex *Antennaria rosea* (pussytoes) that has formed from multiple crosses involving various diploid taxa occurring in specific habitats of the Rocky Mountains (Bayer 1997). The polyploid lineages of this "compilospecies" are gametophytic apomicts that display larger ecological amplitude than their diploid progenitors.

Not all successful allopolyploids have multiple origins. In contrast to *Tragopogon* spp., *Sencecio cambrensis* and *Antennaria rosea*, the invasive saltmarsh species *Spartina anglica* has undergone a severe genetic bottleneck at the time of its formation in the Bay of Southampton (U.K.). This dodecaploid species formed after chromosome doubling of the first generation hybrid *Spartina x townsendii* that has resulted from hybridization between the introduced East-American hexaploid *Spartina alterniflora* and the native hexaploid *Spartina maritima* (Raybould *et al.* 1991). Both parental species lack genetic diversity in the hybridization site; a limited number of *S. alterniflora* genotypes have been introduced in Western Europe (Baumel *et al.* 2003), whereas a striking lack of molecular variation is encountered in populations of the native *S. maritima* (Yannic *et al.* 2004). European populations of *S. anglica* are mostly composed of one major multilocus genotype that has formed in Southampton, and that is identical to the first generation hybrid *S. x townsendii* (Baumel *et al.* 2001; 2002a). Chloroplast DNA analysis has revealed all populations of *S. anglica* in western Europe display identical plastome to *S. alterniflora* which is then considered as the maternal genome donor (Ferris *et al.* 1997; Baumel *et al.* 2001). *Spartina anglica* has rapidly invaded the British saltmarshes since its formation (Thompson 1991) and it has been naturally or deliberately introduced in various continents (such as China or Australia) where it is now considered as a serious threat to native flora and fauna (Aïnouche *et al.* 2004a and references therein). Although it has limited inter-individual genetic diversity, *S. anglica* contains two well-differentiated homoelogous genomes inherited from its hexaploid parents, which provides fixed heterozygosity at homoeologous loci (Baumel *et al.* 2002b; Aïnouche *et al.* 2004b). *Spartina* is an ideal system in which to explore the genetic and genomic consequences of hybridization and gene duplication in successful invasive species; although previous attempts of re-synthesizing experimentally the allopolyploid have failed, time since species formation and the parental species are known; it is possible to differentiate between the effects of hybridization (in. *S. x townsendii*) and genome duplication (in *S. anglica* populations); moreover, two natural replicates of hybridization events between *S. martima* and *S. alterniflora* are available in *S. x towsendii* and *S. x neyrautii* another hybrid that has formed at the same period in southewest France, with no genome doubling (Baumel *et al.* 2003).

The immediate consequence of polyploidy is a greater intra-individual genetic diversity and heterozygosity at duplicated loci than in diploids that results in increased biochemical diversity (Roose and Gottlieb 1976) and confers a greater tolerance to environment variation and may promote successful colonization (Brown and Marshall 1981). Gene duplications resulting from polyploidy are also believed to have a buffering effect against deleterious mutations (Ohno 1970, Lynch and Conery 2000, Lawton-Rauh 2003). Additionally, polyploids of hybrid origin (allopolyploids) may benefit from higher fitness due to heterosis. Similarly, traits which result in the "gigas" effect, or a general increase in morphological characteristics (de Vries 1905), are associated with both polyploidy and invasive plant species (Baker 1974; Rieseberg *et al.* 1999). *Hypericum perforatum* (St. John's wort), a tetraploid thought to be of allopolyploid origin, has high reproductive plasticity (Matzke *et al.* 2001) but reproduces most commonly via facultative apomixis. Field collections of *H. perforatum* in Australia found a lack of within population variation (Mayo and Langridge 2003) but high interpopulational genetic diversity. *Hypericum perforatum* has high levels of hypericin, a chemical that causes photosensitization and reduced herbivory, that are hypothesized to be a direct result of allopolyploidization (Mayo and Langridge 2003). Introduced into North American in the 1920s, the allopolyploid and nearly monomorphic *Setaria faberi* (giant foxtail) is now present throughout the disturbed areas on the continent (Pohl 1951, 1966; Hafliger and Scholz 1980). Dekker (2003) named the allopolyploization of *S. faberi*, the "weed speciation event", as the polyploid has higher fitness in agricultural systems than its diploid ancestors. One of the world's worst weeds is the polyploid *Sorghum halepense* (Johnson grass), a product of hybridization between the cultivar *S. bicolor* and the wild *S. propinquum*, and which responds well to cultivation and is extremely pernicious to due the rhizomatous growth habit (Paterson 2002).

A growing body of evidence continues to accumulate in regard to the dynamic and plastic nature of polyploid genomes that would explain their evolutionary success. Allopolyploid genomes are particularly dynamic at both the structural and expression levels over the long- and also short-term evolutionary time scale (reviewed in Wendel 2000, Liu and Wendel 2002, Osborn *et al.* 2003). Significant advances in revealing the occurrence and nature of the early evolutionary changes in polyploid genomes are possible due to experimentally resynthesized allopolyploids involving well-known model systems such as *Brassica* (oilseed rape), *Arabidopsis*, *Gossypium* (cotton), or *Triticum-Aegilops* (wheat). These model systems allow the exploration of allopolyploid material of known origin with the comparison of their actual parental genotypes, a condition rarely met for most natural allopolyploids. Rapid and biased structural changes have been encountered in the first generations following polyploidization in *Brassica* (Song *et al.* 1995) and wheat (Feldman *et al.* 1997; Liu *et al.* 1998a, 1998b, Ozkan *et al.* 2001). However, Liu *et al.* (2001) did not find consistent structural changes in newly synthesized allopolyploid *Gossypium*. Various evolutionary mechanisms appear to affect allopolyploid genomes over a longer term and include: the independent evolution of duplicated genes in allotetraploid cotton that formed one to two million years ago (Cronn *et al.* 1999, Senchina *et al.* 2003); interaction between the homoelogous subgenomes for repetitive sequences *via* concerted evolution (*e.g.*, *Gossypium*, Wendel *et al.* 1995, *Nicotiana* (tobacco), Volkov 1999); or spread of transposable elements (Zhao *et al.* 1998). The fate of duplicated homoeologous genes has been particularly well-investigated in the *Gossypium* system by Jonathan Wendel and his colleagues (*e.g.*, *ADH* genes, Small and Wendel 2000, 2002; *MYB* genes, Cedroni *et al.* 2003) and reveals various evolutionary patterns including copy number lability, pseudogenization, gene elimination, or accelerated rate of nucleotide substitution (*e.g.*, for *ADH-C* genes). However, a recent analysis of 48 nuclear genes (Senchina *et al.* 2003) indicated that polyploidy led to an overall modest enhancement in rates of nucleotide substitution in *Gossypium*.

When compared to resynthesized allopolyploids, naturally nascent allopolyploids display different patterns of genome evolution; various levels of concerted evolution seem to have affected rDNA sequences of the young allopolyploid populations in *Tragopogon* (Soltis *et al.* 2004), whereas no homogenization of the parental sequences is observed in *Spartina anglica* (Aïnouche *et al.* 2004a). In the latter system, no major change of the parental genomes are observed for various multilocus markers (ISSRs, RAPDs, AFLPs), although some preferential loss of maternal (from *S. alterniflora*) AFLP fragments are observed in *S. x townsendii* and *S. anglica* (Aïnouche *et al.* 2004b, Salmon A. *et al.* unpublished). A transposon display analysis indicates no burst of retro-element activation in *S. anglica* (Baumel *et al.* 2002a) and suggests that different biological systems respond variously to polyploidy (Liu *et al.* 2001). In spite of the structural genomic stasis encountered for most of the markers investigated to date, *Spartina anglica* populations exhibit consistent morphological plasticity (Thompson 1990), and suggest a probable functional plasticity in the expression of the duplicated loci (Aïnouche *et al.* 2004a).

Recent studies have pointed out that there is modulated expression of duplicated loci in polyploids (Comai 2000, Shaked *et al.* 2001, Kashkush *et al.* 2002, Kashkush *et al.* 2003, He *et al.* 2003, Adams *et al.* 2003). These expression changes may involve various mechanisms, including increased variation in dosage-regulated gene expression, altered regulatory networks, and genetic or epigenetic changes (Riddle and Birchler, 2003; Osborn *et al.* 2003). In wheat, polyploidy was accompanied by transcriptional activation of retroelements that led to novel expression patterns (Kashkush *et al.* 2003; Levy and Feldman 2004). Expression changes may have profound impact on fitness when they result in variable phenotypes. Gene silencing resulted in phenotypic variation and instability in experimentally resynthesized *Arabidopsis* allotetraploids that displayed considerable variation in morphology, flowering time, and fertility (Comai *et al.* 2000).

Novel flowering time variation was also observed in resynthesized allopolyploid *Brassica napus* (Schranz and Osborn 2000). Expression of 40 homoeologous gene pairs was analysed by Adams *et al.* (2003) in natural (1 to 2-myr old) and synthetic allotetraploid *Gossypium*. The duplicated genes showed unequal levels of expression and organ-specific reciprocal silencing, suggesting a partitioning of the ancestral functions as both immediate (in synthetic polyploids) and long-term (in natural polyploids) responses to polyploidization. In recently formed natural allopolyploids, such expression changes appear to take place also, as revealed by cDNA AFLP investigations in *Tragopogon* where about 5% of the genes examined in the allopolyploids have been silenced and an additional 4% exhibit novel gene expression relative to their diploid parents (Soltis *et al.* 2004).

Epigenetic changes associated with invasive ability

Epigenetics refers to heritable changes in phenotype that do not result from changes in gene sequence but rather from regulatory mechanisms of gene expression (Wolffe and Matzke 1999). These mechanisms are known to be involved in growth and development (Finnegan *et al.* 2000), and can result in various morphological changes including flower structure (*e.g.*, Cubas *et al.* 1999; Comai *et al.* 2000). Such mechanisms have important evolutionary consequences because they increase phenotypic plasticity, which in turn buffers environmental pressures on genotypes. This may be astonishing to our naïve perspective of genomic interactions, but epigenetic processes are proving to be predictably associated with the structural limitations of genomes. Epigenetics is consequently becoming a very active field of research in evolutionary genomics (reviewed in Finnegean *et al.* 1998; Comai 2000; Liu and Wendel 2003).

Mechanistically, epigenetic changes result from various interacting processes, which include cytosine methylation of DNA (Martienssen and Colot 2001), histone deacetylation (Tian and Chen 2001), and short RNAs (Mette *et al.* 2000) that modulate gene silencing. Epigenetic alterations are known to be triggered by environmental stress (Finnegan 2001), and in some cases are viewed as genome defense mechanisms (Yoder *et al.* 1997; Matzke *et al.* 2000). In introgressed hybrid rice plants, Liu and Wendel (2000) observed retrotransposon activation that was rapidly repressed by cytosine methylation.

The reunion of two divergent genomes in the same nucleus in hybrid and allopolyploid species may be considered as a genomic stress that generates epigenetic changes altering gene expression and phenotypes (Comai *et al.* 2003). In synthetic *Arabidopsis* allotretraploids, Comai *et al.* (2000) observed that about 1% of the genes were silenced compared to their parents. The silenced genes were both normal genes or genes related to transposons. These changes were further found to be related to methylation modifications that were associated to phenotypic instability (Madlung *et al.* 2002). Similar levels of silencing, related to cytosine methylation, were also encountered in the corresponding natural allotetraploid *Arabidopsis suecica* (Lee and Chen 2001). In experimentally re-synthesized allopolyploid wheat, 13% of the loci investigated using Methylation Sentitive AFLP (MSAP) were found methylated (Shaked *et al.* 2001), and transcriptional activation of retrotransposons was shown to alter the expression of adjacent genes (Kashkush *et al.* 2003; Levy and Feldman 2004). Although no changes in methylation patterns were observed in newly synthesized *Gossypium* allopolyploids (Liu *et al.* 2001), the organ-specific and reciprocal gene silencing found by Adams *et al.* (2003) is interpreted as resulting from epigenetic regulation through mechanisms that have yet to be elucidated (Adams and Wendel 2004).

Spartina anglica, is to our knowledge, the first invasive species that has been investigated in the context of epigenetic gene expression. Aïnouche *et al.* (2004b; Salmon *et al.* 2005) have attempted to differentiate between the genomic consequences that result from hybridization and those that result from genome duplication. This was made possible by the comparison of the allopolyploid to the natural F1 hybrid species (*S. x townsendii*, the progenitor of *S. anglica*) and *S. x neyrautii*, the other hybrid that has formed independently in southwest France (Baumel *et al.* 2003). MSAP data analysis revealed consistent methylation changes that contrast with the structural additivity of the parental genomes mentioned above. Most methylation changes were found in both *S. x townsendii* and *S. x neyrautii*, which indicates the reproducibility of the changes in the two different hybridisation events. The methylation alterations found in *S. anglica* were already present or initiated in *S. x townendii*, suggesting that epigenetic changes were triggered by hybridization rather than by genome duplication. The extent of such changes and their variability when plants are facing different environmental conditions need to be explored at the population level, and the sequences that are epigenetically affected have to be identified.

In the context of rapid expansion of invasive species that explore new habitats, epigenetic processes are of major interest as they definitely influence fitness. Because epigenetic cases of gene silencing will not be reflected in sequence data, the assessment of gene expression is key to the understanding of gene function and its importance in species adaptation and invasiveness. Molecular evolutionists are becoming more aware of the necessity to approach adaptive processes at both gene and genome levels.

SELECTION, DEVELOPMENTAL RESPONSE, AND INVASIONS

Genetic change and subsequent alterations in developmental response have been minimally addressed in invasive plant species. Similarly, although the evaluation of developmental quantitative traits has been considered theoretically, there is little empirical evidence for or against their importance in plant invasions. The genetic and ecological interactions that result from changes in developmental genes or from genes of major effect could have profoundly complex repercussions. For example, the release from enemies may allow for decreased allocation to protective mechanisms and a concomitant increase in fitness that results from developmental responses. We know little about the ecological impacts of developmental changes and interactions in predator-prey relationships (Nijhout 2003). Action from just a few particularly important genes such as these may be key to the ability to invade (Paterson *et al.* 1995).

All adaptations important to invasiveness, *e.g.*, high seed production, breeding system changes, and vegetative propagation, are amenable for study in a developmental genetic and ecological context (Puragganan 2000, Gilbert and Bolker 2003). Genes such as *teosinte branched1* in maize illustrate the importance of developmental genes in plant establishment and spread (Q. Cronk pers. comm, Doebley *et al.* 1995). *Teosinte branched1*, which controls tillering in maize, is responsible for the difference between teosinte that is persistent in the environment, and varieties that are not ecologically competitive (Doebley *et al.* 1995). Variation in regulatory genes may be key to the provision of the genetic variation necessary to alter phenological developmental pathways, floral variation, secondary plant compound production, and clonal growth.

Life history strategies, often emphasized as key to understanding invasive ability have been illustrated as phenotypically environmentally dependent (Pigliucci 2001). The developmental response of reaction norms is based on mean phenotypic trait values, thus the variation must be explained by additive genetic variance or genes for "adaptive plasticity" (Sultan 2003). A comparison between *Polygonum* species with broad ecological tolerance to those limited by environmental factors demonstrate that developmental timing of plastic responses is important to environmental breadth (Sultan 2003). Gene expression is altered by both internal and more importantly, within the invasive species context, external cues (Schlichting 2003). Of particular importance to invasive ability may be the developmental plasticity of gender and breeding system flexibility which will impact sex ratios in propagules responding to new range environmental factors (Delph 2003). As Schaal *et al.* (2003) suggest, gene genealogies will be important to test hypotheses for adaptation, and we further suggest that this should include developmental traits that may be closely related to selection and adaptive divergence. Our expectation is that the phenomenon of rapid divergence of developmental traits will be found in other examples of island radiation or in regions similarly susceptible to invasion. In cases of gene duplication and polyploidy, the rate of this process may be increased, because of the "backup"genetic expression and copies with which selective processes can "tinker".

RESEARCH NEEDS IN THE AREA OF THE GENETICS AND EVOLUTION OF INVASIVE SPECIES

The distribution of genetic variation in populations of invasive species, and the relative invasiveness of different genotypes remains ellusive with few definitive studies outside of agricultural (but see Gaskin and Schaal 2002 and Saltonstall 2002). Of particular need in understanding the dynamics of adaptation and spread, is an assessment of the molecular population genetics of regulatory loci (Puragganan 2000).

Soil seedbank analysis as a source of new genetic variation has not been adequately addressed in plant invasions. Seed dormancy is an adaptive trait and although dormancy periods are well-described for many agricultural weeds (Baskin and Baskin 1998; Dekker 1997, 1999), virutally nothing is known about the importance of this trait for the maintainance of genetic diversity in invasive plant species. Seed viability for *Cytisus scoparius* can be as long as 60 years and a two- or three-year old plant can produce up to 18,000 seeds/year (Parker *et al.* 1994). Thus, gene flow from seed banks may be considerable. *Chromolaena odorata* (triffid weed), a shrub native to the neotropics and invasive throughout the Pacific Islands, is quite aggressive and even the occurrence of an occassional seedling six years past an irradication effort may be a potentially important source of genetic variation (Waterhouse and Zeimer 2000).

Quantitative trait loci analyses have proven to be very useful in agricultural genetics and there are numerous examples (Paterson 2002). In *Sorghum halepense*, variation in the number of rhizomes has been associated with three QTLs and the number of vegetative buds with an additional QTL (Paterson *et al.* 1995). The potential importance of this type of work can not be overstated; it has been well established that clonal growth and the ability to form more vegetative buds can be important to invasive ability (Paterson 2002). Continued work with agriculture species will aid the study of those genes that may be important in weediness such as floral or fruit regulation and vegetative growth. For example, the vegetative growth traits found in *S. halepense* correspond with the same QTLs in rice, wheat, corn, and likely other, but invasive, grasses. As so astutely noted by Paterson *et al.* (1995), QTL analysis can lead to the cloning of candidate genes important in vegetative growth. Once these genes are identified it may help in the specific eradication of some invaders that have large amounts of vegetative propagation through genetic or chemical methods.

Important contributions of the quantitative genetics in understanding invasive species evolution will result from studies which consider the adaptive response, or additive genetic variance of a particular trait (Lynch and Walsh 1998; Müller-Schärer and Steinger 2004). Such studies will be further strengthened by the identification of QTLs associated with these traits and their eventual molecular characterization. The integration of quantitative genetics and the molecular basis of these traits will eventually allow an assessment of the large-scale ecological effects of genetic traits.

As previously discussed, homeologous genes, or those genes duplicated by polyploid events, may undergo various evolutionary processes. Of particular interest are those genes that have adaptive significance. Specific questions that can be addressed to assess the dynamics of the genes in polyploid systems include: characterization in sequence and function between the different copies of the genes and in both the parental species and polyploid offspring; characterization of divergence times among the parental species and polyploid offspring.

Most importantly, the determination of the effects of polyploidy on duplicate gene expression and adaptive function is now possible. Much remains to be learned about the dynamics and subsequent genotypic and phenotypic changes due to the hybridization between historically allopatric taxa, homeologous genes, and allopolyploidization.

Although currently limited in their application to model systems or closely related taxa, the exploration of the use of new genomic tools, *e.g.*, microarrays may prove useful for the investigation of genomic and ecological interactions (Jackson *et al.* 2002).

CONCLUSIONS

As evolutionary biologists, many of us have worked toward the day in which we can link gene and genomes to the level of the ecosystem. This day has arrived, and for better or worse, a forum that offers one of the best opportunities for the integration of genetics and ecology in an evolutionary context is the study of invasive species. Although the participants of the "genetics of colonizing species" symposium made this same observation in 1965; only now do we have the tools to address the myriad of questions associated with the evolution of invasive ability. We are well aware that it is perhaps more than a coincidence that many of the examples of invasive species we have cited often share the phenomena of rapid expansion into new ranges, polyploidy, and hybridization.

The integration of molecular genetics, ecology, and large-scale field experimentation is too large for any single investigator to study in invasive species or any other context. Increased knowledge in the respective fields of genomics, proteomics, cell biology, population genetics, phylogenetics, and ecology, will require the cooperation of laboratories with expertise in these respective areas. Perhaps invasive species genetics will be the field that brings together evolutionists from various disciplines for both a better understanding and the management of contemporary environments.

ACKNOWLEDGEMENTS

We thank the National Science Foundation in the U.S.; the Centre National de la Recherche Scientifique, and the Centre Armoricain de Recherches en Environnement in France for their support of collaborative research on genome evolution of invasive plants. We also acknowledge and thank the contributions of our students and many collaborators to our work, particularly Norman Ellstrand. Curt Daehler and one anonymous reviewer are thanked for helpful comments on the manuscript.

REFERENCES

Abbott R. J. 1992. Plant invasions, interspecific hybridization and the evolution of new plant taxa. Trends in Ecology and Evolution **7**, 401-405.

Abbott R. J., J. K. James, R. I. Milne and A. C. M. Gillies. 2003. Plant introductions, hybridization and gene flow. Philosophical Transactions of the Royal Society of London B **358**, 1123-1132.

Abbott R. W. and A. J. Lowe 2004. Origins, establishment and evolution of new polyploid species: *Senecio cambrensis* and *S. eboranensis* in the British isles. Biological Journal of the Linnean Society. In press.

Adams K. L. R. Cronn, R. Percifield and J. F. Wendel. 2003. Genes duplicated by polyploidy show unequal contributions to the transcriptome and organ-specific reciprocal silencing. Proceedings of the National Academy of Sciences **100**, 4649-4654.

Adams K. L. and J. F. Wendel 2004. Exploring the genomic mysteries of polyploidy in cotton. Biological Journal of the Linnean Society. In press.

Aïnouche M. L, A. Baumel and A. Salmon. 2004a. *Spartina anglica*: a natural model system for studying early evolutionary changes that affect allopolyploid genomes. Biological Journal of the Linnean Society. In press.

Aïnouche M. L., A. Baumel, A. Salmon and G. Yannic. 2004b. Hybridization, polyploidy and speciation in *Spartina* (Poaceae). New Phytologist **161**, 151-164.

Allendorf F. W. and L. L. Lundquist. 2003. Introduction: population biology, evolution and control of invasive species. Conservation Biology **17**, 24-30.

Amsellem L., J. L. Noyer, T. Le Bourgeois and M. Hossaert-McKey. 2000. Comparison of genetic diversity of the invasive weed *Rubus alceifolius* Poir. (Rosaceae) in its native range and in areas of introduction, using AFLP markers. Molecular Ecology **1**, 433-455.

Anderson E. 1949. Introgressive hybridization. New York, USA: John Wiley.

Anderson E. and G. L. Stebbins. 1954. Hybridization as an evolutionary stimulus. Evolution **8**, 378-388.

Anttila C. K, R. A. King, C. Ferris, D. R. Ayres and D. R. Strong. 2000. Reciprocal hybrid formation of *Spartina* in San Francisco Bay. Molecular Ecology 9, 765-770.

Antilla C. K., C. C. Daehler, N. E. Rank and D. R. Strong. 1998. Greater male fitness of a rare invader (*Spartina alterniflora*) threatens a common native (*Spartina foliosa*) with hybridization. American Journal of Botany **85**, 1597-1601.

Arnold M. L. 1997. Natural hybridization and evolution. Oxford, UK: Oxford University Press.

Arnold M. L. 2004. Transfer and origin of adaptations through natural hybridization: Were Anderson and Stebbins right? The Plant Cell **16**, 562-570.

Auge H., B. Neuffer, F. Erlinghagen, R. Grupe and R. Brandl. 2001. Demographic and random amplified polymorphic DNA analyses reveal high levels of genetic diversity in a clonal violet. Molecular Ecology **10**, 1811-1819.

Ayres D. R., D. Garcia-Rossi, H. G. Davis, D. R. Strong. 1999. Extent and degree of hybridization between exotic (*Spartina alterniflora*) and native (*S. foliosa*) cordgrass (Poaceae) in California, USA determined by random amplified polymorphic DNA (RAPDs). Molecular Ecology **8**, 1179-1186.

Baker H. G. 1965. Characteristics and modes of origins of weeds. Pages 147-172.

Baker H. G. and G. L. Stebbins, editors. The Genetics of Colonizing Species. Academic Press, New York.

Baker H. G. and G. L. Stebbins. 1965. The Genetics of Colonizing Species. Academic Press, New York.

Baker H. G. 1974. The evolution of weeds. Annual Review of Ecology and Systematics **5**, 1-24.

Bartlett E., S. J. Novak and R. N. Mack. 2002. Genetic variation in *Bromus tectorum* (Poaceae), differentiation in the eastern United States. American Journal of Botany **89**, 602-612.

Barrett S. C. H. and B. J. Richardson. 1986. Genetic attributes of invading species. Pages 21-33 in Groves, R. H. and Burdon, J. J., editors. Ecology of Biological Invasions. Cambridge University Press, Melbourne Australia.

Barrier M., B. G. Baldwin, R. H. Robichaux and M. D. Purugganan. 1999. Interspecific hybrid ancestry of a plant adaptive radiation: allopolyploidy of the Hawaiian silver-sword alliance (Asteraceae) inferred from floral homeotic gene duplications. Molecular Biology and Evolution **16**, 1105-1113.

Barrier M., R. H. Robichaux and M. D. Purugganan. 2001. Accelerated regulatory gene evolution in an adaptive radiation. Proceedings of the National Academy of Sciences **98**, 10208-10213.

Baskin C. C. and J. M. Baskin. 1998. Seeds: Ecology, Biogeography and Evolution of Dormancy and Germination. Academic Press, New York.

Baumel A., M. L. Aïnouche and J. E. Levasseur. 2001. Molecular investigations in populations of *Spartina anglica* C. E. Hubbard (Poaceae) invading coastal Brittany (France). Molecular Ecology **10**, 1689-1701.

Baumel A., M. L. Aïnouche, R. Kalendar, A. H. Schulman. 2002a. Retrotransposons and genomic stability in populations of the young allopolyploid species *Spartina anglica* C. E. Hubbard (Poaceae). Molecular Biology and Evolution **19**, 1218-1227.

Baumel A., M. L. Aïnouche, R. J. Bayer, A. K. Ainouche and M. T. Misset. 2002b. Molecular phylogeny of hybridizing species from the genus *Spartina Schreb.* (Poaceae) Molecular Phylogenetics and Evolution **22**, 303-314.

Baumel A., M. L. Aïnouche, M. T. Misset, J. P. Gourret and R. J. Bayer. 2003. Genetic evidence for hybridization between the native *Spartina maritima* and the introduced *Spartina alterniflora* (Poaceae) in South-West France: *Spartina x neyrautii* re-examined. Plant Systematics and Evolution **237**, 87-97.

Bayer R. J. 1997. *Antennaria rosea* (Asteraceae) — A model group for the study of the evolution of polyploid agamic complexes. Opera Botanica. **132**, 53-65.

Bazzaz F. 1986. Life history of colonizing plants: some demographic, genetic and physiological features. Pages 96-110 in H. A. Mooney and J. A. Drake, editors. Ecology of biological invasions of North America and Hawaii. Springer-Verlag, Heidelberg, Germany.

Birchler J. A., D. L. Auger and N. C. Riddle. 2003. In search of a molecular basis of heterosis. The Plant Cell **15**, 2236-2239.

Blanc G., K. Hokamp and K. H. Wolfe. 2003. A recent polyploidy superimposed on older large-scale duplications in the *Arabidopsis* genome. Genome Research **13**, 137-144.

Brown A. H. D. and D. R. Marshall. 1981. Evolutionary changes accompanying colonization in Plants. Pages 351-363 in G. C. E. Scudder and J. L. Reveal, editors. Colonization, Succession and Stability. Blackwell Scientific Publications, Oxford.

Burke J. M. and M. L. Arnold. 2001. Genetics and the fitness of hybrids. Annual Review of Genetics **35**, 31-52.

Cedroni M. E., R. C. Cronn, K. L. Adams, T. A. Wilkins and J. F. Wendel. 2003. Evolution and expression of MYB genes in diploid and polyploid cotton. Plant Molecular Biology **51**, 313-325.

Chen Z. J., L. Comai and C. S. Pikaard. 1998. Gene dosage and stochastic effects determine direction and severity of uniparental rRNA gene silencing (nucleolar dominance) in Arabidopsis allopolyploids. Proceedings of the National Academy of Sciences **95**, 14891-14896.

Clauss M. J. and T. Mitchell-Olds. 2003. Population genetics of tandem trypsin inhibitor genes in *Arabidopsis* species with contrasting ecology and life history. Molecular Ecology **12**, 1287-1299.

Cohen M. B., M. T. Jackson, B. R. Lu, S. R. Morin, A. M. Mortimer, J. L. Pham and L. J. Wade. 1999. Predicting the environmental impact of transgenic outcrossing to wild and weedy rices in Asia. Pages 151-158 in Symposium Proceedings No. 72., Gene Flow and Agriculture, Relevance for Transgenic Crops. BCPC/University of Keele, Staffordshire.

Comai. L. 2000. Genetic and epigenetic interactions in allopolyploid plants. Plant Molecular Biology **43**, 387-399.

Comai L., A. P. Tyagi, K. Winter, R., Holmes-Davis, S. Reynolds, Y. Stevens and B. Byers. 2000. Phenotypic instability and rapid gene silencing in newly formed *Arabidopsis* allotetraploids. Plant Cell **12**, 1551-1567.

Comai L, A. Madlung, C. Josefsson and A. P. Tyagi. 2003. Do the different parental 'heteromes' cause genomic shock in newly formed allopolyploids? Philosophical Transactions of the Royal Society of London. London B **358**,1148-1159.

Cronn R, Small, R. L. and J. F. Wendel. 1999. Duplicated genes evolve independently following polyploid formation in cotton. Proceedings of the National Academy of Sciences USA **96**, 14406-14411.

Cronn R. C., R. L. Small, T. Haselkorn and J. F. Wendel. 2003. Cryptic repeated genomic recombination during speciation in *Gossypium gossypioides*. Evolution **57**, 2475-1567.

Cronn, R. and J. F. Wendel. 2004. Cryptic trysts, genomic mergers, and plant speciation. New Phytologist, **161**,133-142.

Cubas P., C. Vicent and E. Coen. 1999. An epigenetic mutation responsible for natural variation in floral symmetry. Nature **401**, 157-161.

Doebley J., A. Stec and C. Gustus. 1995. *Teosinte branched1* and the origin of maize: evidence for epistasis and the evolution of dominance. Genetics **141**, 333-346.

Darwin C. 1872. The origin of species by means of natural selection, 6th edn. London, UK: John Murray.

Daehler C. C. and D. R. Strong. 1997. Hybridization between introduced smooth cordgrass *Spartina alterniflora* (Poaceae) and native California cordgrass (*S. foliosa*) in San Francisco Bay, California USA. American Journal of Botany **81**, 307-313.

daSilva J. and B. W. S. Sobral. 1996. Polyploid genetics. Pages 3-35 in The Impact of Plant Molecular Genetics. B. W. S. Sobral, editor. Birkhäuser, Boston, pp. 3-35.

Delph L. F. 2003. Sexual dimorphism in gender plasticity and its consequences for breeding system evolution. Evolution and Development **5**, 34-39.

Dekker J. 1997. Weed diversity and weed management. Weed Science **45**, 357-363.

Dekker J. 1999. Soil weed seed seed banks and weed management. Journal of Crop Production **2**, 139-166.

Dekker J. 2003. The evolutionary biology of the foxtail (*Setaria*) species-group. Pages 65-114 in Principles and Practices in Weed Management: Weed Biology and Management. Inderjit, editor. Kluwer Academic Publishers, The Netherlands.

de Vries H. 1905. Species and Varieties, their Origin by Mutation. Open Court Publishing Company. Chicago, IL.

DeWalt S. 2003. The invasive tropical shrub *Clidemia hirta* (Melastomataceae) in its native and introduced ranges: tests of hypotheses of invasion. PhD Dissertation. Louisiana State University.

Doyle J. J., J. L. Doyle, J. T. Rauscher and A. H. D. Brown. 2004. Diploid and polyploid reticulate evolution throughout the history of the perennial soybeans (*Glycine* subgenus Glycine). New Phytologist **161**, 121-132.

Ellstrand N. and K. A. Schierenbeck. 2000. Hybridization as a stimulus for the evolution of invasiveness in plants? Proceedings of the National Academy of Sciences USA **97**, 7043-7050.

Feldman M., B. Liu, G. Segal, S. Abbo, A. A. Levy and J. M. Vega. 1997. Rapid Elimination of low-copy DNA sequences in polyploid wheat: a possible mechanism for differentiation of homoeologous chromosomes. Genetics **147**, 1381-1387.

Ferris C., R. A. King and A. J. Gray. 1997. Molecular evidence for the maternal parentage in the hybrid origin of *Spartina anglica* C. E. Hubbard. Molecular Ecology **6**, 185-187.

Finnegan E. J. 2001. Epialleles — a source of random variation in times of stress. Current Opinion in Plant Biology **5**, 101-106.

Finnegan E. J., R. K. Genger, W. J. Peacock and E. S. Dennis. 1998. DNA methylation in plants. Annual Review of Plant Physiology. Plant Molecular Biology **49**, 223-247.

Finnegan E. J., W. J. Peacock and E. E. Dennis. 2000. DNA metylation, a key regulator of plant development and other processes. Current Opinion in Genetics and Development **10**, 217-223.

Gaskin J. F. and B. A. Schaal. 2002. Hybrid *Tamarix* widespread from U. S. invasion and undected in native Asian range. Proceedings of the National Academy of Sciences USA **99**, 11256-11259.

Gilbert S. F. and J. A. Bolker. 2003. Ecological developmental biology: preface to the symposium. Evolution and Development **5**, 3-8.

Grant V. 1981. Plant speciation. New York, USA: Columbia University Press.

Gressel J. 2000. Molecular biology of weed control. Transgenic Research **9**, 355-382.

Grotkopp E., M. Rejmanek and T. L. Rost. 2002. Toward a causal explanation of plant invasiveness: seedling growth and life-history strategies of 20 pine (*Pinus*) Species. American Naturalist **159**, 396-419.

Hafliger E. and H. Scholz. 1980. Grass weeds 1. Weeds of the Subfamily Panicoideae. Ciba-Geigy, Ltd., Basel, Switzerland. 123 pp.

Hamrick J. L., M. J. W. Godt and S. L. Sherman-Broyles. 1992. Factors influencing levels of genetic diversity in woody plant species. New Forests **6**, 95-124.

Hamrick J. L., Y. B. Linhart and J. B. Mitton. 1979. Relationships between life history characteristics and electrophoretically detectable genetic variation in plants. Annual Review of Ecology and Systematics **10**, 173-200.

He P., B. Friebe, B. S. Gill B. S. and J. Zhou. 2003. Allopolyploidy alters gene expression in the highly stabble hexaploid wheat. Plant Molecular Biology **52**, 401-414.

Hegde S. G., J. Valkoun and J. G. Waines. 2002. Genetic diversity in wild and weedy *Aegilops*, *Amblyopyrum*, and *Secale* species — A preliminary survey. Crop Science **42**, 608-614.

Huey R. B., G. W. Gilchrist, M. L. Carlson, D. Berrigan and L. Serra. 2000. Rapid evolution of a geographic cline in size in an introduced fly. Science **287**, 308-309.

Huxel G. R. 1999. Rapid displacement of native species by invasive species: effects of hybridization. Biological conservation **89**, 143-152.

Iltis Hugo. *Life of Mendel*, trans. Eden and Cedar Paul. (Hafner Publishing, 1932/1966).

Jackson R. B., C. R. Linder, M. Lynch, M. Purugganan, S. Somerville and S. S. Thayer. 2002. Linking molecular insight and ecological research. Trends in Ecology and Evolution **17**, 409-414.

Johnston J. A., M. L. Arnold and L. A. Donovan. 2003. High hybrid fitness at seed and seedling life history stages in Louisiana irises. Journal of Ecology **91**, 438-446.

Kashkush K., M. Feldman and A. A. Levy. 2002. Gene Loss, Silencing and Activation in a Newly Synthesized Wheat Allotetraploid. Genetics **160**, 1651-1659.

Kashkush K., M. Feldman and A. A. Levy. 2003. Transcriptional activation of retrotransposons alters the expression of adjacent genes in wheat. Nature Genetics **33**, 102–106.

Kim S.-C. and L. H. Rieseberg. 1999. Genetic architecture of species differences in annual sunflowers: implications for adaptive trait introgression. Genetics. **153**, 965-977.

Kolar C. and D. M. Lodge. 2001. Progress in invasion biology: predicting invaders. Trends in Ecology and Evolution **18**, 199-204.

Lawton-Rauh A. 2003. Evolutionary dynamics of duplicated genes in plants. Molecular Phylogenetics and Evolution **29**, 396-409.

Lee C. E. 2002. Evolutionary genetics of invasive species. Trends in Ecology and Evolution **17**, 386-391.

Lee H. and Z. J. Chen 2001. Protein-coding genes are epigenetically regulated in *Arabidopsis* polyploids. Proceedings of the National Academy of Science USA **98**, 6753-6758.

Leger E. A. and K. J. Rice. 2002. Invasive California poppies (*Eschscholzia californica* Cham.) grow larger than native individuals under reduced competition. Ecology Letters **6**, 257-264.

Levy A. A. and M., Feldman 2004. Genetic and epigenetic reprogramming of the wheat genome upon allopolyploidisation. Biological Journal of the Linnean Society. In press.

Lewis W. H. 1980. Polyploidy: biological relevance. Plenum, New York, New York, USA.

Lexer C., M. E. Welch, J. L. Durphy and L. H. Rieseberg. 2003. Natural selection for salt tolerance quanitative trait loci (QTLs) in wild sunflower hybrids: Implications for the origin of *Helianthus paradoxus*, a diploid hybrid species. Molecular Ecology **12**, 1225-1235.

Lexer C., Z. Lai and L. H. Rieseberg. 2004. Candidate gene polymorphisms associated with salt tolerance in wild sunflower hybrids: implications for the origin of *Helianthus paradoxus*, a diploid hybrid species. New Phytologist **161**, 225-233.

Losos J. B., K. I. Warheit and T. W. Schoener. 1997. Adaptive differentiation following experimental island colonization in *Anolis* lizards. Nature **387**, 70-73.

Liu B., Vega, J. M., Segal, G., Abbo S, Rodova, M. and M., Feldman 1998a. Rapid genomic changes in newly synthetized amphiploids of *Triticum* and *Aegilops*. I. Changes in low-copy noncoding DNA sequences. Genome **41**, 272-277.

Liu B, Vega, J. M., Segal, G., Abbo, S., Rodova, M. and M. Feldman 1998b. Rapid genomic changes in newly synthetized amphiploids of *Triticum* and *Aegilops*. II. Changes in low-copy coding DNA sequences. Genome **41**, 535-542.

Liu B. and J. F., Wendel 2000. Retrotransposon activation followed by rapid repression in introgressed rice plants. Genome **43**, 874-880.

Liu B., C. L. Brubaker, G. Mergeai, R. C. Cronn, J. F. Wendel. 2001. Polyploid formation in cotton is not accompanied by rapid genomic changes. Genome **44**, 321-330.

Liu B. and J. F. Wendel. 2002. Non-Mendelian phenomena in allopolyploid genome evolution. Current Genomics **3**, 489–506.

Liu B. and J. F. Wendel. 2003. Epigenetic phenomena and the evolution of plant allopolyploids. Molecular Phylogenetics and Evolution **29**, 369-379.

Linde M., S. Diel and B. Neuffer 2001. Flowering ecotypes of *Capsella bursa-pastoris* (L.) *Medik.* (Brassicaceae) analysed by a cosegregation of phenotypic characters (QTL) and molecular markers. Annals of Botany **87**, 91-99.

Loveless M. and J. L. Hamrick. 1984. Ecological determinants of genetic structure in plant populations. Annual Review of Ecology and Systematics **15**, 65-95.

Lynch M. and B. Walsh. 1998. Genetics and Analysis of Quantitative Traits. Sinauer Associates, Sunderland, Massachusetts.

Lynch M. and J. S. Conery. 2000. The evolutionary fate and consequences of duplicated genes. Science **290**, 1151-1155.

Madlung A., Masuelli, R. W., Watson, B., Reynolds, S. H., Davison, J. and L. Comai 2002. remodeling of DNA methylation and phenotypic and transcriptional changes in synthetic *Arabidopsis* allotetraploids. Plant Physiology **129**, 733-746.

Martienssen R. A. and V. Colot. 2001. Methylation and epigenetic inheritance in plants and filamentous fungi. Science **293**, 1070-1074.

Matzke M. A., Mette, M. F. and A. J. M. Matzke. 2000. Transgene silencing by the host genome defense: implications for the evolution of epigenetic control mechanisms in plants and vertebrates. Plant Molecular Biology **43**, 401-415.

Matzke M. A., Matzke, A. J, Pruss, G. J. and V. B. Vance. 2001. RNAbased silencing strategies in plants. Current Opinions in Genetics and Development **11**, 221-227.

Mayo G. M. and P. Langridge. 2003. Modes of reproduction in Austrailian populations of *Hypericum perforatum* L. (St. John's wort) revealed by DNA fingerprinting and cytological methods. Genome **46**, 573-579.

Mette M. F., W. Aufsatz, J. van der Winden, M. A. Matzke and A. J. M. Matzke. 2000. Transcriptional gene silencing and promoter methylation triggered by double stranded RNA. EMBO Journal **19**, 5194-5201.

Moore N. J. and G. F. Moran. 1989. Microgeographic patterns of allozyme variation in Casuarina cunninghamiana within and between the Murrumbidge and coastal drainage systems. Australian Journal of Botany **37**, 181-192.

Mooney H. A. and E. E. Cleland. 2001. The evolutionary impact of invasive species. Proceedings of the National Academy of Sciences USA **98**, 5446-5451.

Moran G. F., O. Muona and J. C. Bell. 1989. Breeding systems and genetic diversity in Acacia auriculiformis and A. crassicarpa. Biotropica **21**, 250-256.

Moran G. F., J. C. Bell and J. W. Turnbull. 1989. A cline in genetic diversity in River she-oak (Casuarina cunninghamiana) Australian Journal of Botany **37**, 169-180.

Müller-Schärer H. and T. Steinger. 2004. Predicting evolutionary change in invasive, exotic plants and its consequences for plant-herbivore interactions. Pages 137-162 in Genetics, Evolution, and Biological Control. L. E. Ehler, R. Sforza, and T. Mateille, editors. CAB International.

Neuffer B. and H. Hurka. 1999. Colonization history and introduction dynamics of *Capsella bursa-pastoris* (Brassicaceae) in North America: isozymes and quantitative traits. Molecular Ecology 8, 1667-1681.

Nijhout H. F. 2003. Development and evolution of adaptive polyphenisms. Evolution and Development **5**, 9-18.

Novak S. J. and R. N. Mack. 1993. Genetic variation in *Bromus tectorum* (Poaceae): comparison between native and introduced populations. Heredity **71**, 167-176.

Novak S. J., R. N. Mack and P. S. Soltis. 1993. Genetic variation in *Bromus tectorum* (Poaceae): Introduction dynamics in North America. Canadian Journal of Botany: **71**, 1441-1448.

Ohno S. 1970. Evolution by gene duplication. Springer Verlag, New York.

Ohta T. and J. H. Gillespie. 1996. Development of neutral and nearly neutral theories. Theoretical Population Biology **49**, 128-142.

Olden J. D., N. L. Poff, M. R. Douglas, M. E. Douglas and K. D. Fausch. 2004. Ecological and evolutionary consequences of biotic homogenization. Trends in Ecology and Evolution **19**, 18-24.

Osborn T. C., J. C. Pires, J. A. Birchler, D. L. Auger, Z. J. Chen, H.-S. Lee, L. Comai, A. Madlung, R. W. Doerge, V. Colot and R. A. Martienssen. 2003. Understanding mechanisms of novel gene expression in polyploids. Trends in Genetics **19**, 141-147.

Otto S. P. and J. Whitton. 2000. Polyploid incidence and evolution. Annual Review of Genetics **34**, 401-437.

Ozkan H., A. A. Levy and M. Feldman. 2001. Allopolyploidy-induced rapid genome evolution in the wheat (*Aegilops-Triticum*) group. Plant Cell **13**, 1735-1747.

Paneida F. D. and S. A. Carstairs. 1989. Isozymic discrimination of tropical Australian populations of mesquite (*Prosopis* spp.); implications for biological control. Weed Research **29**, 157-165.

Pappert R. A., J. L. Hamrick and L. A. Donovan. 2000. Genetic variation in *Pueraria lobata* (Fabaceae), an introduced, clonal, invasive plant of the southeastern United States. American Journal of Botany **87**, 1240-1245.

Parker B., G. Miller and L. C. Burrill. 1994. Scotch broom, *Cytisus scoparius* (L.) Link. Pacific Northwest Extension Publication 103.

Parker I. M., J. Rodriguez and M. E. Loik. 2003. An evolutionary approach to understanding the biology of invasions: Local adaptation and general-purpose genotypes in the weed *Verbascum thapsus*. Conservation Biology **17**, 59-72.

Paterson A. H., K. F. Schertz, Y-R. Lin, S-C. Liu and Y-L. Chang. 1995. The weediness of wild plants: Molecular analysis of genes influencing dispersal and persistence of johnsongrass, *Sorghum halepense* (L.) Pers. Proceedings of the National Academy of Sciences USA **92**, 6127-6131.

Paterson A. H. 2002. What has QTL mapping taught us about plant domestication? New Phytologist **154**: 591–608.

Petit R. J., C. Bodénès, A. Ducousso, G. Roussel and A. Kremer. 2004. Hybridization as a mechanism of invasion in oaks. New Phytologist **161**, 151-164.

Pigliucci M. 2001. Phenotypic Plasticity: Beyond Nature and Nurture. John Hopkins University Press, Baltimore.

Pohl R. W. 1951. The genus *Setaria* in Iowa. Iowa State Journal of Science. **25**, 501-508.

Pohl R. W. 1966. The grasses of Iowa. Iowa State Journal of Science. **40**, 341-373.

Purugganan, M. 2000. The molecular population genetics of regulatory genes. Molecular Ecology **9**, 1451-1461.

Raybould A., A. J. Gray, M. J. Lawrence, D. F. Marshall. 1991. The evolution of *Spartina anglica* C. E. Hubbard (Gramineae): origin and genetic variability. Biological Journal of the Linnean Society **43**, 111-126.

Reed D. H. and R. Frankham. 2001. How closely correlated are molecular and quantitative measures of genetic variation. A meta-analysis. Evolution **55**, 1095-1103.

Rejmanek M. and D. M. Richardson. 1996. What attributes make some plant species more invasive? Ecology **77**, 1655-1661.

Reznick D. N. and C. K. Ghalambor. 2001. The population ecology of contemporary adaptations: what emprical studies reveal about the conditions that promote adaptive evolution. Genetica **112-113**, 183-198.

Riddle N. C. and J. A. Birchler. 2003. Effects of reunited diverged regulatory hierarchies in allopolyploids and species hybrids. Trends in Genetics **19**, 597-600.

Rieseberg L. H. 1997. Hybrid origins of plant species. Annual review in Ecology and Systematics **28**, 359-389.

Rieseberg L. H. and J. F. Wendel. 1993. Introgression and its consequences in plants. In: Harrisson R., ed. Hybrid zones and the evolutionary process. New York, USA. Oxford University Press 70-109.

Rieseberg L. H. and J. Wendel. 2004. Plant speciation – rise of the poor cousins. New Phytologist **161**, 3-8.

Rieseberg L. H., M. A. Archer and R. K. Wayne. 1999. Transgressive segregation, adaptation, and speciation. Heredity **83**, 363-372.

Rieseberg L. H., O. Raymond, D. M. Rosenthal, Z. Lai, K. Livingstone, T. Nakazato, J. L. Durfy, A. E. Scwarzabach, L. A. Donovan and C. Lexer. 2003. Major ecological transitions in wild sunflowers facilitated by hybridization. Science **301**, 1211-1216.

Roose M. L. and L. D. Gottlieb. 1976. Genetic and biochemical consequences of polyploidy in *Tragopogon*. Evolution **30**, 818-830.

Saltonstall K. 2002. Cryptic invasion by a non-native genotype of the common reed, Phragmites australis, into North America. Proceedings of the National Academy of Sciences USA **99**, 2445-2449.

Schaal B. A., J. F. Gaskin and A. L. Caicedo. 2003. Phylogeography, haplotype trees, and invasive plant species. Journal of Heredity **94**, 194-204.

Schierenbeck K. A., G. L. Stebbins and R. W. Patterson. 1992. Morphological and cytological evidence for polyphyletic allopolyploidy in *Arctostaphylos mewukka* (Ericaceae). Plant Systematics and Evolution **179**, 187-205.

Schierenbeck K. A., J. L. Hamrick, R. N. Mack. 1995. Comparison of allozyme variability in a native and an introduced species of *Lonicera*. Heredity **75**, 1-9.

Schranz M. E. and Osborn T. C. 2000. Novel flowering time variation in the resynthesized polyploid *Brassica napus*. Journal of Heredity **91**, 242-246.

Schlichting C. D. 2003. Origins of differentiation via phenotypic plasticity. Evolution and Development **5**, 98-105.

Schranz M. E. and Osborn T. C. 2000. Novel flowering time variation in resynthesized polyploid *Brassica napus*. Journal of Heredity **91**, 242-246.

Seehausen O. 2004. Hybridization and adaptive radiation. Trends in Ecology and Evolution (in press).

Shaked H., K. Kashkush, H. Ozkan, M. Feldman and A. A. Levy. 2001. Sequence elimination and cytosine methylation are rapid and reproducible responses of the genome to wide hybridization and allopolyploidy in wheat. Plant Cell **13**, 1749-1749.

Siemann E. and W. E. Rogers. 2001. Genetic differences in growth of an invasive tree species. Ecology Letters **4**, 514-518.

Senchina D. S., I. Alvarez, R. C. Cronn, B. Liu, J. Rong, R. D. Noyes, A. H. Paterson, R. A. Wing, T. A. Wilkins and J. F. Wendel. 2003. Rate variation among nuclear genes and the age of polyploidy in *Gossypium*. Molecular Biology and Evolution **20**, 633-643.

Slatkin M. and R. R. Hudson. 1991. Pairwise Comparisons of Mitochondrial DNA Sequences in Stable and Exponentially Growing Populations. Genetics **129**, 555-562.

Small R. L. and J. F. Wendel. 2000. Phylogeny, duplication, and intraspecific variation of Adh sequences in New World diploid cottons (*Gossypium*, Malvaceae). Molecular Phylogenetics and Evolution **16**, 73-84.

Small R. L. and J. F. Wendel. 2002. Differential evolutionary dynamics of duplicated paralogous Adh loci in allotetraploid cotton (*Gossypium*). Molecular Biology and Evolution **19**, 597-607.

Soltis D. E. and P. S. Soltis. 1999. Polyploidy: recurrent formation and genome evolution. Trends in Ecology and Evolution **14**, 348-352.

Soltis P. S. and D. E. Soltis. 2000. The role of genetic and genomic attributes in the success of polyploids. Proceedings of the National Academy of Sciences USA **97**, 7051-7057.

Soltis D. E., P. S. Soltis, J. C. Pires, A. Kovarik and J. A. Tate. 2004. Recent and recurrent polyploidy in *Tragopogon* (Asteraceae): cytogenetic, genomic, and genetic comparisons. Biological Journal of the Linnean Society. In press.

Song K., P. Lu, K. Tang and T. C. Osborn. 1995. Rapid genome change in synthetic polyploids of *Brassica* and its implications for polyploid evolution. Proceedings of the National Academy of Sciences USA **92**, 7719-7723.

Squirrell J., P. M. Hollingsworth, R. M. Bateman, J. H. Dickson, M. H. S. Light, M. MacConaill and M. Tebbitt. 2001. Partitioning and diversity of nuclear and organelle markers in native and introduced populations of *Epipactis helleborine* (Orchidaceae). American Journal of Botany **88**, 1409-1418.

Stebbins G. L. 1950. Variation and evolution in plants. New York; USA: Columbia University Press.

Stebbins G. L. 1965. Colonizing species of the native California flora. Pp. 173-195 in H. G. Baker and G. L. Stebbins, editors. The genetics of colonizing species. Academic Press, New York. 588 pp.

Stebbins G. L. 1971. Chromosomal evolution in higher plants, Addison-Wesley, Reading.

Strauss S. Y., J. A. Rudgers, J. A. Lau and R. E. Irwin. 2002. Direct and ecological costs of resistance to herbivory. Trends in Ecology and Evolution **17**, 278-285.

Sultan S. 2003. Phenotypic plasticity in plants: a case study in ecological development. Evolution and Development **5**, 25-33.

Surles S. C., J. Hamrick and B. C. Bongarten. 1989. Allozyme variation in black locust (*Robinia pseudoacacia*). Candian Journal of Forest Research **19**, 471-479.

Templeton A. R. 2004. Statistical phylogeography: methods of evaluating and minimizing inference errors. Molecular Ecology **13**, 1-21.

Thompson J. D. 1990. Morphological variation among natural populations of *Spartina anglica*. Pages 26-33 in *Spartina anglica*, a research review. A-J. Gray and P. E. M. Benham, editors. Institute of Terrestrial Ecology. Natural Environment Research Council. London.

Thompson J. D. 1991. The biology of an invasive plant: What makes *Spartina anglica* so successful? Bioscience **41**, 393-401.

Tian L. and Z. F. Chen. 2001. Blocking histone deacetylation in *Arabidopsis* induces pleiotropic effects on plant gene regulation and development. Proceedings of the National Academy of Sciences USA **98**, 200-205.

Urbanska K. M., H. Hurka, E. Landolt, B. Neufler and K. Mummenhoff. 1997. Hybridization and evolution in *Cardamine* (Brassicaceae) at Urnerboden, Central Swtitzerland: biosystematic and molecular evidence. Plant Systematics and Evolution **204**, 233-256.

Volkov R. A., Borisjuk, N. V., Panchuk I. I., Schweizer, D. and V. Helmleben 1999. Elimination and rearrangement of parental rDNA in the allotetraploid *Nicotiana tabacum*. Molecular Biology and Evolution **16**, 311-320.

Vitousek P. 1986. Biological invasions and ecosystem properties : can species make a difference? Pages 163-176 in Ecology of Biological Invasions in North America and Hawaii. H. A. Mooney and J. A. Drake, editors. Springer-Verlag, New York.

Wang R.-L., J. F. Wendel and J. H. Dekker. 1995. Weedy adaptation in *Setaria* spp. I. Isozyme analysis of genetic diversity and population genetic structure in Setaria viridis. American Journal of Botany **82**, 308-317.

Warwick S. I. 1990. Allozyme and life-history variation in five northwardly colonizing North American weed species. Plant Systematics and Evolution **169**, 41-54.

Waterhouse B. M. and O. Zeimer. 2002. 'On the brink': the status of *Chromolaena odorata* in Northern Australia. Pages 29-33 in Proceedings of the fifth internation workshop on Biological Contral and Management of *Chormolaena odorata*, Durban, South Africa. Zachariades, C., R. Muniappan and L. W. Strathie, editors.

Weber E. and B. Schmid. 1998. Latitudinal population differentiation in two species of *Solidago* (Asteraceae) introduced into Europe. American Journal of Botany **85**, 1110-1121.

Wendel J. F., A. Schnabel and T. Seelanan. 1995. Bi-directional interlocus concerted evolution following allopolyploid speciation in cotton (*Gossypium*). Proceedings of the National Academy of Sciences of the USA **92**, 280-284.

Wendel J. F. 2000. Genome evolution in polyploids. Plant Molecular Biology **42**, 225-249.

Wendel J. F. and J. J. Doyle. 2004. Polyploidy and evolution in plants. In: Diversity and Evolution in Plants. R. Henry Ed. CAB Publishing, Oxon U. K.

Wetzel D. K., M. J. Horak, D. Z. Skinner and P. A. Kulakow. 1999. Transfer of herbicide resistance traits from *Amaranthus palmeri* to *Amaranthus rudis*. Weed Science **47**, 538-543.

Williams D. G. 1992. Physiological ecology of the invasive grass *Pennisetum setaceum* on Hawaii. PhD Dissertation. Washington State University.

Wolfe K. H. 2001. Yesterday's polyploids and the mystery of diploidization. Nature Reviews Genetics **2**, 333-341.

Wolffe A. P. and M. A. Matzke. 1999. Epigenetics: regulation through expression. Science **286**, 481-486.

Xu C.-Y., W.-J. Zhang, C.-Z. Fu and B. Rong. 2003. Genetic diversity of alligator weed in China by RAPD analysis. Biodiversity and Conservation **12**, 637-645.

Yannic G., A. Baumel and M. L. Aïnouche. 2004. Uniformity of the nuclear and chloroplast genomes of *Spartina maritima* (Poaceae) a salt marshes species in decline along the Western European Coast. Heredity (In press.)

Ye W. H., J. Li, H. L. Cao and X. J. Ge. 2003. Genetic uniformity of *Alternanthera philoxeroides* in South China. Weed Research **43**, 297-302.

Yoder J. A., Walsh, C. P., Bestor, T. H. 1997. Cytosine methylation and the ecology of intragenomic parasites. Trends in Genetics **13**, 335-340.

Zhao X. P., Si, Y., Hanson, R. E., Crane, C. R., Price, H. J., Stelly, D. M., Wendel, J. F. and A. H. Paterson. 1998. Dispersed repetitive DNA has colonized new genomes since polyploid formation in cotton. Genome Research **8**, 479-492.

Unwelcomed visitor: species interactions

Contact experience, alien-native interactions, and their community consequences: a theoretical consideration on the role of adaptation in biological invasions

M. Kondoh

INTRODUCTION

Ecological theory suggests that interspecific interactions arising between an alien species and native species hold the key to invasion success by an alien species. The well-known hypothesis that an alien species that has few natural enemies is more likely to be a successful invader (e.g., Blossey and Nötzold 1995, Mitchell and Power 2003) implies the importance of a trophic interaction. The hypothesis that a community with higher species richness is more resistant to a biological invasion (Stachowicz *et al.* 1999, Kennedy *et al.* 2002) is strongly based on the competition theory (Tilman 1982). Further, the way through which a biological

M.W. Cadotte et al. (eds.), Conceptual ecology and invasion biology, 225–242.

invasion affects the local community is an alien-native interspecific interaction such as resource competition, trophic interaction and allelopathy. The important role of interspecific interactions in a biological invasion suggests the detailed evaluation of the interspecific interactions between alien and native species as an interesting approach to the biological invasion issues.

There are two different, but not mutually exclusive, views of what characterizes an alien-native interaction: the "species-identity" and "contact-experience" views. The species-identity view focuses on the identity of the focal species (e.g., Baker 1974, Roy 1990, Kolar and Lodge 2001). The alien and native species are characterized by fixed sets of species-specific traits such as competitive ability (Herbold and Moyle 1986, Petren and Case 1996, Holway 1999, Byers 2000), which influence their interaction and thus the invasion success and its impact on the local community (e.g., Williams 1996). In this view an interspecific interaction that arises between an alien and native species is explained mainly by their inherent species-specific traits and so are its population-level consequences (Mack *et al.* 2000).

The contact-experience view (Sakai *et al.* 2001) focuses on the lack of "contact experience" between alien and native species as the major factor characterizing their interaction. Native species that coexist in the local habitat have the experience of encounters or interactions with each other over varying timescales, which provides the opportunity for local species to adapt to each other (e.g., Lawlor and Maynard Smith 1976). In contrast, an alien species and a native species, due to temporal or spatial separation, share a relatively short history and are less likely to adapt to each other. The lack of adaptation should have major impact on the alien-native interspecific interaction (e.g., Ware 1971, Dill 1974, Marurran 1989, Croy and Hughes 1991, Hughes *et al.* 1992, Maloney and Mclean 1995, Callaway and Aschehoug 2000, Rehage *et al.* 2004) and the resultant community structure (Griffin *et al.* 2000, Ricciardi *et al.* 1998, Ricciardi and Atkinson 2004). However, only a few studies have comprehensively considered the contact experience effect in the context of biological invasion issues (but see Sakai *et al.* 2001). Now, we need a strong conceptual basis for the issue of how contact experience influences a biological invasion for guiding further research.

In this chapter I present a theoretical framework for applying the contact experience view to alien-native trophic interactions. I chose this particular interaction as it is known to play the essential role in shaping community structure (e.g., Hairstone *et al.* 1960, Fretwell 1977, Power 1992). It drives diverse prey-predator population dynamics, influences population density and distribution (Hairstone *et al.* 1960), generates two major interspecific indirect interactions {resource competition (Tilman 1982) and apparent competition (Holt 1977, Bonsall and Hassell 1997; but see Chaneton and Bonsall 2000)} that limit the number of coexisting species and forms a major component of the natural selection pressure (Endler 1986). Further, the macroscopic food-web topology can be an essential determinant of population stability (Pimm 1991) and ecosystem function (Thébault and Loreau 2003).

The two major questions of the present chapter are of (1) how the lack of contact experience influences a trophic interaction, and (2) what is the population to community-level consequence of a contact experience-free interaction. To answer these questions, I first explain the behavioral basis of a trophic interaction. Then I will provide several hypotheses on how alien-native interspecific trophic interactions differ from interactions between natives and how this difference affects the population or community following a biological invasion.

THE CONTACT-EXPERIENCE VIEW OF ALIEN-NATIVE TROPHIC INTERACTIONS

Failure of recognition

Consider a dog and a piece of meat that is releasing an odor. If hungry enough, the dog will follow the odor and eat the meat; there is an individual-level trophic interaction. The dog's foraging behavior consists of three sub-processes (e.g., Griffin *et al.* 2000): "information reception", "information processing", and "behavioral response". The dog detects the odor using chemical receptors and the nervous system (information reception). The odor is associated with meat based on the dog's previous experience or an inborn association (information processing). Using this information, the dog makes a decision to follow the odor and locates and eats the meat (behavioral response).

Information processing is the sub-process by which a meaning (e.g., "there is meat") that is conveyed by the cue (the odor) is "understood". In general, cue alone often have little to do with the information that is conveyed. This is analogous to the fact that how a word sounds has little to do with what the word actually means. Thus successful information processing, or whether the meaning of the cue is correctly understood, depends on the capability of cue-information association (e.g., Thorndike 1911, Pavlov 1927). In the case of the dog and the meat, for example, the chemical composition of the meat's odor does not necessarily mean a block of meat or potential food without the correct association.

The cue-information association, the core of information processing, is shaped through "experience" on various time scales. Natural selection might create the correct association if the capability of information processing affects reproductive success (Roeder and Treat 1961), while an individual might learn the visual, chemical, or phonic cues of potential prey through continuous encounter experiences with the prey on the shorter time scale (Domjan and Burkhard 1986, Foster 1999). This experience dependence implies that the lack of contact experience between a prey and predator influences the behavioral responses of one to the other by altering information processing and, consequently, affects their trophic interaction.

In trophic interactions it is convenient to consider two different levels of recognition, "trophic role recognition" and "trophic species recognition". The former is recognition of whether the individual is potential prey, potential predator or neither, while the latter is related to more detailed recognition within a trophic role such as distinction between species or organism types that require different strategies to cope with. If trophic role recognition fails, all the behaviors, which should follow a contact with potential prey/predator, does not follow. Foraging behavior, which normally follows contact with prey, might not occur between two species that have no contact experience with each other (e.g., Ware 1971, Croy and Hughes 1991, Bond and Kamil 1998). A prey might fail to recognize (Curio 1993), and can be more vulnerable to, a novel predator (e.g., Dill 1974, Magurran 1989, Maloney and Mclean 1995). If trophic species recognition fails, an incorrect behavioral response, which should follow a contact with another trophic species, may follow the encounter. Possible consequences of this will be discussed later.

Failure of the behavioral response

Another possible mechanism through which the lack of contact experience alters a trophic interaction is a failure in the third sub-process, behavioral response. The strategy required to consume a prey is often prey-species specific. A predator has to choose the correct place and time to find a prey, or might need a specific technique to capture or consume the prey. In this situation, whether a predator can use a novel prey depends on the applicability of the predator's foraging strategy, which is shaped through previous interactions with other prey species, to the novel prey species (skill transfer theory; Ellis 1965, Hazlett 1995). If the novel prey behaves or responds like a prey familiar to the predator, the predator might be able to handle the novel prey (Ricciardi and Atkinson 2004), and a trophic interaction would occur normally. If previous foraging behavior is not applicable to the novel prey, the predator is unlikely to consume the novel prey successfully. Thus the lack of contact experience can prevent successful consumption behavior (Ware 1971, Croy and Hughes 1991, Hughes *et al.* 1992). The same should be true for the prey's anti-predation behavior (Soluk 1993; Sih *et al.* 1998), as a behavior required to avoid a predator is predator-specific. If the prey has no contact experience with the focal predator, and if the prey's anti-predation strategy is not applicable to the novel predator, the correct predator-specific defence might not follow the contact.

In an alien-native interaction, the behavioral response (either prey consumption or predator avoidance) might fail even if the counterpart species is not novel to the focal species. This is because a habitat-specific history can lead to divergence in the way in which prey is used or predators are avoided in different habitats (e.g., Brodie and Brodie 1991, Thompson 1994). Consider two separate habitats, A and B, inhabited by the same set of prey and predator species. Prey-predator co-adaptation occurs independently in the two habitats, implying that the prey and predator from different habitats are not experienced with each other. Therefore, the prey-predator interaction arising from the opposite combination (prey from A and predator from B or *vice versa*) can differ from the original interaction within Habitat A or B.

CONSEQUENCES OF THE CONTACT EXPERIENCE-FREE TROPHIC INTERACTIONS

High variance in alien-native interaction strength

According to biological invasion theory, an alien prey species can have more advantages than a native prey species in the local habitat because the alien species is less likely to have natural enemies in the new habitat (e.g., Elton 1958, Blossey and Nötzold 1995, Mack *et al.* 2000, Keane and Crawley 2002, Mitchell and Power 2003, Torchin *et al.* 2003). This is known as enemy release hypothesis. This hypothesis suggests that the lack of a native consumer's contact experience with the alien prey species lowers the strength of the trophic interaction between them. From the contact experience point of view, there are three potential reasons for this hypothetical pattern. First, a native consumer does not recognize the alien species as potential prey. Second, the native consumer does not "know" how to handle the novel prey species. Third, even if the alien prey species is not novel, it might behave in a novel way due to a habitat-specific evolutionary process.

However, I suggest that the enemy release hypothesis only considers a part of the total picture of alien-native trophic interactions. First, the hypothesis overlooks the prey's anti-predatory behavior. As the lack of contact experience may lower the alien's defence efficiency against the native predator (Rehage *et al.* 2004), then their interaction strength can be stronger. Taking this effect into account, the lack of contact experience has two opposite consequences: low predation efficiency, which decreases the interaction strength, and low defence efficiency, which increases the strength. Second, the enemy release hypothesis overlooks alien's prey. If the lack of contact experience affects the alien's vulnerability to native natural enemies, this should be also applicable to the interaction of the focal alien as a predator and a potential prey in the local habitat. In this interaction, the behaviors influenced by the contact experience are the alien's foraging behavior and the native's defence behavior.

In summary, four different behaviors must be considered to evaluate the effect of the lack of contact experience on interaction strength and invasion success: the alien's foraging behavior and anti-predatory behavior and the native's foraging behavior on the alien and anti-predatory behavior against the alien (Fig. 1). The complex interaction between the four inter-related behaviors makes consistent prediction more difficult than previously thought. Assume that Species A is introduced to a local habitat in which it interacts with a native natural enemy, Species E, and a native prey, Species R; and assume that the strength of a trophic interaction is determined by the balance between the prey's anti-predatory defence and the predator's foraging behavior. Whether the lack of contact experience

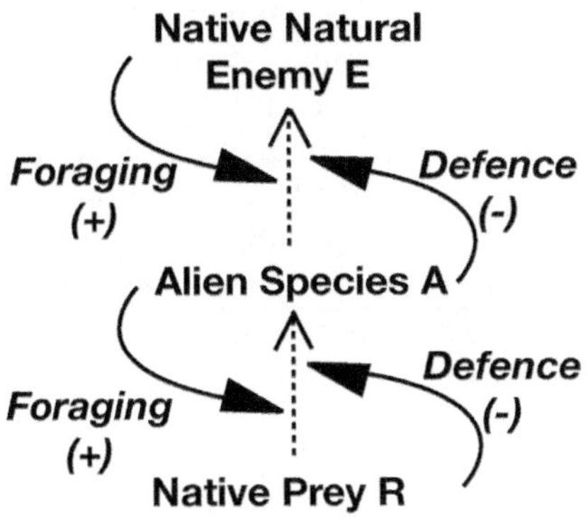

Fig. 1 Four behaviors affecting the trophic interactions between an alien species and a native species. If predator avoidance is more dependent on contact experience than is prey consumption, an alien species will have a stronger interaction with the native species. If prey consumption is more dependent on contact experience than is predator avoidance, an alien species will have a weaker interaction with the native species. If there is no consistent trend, the interaction strength will be characterized by a higher variance.

benefits the alien species depends on the level of each species' contingent pre-adaptation (Agrawal and Kotanen 2003, Ricciardi and Atkinson 2004) to an interacting species. The alien species would be at an advantage if its pre-adaptive defence level to Species E were higher, the pre-adaptive defence level of Species R to Species A were lower, the pre-adaptive foraging level of Species E to Species A were lower, or the pre-adaptive foraging level of Species A to Species R were higher.

If there is a consistent pattern in which behavior (foraging or antipredatory defence) or trophic role is more dependent on contact experience, then it is possible to make a consistent prediction on how the contact experience affects the interaction strength. For example, if predator relies more on the experience than prey does, a trophic interaction in the absence of contact experience will be weaker than an in its presence. If the prey is more dependent on the experience, a trophic interaction between novel species would be stronger than the normal one. If there is no consistent pattern in which one trophic role is more dependent on contact experience, the disruption of the predation-defence balance should lead to more variable trophic interaction strength. Which is the real pattern in nature? It would be interesting to examine how the frequency distribution of the interaction strength of alien-native interactions differs from that of native-native interactions.

It is well recognized that competitive ability is affected by the interacting effects of resource availability (bottom-up effect) and vulnerability to natural enemies (top-down effect) (e.g., Power 1992, Leibold 1996). Theory suggests that the coexistence of species sharing a similar resource and natural enemy requires a trade-off between resource competition and natural-enemy avoidance (Lubchenco 1978, Leibold 1996). If the strength of an alien-native trophic interaction is characterized by a high variance, it might lead to the prediction that the competitive superiority of an alien species should be abnormally high or low. If the trophic link to the consumption of the alien species is weak and the link from the alien species is strong, the alien will have a high population density. If the trophic link to the consumption of the alien species is strong and the link from the alien species is weak, the alien will have a low population density. This implies that, although there are many alien species that cannot become established, when an alien species succeeds in becoming established, it tends to be very abundant.

Alien-native mismatching in adaptive speed and "boom-and-bust" pattern of alien populations

Although the lack of contact experience may, as I have argued in the last section, lower the foraging efficiency or defence efficiency at the initial stage of invasion, the efficiency is likely to be gradually improved as an alien species and resident species learn or evolve to adapt each other (Carroll and Dingle 1996, Maron and Vila 2001). The previous studies that consider the coadaptive process have stressed the disadvantage of an alien species by suggesting that an alien species is often characterized by a low population density and low genetic diversity, which lowers evolution speed (Tsutsui *et al.* 2000). This is why multiple introductions, which increase genetic diversity, enhance alien's invasion success (Kolbe *et al.* 2004).

However, the disadvantage of alien species can be compensated. An example is alien's behavioral plasticity (Gray 1986). Sol and Lefebvre (2000) analyzed the relationship between brain size in birds and invasion success in New Zealand; they found that species with larger brains had a higher probability of introduction success than did species with smaller brains. This pattern is explained by the effect of brain size on the bird's interaction with resources: a bird with a larger brain is better at discovering new foods in the new habitat (Sol and Lefebvre 2000). This view provides a species identity explanation for invasion success, i.e., a species with high plasticity is a successful invader.

Here, from the viewpoint of contact experience, I show another factor that may overcome the alien's disadvantage. Consider that a small number of Species A is introduced to a community where a potential predator, Species E, exists (Figs. 2, 3). After its introduction, the alien Species A would learn or evolve to lower the predation pressure by Species E, while Species E would adapt to increase the predation pressure on Species A. Note that, in this co-adaptation process, the adaptive speeds would differ due to differences in the contact opportunities between alien and native species. When an alien species is introduced, its abundance should be much lower that that of native species. This implies that the alien prey has more opportunity to gain experience in handling the native predator

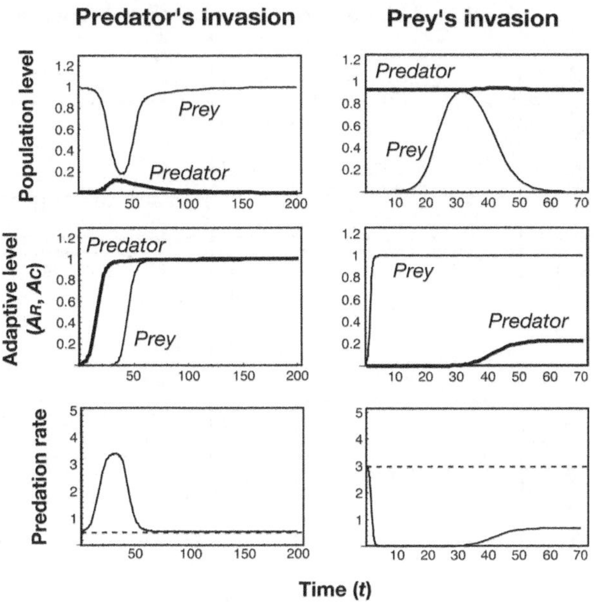

Fig. 2 Alien-native mismatching in evolutionary speed leads to boom-and-bust pattern of alien populations. Each column is for an alien predator and its prey (left column) and an alien prey and its predator (right column) in the presence of prey-predator coevolution. Each panel shows population dynamics (top; $C_N + C_A$, $R_N + R_A$), adaptive level (middle; the fraction of adapted individuals, $C_A/(C_N+C_A)$ for the predator and $R_A/(R_N+R_A)$ for the prey) and per-capita predation rate (bottom; (ann $C_N R_N + a_{na} C_A R_N + a_{an} C_N R_A + a_{aa} C_A R_A)/[(C_N + C_A)(R_N + R_A)]$). In the top and middle panels the thick lines and thin lines are for predator and prey, respectively. Consider populations of an alien species and native species, each of which consists of many non-adapted individuals (with low defence ability or low foraging ability) and a few adapted individuals (with high defence ability or high foraging ability). The population dynamics is described by: $d C_N / d t = r_c (K_c - C_N - C_A) C_N + e a_{nn} C_N R_N + e a_{an} C_N R_A - m_c C_N$; $d C_A / d t = r_c (K_c - C_N - C_A) C_A + e a_{aa} C_N R_N + e a_{an} C_N R_A - m_c C_A$; $d R_N / d t = r_r (K_r - R_N - R_A) R_N - a_{nn} C_N R_N - a_{na} C_A R_N - m_r R_N$; $d R_A / d t = r_r (K_r - R_N - R_A) R_A - a_{an} C_N R_A - a_{aa} C_A R_A - m_r R_A$, where C_N, C_A, R_N and R_A are population abundances of non-adapted predator, adapted predator, non-adapted prey and adapted prey, respectively, r_i is the intrinsic growth rate, K_i is carrying capacity supported by resources that are not explicitly represented by the equations, m_i is the mortality rate of the prey ($i = r$) and predator (c), e is the assimilation efficiency. The per-capita predation rates hold that $a_{na} > a_{nn} = a_{aa} > a_{an}$, implying that the predation rate of the non-adapted predator on the non-adapted prey is same with that of the adapted predator on the adapted prey. Parameters a_{nn} and a_{aa} are set to a value with which an alien species cannot persist. This setting confirms that a temporal increase of an alien species is due to the alien-native mismatch in evolutionary speed. The initial condition is that $(C_N(0), C_A(0), R_N(0), R_A(0)) = (0.0099, 0.0001, 0.999, 0.001)$ for predator's invasion and (0.924, 0.001, 0.099, 0.001) for prey's invasion. Parameters that are used are: (rc, rn, Kc, Kn, mc, mn, ann, ana, aan, aaa, e) = (0, 0.4, 0, 1, 0.075, 0, 0.5, 3.5, 0, 0.5, 0.1) for predator's invasion and (1, 0.4, 1, 1, 0.075, 0, 3, 5, 0, 3, 0.1) for prey's invasion.

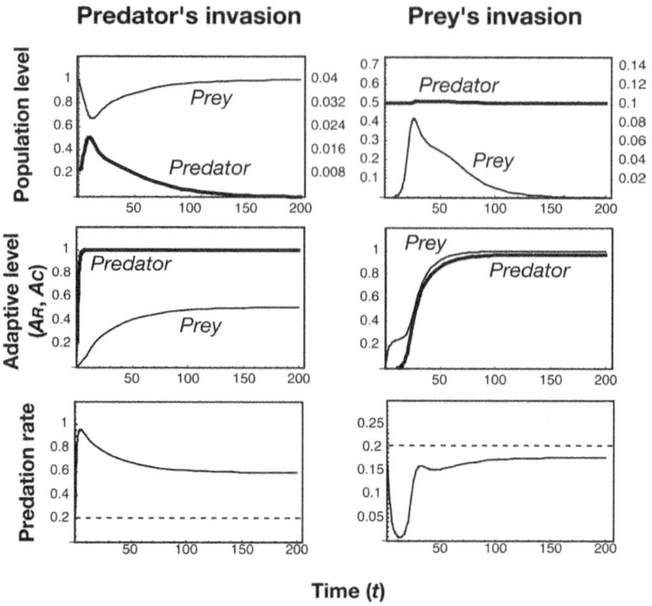

Fig. 3 Alien-native mismatching in speed of adaptive behavioral plasticity leads to boom-and-bust pattern of alien populations. Each column is for an alien predator species and its predator prey (left column) and an alien prey species and its predator (right column) in the presence of the learning process. Each panel shows population dynamics (top; *C*, *R*), adaptive level (middle; A_C) for the predator, A_R for the prey) and interaction strength (bottom; $A_C - a\ A_R$)). In the top and middle panels the thick lines and thin lines are for predator and prey, respectively. A scale for prey population level is indicated on the right of each panel. Consider populations of a prey species and a predator species, whose population dynamics is described by: $d\ C\ /\ d\ t = r_c\ (K_c - C)\ C + e\ F\ C\ R - m_c\ C$ and $d\ R\ /\ d\ t = r_r\ (K_r - R)\ R - F\ C\ R - m_r\ R$, where C and R are population abundances of the predator and prey, respectively, r_i is the intrinsic growth rate, Ki is carrying capacity supported by resources that are not explicitly represented by the equations, mi is the mortality rate of the prey ($i = r$) and predator (c), e is the assimilation efficiency. The per-capita predation rate, F, is a temporal variable and determined by predator's foraging-related adaptive levels, A_C, and prey's defence-related adaptive level, A_R: that is, $F = b\ (A_C - a\ A_R)$. The adaptive dynamics is given by: $d\ A_C\ /\ d\ t = G_c\ (A_{Cmax} - AC)\ R$; and $d\ A_R\ /\ d\ t = Gr\ (A_{Rmax} - A_R)\ (A_C - a\ A_R)\ C$, where G_c and G_r are the learning rates, A_{Cmax} and A_{Rmax} are the maximum limits of adaptive levels for the predator and prey, respectively. I used the parameters, $(G_c, G_r, A_{Cmax}, A_{Rmax}) = (1, 1, 1, [1 - A_{CO}]/a)$ and initial condition, $(A_C(0), A_R(0)) = (A_{CO}, 0)$. This assumption implies that the predation rate of the naive prey (whose $A_R = 0$) and naive predator ($A_C = A_{CO}$) is same with that of the maximally adapted predator ($A_C = A_{Cmax} = 1$) on the maximally adapted prey ($A_R = [1 - A_{CO}]/a$). Parameter A_{CO} is set to a value with which an alien species cannot persist to confirm that a temporal increase of an alien species is due to the alien-native mismatch in adaptive speed. The initial condition is that $(C(0), R(0), A_{CO}) = (0.01, 1, 0.2)$ for predator's invasion and $(0.5, 0.0001, 0.2)$ for prey's invasion. Parameters that are used are: $(r_c, r_n, K_c, K_n, m_c, m_n, a, b, e) = (0, 0.4, 0, 1, 0.5, 0, 1, 8, 0.1)$ for predator's invasion and $(0.4, 0.4, 1, 1, 0.2, 0, 0.2, 5, 0.1)$ for prey's invasion.

than *vice versa*. Therefore, the adaptation process is alien-biased, and the trophic interaction between a native predator and an alien prey is likely to be kept low after the introduction (Figs. 2, 3). The same is true for the interaction between the alien species (Species A) as a predator and the native prey (Species R) (Figs. 2, 3). Because the opportunity for the alien predator to gain experience with the native prey is greater than for the opposite situation, an imbalance can exist between the alien's predation efficiency and the native's defence efficiency. Therefore, the trophic interaction between the alien predator and native prey would tend to be relatively high in the transient dynamics.

This alien-native mismatch in adaptive speed suggests that the alien species has an advantage in not being experienced by interacting species. This makes an alien species less vulnerable to natural enemies and lowers the prey's defence level against the alien species, allowing the alien species to increase its population quickly after introduction. This advantage will last until the alien species becomes abundant, at which point it becomes easier for the native species to have contact experience with the alien species. After the alien species becomes abundant, the native species will catch up with the alien species (defence and predation behavior against the alien species will develop), and the alien species will be controlled more efficiently. This successfully generates the "boom-and-bust" pattern of alien populations that is often observed in nature (Figs. 2, 3, 4; Williamson and Fitter 1996).

Disruption of switching behavior and community instability

An organism can change the species (or trophic species) to interact in response to a change in the relative abundance of potential interacting species (Abrams 1984, Bolker *et al.* 2003). This is called "switching behavior" (Murdoch 1969). There are two trophic interaction-related switching behaviors, foraging (Murdoch 1969) and defence (Fryxell and Lundberg 1997) switches. A foraging switch refers to a choice of diet with higher quality or quantity from a set of nutritionally substitutable potential diets (a foraging switch or foraging shift; Stephens and Krebs 1986). It is a simple strategy to maximize the energetic or material gain per unit effort when different strategies are required to find or capture different diets; the use of a less profitable resource lowers the gain per unit effort. There are a number of examples in which organisms switch to more valuable or abundant diets as the relative abundance or quality of the potential diets change (see Stephens and Krebs 1986). The other switching behavior is a defence switch. Consider a prey species with multiple predators and assume that the avoidance of different predators requires different strategies with trade-offs between them. This might occur when predators attack in different manners or the total predator avoidance effort is limited (Soluk 1993; Sih *et al.* 1998). In such a case, a prey must allocate its defence effort among the potential predators, with more effort being allocated to the more risky predator (Lima 1992).

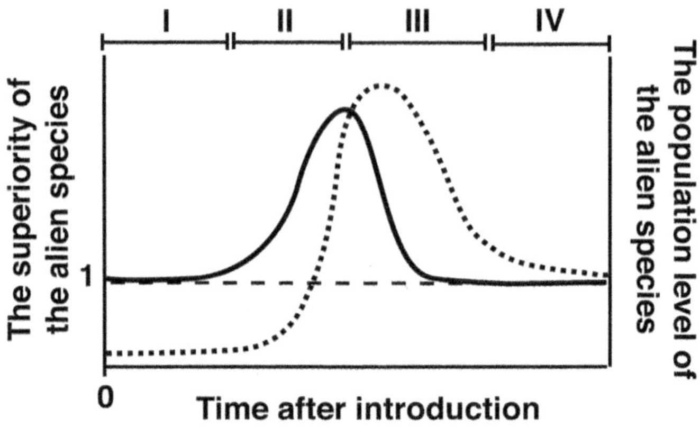

Fig. 4 From the viewpoint of contact experience, a biological invasion is divided into four phases: the (I) pre-adaptation, (II) adaptive-mismatching, (III) native-adaptation, and (IV) post-adaptation phases. First, just after the introduction, the alien and native species are not experienced with each other, and their trophic interaction, which depends on the pre-adaptation level of either species, is unlikely to have a consistent pattern (i.e., Phase I: pre-adaptation phase). Second, the alien species starts to gain experience with the native species; however, the native species cannot gain experience with the alien species because the abundance of the alien species is low. This alien-biased adaptive process provides the alien species with a minority advantage, resulting in the rapid growth of the alien species (i.e., Phase II: adaptive-mismatching phase). Third, after the alien species becomes abundant, the native species begins to gain experience with the alien species (i.e., Phase III: native adaptation phase). During this third phase, the local community can efficiently suppress an alien species. In the last phase, the alien and native species are well adapted to each other (Phase VI: post-adaptation phase), and the population of the alien species will become comparable to that of the native species. The solid and dotted lines represent the superiority of an alien species to a native species and the population abundance of the alien species, respectively. [Superiority] = 1 means no superiority; [population level] = 1 means comparable abundance of the alien species to that of native species.

Theory suggests that a foraging switch enhances species coexistence (e.g., Tansky 1978, Gleeson and Wilson 1986). Consider a simple trophic system consisting of two prey species and one predator species. In the absence of a foraging switch, the two prey species might not coexist because the species with low vulnerability to predation tends to out-compete the other through the negative, indirect effect mediated by the shared predator (apparent competition; Holt 1977). By contrast, in the presence of a foraging switch, the two prey species can coexist because the species that is less abundant tends to be attacked less often by the switching predator. One prey species can provide another prey species with an opportunity to avoid the predator. The same is true for more complex systems (Kondoh 2003).

A defence switch also enhances species coexistence (Lima 1992; Matsuda *et al.* 1993, 1996). In a simple system consisting of two predator species and one prey species, the two predator species tend not to coexist because resource competition leads to the extinction of the inferior predator species (Tilman 1982). In the presence of an adaptive defence switch, however, a completely different picture arises. An adaptive prey is more likely to be defensive against the more abundant predator species, creating a mechanism of minority advantage. Therefore, the switching prey can enhance the predator's coexistence. Matsuda *et al.* (1996) showed that an adaptive defence prevents the extinction of the predator species in a two trophic-level system with higher species richness, suggesting that the defence switch can enhance species coexistence in a complex food web. This is also true in an even more complex food web (Kondoh, unpublished manuscript).

In general, adaptive switching behavior requires the ability to discriminate multiple prey or predator species and knowledge of the relative quantity or quality of the potentially interacting species (Stephens and Krebs 1986). Consider one predator and multiple prey species, for example. The forager is unlikely to be able to choose the more abundant or more nutritious prey species if this prey species is not differentiated from other prey species. Similarly, in the absence of species recognition ability, a prey species is unlikely to make a correct switch in its defence behavior against predators.

The fact that trophic interaction-related switching behavior, which requires correct recognition of the interacting species, tends to enhance species coexistence suggests another important role of contact experience in maintaining biodiversity. An alien-native interaction, in which species recognition is less likely to occur, might affect biodiversity maintenance in a completely different way from that of a native-native interaction. More specifically, an alien-native interaction might be more likely to lower the probability of species coexistence than would a native-native interaction.

This interpretation is based on a model analysis by Kondoh (2003). Using a dynamic mathematical model of a food web (Fig. 5), Kondoh (2003) revealed an important role of the foraging switch in biodiversity maintenance. Consider a food web of N species, each of which is connected by a directed trophic link with probability C (connectance). In this model, C determines the number of potential prey species that a consumer can use (*i.e.*, $[N\text{-}1]C/2$ on average). A consumer species is capable of choosing the more abundant prey species from the potential diet with probability F. By contrast, a fraction $(1\text{-}F)$ of predators does not discriminate between prey species and uses every potential prey. Kondoh (2003) analyzed the interactive effect of the predator's capability of an adaptive foraging switch and the food-web complexity on community persistence. The analysis showed that an adaptive foraging switch potentially inverts the complexity-stability relationship of a food web. In the absence of adaptive switching behavior, increasing food-web complexity lowers community persistence, while in the presence of adaptive switching, a community is more likely to be persistent when the food web structure is more complex (Kondoh 2003). Further, the relationship becomes

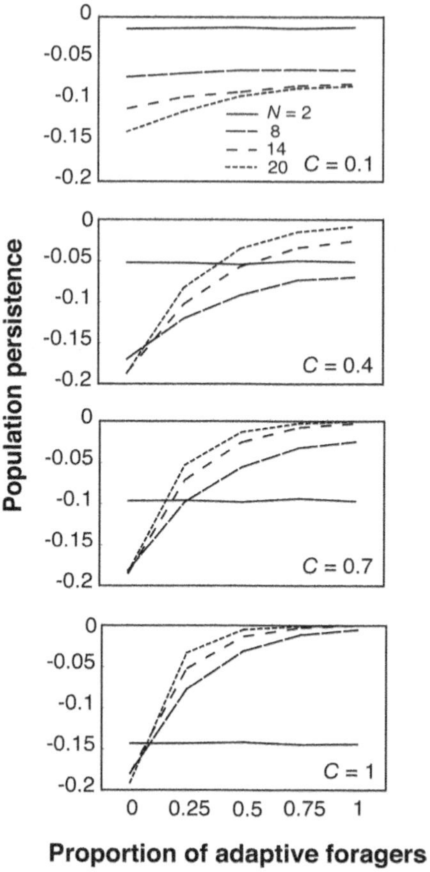

Fig. 5 Population persistence changing with changing fraction of adaptive foragers for varying food-web complexity (C = 0.1–1, N = 2–20). Consider a food web with N species and connection probability C. Population dynamics of Species i is given by:

$$\acute{Y} = X_i \left[r_i - s_i X_i + \sum_{k \in \text{sp. i's } prey} e_{ik} f_{ki} a_{ki} X_k - \sum_{k \in \text{sp. i's } predator} f_{ki} a_{ki} X_k \right],$$

where r_i is the intrinsic growth rate; s_i the self regulation strength; e_{ij} the conversion efficiency; f_{ij} the foraging efficiency of predator i on prey j; a_{ij} the foraging effort of predator i allocated to prey j. The foraging effort of adaptive forager changes to increase the energy gain per unit effort:

$$\acute{Y}_{ij} = G a_{ij} \left[e_{ij} f_{ij} X_j - \sum_{k \in \text{sp. i's } prey} e_{ik} f_{ki} a_{ki} X_k \right].$$

See Kondoh (2003) for more detail. I used the following parameters to create this figure: ($X_i(0)$, a_{ij} (0), r_i, s_i, e_{ij}, f_{ij}, G) = (ran[0, 0.1], 1/(the number of potential prey species, ran[0, 0.1], 1, 0.15, ran[0, 1], 0.25). Random model is used as a food web topology. Population persistence is defined by ln[the probability that no species go extinct]/N.

closer to negative as the fraction of adaptive foragers decreases or the speed of the adaptive switch decreases (Kondoh 2003).

A possible interpretation of Kondoh's result (2003) is that a community with a greater fraction of alien species is more likely to lose a species especially in a more complex food web (Fig. 5). Because an alien species is less likely to discriminate between novel native species, it might not be able to switch diets correctly, implying that a correct foraging switch is less likely to occur as the fraction of alien species increases in the focal food web. Further, if a native predator confuses an alien species with a native prey, the predator might fail to evaluate the relative abundance or quality of the native species. Thus, the introduction of an alien species interrupts switching behavior and can threaten local biodiversity by lowering the population stability.

In this scenario, the effect of an alien species is analogous to the confusion of the body's signaling system caused by endocrine disruptors (Colborn *et al.* 1996). Endocrine disruptors occupy a hormone receptor and are mistakenly recognized as the correct hormone, resulting in a body disorder (Colborn *et al.* 1996). The occupation interferes with the proper hormone signal and produces an excessive or insufficient cellular response (Colborn *et al.* 1996). If biodiversity is maintained by adaptive switching behavior, as Kondoh (2003) suggests, an alien species that confuses species recognition might be regarded as an endocrine disruptor at the community level.

SUMMARY

This chapter presents a hypothesis of the way in which the lack of contact experience influences the trophic interaction between alien and native species. A lack of contact experience can lead to a failure of trophic role recognition, prey handling, or predator avoidance, which can result in a trophic interaction with abnormally high or low strength and destabilized prey-predator interactions. A mismatch in the adaptation speed between alien and native species can produce a temporal explosion in the alien species abundance. A failure of species discrimination in an alien-native interaction can disrupt the adaptive response to qualitative and quantitative changes in the prey or predator and can thereby threaten biodiversity. These hypotheses imply that an alien-native interaction is qualitatively different from a native-native interaction and create a general pattern of alien species invasions and their effects on local communities.

Nevertheless, these hypotheses still rely on a number of simplifying assumptions concerning the effect of experience on a trophic interaction. These assumptions tions are required because we know little about how contact experience affects such macroscopic properties as trophic interaction strength, trophic link flexibility, and population stability. We need to examine the behavioral basis of a trophic interaction and test the possibility that an alien-native interaction is qualitatively different from a native-native interaction. In so doing, it is important to isolate the effect of the absence of experience from that of species identity. A swapping experiment between two communities with the same species composition and different histories may provide an interesting opportunity this test.

ACKNOWLEDGMENT

This study was partially supported by a JSPS Research Fellowship for Young Scientists and an MEXT Grant-in-Aid for the 21st Century COE Program (A2 to Kyoto University).

REFERENCES

Abrams, P. A. 1984. Foraging time optimization and interactions in food webs. The American Naturalist **124**, 80-96.

Agrawal, A. A. and P. M. Kotanen. 2003. Herbivores and the success of exotic plants: a phylogenetically controlled experiments. Ecology Letters **6**, 712-714.

Baker, H. G. 1974. The evolution of weeds. Annual Review of Ecology and Systematics **5**, 1-24.

Blossey, N. and R. Nötzold. 1995. Evolution of increased competitive ability in invasive non-indigenous plants: a hypothesis. Journal of Ecology **83**, 887-889.

Bolker, B., M. Holyoak, V. Křivan, L. Rowe and O. Schmitz. 2003. Connecting theoretical and empirical studies of trait-mediated interactions. Ecology **84**, 1101-1114.

Bond, A. B. and A. C. Kamil. 1998. Apostatic selection by blue jays produces balanced polymorphism in virtual prey. Nature **395**, 594-596.

Bonsall, M. B. and M. P. Hassell. 1997. Apparent competition structures ecological assemblages. Nature **388**, 371-373.

Brodie, E. D. and E. D. Brodie. 1991. Evolutionary response of predators to dangerous prey — reduction of toxicity of newts and resistance of garter snakes in island populations. Evolution **45**, 221-224.

Byers, J. E. 2000. Competition between two estuarine snails: implications for invasions of exotic species. Ecology **81**, 1225-1239.

Callaway, R. M. and E. T. Aschehoug. 2000. Invasive plants versus their new and old neighbors: a mechanism for exotic invasion. Science **290**, 521-523.

Carroll, S. and H. Dingle. 1996. The biology of post invasion events. Biological Conservation **78**: 207-214.

Chaneton, E. J. and M. B. Bonsall. 2000. Enemy-mediated apparent competition: empirical patterns and the evidence. Oikos **88**, 380-394.

Colborn, T., D. Dumanoski and J. P. Myers. 1996. Our Stolen Future. Penguin Books, New York, NY.

Croy, M. I. and R. N. Hughes. 1991. The role of learning and memory in the feeding behaviour of the fifteen-spined stickleback (*Spinachia spinachia* L.). Animal Behaviour **41**, 149-160.

Curio, E. 1993. Proximate and developmental aspects of antipredator behavior. Advances in the Study of Behavior. **22**, 135-238.

Dill, L. M. 1974 The escape response of the zebra donio (*Brachydanio rerio*). II. The effect of experience. Animal Behaviour **22**, 723-730.

Domjan, M. and B. Burkhard. 1986. The principles of learning behavior. 2nd edition. Brooks Cole, Montery, CA.

Ellis, H. C. 1965. The transfer of Learning. Macmillan, New York, NY.

Elton, C. S. 1958. The Ecology of Invasions by Animals and Plants. Methuen, London, UK.

Endler, J. A. 1986. Defence against predators. Pages 109-134 in Predator-Prey Relationships. M. E. Feder and G. V. Lauder, editors. University of Chicago Press, Chicago, IL.

Foster, S. A. 1999. The geography of behavior: an evolutionary perspectives. Trends in Ecology and Evolution **14**, 190-195.

Fretwell, S. D. 1977. The regulation of plant communities by food chains exploiting them. Perspective in Biology and Medicine **20**, 169-185.

Fryxell, L. M. and P. Lundberg. 1997. Individual Behavior and Community Dynamics. Chapman & Hall, New York, NY.

Gleeson, S. K. and D. S. Wilson. 1986. Equilibrium diet: optimal foraging and prey coexistence. Oikos **46**, 139-144.

Gray, A, J. 1986. Do invading species have definable genetic characteristics? Philos. Trans. R. Soc. London Ser. B **314**, 655-672.

Griffin, A. S., D. T. Blumstein and C. S. Evans. 2000. Training animals about predators: a critical review and synthesis. Conserv. Biolo. **14**, 1317-1326.

Hairstone, N. G., F. E. Smith and L. B. Slobodkin. 1960. Community structure, population control and competition. The American Naturalist **94**, 421-425.

Hazlett, B. A. 1995. Behavioural plasticity in crustacea: why not more? Journal of Experimental Marine Biology and Ecology **193**, 57-66.

Herbold, B. and P. Moyle. 1986. Introduced species and vacant niches. The American Naturalist **125**, 751-760.

Holt, R. D. 1977. Predation, apparent competition, and the structure of prey communities. Theoretical Population Biology **12**, 197-229.

Holway, D. A. 1999. Competitive mechanisms underlying the displacement of native ants by the invasive argentine ants. Ecology **80**, 238-251.

Hughes, R. N., M. J. Kaiser, P. A. Mackney and K. Warburton. 1992. Optimizing foraging behaviour through learning. Journal of Fish Biology **41**, 77-91.

Keane, R. M. and M. J. Crawley. 2002. Exotic plant invasions and the enemy release hypothesis. Trends in Ecology and Evolution **17**, 164-170.

Kennedy, T., S. Naeem, K. Howe, J. M. H. Knops, D. Tilman and P. B. Reich. 2002. Biodiversity as a barrier to ecological invasion. Nature **417**, 636-638.

Kolar, C. and D. M. Lodge. 2001. Progress in invasion biology: predicting invaders. Trends in Ecology and Evolution **16**, 199-204.

Kolbe, J. J., R. E. Glor, L. Rodriguez Schettino, A. C. Lara, A. Larson, A. and J. B. Losos. 2004. Genetic variation increases during the biological invasion of a Cuban lizard. Nature **431**, 177-181.

Kondoh, M. 2003. Foraging adaptation and the relationship between food-web complexity and stability. Science **299**, 1388-1391.

Lawlor, L. R. and J. Maynard Smith. 1976. The coevolution and stability of competing species. The American Naturalist **110**, 79-99.

Leibold, M. A. 1996. A graphical model of keystone predators in food webs: trophic regulation of abundance, incidence and diversity patterns in communities. The American Naturalist **147**, 784-812.

Lima, S. L. 1992. Life in a multi-predator environment: some considerations for antipredatory vigilance. Annales Zoologici Fennici **29**, 217-226.

Lubchenco, J. 1978. Plant species diversity in a marine intertidal community: importance of herbivore food preference and algal competitive abilities. The American Naturalist **112**, 23-39.

Mack, R. N., D. Simberloff, W. M. Lonsdale, H. Evans, M. Clout and F. A. Bazzaz. 2000. Biotic invasions: causes, epidemiology, global consequences, and control. Ecological Applications **10**, 689-710.

Maloney, R. F. and I. G. Mclean. 1995. Historical and experimental learned predator recognition in free-living New Zealand Robins. Animal Behaviour **59**, 1193-1201.

Maron, J. L. and M. Vila. 2001. When do herbivores affect plant invasion? Evidence for the natural enemies and biotic resistance hypotheses. Oikos: **95**, 361-373.

Magurran, A. E. 1989. Acquired recognition of predator odour in the European minnow (*Phoxinus phoxinus*). Ethology **82**, 216-223.

Matsuda, H., P. A. Abrams and M. Hori. 1993. The effect of adaptive antipredator behavior on exploitative competition and mutualism between predators. Oikos **68**, 549-559.

Matsuda, H., M. Hori and P. A. Abrams. 1996. Effects of predator-specific defense on biodiversity and community complexity in two-trophic-level communities. Evolutionary Ecology **10**, 13-28.

Mitchell, C. E. and A. G. Power. 2003. Release of invasive plants from fungal and viral pathogens. Nature **421**, 625-627.

Murdoch, W. W. 1969. Switching in general predators: experiments on predator specificity and stability of prey populations. Ecological Monographs **39**, 335-354.

Pavlov, I. P. 1927. Conditioned relaxes. Oxford University Press, New York, NY.

Petren, K. and Case, T. J. 1996. An experimental demonstration of exploitative competition in an ongoing invasion. Ecology **77**, 118-132.

Pimm, S. L. 1991. The Balance of Nature? University of Chicago Press, Chicago.

Power, M. E. 1992. Top down and bottom up forces in food webs: Do plants have primacy? Ecology **66**, 1448-1456.

Rehage, J. S., Barnett, B. K. and Sih, A. 2004. Behavioral responses to a novel predator and competitor of invasive mosquitofish and their non-invasive relatives (*Gambusia* sp.) Behavioural Ecology and Sociobiology (**online**) DOI: 10. 1007/s00265-004-0850-1.

Ricciardi, A., R. J. Neves and J. B. Rasmussen. 1998. Impending extinctions of North American freshwater mussels (Unionoida) following the zebra mussel (*Dreissena polymorpha*) invasion. Journal of Animal Ecology **67**, 613-619.

Ricciardi, A. and S. K. Atkinson. 2004. Distinctiveness magnifies the impact of biological invaders in aquatic ecosystems. Ecology Letters **7**, 781-784.

Roeder, K. D. and A. E. Treat. 1961. The detection and evasion of bats by moths. American Scientist **49**, 135-148.

Roy, J. 1990. In search of the characteristics of plant invaders. Pages 335-352 in F. di Castri, A. J. Hansen and M. Debussche, editors. Biological invasions in Europe and the Mediterranean Basin. Kluwer, Dordorecht, The Netherlands.

Sakai, A. K., F. W. Allendorf, J. S. Holt, D. M. Lodge, J. Molofsky, K. A. With, S. Baughman, R. J. Cabin, J. E. Cohen, N. C. Ellstrand, D. E. McCauley, P. O'Neil, I. M. Parker, J. N. Thompson and S. G. Weller. 2001. The population biology of invasive species. Annual Review of Ecology and Systematics **32**, 305-332.

Sih, A., G. Englund and D. Wooster. 1998. Emergent impacts of multiple predators on prey. Trends in Ecology and Evolution **13**, 350-355.

Sol, D. and L. Lefebvre. 2000. Forebrain size and foraging innovations predict invasion success in birds introduced to New Zealand. Oikos **90**, 599-605.

Soluk, D. A. 1993. Multiple predator effects: predicting combined functional response of stream fish and invertebrate predators. Ecology **74**, 219-225.

Stachowicz, J. J., R. B. Whitlatch and R. W. Osman. 1999. Species diversity and invasion resistance in a marine ecosystem. Science **286**, 1577-1579.

Stephens, D. W. and J. R. Krebs. 1986. Foraging Theory. Princeton University Press, Princeton, NJ.

Tansky, M. 1978. Switching effect in prey-predator system. Journal of Theoretical Biology **70**, 263-271.

Thébault, E. and M. Loreau. 2003. Food-web constraints on biodiversity–ecosystem functioning relationships. Proceedings of the National Academy of Sciences **25**, 14949-14954.

Thompson, J. N. 1994. The coevolutionary process. University of Chicago Press, Chicago.

Thorndike, E. L. 1911. Animal intelligence: experimental studies. Macmillan, New York, NY.

Tilman, D. 1982. Resource Competition and Community Structure. Princeton University Press, Princeton, NJ.

Torchin, M. E., K. D. Lafferty, A. P. Dobson, V. J. McKenzie and A. M. Kuris. 2003. Introduced species and their missing parasites. Nature **421**, 628-630.

Tsutsui, N. D., A. V., Suarez, D. A. Holway and T. J. Case. 2000. Reduced genetic variation and the success of an invasive species. Proceedings of the National Academy of Science, USA **97**, 5948-5953.

Ware, D. M. 1971. Predation by rainbow trout: the effect of experience. Journal of the Fisheries Research Board of Canada **28**, 1847-1852.

Williams, D. F. 1996. Exotic Ants. Biology, Impact, and Control of Introduced Species. Westview Press, Boulder, CO.

Williamson, M. H. and A. Fitter. 1996. The characters of successful invaders. Biological Conservation **78**, 163-170.

Chapter eleven

Use of biological invasions and their control to study the dynamics of interacting populations

F. Courchamp and S. Caut

INTRODUCTION

One of the difficulties of conservation biology is the general lack of experimental approaches. Because it is often unethical, or simply because this new discipline deals with small and/or fragile populations, experiments on those populations are not always feasible. As a result, the knowledge on population dynamics, when not dealing with laboratory populations of caged invertebrates, has often come from theoretical studies, with notable exceptions such as those based on of some populations isolated on particular islands (e.g., (Clutton-Brock and Coulson, 2002; Grenfell *et al.* 1998). However, one aspect that is often lacking from theoretical studies, as well as from natural isolated populations, is the interspecific dimension: in the above cases, it is rather exceptional to take into account more than two interacting populations. Yet, as we hope to show in this chapter, direct and indirect "complex" interspecific relationships may be the major ecological forces in some communities. They can thus be crucial for applied ecology as well

M.W. Cadotte et al. (eds.), Conceptual ecology and invasion biology, 243–269.

as represent heuristic tools for students, and intellectual candies for functional ecologists.

However, there is an enormous set of ecological events that can be viewed as natural, large-scale experiments: biological invasions. Several aspects make biological invasions an interesting tool for the study of interspecific interactions: they are of various types, involve many different organisms, and happen in contrasted ecosystems. Biological invasions are often a very rich source of information for the understanding of ecosystem functioning, as they originate from introductions that are generally relatively well documented. In addition, in many cases, the invaded ecosystem is an island, with all the advantages that insular ecosystems provide for fundamental research: closed ecosystems, with limited size and of relatively simple and non-redundant trophic webs. In the same way that the physiology of an individual may be better understood during illness, the disfunctioning of an ecosystem may help gain knowledge about its normal functioning. In this regard, the changes generated by the simple modifications that are species introductions and their consecutive spread provide many different types of information. Thus, biological introductions represent simple experiments of species addition into a new trophic web. Such experiments can benefit from controls and replicas in the case of archipelagoes.

Similarly, species deletions can be studied in large-scale experiments that are even more accessible to population or community biologists, as they can be designed by them: the control or eradication of the alien species. With this new tool of species addition and deletion from the comparatively simple island ecosystems, one may gain more knowledge of basic processes such as colonization, dispersion, spatial spread, as well as the dynamics of interacting populations. We will here focus on this latter aspect, restricting our analysis on three- and four-species interactions, and shamelessly basing it on our previous studies.

In this chapter, we will present a number of mathematically simple models that depict some "complex" interspecific relationships, with the aim of showing how the study of biological invasions and their control can be useful for the study of fundamental ecological processes that are more problematical to understand in other contexts. Here, complex relationships are defined as interactions within trophic webs that encompass more than two populations (with possible indirect processes), and that may not be really complex in a biological sense, but that are more demanding to study analytically. We also use 'control' in a somewhat lenient style. This term can have two meanings: it can be a general term of action against an alien species ranging from simple reduction up to eradication, and it can more specifically mean reduction of the population size down to acceptable levels, in ecological or economic terms. The latter is called 'mitigation' or 'reduction' and is opposed to 'eradication'. In this Chapter, we will use 'mitigation' for partial population removal, 'eradication' when removal is total, and 'control' as a general term. We will articulate our presentation in two parts; the first part depicts systems where species are added to a trophic web (the biological invasions), and the second part depicts systems where species are removed (conservation programs), focusing on the possible associated indirect processes in each case. A secondary objective of this chapter is to convince the readers, be they students in biology or conservation managers, that mathematical modeling is a powerful tool to understand, and in some case to predict, ecosystem functioning and reactions. Yet for pedagogic purposes, we will present our analyses based on the description of several concrete examples with little or no emphasis on the technical aspects of the mathematical models. We provide references for more details about the models and their analysis.

All the models presented here are deterministic coupled differential equations based on classical Lotka-Volterra predation or competition models. Each population is described by a simple logistic equation, modified to take into account its relationship with the other population(s). Although biologically simple, the models presented here can show a relatively high mathematical complexity when it comes to, for example, determining equilibrium points. Confident that the simplest models are the most useful (Ginzburg and Jensen, 2004), we systematically refrained from unduly adding complexity to our equations, which results in a lack of predictive power. These models are therefore not aimed at providing precise values of population trends or of control measures in the field; neither the nature of the models, nor the state of current knowledge in the field would allow useful quantitative predictions. Rather, the aim of this exercise is to emphasize the link between species, the importance of indirect interactions, and the unexpected outcome of control actions if they are not thoroughly taken into account. We believe that the qualitative information provided by our mechanistic models is suitable to offer the information we seek in this context. Also, while parameterising models to reproduce field results is a very useful way of identifying plausible mechanisms of trophic interactions, it does not in itself provide a direct test of the importance of those plausible mechanisms. The information they generate should always be completed by information coming from empirical and experimental studies. In this Chapter, we will only deal with the modeling part.

In order to render the reading of this chapter less tedious, we will describe in detail the process leading to the model for the first example only, and will only give the model for the other examples. Although some will differ in details, all models are based on similar principles. Apart from the more complete description of the first case, all examples will be presented in a similar way, to allow easy comparisons between cases.

NATURAL ECOSYSTEM EXPERIMENTS: ADDITIONS AND DELETIONS OF SPECIES

Biological invasions as a species addition experiment

The hyperpredation process: three-species interactions

a - The case

Introduced species are notorious for their deleterious impact on invaded communities and their direct effects on trophic systems, such as decrease of prey (or competitor) populations. While they represent catastrophic events in terms of biodiversity conservation, those effects are not of major interest to theoretical ecology. For this reason, we will focus on less obvious effects, starting with an example concerning the extinction of an endemic parrot caused by the introduction of rabbits to an insular system.

The rabbit is one of the most documented introduced mammal species, often associated with a dramatic impact on endemic plant species. To date, this herbivore has been introduced (most of the time purposefully) to more than 800 islands (Flux and Fullagar, 1992). Rabbits have a high ecological adaptability, and as such easily succeed when introduced in to ecosystems where indigenous grazers are less numerous and competitive (Flux, 1993). The very rapid increase of their populations can lead to a dramatic quantitative and qualitative impoverishment of the vegetation (Chapuis *et al.* 1994; Selkirk *et al.* 1983), resulting in dramatic denudation of the soil (Scott, 1988) and have an impact on animal species which depend on the vegetation (Gillham, 1963).

Effects of rabbits on indigenous vertebrate species can also be more complex. These mammals are preyed upon by other introduced vertebrates, in particular by feral domestic cats. Cats, for example, are opportunistic predators which switch prey according to relative spatial and/or temporal availability (Fitzgerald, 1988). When rabbits are abundant, domestic cats are known to prey largely upon them. However, rabbits can constitute a smaller part of the cat diet when birds, reptiles or other mammals are relatively more abundant. In several sub-Antarctic islands, rabbits are only a secondary prey item in months when seabirds are present, but appear to enable cats to subsist over winter when seabirds are absent (Chapuis, 1995a, b). A similar effect is documented in the spatial dimension: rabbits often enable cats to reach remote colonies or populations of indigenous prey in islands with heterogeneous indigenous prey distribution (Brothers and Copson, 1988). In these cases, the presence of rabbits has an indirect effect on other prey species used by introduced cats.

Predation by cats introduced to Macquarie Island caused the decline of burrow-nesting petrels (Brothers, 1984) and the extinction of an endemic parakeet and a banded rail (Taylor, 1979). Cats were introduced to the island 60 years before the introduction of rabbits, however the cat driven extinction of birds dates back to just 10 years following the introduction of rabbits (Taylor, 1979). Rabbits were not observed as having any direct effects on the land birds. In fact, it is believed that the rabbit population allowed a significant increase in the cat population, resulting in an increased predation pressure on the land bird species. This process, related to the more general "apparent competition" (e.g., (Abrams, 1987; Abrams *et al.* 1998; Holt, 1977) has been termed hyperpredation (Courchamp *et al.* 2000; Smith and Quin, 1996).

It is generally assumed that life history traits and behavior of the introduced prey make it resistant to high levels of predation pressure. A higher reproductive rate, high density and efficient anti-predator responses (which are often lacking in the indigenous species) exhibited by the introduced prey could enable an increase in the predator population without a large decrease in the introduced prey population. Furthermore, the lack of serious competitors and the relatively few parasites typically found in introduced populations can also increase their potential for dramatic population growth (the enemy release hypothesis: (Keane and Crawley, 2002). These features imply the ability to sustain high predation pressure, as cats are supposed to remove only the individuals with low survival (dispersing young, sick and dead, (Smith and Quin, 1996). The resulting increased population of predators cannot be sustained by the indigenous prey species which, compared with the introduced prey species, has inferior (less well adapted) reproductive and anti-predator characteristics. The conjunction of a low adaptation to predation and an artificially high predation pressure can lead to a dramatic decrease in an indigenous prey population, up to total extirpation.

b - The model

We will thus present a simple model of hyperpredation to illustrate how the extinction of the bird population on Macquarie Island can be explained by an indirect effect of the introduced rabbits. For heuristic purposes, we first present a two-species model, and then modify it into a three-species model. We hope that this will help the understanding of this model and of the other sets of equations in this chapter (which are all based on the same principle). We use the example of bird (local prey), rabbit (introduced prey) and cat (introduced predator), and will refer to these species for the sake of simplicity, but other species can present similar relationships.

The bird-cat model can be given the following form:

$$\begin{cases} \dfrac{dB}{dB} = r_b B \left[1 - \dfrac{B}{K_b}\right] - \mu_b CB \\ \\ \dfrac{dC}{dt} = \lambda_b \mu_b BC - \nu C \end{cases},$$

where the number of individuals at time *t* in the bird, rabbit and cat populations are *B*, *R* and *C*, respectively. The rabbit-cat model is the same. The intrinsic growth rates of the bird and the rabbit populations are r_b and r_r, respectively. The predation rate is μ_b on the bird population and μ_r on the rabbit population. The carrying capacity of the environment for the bird population is K_b and the carrying capacity of the environment for the rabbit is K_r. The rate at which eaten prey are turned into new predators is λ_b for birds and λ_r for rabbits, and ν is the predator mortality rate. In a more general manner, the same parameters will be used for the next models, with indexes corresponding to the first letter of the species considered.

If two prey species are to be considered simultaneously, the formulation of the predation rates and of the growth rate of the predator must be changed accordingly: instead of $\mu_b C$ and $\mu_r C$, the predation rates are given the form

$$\frac{B}{B+R}\mu_b C \quad \text{and} \quad \frac{R}{B+R}\mu_r C$$

for the bird and the rabbit predation rates respectively, so that the predation rate is still a function of the availability of the prey but varies with relative prey proportions. The adaptation of the introduced prey in terms of an anti-predator behavioral response is given by a preference of the predator for the indigenous prey (which is more easily detected and/or caught) over the introduced prey. This preference is a ratio (α) with a simple biological meaning: given equal availability, the predator will prey upon the indigenous prey α times more often than on the introduced prey. We assume that $\alpha \geq 1$, and that one rabbit and one bird prey items are energetically equally valuable to the cat. The predation terms are now given by

$$\frac{\alpha B}{\alpha B+R}\mu_b C \quad \text{and} \quad \frac{R}{\alpha B+R}\mu_r C$$

on the indigenous and introduced prey respectively. This change is reflected in a similar way in the predator growth rate:

$$\lambda_b B\left(\frac{\alpha B}{\alpha B+R}\right)\mu_b C + \lambda_r R\left(\frac{R}{\alpha B+R}\right)\mu_r C:$$

it depends on both the numbers and the proportion of prey. We have now the following system:

$$\begin{cases} \dfrac{dB}{dt} = r_b B \left(1 - \dfrac{B}{K_b}\right) - \dfrac{\alpha B}{\alpha B + R} \mu_b CB \\ \dfrac{dR}{dt} = r_r R \left(1 - \dfrac{R}{K_r}\right) - \dfrac{R}{\alpha B + R} \mu_r CR \\ \dfrac{dC}{dt} = \dfrac{(\lambda_b \mu_b \alpha B^2 + \lambda_r \mu_r R^2)C}{\alpha B + R} - \nu C \end{cases}$$

Note that similar formulations of the model could be used and that the aforementioned choices are arbitrary. In particular, we used a predation term proportional to the number of prey. We did so to keep the formulation of the original paper (Courchamp *et al.* 2000) but alternative models can reproduce the hyperpredation process. Also, like for the rest of the models presented in this chapter, we do not take into account the further risks encountered by populations when at small sizes (environmental and demographic stochasticity, Allee effects etc.). Most of those would anyway only strengthen our point. The classical compartmental representation of the model is presented in Fig. 1A, the corresponding set of equations is shown in Fig. 1B, and a selected representation of the population trends with time is given in Fig. 1C. The other examples will be illustrated with figures following the same format.

The study of this set of equations (both analytically and numerically, see Courchamp *et al.* 2000) shows that the indirect effect of the introduced prey may be very important. Indeed, according to the values of the parameters, the increase of the predator population triggered by the presence of the introduced prey can drive the indigenous prey to very low numbers and potentially to extinction. The effect of hyperpredation is the strongest for species with low intrinsic growth rate and low environmental carrying capacity. This model also illustrates that the hyperpredation process may be due to a combination of well-adapted life history traits and efficient behavioral response of prey, but that the "better-adapted" behavioral response may have more importance than "better-adapted" life history traits, at least for the cases considered (Courchamp *et al.* 2000). Thus, a prey species introduced into an environment in which a predator has also been introduced is likely to allow so high an increase of this predator, that local prey, less adapted to high levels of predation, could suffer a population decline and possibly even extinction. Such a process has consequences when it comes to management actions.

c - Conservation consequences

Historically, conservation programs for many islands have processed case by case. Until recently, introduced species were always considered separately, with, at best, separate programs for each species, and timings depending mostly on funding and logistics, or, more frequently, one single program for the visibly most

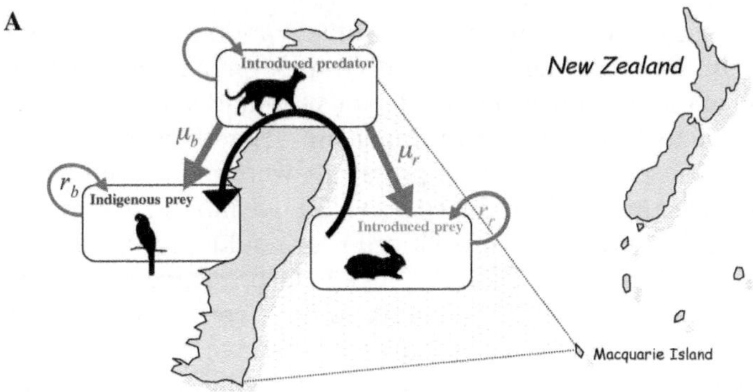

B

$$
\begin{cases}
\dfrac{dB}{dt} = r_b B\left(1 - \dfrac{B}{K_b}\right) - \dfrac{\alpha B}{\alpha B + R}\,\mu_b C B \\[2ex]
\dfrac{dR}{dt} = r_r R\left(1 - \dfrac{R}{K_r}\right) - \dfrac{R}{\alpha B + R}\,\mu_r C R \\[2ex]
\dfrac{dC}{dt} = \dfrac{\left[\lambda_b \mu_b \alpha B^2 + \lambda_r \mu_r R^2\right] C}{\alpha B + R} - \nu C
\end{cases}
$$

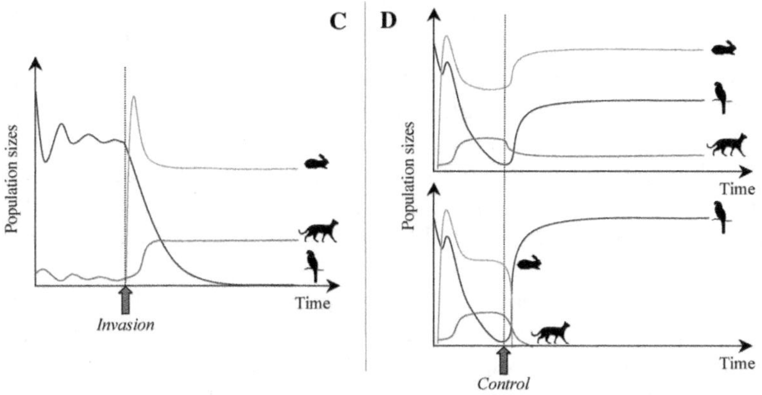

Fig. 1 Representation of the hyperpredation model, with the introduced prey, indigenous prey and introduced predator being illustrated by rabbits (R), birds (B) and cats (C), respectively. (A): compartmental representation and illustration of the island from where the example is taken, here Macquarie Island, off New Zealand. Each box represents a population, and the arrows represent fluxes between them. The large, curved arrow represents an indirect effect. Each species is illustrated in a color that is also used for the set of equations (B) and for the drawings that represent the population dynamics of the interacting species after introduction of a species (C) or control of an introduced species (D). In this example, following the cat introduction (red arrow), the increase of the cat population allowed by the large population of rabbits leads to bird extinction (C). Following control (green arrow in D), the bird population only partially recovers if only the cat is controlled (top panel of D). However, the same cat control level leads to cat eradication and full recovery of birds if both the rabbits and the cat are controlled (bottom panel of D).

devastating species. As a result, when a conservation program involved an island such as the one we just mentioned, with a hyperpredation process taking place, the key role of the introduced prey was not systematically obvious. Predators are often perceived as having the most important deleterious effects on invaded ecosystems, and consequently control programs were more often directed at them, sometimes neglecting the introduced prey. Yet, we have just seen that in the presence of introduced predators, introduced prey could have an indirect impact on indigenous prey. Basing our efforts on the model presented in the previous section of this chapter, we studied the relative efficiency of control programs aiming either at the predator only, or at the introduced prey and predator simultaneously. The model on which we based this has been published in (Courchamp *et al.* 1999), and this model is only slightly different from the associated model without control (Courchamp *et al.* 2000). We will not reproduce the model here, as the interest lies not in the details of the equations. The only important point is the addition of a control effort on either the alien prey or the alien predator (or both). This control is added to the corresponding equation by a simple linear term. We emphasize the fact that, in the model, the introduced prey (rabbit) still has no direct effect on the local prey (bird).

Some possible population trends of the system in presence of control are shown in Fig. 1D. The study of this model shows that control of rabbits can facilitate the eradication of cats. Indeed, when no control is undertaken, the cat population stays large, mainly because of the presence of rabbits, and can eliminate the birds in the long term. When cats only are controlled, the presence of rabbits can preclude cat eradication, and the bird population recovery is only partial. In contrast, for the same cat control effort, eradication of rabbits allows eradication of cats and total recovery of birds. Actually, if the control of introduced prey is not sufficient, the indigenous prey will be destroyed, even if the predator population is being controlled.

Obviously we argue here, that even in absence of visible direct effect, introduced prey should be controlled when a predator has been introduced, in order to prevent an artificial predator population increase. In addition, removing an introduced predator population without controlling the introduced prey may be difficult to achieve since they constitute a constant source of food to the predator. Also, it would not be an appropriate solution because removing the predation pressure would increase the difficulties of later coping with introduced prey, which are often characterized by high reproductive rates. On the other hand, controlling only the introduced prey is unsatisfactory in the long term because predators could report high predation pressure on the indigenous prey. Combined control of both species seems here to be the best restoration strategy. In addition, starting both control programs together would also result in advantages due to synergetic effects: costs may be reduced (if costs related to transportation, or hunting and trapping can be shared by the two programs) and efficiency might be increased (e.g., through the additive effects of primary and secondary poisoning of predators; Flux, 1993; Rammell *et al.* 1984; Robertson *et al.* 1994).

To conclude, it is worth reiterating that the higher efficiency of dual control is not due to direct effects of rabbits on birds (habitat destruction and competition for food and shelter), since they are not taken into account here. Nor is the predicted success of dual control due to the preference of the predator, since this preference is set in favor of the indigenous prey in the model. This success is due to the addressing of the hyperpredation process.

The hyperpredation process: four-species interactions

a - The case

The cat is a well-known predator of both insular birds and small introduced mammals, so the example above should be relatively easy to spot. Whenever a local population is threatened by an introduced predator, a diet study of the predator in question should be conducted in order to assess the importance of the impact on the local population, but also potential hyperpredation processes. However, there are cases where this strategy is not obvious, because the cause of a prey population decline may not be spotted as easily. An interesting illustration of this is the severe decline of the insular fox on the Channel Islands in the 1990's.

The Channel Islands group is made up of eight small islands off the Californian coast, USA. The island grey fox (*Urocyon littoralis*) is a small carnivore that arrived on the first island 16,500 years ago. The fox now inhabits the six largest islands and has evolved on each of these in isolation, resulting in six populations representing six different subspecies, all endemic to these islands. On the three northern islands, the insular foxes and their main competitor, the endemic spotted skunks (*Spilogale gracilis amphiala*), were the two terrestrial top-predators of the Channel Islands. In the early 1990's, a study conducted on the home range of the foxes witnessed a considerable decline in the three northern island populations (Roemer, 1999). This severe decline had no obvious cause at first, and many classical ecological forces were investigated: lack of sufficient resources, competition with the spotted skunk and diseases. Predation was also investigated, although the insular fox was the top terrestrial predator of these ecosystems. Suspicion that foxes were killed by golden eagles led to a new effort of research in this direction. Golden eagles have historically been seen visiting the islands, but they never stayed long enough to constitute a threat to the local prey. A study combining metabolic and energetic approaches with population modeling demonstrated that the local prey were too few to allow a pair of dispersing eagles to breed on and colonize the islands. Yet it became obvious that eagles were killing foxes as well as spotted skunks. On one of these islands, Santa Cruz, field workers eventually discovered a golden eagle nest, in which fox remains attested for the suspected predation on this species. But the problem remained. How would the eagle threaten fox survival through predation, if there was not enough local prey on the island to allow the continuous presence of the eagles? The discovery of the nest provided the answer: remains of piglets were also found in the nest. Feral pigs (*Sus scrofa*) were introduced on to the three northern islands where the foxes are declining (they were also present on two of the southern islands, but have already been or are almost eradicated there). In addition to the direct damages that introduced pigs are known to cause to the flora and fauna they invade (Hone, 1995), this alien species also threatened some local species through an indirect process. By producing piglets all year round, they provided visiting eagles with enough resources for them to colonize the islands. Eagles also irregularly depredated other local prey such as foxes or skunks. However, even this low predation rate on a species that is ill-adapted to avian predation, behaviorally as well as at the population level, was sufficient to drive the fox population towards extinction. The decline was all the more dramatic that the breeding eagle population rapidly grew in numbers, thereby increasing the pressure on the fox population. In the mean time the nocturnal skunks benefited from the arrival of the eagle because they benefited from the release of competition pressure from the declining fox populations while also being killed less often than them. The hypothesis was thus that the arrival of pigs had allowed visiting eagles to stay and breed, and thereby they attracted a shared predator to insular prey. Pigs are well adapted to predation — they produce numerous piglets that can escape eagle predation once they reach three months of age. Therefore, there were less consequences for the pig population than the local prey due to the arrival of the eagle (Roemer *et al.* 2001).

b - The model

To test this hypothesis, a model of the population dynamics of the interacting species was constructed and parameterized with data obtained from the field. The model was based on a simple combination of two classical Lotka-Volterra models: one of competition and one of predation. The skunk and the fox population dynamics were described by a competition model, the pig and the eagle population dynamics were described by a predation model, and a predation term of the eagle was added on both fox and skunk populations. Using a correction term for proportions and preference coefficients (ϕ and σ respectively) as in the previous example, we end up with a system of four equations, one predator and its three prey, two of which are competitors. The system and illustrations of population trends are shown in Fig. 2. More details can be found in Roemer *et al.* (2002).

Simple simulations show that in absence of the pigs (if the system is run with an initial number of pigs set at zero), any introduction of eagles, however large, will eventually lead to colonization failure and fox population persistence. However when pigs are present, a single pair of eagles will be able to colonize the island and build a population that is so large that foxes will go extinct while pigs will remain at moderate densities.

It is also interesting to note that the decline in fox numbers, consecutive to the hyperpredation process triggered by the introduction of pigs, is concomitant with an increase of the endemic skunk. In fact, the arrival of eagles reversed the

A

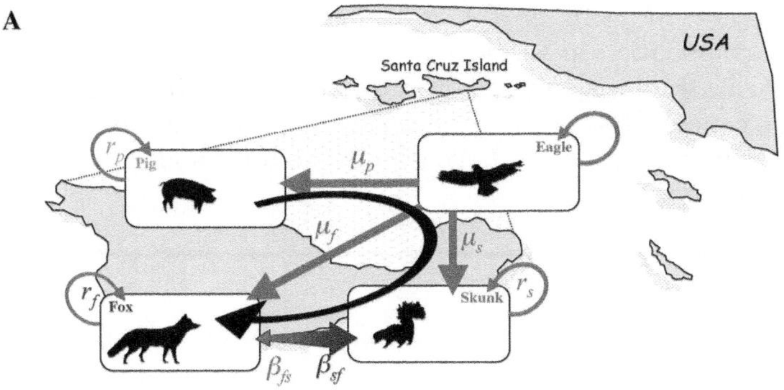

B

$$\begin{cases} \dfrac{dF}{dt} = r_f F\left(1 - \dfrac{F + \beta_{fs} S}{K_f}\right) - \mu_f \dfrac{\phi F}{\phi F + \sigma S + P} EF \\ \dfrac{dS}{dt} = r_s S\left(1 - \dfrac{S + \beta_{sf} F}{K_S}\right) - \mu_S \dfrac{\sigma S}{\phi F + \sigma S + P} ES \\ \dfrac{dP}{dt} = r_p P\left(1 - \dfrac{P}{K_p}\right) - \mu_p \dfrac{P}{\phi F + \sigma S + P} EP \\ \dfrac{dE}{dt} = \dfrac{\left[\lambda_f \mu_f \phi F^2 + \lambda_s \mu_s \sigma S^2 + \lambda_p \mu_p P^2\right] E}{\phi F + \sigma S + P} - \nu E \end{cases}$$

C D

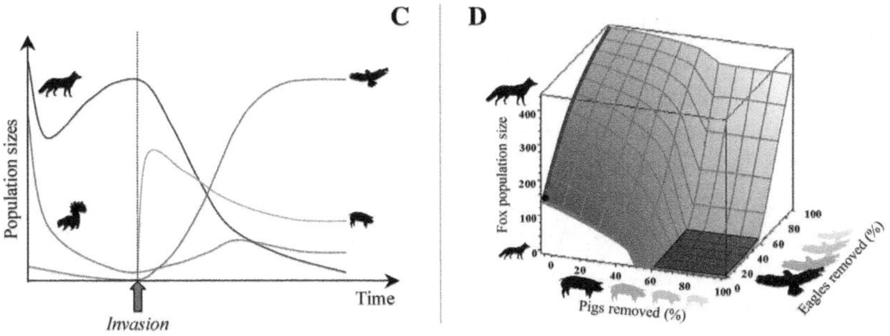

Fig. 2 Representation of the hyperpredation model example with four species: fox (F, blue), skunk (S, green), pig (P, yellow) and eagle (E, red). As in Fig. 1: (A) is the compartmental representation, (B) is the resulting set of equations, (C) is the illustration of populations trends given by (B) following pig introduction and eagle colonization, and (D) is the populations trends following pig and/or eagle control. The parameters are the same than previously, with ϕ and σ being the preference parameters of the eagle for the fox and the skunk over the pig, respectively (same as α in Fig. 1). The control strategy (D) is represented in three dimensions. To help visualize the 3D effect, the colors do not refer to species, but to different population sizes. This graph shows that the population size of foxes is proportional to eagle control, but inversely proportional to pig control. As a result, if eagles are not controlled simultaneously foxes will decline following pig control only. In absence of significant eagle mitigation, high levels of pig mitigation can result in fox extinction (dark area).

competitive outcome between the two top terrestrial predators, shifting forces from direct competition in favor of the fox to apparent competition in favor of the skunk. This second apparent competition process, embedded in the first one, renders any conservation strategy at the least complicated, as the insular spotted skunk is endemic from the northern Channel Islands and is currently benefiting from the fox decline.

c - Conservation consequences

Foxes are now extinct in the wild on two of the three northern islands, with the population on the third island, Santa Cruz, on the verge of extinction (62 individuals in the wild at the end of 2004). Our modeling exercise suggests that the extinction of two populations of the top predator in two insular ecosystems is likely due to an indirect process: a process where an introduced prey attracted a shared predator and eliminated an endemic prey through apparent competition only.

Obviously, the solution to this problem lies with the pigs. The most evident plan of action was to remove the pigs from the northern islands. This would have the double advantage of stopping their direct deleterious effects on the local flora and fauna, as well as eliminating the prey basis for the eagles, leading them with little more choice than starvation or emigration. In fact, several conservation strategies were implemented simultaneously. Among them, eagle live-trapping was quite successful, with less than ten individuals proving impossible to trap or that kept coming back from the translocation area. However, it was easy to see that as long as pigs remained, eagles would start breeding on the island again and thus start a new population. The solution therefore seemed to be the complete removal of pigs from Santa Cruz Island. Yet, the study of a model based on the previous one showed once more that indirect interactions may lead to counter-intuitive results (Courchamp *et al.* 2003b). As for the previous example, the basic model shown in Fig. 2B was changed simply by adding a linear control term to the pig and to the eagle equations.

By varying the control rate of pigs and eagles from zero (no control) to one (eradication), we can mimic different control strategies (control of pigs only, of eagles only or of both species, with different strength) and compare their relative efficiency with no risk to the local populations. Doing so revealed that mitigation of pigs would in fact lead to a decrease in the fox population (Fig. 2D). Eradication of pigs, the intended course of action on Santa Cruz, would lead to fox extinction. Due to the low fox population and the large eagle population, the foxes would be entirely destroyed before the eagles died or emigrated. In theory, the solution is thus simple: remove both the eagles and the pigs. Howerver in practice, the removal of such a large pig population is logistically difficult. In addition, the removal of the eagle would be impossible through live trapping only, and ethically and legally challenging, because golden eagles are protected species in the USA and therefore cannot be killed.

To conclude this part, it may be interesting to note that when more species and more interactions are taken into account, new processes may be unveiled that could not be perceived with only two-species studies. Yet, if better understanding a system is undoubtedly useful for conservation managers, it is not always sufficient for them to be able to know how to act. In the present case, a study taking all species into account revealed that a seemingly obvious line of action (pig removal) would indeed likely achieve results opposite to those desired, and poses a difficult decision to make: remove a protected population in order to save an endangered subspecies. If it is obviously not trivial to make conservation choices, even in the simplest situations, it can sometimes become a challenging dilemma theoretically, logistically, legally and morally. In the present case, the only remaining populations of several fox subspecies were threatened with imminent extinction. However, the proximate cause of this threat is the presence of a protected bird. Furthermore, the decline of the fox benefits the only populations of an endemic skunk. As we have seen, when a (difficult) choice is made, the opposite outcome may well arise. Furthermore, all this is without considering species outside this simplistic system. One could also consider the question under a wider angle, for example including the San Clemente loggerhead shrike, *Lanius ludovicianus mearnsi*, a critically endangered bird, to which the insular fox is the main predator. On San Clemente Island, the fox population has been, ironically, adversely impacted by a US Navy program to protect this bird: attempting to thwart any predation of shrikes, fox were initially trapped and shipped off island or euthanized during the shrike-nesting season (95 foxes removed in 1999, 46 of which permanently). But this is another story...

Along the same line of this conservation riddle, the next part of this chapter investigates the importance of direct interactions in control programs, with the aim of showing that the removal of the primary cause of a ecosystem disfunction will not always help restore the initial conditions. In some cases, not taking into account indirect interaction may lead to even further damage, to the point that it may be wiser to advocate not to remove populations that are known to cause direct negative impacts on invaded communities, at least until adequate knowledge is gained and relevant control strategies are inferred.

Control of invaders as a species removal experiment

Release from introduced herbivores

One concept that is relatively new in the study of invading species, and that has been the core principle of our own studies, is that even if a species is proven to be inflicting important damages to a community it invades, the mere removal of that species may not systematically be the solution to restoring the community. As we have shown with the California Channel Islands example, unconsidered control may even lead to the opposite outcome, that is, further damage, including possible extinction of the species intended to be protected. This fact highlights the need to have a clear and complete view of the relationships among species that are connected directly or indirectly with the introduced species that is subject to control. It nowadays sounds trivial to state that all the species that interact with a population for which removal is planned, are likely to be affected in diverse ways by any such actions. Thus, the success of an eradication program is measured not only by the complete removal of the controlled species, but also by the absence of further dysfunction. Yet, such errors still occur regularly during conservation programs, sometimes simply because conservation programs have insufficient funds to allow thorough pre-control studies of the invaded community, as well as long term post-control monitoring. Sometimes, simply because conservation action is urgently needed and there is no time for such pre-control study.

The importance of knowing the relationships between invading species and those in the invaded community is well illustrated by the goat and pig eradication on the Sarigan Island. This island is part of the Commonwealth of the Northern Mariana Island, in the Pacific Ocean. Introduced pigs and goats threatened the local flora and fauna, triggering a conservation program consisting mainly of goat and pig eradication (Kessler, 2001). As the island is isolated, and hence difficult to access, the program designed included only a minimal pre-eradication study. The program was considered a full success in terms of removing the introduced mammals, however it failed in its ability to detect that the island had also been colonized by an introduced vine, *Operculina ventricosa*, which appeared to be a preferential food item for the goats. It is perhaps not surprising that this vine was not found in the pre-control study as it was likely to have been at a low density due to selective grazing by goats. Even a very thorough study might have failed to see it. Yet, it may be valuable to point out that, when possible, simple fenced exclosure studies prior to eradications can often help land managers see if unwanted results will arise after an eradication of herbivores. If so, then appropriate control of non-native plants can be planned along with the herbivore removal. Unfortunately, this was not done and the control program, which aimed at releasing plant species from goat grazing, had a different impact on the overall plant community than the one expected. As the pressure of grazing was removed from all grazed plants, introduced plants were able to fully express their competitive superiority with regards to native plants, resulting in the rapid invasion of the community. Fig. 3 shows how, within only two years, the removal of an exotic grazer led to a complete invasion of the island community by an exotic plant that appears to have a competitive superiority over local plants. As most of the Sarigan Island ecosystem is now covered by vines, one can easily imagine how the indirect effect of having removed goats is now deleterious for the local plants as well as animals that depend upon them.

The mesopredator release effect

The process we have seen with the release of an exotic plant maintained at low density by a browser can be generalized to releases from almost any other type of

Fig. 3 Evolution of the landscape of Sarigan Island, following the eradication of goats. As this introduced herbivore no longer held in check the expansion of the introduced vine, a favored food item, the vine rapidly increased, covering most of the insular plant communities within two years. This illustrates how an exotic grazer affected the competition relationships between local and introduced plants, and how its removal can lead to dramatic and unexpected outcome for the communities which protection was aimed at.

natural enemy. We will illustrate this with the next two examples. This is, however, not to be mistaken with the enemy release hypothesis (Keane and Crawley, 2002), which proposes that invading species are so successful partly because they are released in the invaded habitat from the pressure of their natural enemies (rarely introduced with them).

On Stewart Island, New Zealand, a population of introduced cats was threatening one of the last populations of kakapo (*Strigops habroptilus*), an endemic flightless parrot. A diet study revealed kakapo remains in 5.1% of 118 collected cat scats (Karl and Best, 1982). This seemingly low predation pressure can have a dramatic effect on insular populations which evolved in the absence of such predators and are therefore not adapted to even low levels of predation. Moreover, the kakapo population was already small and fragile, adding to the weight that introduced cats could have on its fate. This could have been sufficient to trigger a program of cat control on the island.

However, rats (known to be important bird predators) had also been introduced on Stewart Island. In the same diet study (Karl and Best, 1982), rat remains were found in 93.0% of these cat faeces. This shows the indirect role cats might play in preserving native fauna, through reduction of rat predation pressure on kakapo. In fact it is easy to see that in some cases the indirect positive effect of cat predation on rats is more beneficial for the local prey than the direct negative effects of cat predation on the prey themselves. In such cases, the elimination of the feral cat population could lead to a more severe negative impact on the local species through an increase in the rodent population as a consequence of the removal of their predators. The attempted reduction of the cat population on Amsterdam Island has been abandoned as it is alleged to have caused a compensating rise in the number of rats and mice (Holdgate and Wace, 1961). This process, termed "mesopredator release", has been described in fragmented insular ecosystems (Soulé *et al.* 1988) and applies well to many insular food webs (e.g., Schoener and Spiller, 1999).

Conversely, the eradication of rodents first (which has now proven feasible, even on relatively large islands) might induce cats to switch prey, resulting in a brutal increase in predation pressure on the threatened indigenous species, as experienced for stoats and rats in New Zealand (Murphy and Bradfield, 1992). This is a similar process to the one described above for the pig control in presence of eagles on Santa Cruz Island. As the optimal control strategy is neither simple to find, nor intuitive, it is convenient to study it through the analysis of a mathematical model which mimics the dynamics of the three species in this system. The main results of such a study (Courchamp *et al.*, 1999a) are reproduced below to illustrate how the control of an invading species can provide an ideal opportunity to progress in the understanding of the numerous and often complex interactions among populations.

For the sake of clarity, we will not present the model equation in any detail. Suffice to say that the system is described by a set of three coupled equations: one prey (parrot), its predator (the rat, in this case a mesopredator) and one superpredator (cat) which eats both the prey and the mesopredator. The model, and the resulting population trends are represented in Fig. 4.

The study of the above system leads to the quite obvious conclusion that both predators should be eradicated at the same time. However, not only is such a strategy challenging from a logistical point of view, but in addition it may not always be the best solution, especially if the system includes more introduced predators. Indeed, the prey-mesopredator-superpredator system that we have just described can be further complicated. There are cases where the presence of a third predator can render the eradication of the mesopredator problematic if it also acts as a predator to this third predator (and thus at the same time as a mesopredator and a superpredator). Being aware of the potential dangers of removing a superpredator when a mesopredator is present, the managers of the conservation program of Bird Island, Seychelles, decided, rightly, to simultaneously remove introduced cats and rats in order to protect the local bird colonies. They had, however, overlooked the presence of the introduced crazy ant (*Anoplolepis longipes*), which were present in very low numbers on the island (Feare, 1999). The larvae of these ants seem to have been an important prey item of the introduced rodents, such that the rat eradication led to a demographic explosion of the ants. This resulted in the ants covering a large part of the island, with a heavy impact on land crabs and bird colonies. In fact, this problem of chain reaction following the removal of an introduced species, also called surprise effect or Sysiphus effect (Mack and Lonsdale, 2002), can be generalized to other natural enemies such as herbivores or competitors.

The competitor release effect

The control of an invader has the potential to release any species interacting with the controlled invaders from its pressure, be it exploitation or interference. Therefore, one can imagine very similar processes with a browser or a competitor. Let us consider, as a final example, a "competitor release effect". Let us set the scene: an island, invaded by, say, a rat species. The island is north of New Caledonia, in the Entrecasteaux Reef. Let us call it Surprise Island, which suits very well a study on surprise effects. A thorough study of the invaded ecosystem, completed to characterize the impact of introduced rats on that island, revealed the presence of a small isolated population of introduced domestic mice. The population seems small and restricted, so that their impact on the ecosystem is likely to be negligible. In fact, in similar situations, they have been in the past neglected, partly because mouse populations are difficult to eradicate, partly because such small populations were not viewed as a threat, and partly because the conservation program concerned another species and funds and protocols were not available to deal with mice. As an example, the rat and rabbit control of Saint Paul Island, in the Antarctic ocean, has been very successful in eradicating these two introduced mammals (it was even at the time the greatest area ever cleaned up from introduced rabbits), but the program did not focus on the small

A

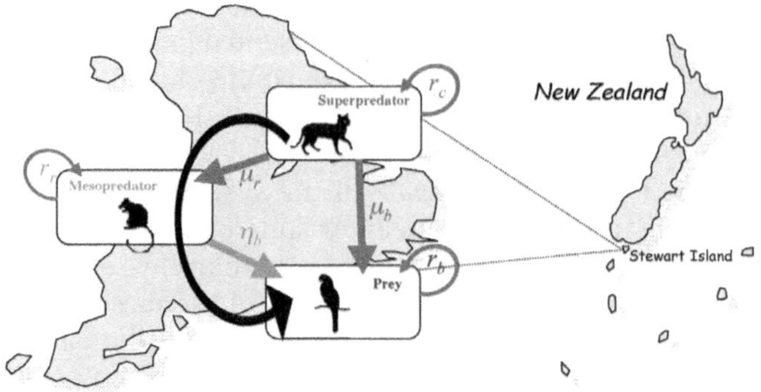

B

$$
\begin{cases}
\dfrac{dB}{dt} = r_b B\left(1 - \dfrac{B}{K_b}\right) - \dfrac{B}{S + B}\,\eta_b R - \dfrac{B}{B + R}\,\mu_b C \\
\dfrac{dR}{dt} = r_r R\left(1 - \dfrac{\eta_b \eta_s R}{\eta_b S + \eta_s B}\right) - \dfrac{R}{B + R}\,\mu_r C \\
\dfrac{dC}{dt} = r_c C\left(1 - \dfrac{\mu_b \mu_r C}{\mu_r B + \mu_b R}\right)
\end{cases}
$$

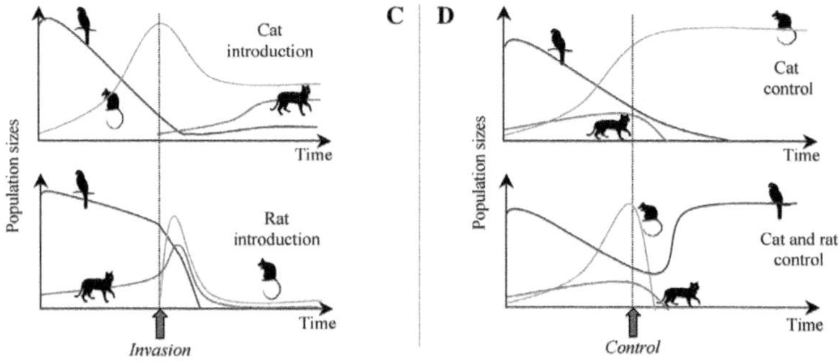

Fig. 4 Representation of the mesopredator release process, with the introduced superpredator being the cat (C, red), the introduced mesopredator being the rat (R, yellow), and the indigenous prey being the bird (B, blue). As in Fig. 1, (A) is the compartmental representation, the example coming from Stewart Island, off the New Zealand mainland, (B) is the resulting set of equations, and the two following panels are typical population trends following introductions (of the superpredator, (C) top and of the mesopredator, (C) bottom) or control (of the superpredator only, (D) top and of both the superpredator and the mesopredator, (D) bottom). Panel (C) shows that in some cases, a superpredator introduction will decrease the predation pressure on local prey, via its predation on the mesopredator (top), while a mesopredator introduction will allow an increase of the superpredator (process similar to the hyperpredation), leading to a further decrease of the prey, via a combination of increased predation of the increased superpredator population, and additional predation from the newly introduced mesopredator (bottom). Panel (D) shows that control of both introduced predator needs to be done to protect the prey (bottom), as the control of only the superpredator can trigger a mesopredator release, which eventually leads to the local prey extinction (top).

mouse population that was known to occur on the island (Micol and Jouventin, 2002).

The study of the Surprise ecosystem suggests that resources are abundant enough for the introduced mice to develop a larger population than they have done. This suggests that the mouse population is restricted by a natural enemy, and it comes naturally to mind that it is the competing rat that restricted them from further expanding their range. Although it is quite difficult to unambiguously demonstrate such competition relationships, field specialists seem unanimous in the view that rats are strong competitors of mice, to the point of often excluding them when common resources are few. It seems then quite predictable, especially after having read the previous examples of “surprise effects” in this chapter, that the planned eradication of the introduced rats on Surprise Island is likely to release the mice from rat competition, and thus to allow them to increase in numbers. Such an outcome is not only intuitively logical, it is also very easy to demonstrate through the analysis of a basic Lotka-Volterra competition model, to which a control term is added to one of the competitors. For the sake of simplicity, we study this three-species system (one prey and two competing predators) through only a two competitors model: the shared prey is ignored here, which also allows a generalization of the system to non-predator competitors.

Analysis of this system clearly shows that the mitigation of the higher competitor (the rat), will lead to an increase of the lower competitor as pressure from competition is lifted. The higher the mitigation, the larger the mouse population. A sudden, complete removal of the rat population is likely to result in a demographic explosion of the mouse population. This was the case on Saint Paul Island following the removal of rats in 2000 (Micol and Jouventin, 2002): released from their competitors, mice numbers increased dramatically, to such a point that for a time they far exceeded the carrying capacity of the habitat. Obviously, mice are less harmful than rats, and thus in some cases the end benefit of the rat removal is positive, even if the mouse population increases. Yet, mouse outbreaks can be very problematic, as mice have been shown to be active predators of invertebrates, reptiles and even birds that can be 300 times their weight (Campos and Granadeiro, 1999; Cuthbert and Hilton, 2004; Fitzgerald *et al.* 1996; Le Roux *et al.* 2002; Newman, 1994; Smith *et al.* 2002).

But this competitor release effect is in fact neither surprising nor very interesting for the study of interspecific relationships. The obvious approach to such situations seems simply to apply a simultaneous control to both competitors. And it is all the better that we are dealing with competing rodents, as a simultaneous control is easily feasible with a common rodenticide, for example. End of story, or so it seems. In fact, just to make sure no more surprises are going to emerge from the system, it is possible to complete the model by adding a simultaneous control term to the inferior competitor as well (see Figs 5A and 5B). It seems logical to link the two control rates, for they will often be (at least in the case of rodents) controlled in the same program. For example, one can have $\omega_r/\delta = \omega_m$, with ω_m being the control rate of the mice, the lower competitor (the mouse) and δ the control specificity. A value of, say, 2 for δ means that superior competitors are controlled twice as much as inferior competitors. This is very likely if, for example, baits are accessible in priority to higher competitors, or if the trapping design is aimed at the higher competitor, but they also allow to trap the lower competitor, although with a lower efficiency (the program aim is to remove the rats, but rat traps also can catch mice). Studying this very simple system reveals in fact a subtler and less expected competitor release effect. As shown in Fig. 5C, the simultaneous mitigation of both competitors can lead to a release of the lower competitor. In some cases, this release can amount to actual population explosions. It is possible that aiming at controlling, say an introduced rodent, a control program will lead to a dramatic increase of another rodent, even if that one is controlled too.

Moreover, Fig. 5D shows that the competitor release is directly proportional to the control rate. This means that the more the targeted species is controlled, the more important the competitor release effect will be. This is not so obvious to predict, as it implies that the inferior competitor is controlled too, with a control effort that increases at the same rate as that of the superior competitor. In other words, the more rodents are controlled, the more mice appear. Even if mice are actually caught in traps, and killed by poison, the more we kill them, the more their population will increase. Quite the opposite of an expected outcome.

Although this process may be less intuitive and therefore less often foreseen, it is *a posteriori* quite easy to understand. This is likely to occur as soon as the lower competitor benefits from the differential effect of the simultaneous control of both competitors; when its indirect positive effect (the removal of their competitors) exceeds its direct negative effect (their own removal).

Obviously, such a process can be interesting if the lower competitor is a local species, which survival was threatened by the controlled population. In this case, the dramatic increase of its population following competition release is nothing less than the program objectives. This can be considered in cases where an introduced species is to be removed because it threatens a local population through competition, but where control programs were not implemented for fear of damage to non-target species. In those cases, any unintended non-target death should be more than balanced by the death of the introduced competitors. There is no happy ending, however, if the lower competitor appears to be another introduced species, which increase can inflict further damages to the invaded ecosystem, especially if that increase is dramatic. Again, in many conservation situations, managers have to make trade-off choices, and it may appear that a competitor release be eventually less detrimental than allowing the presence of the introduced predator. Yet, dramatic increases of alien species are often very harmful to ecosystems, and this eventually should always be assessed.

There is much more that could be said about such a system, even as simple as it is, and about the case studies in which the overlooking of an enemy release led to surprise effects that eventually caused further damages to invaded ecosystems. However, being that the aim of this chapter is to convince the reader that biological invasions and their control can provide an excellent model system for

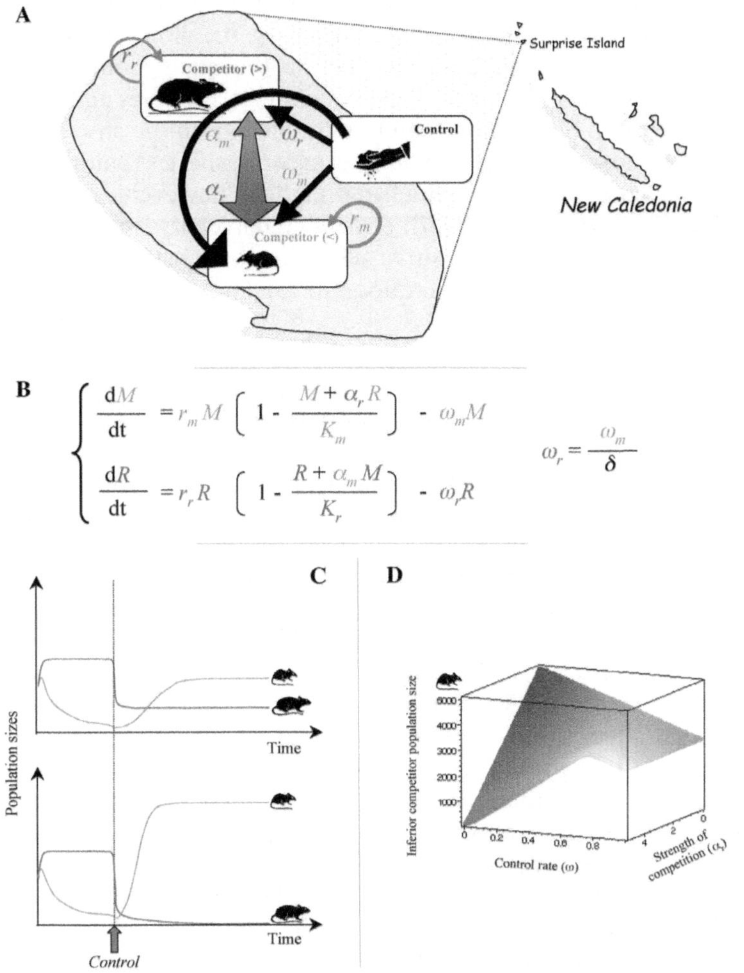

Fig. 5 Illustration of the competitor release effect. Panel (A) shows the compartmental representation of the example taken from Surprise Island, off New Caledonia, where two rodents are competitors: rats (R, red) being a superior competitor over the mice (M, yellow). The equation set in the presence of simultaneous control is given in panel (B). Here, both (C) and (D) represent population trends following control. Panel (C) shows the population trends of the two competitors as a function of time, for two different combinations of control effort and control specificity (see text). In both cases, the rodent control can lead to the demographic explosion of one of the rodent populations. Panel (D) shows the population trends of the lower competitor, as a function of the control effort and the control efficiency. As in Fig. 2D, the colors are given here to facilitate the 3D effect. This panel shows that if the competition is strong, the demographic explosion of the lower competitor is proportional to its control. As the competition pressure from the superior competitor is lifted by the control, the resulting gain in population growth is compensating the losses occurred by the control, resulting in a larger population than in absence of control.

the study of interspecific relationships, we will only close this example by an obvious, if overlooked, statement. In some cases, the direct negative impact of a species can hide an indirect positive effect on the same community, sometimes on the very same species. In a classical system where species addition or removal is not the rule, such interactions can remain hidden. In the case of invaded communities and of associated conservation programs, species removal can highlight these undetected relationships. Such highlight is however done in the form of further damage in the system, with a risk of biodiversity loss, and should therefore be anticipated. If indirect positive effects are greater than the direct negative effects, one must be extremely cautious in any restoration action intended to protect the affected species.

CONCLUSION

Biological invasions are primarily considered for their harmful effects on biodiversity in invaded ecosystems, especially on islands. This is far from surprising as islands are places of major biological diversity (and are often included in ecological hotspots, Myers *et al.* 2000; Reid, 1998) and of high probability of exotic species introduction. For example, 644 mammal introductions have been documented on islands (Ebenhard, 1988), while Gargominy *et al.* recorded more than 800 exotic species of flowering plants in the wild in New Caledonia alone (Gargominy *et al.* 1996). These figures have undoubtedly increased nowadays. The number of species that went extinct as a result of these invasions is imposing too. Consequently, the number of control programs has been on the rise this last decade, and here again there are a great number of documented examples, with, for example, over 150 eradication programs just for exotic mammals in New Zealand (C.R. Veitch, pers. comm.).

We have so far insisted much on caution and planning in any mitigation/eradication effort in management programs. We feel it is nevertheless essential to start our conclusion by rewording the obvious: the best response to biological invasion is almost always mitigate, and when possible eradicate, the alien population. In many cases a hesitancy to proceed with this has caused more damage to biodiversity than have the unexpected results of poorly planned or simply unlucky eradications.

However, the aim of this chapter was to take a different point of view regarding biological invasions, and to show that such events, however not systematic, can be viewed as opportunities to increase fundamental knowledge in ecology. Additions and deletions of species in trophic webs that are comparatively much simpler than usual should greatly enhance our ability to discern intrinsic dynamical processes as well as direct and indirect interactions between species (and here we view population dynamics as good markers of such mechanisms). This approach also has the dual advantage of addressing mechanisms on a real-scale (something impossible in laboratory based studies) and of encompassing all the charming constraints of biological reality (something missing in purely theoretical ecology).

It is, of course, not our purpose to minimize the contribution of theory in this regard. Despite their indubitable advantage in this domain, field-based studies that use invasions as an experiment have much to gain from a synergy with concomitant laboratory and theoretical approaches. We even urge conservation biologists and biodiversity managers to use theoretical-based results to found their conservation program design, and to look into historical cases in order not to endlessly repeat mistakes of the past. An idea associated to this is the need for conservationists as well as biodiversity managers to know when to stop a program and rethink the strategy (Courchamp *et al.* 2003a). To know when we are starting to play the witch's sorcerer and when to stop is an essential, yet difficult faculty. History has shown us that often more harm is generated when trial and error processes are used indiscriminately. One good example of this is given by the attempted mitigation of rats in sugarcane fields in Jamaica. There, cane growers introduced ants (*Formica omnivora*), which did not reduce rat numbers but soon became a problem themselves. To remove rats and ants together, it was then decided to introduce toads (*Bufo marinus*). But toads still did not control rats, and became a pest themselves. Finally, small Indian mongooses were introduced to control rats and toads. Mongooses failed to control either, and began preying on native birds, posing new threats to wildlife (Silverstein and Silverstein, 1974).

We recognize that there has been much to gain from these kinds of historical mistakes and that these hard-won lessons might even turn out to be globally beneficial for the biodiversity in the long run. However, we believe the trial-and-error approach time has now passed, and biodiversity managers should persevere in the current trend of basing restoration strategies on sound scientific grounds. This can only be done in developing specific studies on populations interacting in invaded communities as well as those that use such invasions as a didactic tool to understand basic processes in population ecology.

ACKNOWLEDGMENTS

This work was funded by the Institut Français de Biodiversité. We thank Leigh Bull for her help with the English.

REFERENCES

Abrams, P. A. 1987. Indirect interactions between species that share a predator: varieties of indirect effects. In Predation: Direct and Indirect Impacts on Aquatic Communities (eds. W. C. Kerfoot & A. Sih), pp. 38-54. University Press of New England, Hanover.

Abrams, P. A., Holt, R. D. and Roth, J. D. 1998. Apparent competition or apparent mutualism? Shared predation when populations cycle. Ecology, **79**, 201-212.

Brothers, N. P. 1984. Breeding, distribution and status of burrow-nesting petrels at Macquarie Island. Austrialian Wildlife Research, **11**, 113-131.

Brothers, N. P. and Copson, G. R. 1988. Macquarie Island flora and fauna management - interpreting progress and predictions for the future. Papers and Proceedings of the Royal Society of Tasmania, **122**, 129-135.

Campos, A. R. and Granadeiro, J. P. 1999. Breeding biology of the White-faced Storm-petrel on Selvagem Grande Island, North-east Atlantic. Waterbirds, **22**, 199-206.

Chapuis, J.-L. 1995a. Alien mammals in the French Subantarctic Islands. In Progress in conservation of the Subantarctic Islands (ed P. R. Dingwall), Vol. 2, pp. 127-132. IUCN - The World Conservation Union - Conservation of the Southern Polar Region.

Chapuis, J.-L. 1995b. Restoration of two islands in the Kerguelen Archipelago by eradication of the rabbit (Oryctolagus cuniculus). In Progress in conservation of the Subantarctic Islands (ed P. R. Dingwall), Vol. 2, pp. 157-160. IUCN - The World Conservation Union - Conservation of the Southern Polar Region.

Chapuis, J.-L., Boussès, P. and Barnaud, G. 1994. Alien mammals, impact and management in the French sub-Antarctic islands. Biological Conservation, **67**, 97-104.

Clutton-Brock, T. H. and Coulson, T. 2002. Comparative ungulate dynamics: the devil is in the detail. Philosophical Transactions of the Royal Society of London Series B-Biological Sciences, **357**, 1285-1298.

Courchamp, F., Chapuis, J. L. and Pascal, M. 2003a. Mammal invaders on islands: impact, control and control impact. Biological Reviews, **78**, 347-383.

Courchamp, F., Langlais, M. and Sugihara, G. 1999. Control of rabbits to protect island birds from cat predation. Biological Conservation, **89**, 219-225.

Courchamp, F., Langlais, M. and Sugihara, G. 2000. Rabbits killing birds: modelling the hyperpredation process. Journal of Animal Ecology, **69**, 154-164.

Courchamp, F., Woodroffe, R. and Roemer, G. 2003b. Removing protected populations to save endangered species. Science, **302**, 1532-1532.

Cuthbert, R. and Hilton, G. 2004. Introduced house mice Mus musculus: a significant predator of threatened and endemic birds on Gough Island, South Atlantic Ocean? Biological Conservation, **117**, 483-489.

Ebenhard, T. 1988. Introduced birds and mammals and their ecological effects. Swedish Wildlife Research, **13**, 1-107.

Feare, C. 1999. Ants take over from rats on Bird Island, Seychelles. Bird Conservation International, **9**, 95-96.

Fitzgerald, B. M. 1988. Diet of domestic cats and their impact on prey populations. In The domestic cat, the biology of its behaviour (eds. D. C. Turner & P. Bateson), pp. 123-147. Cambridge University Press, Cambridge.

Fitzgerald, B. M., Daniel, M. J., Fitzgerald, A. E., Karl, B. J., Meads, M. J. and Notman, P. R. 1996. Factors effecting the numbers of house mice (*Mus musculus*) in hard beech (*Nothofagus truncata*) forest. Journal of the Royal Society of New Zealand, **26**, 237-249.

Flux, J. E. C. 1993. Relative effect of cats myxomatosis traditional control or competitors in removing rabbits from islands. New Zealand Journal of Zoology, **20**, 13-18.

Flux, J. E. C. and Fullagar, P. J. 1992. World distribution of the rabbit *Oryctolagus cuniculus* on islands. Mammal Review, **22**, 151-205.

Gargominy, O., Bouchet, P., Pascal, M., Jaffre, T. and Tourneur, J. C. 1996. Consequences for biodiversity of plant and animal species introductions in New Caledonia. Revue D Ecologie-La Terre Et La Vie, **51**, 375-402.

Gillham, M. E. 1963. Some interactions of plants, rabbits and seabirds on South African islands. Journal of Ecology, **51**, 275-294.

Ginzburg, L. R. and Jensen, C. X. J. 2004. Rules of thumb for judging ecological theories. Trends in Ecology & Evolution, **19**, 121-126.

Grenfell, B. T., Wilson, K., Finkenstadt, B. F., Coulson, T. N., Murray, S., Albon, S. D., Pemberton, J. M., Clutton-Brock, T. H. and Crawley, M. J. 1998. Noise and determinism in synchronized sheep dynamics. Nature, **394**, 674-677.

Holdgate, M. W. and Wace, N. M. 1961. The influence of man on the floras and faunas of Southern Islands. Polar Record, **10**, 475-493.

Holt, R. D. 1977. Predation, apparent competition, and the structure of prey communities. Theoretical Population Biology, **12**, 197-229.

Hone, J. 1995. Spatial and Temporal Aspects of Vertebrate Pest Damage with Emphasis on Feral Pigs. Journal of Applied Ecology, **32**, 311-319.

Karl, B. J. and Best, H. A. 1982. Feral cats on Stewart Island: their foods and their effects on kakapo. New Zealand Journal of Zoology, **9**, 287-294.

Keane, R. M. and Crawley, M. J. 2002. Exotic plant invasions and the enemy release hypothesis. Trends in Ecology & Evolution, **17**, 164-170.

Kessler, C. C. 2001. Eradication of feral goats and pigs from Sarigan Island, Commonwealth of the Northern Mariana Islands; methods and results. In Eradication of islands Invasives: practical actions and results achieved (ed. M. Clout), pp. 18, University of Auckland, New-Zealand.

Le Roux, V., Chapuis, J. L., Frenot, Y. and Vernon, P. 2002. Diet of the house mouse (Mus musculus) on Guillou Island, Kerguelen archipelago, Subantarctic. Polar Biology, **25**, 49-57.

Mack, R. N. and Lonsdale, W. M. 2002. Eradicating invasive plants: hard-won lessons for islands. In Turning the Tide: the Eradication of Invasive Species (eds. C. R. Veitch & M. Clout). Invasive Species Specialty Group of the World Conservation Union (IUCN), Auckland, New-Zealand.

Micol, T. and Jouventin, P. 2002. Eradication of rats and rabbits from Saint-Paul Island, French Southern Territories. In Turning the tide: the eradication of invasive species (eds. C. R. Veitch & M. N. Clout), pp. 199-205. IUCN SSC Invasive Species Specialist Group. IUCN, Gland, Switzerland and Cambridge, UK.

Murphy, E. and Bradfield, P. 1992. Change in diet of stoats following poisoning of rats in a New-Zealand forest. New Zealand Journal of Ecology, **16**, 137-140.

Myers, N., Mittermeier, R. A., Mittermeier, C. G., da Fonseca, G. A. B. and Kent, J. 2000. Biodiversity hotspots for conservation priorities. Nature, **403**, 853-858.

Newman, D. G. 1994. Effects of a Mouse, *Mus musculus*, Eradication Program and Habitat Change on Lizard Populations of Mana Island, New- Zealand, with Special Reference to Mcgregor Skink, *Cyclodina macgregori*. New Zealand Journal of Zoology, **21**, 443-456.

Rammell, C. G., Hoogenboom, J. J. L., Cotter, M., Williams, J. M. and Bell, J. 1984. *Brodifacoum* residues in target and non-target animals following rabbit poisoning trials. New Zealand Journal of Experimental Agriculture, **12**, 107-111.

Reid, W. V. 1998. Biodiversity hotspots. Trends in Ecology & Evolution, **13**, 275-280.

Robertson, H. A., Hay, J. R., Saul, E. K. and McCormack, G. V. 1994. Recovery of the Kakerori - an Endangered Forest Bird of the Cook-Islands. Conservation Biology, **8**, 1078-1086.

Roemer, G. W., Donlan, C. J. and Courchamp, F. 2002. Golden eagles, feral pigs, and insular carnivores: How exotic species turn native predators into prey. Proceedings of the National Academy of Sciences of the United States of America, **99**, 791-796.

Schoener, T. W. and Spiller, D. A. 1999. Indirect effects in an experimentally staged invasion by a major predator. American Naturalist, **153**, 347-358.

Scott, J. J. 1988. Rabbit distribution history and related land disturbance Macquarie Island Australia. Papers and Proceedings of the Royal Society of Tasmania, **122**, 255-266.

Selkirk, P. M., Costin, A. B., Seppelt, R. D. and Scott, J. J. 1983. Rabbits, vegetation and erosion on Macquarie Island. Proceedings of the Linnean Society N. S. W., **106**, 337-346.

Silverstein, A. and Silverstein, V. 1974. Animal invaders. The story of imported wildlife Athenum, New York.

Smith, A. P. and Quin, D. G. 1996. Patterns and causes of extinction and decline in Australian conilurine rodents. Biological Conservation, **77**, 243-267.

Smith, V. R., Avenant, N. L. and Chown, S. L. 2002. The diet and impact of house mice on a sub-Antarctic island. Polar Biology, **25**, 703-715.

Soulé, M. E., Bolger, D. T., Alberts, A. C., Wright, J., Sorice, M. and Hill, S. 1988. Reconstructed dynamics of rapid extinctions of chaparral-requiring birds in urban habitat islands. Conservation Biology, **2**, 75-92.

Taylor, R. H. 1979. How the Macquarie island parakeet became extinct. New Zealand Journal of Ecology, 2, 42-45.

Chapter twelve

Invasibility of seed predators on synchronized and intermittent seed production of host plants

A. Satake, O. N. Bjørnstad and Y. Iwasa

INTRODUCTION

The spatiotemporal distribution of resources that invaders require for their growth and reproduction is a key factor in controlling success of invasion (Davis *et al.* 2000). The abundance of abiotic resources, such as water supply and nutrient levels, may fluctuate due to variable climatic condition and disturbance regime (e.g., drought, flood, or fire regimes). Therefore the intensity of competition for these limited resources varies resulting in temporal fluctuation of a community's susceptibility to invasion (Davis and Pelsor 2001).

When invading species are consumers of biotic resources (e.g., hervivores and seed predators), successful invasion may trigger the evolution of traits of host species that enhance resistance to invaders. For example, mast seeding or masting, the intermittent and synchronized reproduction by plant populations (Kelly 1994) reported across a broad group of tree species (Herrera *et al.* 1998), is often explained to have evolved to reduce seed losses to (and hence to prevent

M.W. Cadotte et al. (eds.), Conceptual ecology and invasion biology, 271–288.

invasion of) seed predators (the "predator satiation hypothesis", Janzen 1971; Silvertown 1980). Essentially, larger seed crops are synchronized among individuals and satiate seed predators and therefore experience a lower percentage of seed predation. At the same time, years with low seed production will reduce the growth of the seed predator populations. Many studies confirm predator satiation by showing lower predation rates in high seed years (Smith *et al.* 1990; Sork 1993; Kelly and Sullivan 1997; Sperens 1997; Shibata *et al.* 1998; Kelly *et al.* 2000; McKone *et al.* 2001; Satake *et al.* 2004).

The properties or adaptations of the invading consumers also play an important role, particularly the traits associated with dispersal. Even if food resources show unpredictable fluctuation at a single patch (or plant), dispersal between patches buffers variation in fitness by spreading risk over space (Hopper 1999). Dispersal among patches is advantageous if fluctuations of local conditions are independent across space (Levin *et al.* 1984), but if there is positive correlation either in time or across space, dispersal is less favored (Ellner and Shmida 1981; Cohen and Levin 1991; Holt and McPeek 1996).

In this chapter, we discuss a resource-consumer model to understand the spatiotemporal dynamics of a seed predator on a masting resource and explore the invasibility of the predators (Satake and Bjørnstad 2004). Spatiotemporal fluctuations in seeding are represented by the pollen coupled tree model (Isagi 1997; Satake and Iwasa 2000, 2002a, 2002b) in which reproductive behavior of individual plants is constrained by the level of internal energy reserves and local pollen production. We consider specialized insect seed predators. The dynamics of the predator at a single host plant is assumed to follow a spatially-extended Nicholson-Bailey model (Hassell *et al.* 1991; Satake and Bjørnstad 2004), and the adults having a semelparous lifecycle disperse to neighboring host plants to search randomly for seeds, berries, or flowerheads on which to lay their eggs, and the larvae subsequently complete the lifecycle through feeding on the seeds. We calculate an approximate invasion criterion for the predators, which illustrates how variable and synchronized seed production of plants may prevent invasion of seed predator's population.

RESOURCE DYNAMICS OF SEED PRODUCTION

Spatiotemporal fluctuation of food resource for seed predators is modeled as follows. We consider a forest of out-crossing plants each with an internal energy reserve that changes with net energy gain through photosynthesis and energy expenditure through reproduction (and maintenace/growth). Each plant gains net energy, P_s, every year from photosynthesis. Let $S_i(t)$ be the energy reserves of a plant located at site i at the beginning of year t. A plant will not reproduce unless its accumulated energy reserves, $S_i(t) + P_s$, exceeds a critical threshold, L_T. Otherwise the plant produces flowers investing energy ("flowering cost") proportional to the excess given by $a(S_i(t) + P_s - L_T)$, where a is constant of proportionality (Satake and Iwasa 2000). All flowers are pollinated and the plant then invests additional energy due to the cost of fruit production, $R_c\, a(S_i(t) + P_s - L_T)$, where R_c is the ratio of fruiting cost to flowering cost. The energy reserve level thus falls to $S_i(t) + P_s - a(R_c + 1)(S_i(t) + P_s - L_T)$, so that the overall resource dynamics are

$$S_i(t+1) = \begin{cases} S_i(t) + P_s & \text{if } S_i(t) + P_s \le L_T \\ (S_i(t) + P_s - a(R_c + 1)(S_i(t) + P_s - L_T) & \text{if } S_i(t) + P_s > L_T \end{cases}. \tag{1}$$

Considering the non-dimensionalized variable $Y_i(t) = (S_i(t) + P_s - L_T)/P$, equation (1) is rewritten as,

$$Y_i(t+1) = \begin{cases} Y_i(t) + 1 & \text{if } \; Y_i(t) \le 0 \\ -kY_i(t) + 1 & \text{if } \; Y_i(t) > 0 \end{cases}. \tag{2}$$

in which $k = a(R_c + 1) - 1$ is a depletion constant. If $a = 1$, k is equal to the ratio of fruiting cost to flowering cost, R_c. Equation (2) represents a "tent-map" that predicts stable annual reproduction for $k < 1$ and chaotic reproduction for $k > 1$ (Satake and Iwasa 2000). Such temporal energy dynamics at the level of the individual plant is easily illustrated through bifurcation diagrams (Fig. 1). The long-term trajectories of energy level defined in eq. (2) are plotted across a range of values of k. When $k < 1$, trajectories visit a single positive value, which represents annual and constant reproduction. For $k > 1$, the orbit eventually represents chaotic fluctuation. As k increases, the interval between non-reproductive years becomes longer because the plant needs more years to rebuild energy reserves to the threshold level.

Pollen limitation

If plants are self-incompatible, fruiting rates may depend on the flowering activity of other trees in a forest because pollination efficiency changes with the number of flowering plants (Nilsson and Wästljung 1987; Smith *et al.* 1990). To model the pollen limitation of plant reproduction, the pollen availability, $P_i(t)$, to individual i was introduced in the non-dimensionalized energy dynamics (Isagi *et al.* 1997; Satake and Iwasa 2000):

$$Y_i(t+1) = \begin{cases} Y_i(t) + 1 & \text{if } \; Y_i(t) \le 0 \\ -kP_i(t)Y_i(t) + 1 & \text{if } \; Y_i(t) > 0 \end{cases}, \tag{3a}$$

where $P_i(t)$ is determined by the average flowering intensity of the plants in the neighborhood around the focal plant, U_i. It is given by

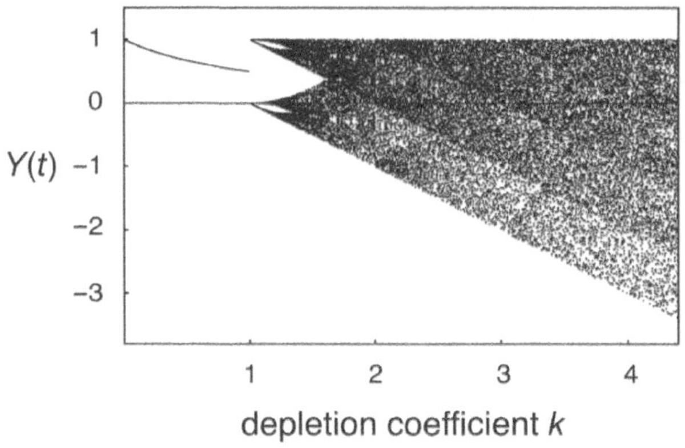

Fig. 1 Bifurcation diagram of an individual plant. Horizontal axis is depletion coefficient k, and the vertical axis is the non-dimensinalized energy level. Parameters: $\beta = 0$, $w_1 = 100$.

$$P_i(t) = \left(\frac{1}{w_1} \sum_{j \in U_i} [Y_j(t)] \right)^{\beta}, \tag{3b}$$

where $[Y]_+ = Y$ if $Y > 0$ and zero otherwise. The parameter w_1 is the number of plants included in the neighborhood U_i (note that the neighborhood does not include oneself), and β determines the degree of neighbor dependence on the pollen limitation. If β is close to 0, fruit production is almost independent of the flowering intensity of other plants. In contrast, a large β implies a strong dependence of seed and fruit production on the pollen production of other plants. Thus, we call β the pollen coupling strength. In this chapter, we consider a situation of "local pollen coupling" in which pollination is limited to the eight nearest neighbors around the focal plant ($w_1 = 8$).

POPULATION DYNAMICS OF SEED PREDATORS

A specialized insect seed predator with a semelparous lifecycle is considered in an all-females model. Adults disperse to nearby host plants to search randomly for seeds and berries on which to lay their eggs. The larvae hatch to complete their lifecycle feeding on the seed, leave the seeds to pupate, and emerge as adults at the beginning of next generation. These assumption are motivated by studies of the apple fruit moth, *Argyresthia conjugella* Zeller, which is a pre-dispersal seed predator of rowan, *Sorbus aucuparia* L. The larvae of the apple fruit moth relies heavily on rowan berries that varies substantially between years in a geographically synchronized fashion (Kobro *et al.* 2003; Satake *et al.* 2004).

We assume that host plants are arranged on lattice points of a two-dimensional square grid. The number of emerged adults at site i at the beginning of year t is denoted by $Z_i(t)$. A fraction, μ, of the adults leaves the plant where they were born and disperses to adjacent plants. The post-dispersal number of adults at site i, denoted as $Z'_i(t)$, is given by:

$$Z'_i(t) = (1 - \mu)\, Z_i(t) + \frac{\mu}{w_2} \sum_{j \in R_i} Z_j(t). \tag{4}$$

The parameter R_i is the neighborhood of site i and w_2 is the number of host plants included in R_i. For simplicity, we also assume that seed predator disperses to the eight nearest neighbors ($w_2 = 8$). When a single plant at site i produces a certain amount of seeds, $F_i(t)$, in year t, the number of pre-dispersal adults in the next generation $(t + 1)$ is given by:

$$Z_i(t + 1) = \gamma F_i(t)(1 - e^{-\alpha Z'_i(t)}), \tag{5}$$

where γ is the per capita emergence rate and α is the per capita attack rate (the searching efficiency). $F_i(t)$ itself is governed by the pollen coupled tree model (equations (1)–(3)) according to:

$$F_i(t + 1) = \begin{cases} 0 & \text{if } \; Y_i(t) \le 0 \\ ckP_i(t)Y_i(t) & \text{if } \; Y_i(t) > 0 \end{cases}, \tag{6}$$

where c is a constant, k is the depletion coefficient, and $P_i(t)$ is the pollen availability given by eqn. (3b).

SEED PREDATION ON MASTING RESOURCE

The reproductive behavior predicted by the pollen coupled tree model and severity of seed predation is illustrated in Fig. 2. When the depletion coefficient, k, is less than 1, each plant produces seeds every year, which is annual reproduction (Fig. 2A). Annual reproduction can be further classified into "annual and constant reproduction" (shaded area in Fig. 2A) in which each plant produces seeds constantly and "annual and 2-year cyclic reproduction" (checker region in Fig. 2A) in which produced seed crop size alternates between high and low (but positive) reproductive levels. Annual and constant seed production results in severe attacks (Fig. 2B), while annual and 2-yr cyclic seed production can cause extinction of the predator (Fig. 2B). This extinction is due to the satiation of numerical response of

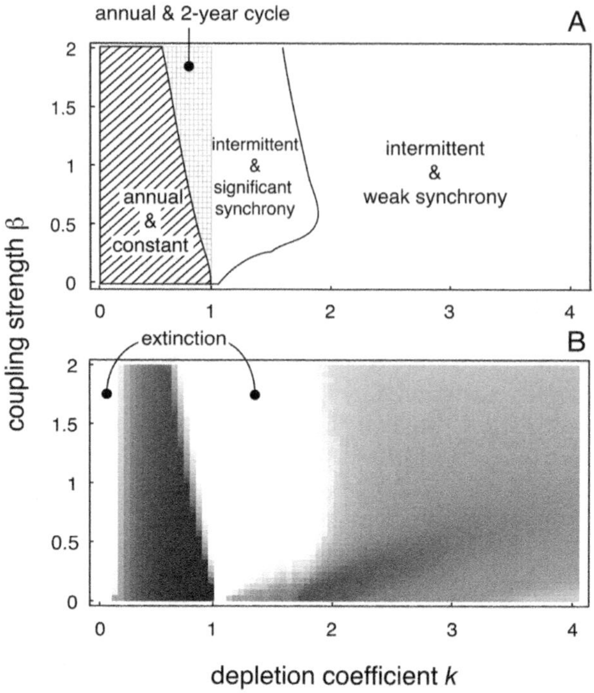

Fig. 2 (A) Classification of reproductive pattern of plants predicted by the pollen coupled tree model with an assumption of local pollen coupling. We classified four types of reproductive pattern as explained in the text. (B) The density plot for proportion of seeds attacked calculated by numerical simulation of the model. The darker area represents more severe seed predation. Parameters: $\mu = 0.8$; $\alpha = 10$. (From A. Satake and O. N. Bjørnstad, Spatial dynamics of specialist seed predators on synchronized and intermittent seed production of host plants, American Naturalist 163, (2004), 591-605. Reprinted with permission from The University of Chicago Press).

the predator — the number of the predators is heavily reduced by a very small seed production in a low reproductive year, and the predator population cannot cope with a great increase in seed production in a high reproductive year, which decreases the population size and finally leads the population to extinction.

When individual plants produce seeds intermittently with numerous years with no seed set (parameter region of $k > 1$ in Fig. 2A), "local" extinction of seed predators is inevitable. However they may nevertheless persist globally due to dispersal when mast synchrony is weak (Fig. 2). In Fig. 2, significant synchrony ($\bar{\rho} \geq 0.8$) is distinguished from weak synchrony ($\bar{\rho} < 0.8$) by calculating the degree of spatial synchrony among plants, $\bar{\rho}$. In a forest exhibiting weak synchrony in seeding, there is always some fraction of plants producing seeds. This enables seed predators to persist globally through dispersal to any asynchronous nearby plants.

In contrast, if plants in a forest produce seeds in a significantly synchronized fashion, seed predators cannot find resources elsewhere within the dispersal range and then go globally extinct (Fig. 2). For masting plants ($k > 1$), the proportion of attacked seeds is inversely related to the degree of spatial synchrony among plants, $\overline{\rho}$ (Fig. 3), meaning that spatial synchrony in seed production effectively reduces losses to seed predators.

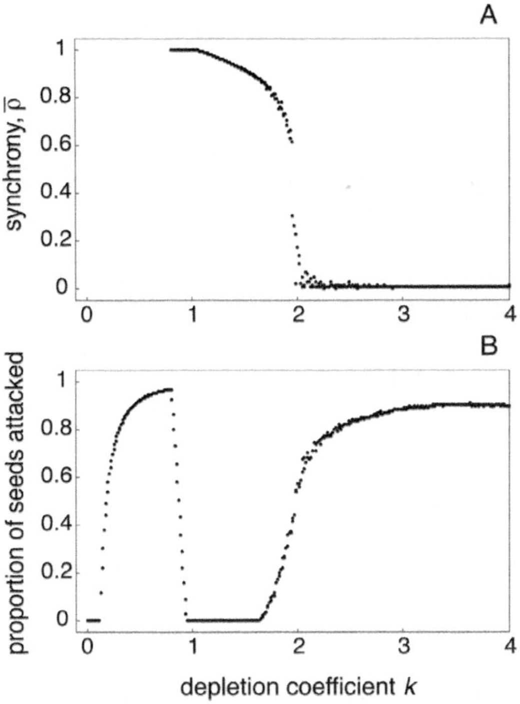

Fig. 3 (A) Degree of synchrony in seed production (ρ). (B) Proportion of seeds attacked. The horizontal axis is the depletion coefficient, k. Parameters: $\beta = 0.8$; $\mu = 0.8$; $\alpha = 10$.

INVASION

We developed an approximate invasion criterion, $\hat{Q}$, that approximately describes when seed predators can invade a system where the host plants produce seeds synchronously and intermittently (see Appendix in Satake and Bjørnstad 2004). Though we have been unable to obtain exact invasion criteria due to the nature of masting resources (i.e., fluctuation across time and space in a cyclic or chaotic manner; Satake and Iwasa 2000, 2002a). The approximate invasion criterion appears to give a good prediction of the invasion and provides key insights into how predator invasion is influenced by the manner of fluctuations of the resource. The approximate invasion criterion $\hat{Q}$ is the average growth or decline in the predator population during two generations following introduction, given as follows:

$$\hat{Q} = \frac{1}{T}\sum_{t=0}^{T-1} Q(t), \tag{7a}$$

where

$$Q(t) = \frac{\alpha^2\gamma^2}{N}\left\{(1-\mu)\sum_{i=1}^{N} F_i(t)\,F_i(t+1) + \frac{\mu}{w_2}\sum_{i=1}^{N}\sum_{j\in R_i} F_i(t)\,F_j(t+1)\right\}. \tag{7b}$$

In the above equation, N is the number of plants in the forest, and w_2 is the number of plants included in the neighborhood, R_i. $F_i(t)$ is the amount of seed produced by a plant at site i in year t given by eqn. (6). Other parameters are as defined in equations (4) and (5). The first term in the right hand side in eqn. (7b) represents the average population increase (or decline) through two generations for the seed predators that do not disperse — their next-year offspring are left to consume seeds produced by the plant where their parents were born. The second term represents the corresponding average for the predators that dispersed to lay eggs on neighboring plants.

By a simple calculation (Stake and Bjornstad 2004), $\hat{Q}$ is related to the time lag-1 auto-correlation $R(1)$ in seed production of individual plants and time lag-1 cross-correlation $C(1)$ between neighboring plants as follows:

$$\hat{Q} = \alpha^2\gamma^2\{(1-\mu)\,R(1)\sigma^2 + \mu C(1)\,\sigma^2 + 2\hat{F}^2\} \tag{8}$$

where σ^2 is the variance and $\hat{F}$ is the temporal mean seedset of each plant. Eqn. (8) shows that invasibility is prompted if seed production is positively correlated in time both for a given plant and for plants within the dispersal neighborhood of the predator. In the case of mast seeding, negative time lag 1 or 2 autocorrelations in seed production are often realized (Norton and Kelly 1988; Sork *et al.* 1993; Koenig *et al.* 1994; Kelly and Sork 2002). In addition, negative cross correlation at time lag 1 often result because of spatially synchronized reproduction. Thus invasibility to the predators is reduced on masting resources.

If seed production with intermittence is synchronized at a "local" spatial scale, nearby plants show a stronger negative cross correlation in seed production at time lag 1 than distantly located plants. In a forest showing regionally (not locally) self-organized seed production, cross correlations calculated at the local spatial scale will be similar to the average across the regional spatial scale. In this case, $Q(t)$ simplifies to:

$$Q'(t) = \frac{\alpha^2\gamma^2}{N}\left\{(1-\mu)\sum_{i=1}^{N} F_i(t)\,F_i(t+1) + \frac{\mu}{(N-1)}\sum_{i=1}^{N}\sum_{j\neq i} F_i(t)\,F_j(t+1)\right\}. \quad (9)$$

Here, the second term in the right hand side in eqn. (7b) is replaced by the average of the product for all possible pairs (i, j) of plants. By taking the temporal average of $Q'(t)$, we have an approximate invasion criterion in the absence of local spatial structure, $\hat{Q}'$.

We evaluated $\hat{Q}$ and $\hat{Q}'$ numerically by generating the spatiotemporal patterns in seed production from a numerical simulation of the pollen-coupled tree model with a range of parameters resulting in varying degree of masting (Satake and Bjornstad 2004). We confirm the accuracy of the approximate invasion criteria, through numerical simulation of the full consumer-resource model. In theory the seed predators should only be able to invade when $\hat{Q} > 1$. To investigate this, we introduced a minute density of the predator (10^{-4}) on each host plant in the model. We subsequently traced their population growth for 100 generations. If the predator's population is still extant at the end of the simulation, the plant reproductive strategy is considered invasible.

The results including the two approximate invasion criteria are illustrated in Fig. 4. Both criteria are seen to decrease with increasing synchrony of seed set, $\bar{\rho}$, meaning that invasibility is inhibited, partly or entirely, when plants show reproductive synchrony. In simulations, establishment fails when $\hat{Q}$ is below 1 and

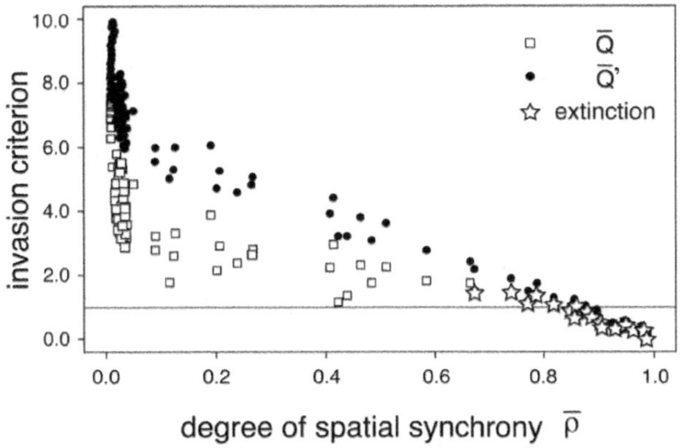

Fig. 4 Plot of the approximate invasion criteria, $\hat{Q}$ (open squares) and $\hat{Q}'$ (solid circles), along the degree of synchrony in seed production among different plants (ρ). Stars represent the values of $\hat{Q}$ in which extinction of seed predators occurred. Parameters: $\mu = 0.8$; $\alpha = 10$. (From A. Satake and O. N. Bjørnstad, Spatial dynamics of specialist seed predators on synchronized and intermittent seed production of host plants, American Naturalist 163, (2004), 591-605. Reprinted with permission from The University of Chicago Press).

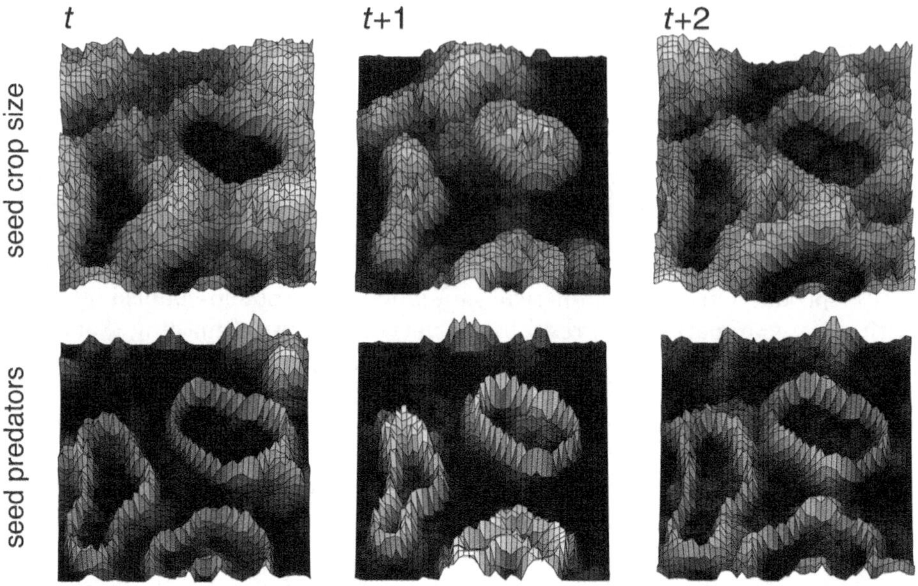

Fig. 5 Spatial patterns in seed production (upper row) and amount of seed predators (bottom row). x and y-axes represent spatial locations, and z-axis represents seed crop size (upper three figures) or seed predator abundance (bottom three figures). Time flows left to right. Parameters: k = 1.8; β = 1.0; μ = 0.8; α = 10. The total number of host plant is 50 × 50. In order to show clear spatial patterns, we apply the pollen dispersal range larger than nearest neighbor: plants located at ith location, (xi, yi), and jth location, (xj, yj) can exchange pollen if |xi − xj|2 + |yi − yj|2 ≤ 52. The overall spatial pattern is similar, but the spatial scale of the pattern is larger and becomes coarse grained as pollen dispersal range increases (for details, see Satake and Iwasa 2002a). (From A. Satake and O. N. Bjørnstad, Spatial dynamics of specialist seed predators on synchronized and intermittent seed production of host plants, American Naturalist 163, (2004), 591-605. Reprinted with permission from The University of Chicago Press).

succeeds for $\hat{Q} > 1$. This suggests that the approximate invasion criteria based on the population growth during two generations appears to give an excellent prediction of the invasion in the full spatiotemporal simulations. This success originates in part from how the entire forest exhibits a 2-year cycle in seed production when the plant's reproduction is highly synchronized (Fig. 2). The 2-year cyclic seed production becomes unclear at the population level as the degree of synchrony decreases. Thus longer intervals between mast years (e.g. 3 or 4 years) should be taken into account to correct the slight overestimation of the invasion criterion (Fig. 4). Interestingly, $\hat{Q}$ is almost always less than $\hat{Q}'$. This difference highlights that seed predators suffer lower invasibility when reproduction among plants is synchronized at a local spatial scale.

The spatial persistence of predators in the presence of highly variable mast seeding is visualized in Fig. 5. The upper three figures represent the temporal change of spatial distributions for seed crop size produced by individual plants. The figure illustrates how localized spatial clusters can result form local pollen coupling. Within a given cluster, plants show synchronized reproduction with a 2-year cycle visiting high and very low reproductive level. The corresponding spatial distributions of seed predators (bottom panel in Fig. 5) show how the seed predator's persistence is facilitated at the boundaries of the clusters. Seed predators can find resources to consume only when host plants showing opposite reproductive phases are within the dispersal range.

COMMUNITY LEVEL MASTING AND INVASION

In this section, we introduce a simple case study of plant-consumer system, and briefly discuss that when seed predators are generalists, masting at community level is necessary to reduce the impact of the predator. Rowan, *Sorbus aucuparia* L., is a deciduous tree, which is common in much of Europe (Tutin *et al.* 1968). The fleshy berries produced by rowan trees suffer from attack by the apple fruit moth, *Argyresthia conjugella*, that is a primary pre-dispersal seed predator whose larvae rely heavily on rowan berries (Sperens 1997a; 1997b; Kobro *et al.* 2002; Satake *et al.* 2004). The spatiotemporal data obtained from a census in southern Norway show that berry production in rowan is variable across years (mean CV = 1.02) and is spatially synchronized ($\overline{\rho} = 0.67$: bootstrapped 95% confidence interval 0.63-0.70). The degree of spatial synchrony in rowan population is lower than the level of synchrony leading to seed predator extinction (i.e., $\overline{\rho} > 0.8$; Fig. 4), and hence the apple fruit moth successfully invade and is persistent in this system.

However relatively high degree of spatial synchrony in rowan masting efficiently reduces seed losses to the apple fruit moth and suppresses the population growth rate of seed predators at a low level as illustrated in Fig. 7. The proportion of seeds predated was negatively correlated with the total number of seeds, evidencing a satiated functional response of the seed predators by masting. In addition, a negative relationship was observed between the seed predation and the ratio of berry production in successive years. This means that seed crops following poor seed crops tended to escape predations, testifying to a satiation through numerical response of the predators (Satake *et al.* 2004). The predator satiation hypothesis further predicts that the more variable populations should attain higher overall seed survival (Janzen 1971, Waller 1979, Silvertown 1980). To assess this, the total proportion of seeds lost to predation was examined as a function of temporal variability in seed production, measured by the CV. The resultant logistic regression revealed a significantly negative relationship between the proportion of berries predated and CV (the slope = −1.03; $p < 0.001$),

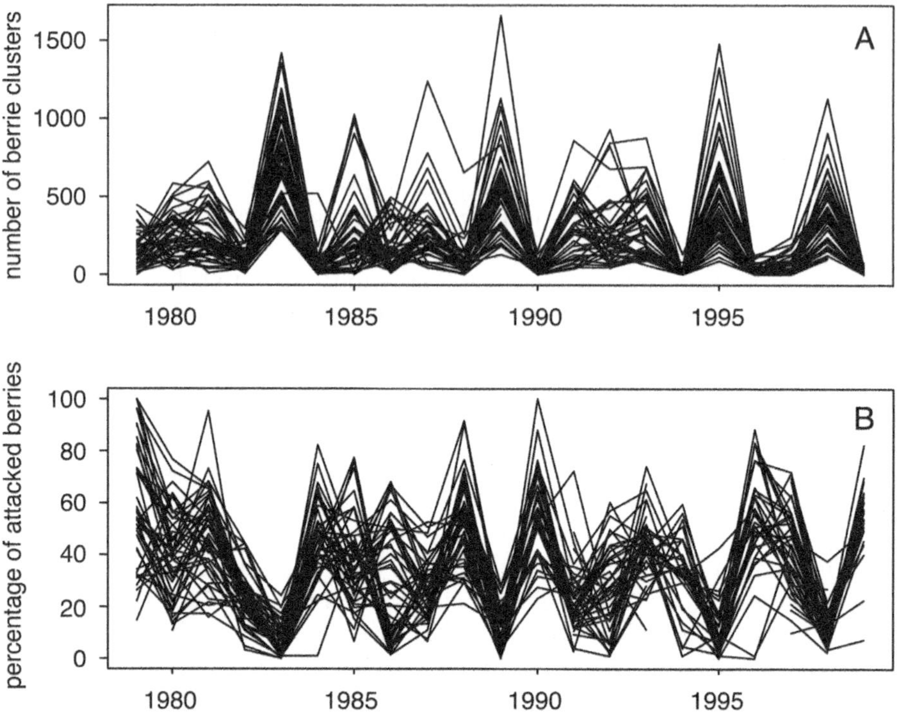

Fig. 6 Time series data of (A) annual berry production of rowan trees, and (B) percentage of attacked berries by the apple fruit moth. Each line represents the time series from one study site. (From A. Satake, O. N. Bjørnstad, and S. Kobro, Masting and trophic cascades: interplay between rowan trees, apple fruit moth, and their parasitoid in southern Norway, Oikos 104, (2004), 540-550. Reprinted with permission from Blackwell Publishing).

indicating that larger inter-annual variation in berry production resulted in smaller losses to predation. These results show that rowan masting has an adaptive foundation, which reduce seed losses to and prevent a rapid expansion in a plant population.

The persistence of the moth in the rowan-moth system is in part because of the fact that the apple fruit moth is not a strict specialist — the primary and preferred host of the apple fruit moth is rowan, but when too few rowan berries are available for egg-laying, many female moths switch the host to apple (Ahlberg 1927). If there are generalist seed predators in a plant community, variable and synchronized seed production only within a single species may not enough to prevent invasion of the seed predator and even masting species showing a significant synchrony in seed production may suffer heavy seed predation. Silvertown

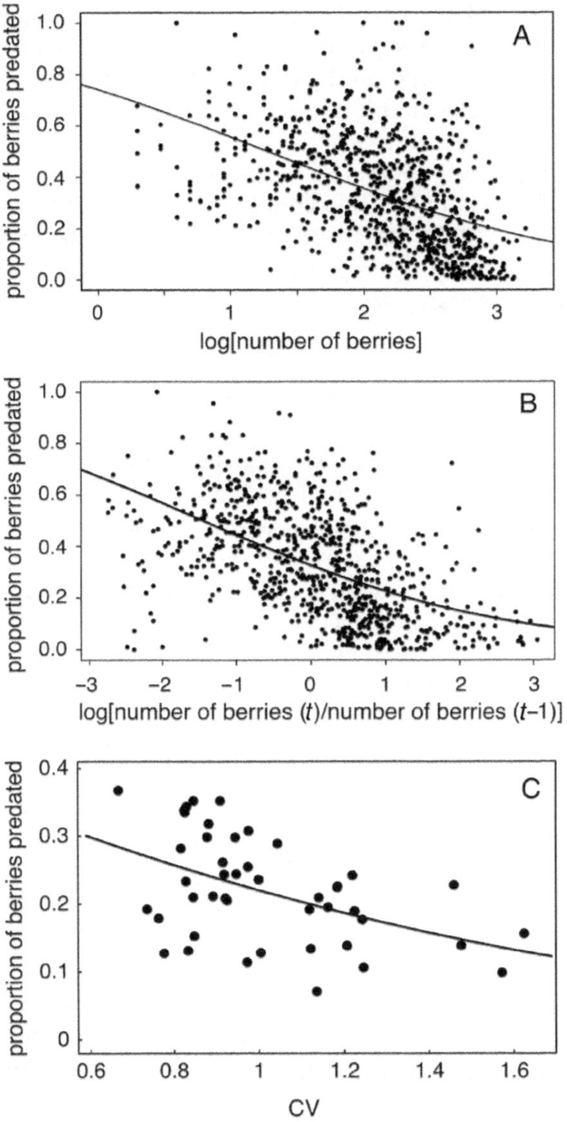

Fig. 7 Plot of proportion of seed predation in rowan as a function of (A) seed crop size and (B) the ratio of seed crop size in year t and that in year $t - 1$ (lag1). Log transformed variables are used for calculation. Lines represents the logistic regression model which have intercepts and slopes of (A) 1.06 (SE = 0.097, $p < 0.001$), –0.82 (SE = 0.055, $p < 0.001$) and (B) –0.73 (SE = 0.030, $p < 0.001$), –0.50 (SE = 0.029, $p < 0.001$). (C) Relationship between the coefficient of variation (CV) of annual seed production and the total proportion of seed predation. Each point represents the result calculated at one study site. Logistic regression analysis gave the regression line as intercept of –0.23 (SE = 0.29, $p = 0.44$) and slope of –1.03 (SE = 0.29, $p < 0.001$).

(1980) stated that synchronous seed production between population of different species sharing the same seed predators reduces the growth rate of the predator population by predator satiation. Spatial synchrony in seed production among species has been reported in animal-pollinated species in a relatively closed community in the rain forest (Momose *et al.* 1999; Inoue and Hamid 1997; Sakai *et al.* 1999), and in wind-pollinated species in open habitats such as temperate forest and savanna (Koenig and Knops 1998, 2000; Schauber *et al.* 2002; Shibata *et al.* 1998, 2002). Such a synchronized and episodic reproduction in a community provides an interesting insight into the evolutionary process how plant communities enhance resistance against generalist seed predators.

CONCLUSION

This chapter focuses on the spatially-extended dynamics of masting plants and their seed predators. Plants within a forest may produce seeds annually or intermittently in time and synchronously or asynchronously across space. This range of dynamical behavior is captured by the pollen coupled tree model (eqns. (1)-(3); Isagi 1997; Satake and Iwasa 2000, 2002a, 2002b). The effect of mast seeding on dynamics and invasibility of specialist consumers was explored by examining the population dynamics of seed predators in a bottom-up fashion (eqns. (4)-(6)). When plants produce seeds intermittently, persistence and invasion of the predator is most critically dependent on the degree of synchrony of seed set. The predator population can only persist through dispersal to adjacent host plants in a forest showing weak synchrony because there is then some fraction of asynchronously reproducing plants within the dispersal range of the predators. In contrast, extinction of seed predators is likely when intermittent reproduction is highly synchronized among different plants (Fig. 2). Spatial synchrony in seed production with intermittence creates negative time lag-1 auto-correlation and cross-correlation in seeding, which most effectively reduce invasibility of the seed predators (eqn. 8). This means that "classic" mast seeding, exhibiting seed set that is negatively correlated in time but positively correlated across space, is a good strategy to reduce seed loss to seed predators. In addition, analysis of invasion criterion suggest that spatial synchrony at local spatial scales may further reduce losses by preventing invasion of seed predators (Fig. 4).

The spatial scale at which synchrony in seed production reduces seed losses to seed predators depends on the mobility of the predators (Kelly and Sork 2002). Insect predators considered in this chapter may disperse over relatively short distances. Birds or mammals in contrast may be highly mobile (Curran and Leighton 2000). In order to satiate such predators so as to reduce seed losses, larger spatial scales of reproductive synchrony must be maintained. The analysis of approximate invasion criteria given in equations 7 and 8 applies also to highly mobile specialist consumers because the formulations can be written for arbitrary neighborhood sizes (not restricted to nearest neighbor dispersal).

The community consequences of masting are only known for selected case studies. Mice and squirrels, for instance, experience enhanced survival and rapid population growth in years of oak mast. Such population may conversely crash to low levels in years of little acorn because of a shortage of food resources (Wolff 1996; McShea 2000). Masting has also been shown to have cascading effects through food webs and ecological communities, in systems ranging from outbreaks of gypsy moth in oak forests to human risk of Lyme exposure (Ostfeld *et al.* 1996; Ostfeld and Keesing 2000). Such studies on the trophic cascades following variable masting may provide key insights into invasion as a community process (Shea and Chesson 2002).

In this chapter, we focus on the dynamics of a single seed predator species on a single masting resource. In the future it will be of interest to consider invasibility to additional seed predators (and possibly exotic species). When exotic seed predators are introduced into a plant population that show masting, a potential strategy the predator may employ to cope with variable resource availability may be an extended diapause. Kelly *et al.* (2000) and McKone *et al.* (2001) demonstrated that plant populations suffering seed predation by the predators who have extended diapause may need to show extremely high levels of mast seeding. This demonstration has been recently supported by a theoretical study of Satake and Bjørnstad 2004. Therefore, how pre-existing masting might mitigate the impact of invading predators is dependent on the level both of pre-existing masting of plant population and the extended diapause of invader. When the resident seed predator persists on a masting resource, the competitive intensity between the resident and invading species will fluctuate due to spatiotemporal variability in resource availability. Invasion success of an exotic species may be enhanced when the resident population is satiated through large seed crops because unconsumed resources may result in a competitive release. This increase in invasibility may occur through temporal competitive release, or spatially through the formation of competition-free spatial clusters. Invasibility of exotic seed predators is determined by the interwined interaction between temporal process and spatial pattern illustrating the complex community consequences of masting.

ACKNOWLEDGEMENT

This work was supported in part by a fellowship and a grant-in aid from the Japan Society for the Promotion of Science (AS) and USDA/NRI grant no. 2002-35302-12656 (ONB). The authors thank the following people: U. Dieckmann, R. A. Ims, D. Kelly, A. Liebhold, J. A. J. Metz, and R. S. Ostfeld for their helpful comments.

REFERENCES

Ahlberg, O. 1927. Rönnbärsmalen, Argyresthia conjugella Zell. En redogörelse för undersökningaråren 1921-1926 (in Swedish, with English summary). Lantbruksentomologiska avdelningen, Stockholm.

Comins, H. N., M. P. Hassell and R. M. May. 1992. The spatial dynamics of host parasitoid systems. Journal of Animal Ecology **61**, 753-748.

Cohen, D. and S. A. Levin. 1991. Dispersal in patchy environments: the effects of temporal and spatial structure. Theoretical Population Biology **39**, 63-99.

Curran, L. M. and M. Leighton. 2000. Vertebrate responses to spatiotemporal variation in seed production of mast-fruiting Dipterocarpaceae. Ecological Monograph **70**, 101-128.

Davis, M. A., J. P. Grime and K. Thompson. 2000. Fluctuating resources in plant communities: a general theory in invasibility. Journal of Ecology **88**, 528-534.

Davis, M. A. and M. Pelsor. 2001. Experimental support for a resource-based mechanistic model of invasibility. Ecology Letters **4**, 421-428.

Ellner, S. and A. Shmida. 1981. Why are adaptations for long-range seed dispersal rare in desert plants? Oecologia **51**, 133-144.

Hassell, M. P., H. N. Comins and R. M. May. 1991. Spatial structures and chaos in insect population dynamics. Nature **353**, 255-258.

Herrera, C. M., P. Jordano, J. Guitian and A. Traveset. 1998. Annual variability in seed production by woody plants and the masting concept — reassesment of principles and relationship to pollination and seed dispersal. American Naturalist **152**, 576-594.

Holt, R. D. and M. A. McPeek. Chaotic population dynamics favors the evolution of dispersal. American Naturalist **148**, 709-718.

Hopper, K. R. 1999. Risk-spreading and bet-hedging insect population biology. Annual Review of Entomology **44**, 535-560.

Inoue T. and A. A. Hamid. 1997. General flowering of tropical rainforests in Sarawak. Center for Ecological Research, Kyoto University, Japan, Kyoto

Isagi, Y., K. Sugimura, A. Sumida and H. Ito. 1997. How does masting happen and synchronize? Journal of Theoretical Biology **187**, 231-239.

Janzen, D. H. 1971. Seed predation by animals. Annual Review of Ecology and Systematics **2**, 465-492.

Kelly, D. 1994. The evolutionary ecology of mast seeding. Trends in Ecology & Evolution **9**, 465-470.

Kelly, D. and V. L. Sork. 2002. Mast seeding in perennial plants: why, how, where? Annual Review of Ecology and Systematics **33**, 427-447.

Kelly, D. and J. J. Sullivan. 1997. Quantifying the benefit of mast seeding on predator satiation and wind pollination in *Chionochloa pallens* (Poaceae). Oikos **78**, 143-150.

Kelly, D., A. L. Harrison, W. G. Lee, I. J. Payton, P. P. Wilson and W. M. Schauber. 2000. Predator satiation and extreme mast seeding in 11 species of *Chionochloa* (Poaceae). Oikos **90**, 472-488.

Kobro, S., L. Søreide, E. Djønne, T. Rafoss, G. Jaastad. and P. Witzgall. 2002. Masting of rowan, *Sorbus aucuparia* L., and consequences for the apple fruit moth, *Argyresthia conjugella* Zeller. Population Ecology **45**, 25-30.

Koenig W. D. and J. M. H. Knops. 1998 Scale of mast-seeding and tree-ring growth. Nature **396**, 225-226.

Levin, S. A., D. Cohen and A. Hastings. 1984. Dispersal strategies in patchy environments. Theoretical Population Biology **26**, 165-191.

McKone, M. J., D. Kelly, A. L. Harrison, J. J. Sullivan and A. J. Cone. 2001. Biology of insects that feed in the inflorescences of *Chinochloa* (Poaceae) in New Zealand and their relevance to mast seeding. New Zealand Journal of Zoology **28**, 89-101.

McShea, W. J. 2000. The influence of acorn crops on annual variation in rodent and bird populations within oak dominated forests. Ecology **81**, 228-238.

Momose, K, T. Nagamitsu and T. Inoue. 1996 The reproductive ecology of an emergent depterocarp in a lowland rain forest in Sarawak. Plant Species Biology **11**, 189-198.

Nilsson, S. G. and U. Wätljung. 1987. Seed predation and cross-pollination in mast-seeding beech (*Fagus sylvatica*) patches. Ecology **68**, 260-265.

Norton, D. A. and D. Kelly. 1988. Mast seeding over 33 years by *Dacrydium cupressinum* Lamb. (rimu) (Podocarpaceae) in New Zealand: the importance of economies of scale. Functional Ecology **2**, 399-408.

Ostfeld, R. S., Jones, C. G. and Wolff, J. O. 1996. Of mice and mast: ecological connections in eastern deciduous forests. BioScience **46**, 323-330.

Ostfeld, R. S. and F. Keesing. 2000. Pulsed resources and community dynamics of consumers in terrestrial ecosystems. Trends in Ecology & Evolution **15**, 232-237.

Rees, M., D. Kelly and O. N. Bjørnstad. 2002. Snow tussocks, chaos, and the evolution of mast seeding. American Naturalist **160**, 44-59.

Sakai S, K. Momose, T. Yumoto, T. Nagamitsu, H. Nagamasu, A. A. Hamid, T. Nakashizuka. 1999. Plant reproductive phenology over four years including an episode of general flowering in a lowland dipterocarp forest, Sarawak, Malaysia. American Journal of Botany **86**, 1414-1436.

Satake, A. and Y. Iwasa. 2000. Pollen-coupling of forest trees: forming synchronized and periodic reproduction out of chaos. Journal of Theoretical Biology 203, 63-84.

Satake, A. and Y. Iwasa. 2002a. Spatially limited pollen exchange and a long-range synchronization of trees. Ecology **83**, 993-1005.

Satake, A. and Y. Iwasa. 2002b. The synchronized and intermittent reproduction of forest trees is mediated by the Moran effect, only in association with pollen coupling. Journal of Ecology **90**, 830-838.

Satake, A., O. N. Bjørnstad and S. Kobro. 2004. Masting and trophic cascades: interplay between rowan trees, apple fruit moth, and their parasitoid in southern Norway. Oikos **104**, 540-550.

Satake A. and O. N. Bjørnstad. 2004. Spatial Dynamics of Specialist Seed Predators on Synchronized and Intermittent Seed Production of Host. American Naturalist **163**, 591-605.

Selås, V., O. Hogstad, G. Andersson and T. von Proschwitz. 2001. Population cycles of autumnal moth, *Epirrita autumnata*, in relation to birch mast seeding. Oecologia **129**, 213-219.

Schauber, E. M., D. Kelly, P. Turchin, C. Simon, W. G. Lee, R. B. Allen, I. J. Payton, P. R. Wilson, P. E. Cowan and R. E. Brockie. 2002. Masting by eighteen New Zealand plant species: the role of temperature as a synchronizing cue. Ecology **83**, 1214-1225.

Shibata, M, H. Tanaka and T. Nakashizuka. 2002. Synchronized annual seed production by 16 principal tree species in a temperate deciduous forest, Japan. Ecology **83**, 1727-1742.

Schnurr, J. L., R. S. Ostfeld and D. Canham. 2002. Direct and indirect effect of masting on rodent populations and tree seed survival. Oikos **96**, 402-410.

Shea, K. and P. Chesson. 2002. Community ecology theory as a framework for biological invasions. Trends in Ecology and Evolution **17**, 170-176.

Shibata, M., H. Tanaka and T. Nakashizuka. 1998. Cause and consequences of mast seed production of four co-occurring *Carpinus* species in Japan. Ecology **79**, 54-64.

Silvertown, J. W. 1980. The evolutionaly ecology of mast seeding in trees. Biological Journal of the Linnean Society **14**, 235-250.

Smith, C. C., J. L. Hamrick and C. L. Kramer. 1990. The advantage of mast years for wind pollination. American Naturalist **136**, 154-166.

Sork, V. L., J. Bramble and O. Sexton. 1993. Ecology of mast-fruiting in three species of north American deciduous oaks. Ecology **74**, 528-541.

Sperens, U. 1997a. Fruit production in *Sorbus aucuparia* L. (Rosaceae) and pre-dispersal seed predation by the apple fruit moth (*Argyresthia conjugella* Zell.). Oecologia **110**, 368-373.

Sperens, U. 1997b. Long-term variation in, and effects of fertilized on, flower, fruit and seed production in the tree Sorbus aucuparia (Rosaceae). Ecography **20**, 521-534.

Tutin, T. G., V. H. Heywood, N. A. Burges, D. M. Moore, D. H. Valentine, S. M. Walters and D. A. Webb. 1968. Flora europaea, 2nd edn. Cambridge University Press, Cambridge.

Wolff, J. O. 1996. Population fluctuations of mast-eating rodents are correlated with production of acorns. Journal of Mammalogy **77**, 850-856.

Chapter thirteen

Invasions and the regulation of plant populations by pathogens

G. S. Gilbert and I. M. Parker

INTRODUCTION

The potential of pathogens to have dramatic impacts on plant populations is made clear by familiar stories of the Irish potato famine (Fry and Goodwin 1997), the ecological extinction of chestnuts caused by chestnut blight (Anagnostakis 1987), and the transformation of Australian Jarrah forests to scrubland by *Phytophthora cinnamomi* (Weste and Marks 1987). Similarly, the annual worldwide expenditure of over $6.6 billion in fungicide application (Donaldson *et al.* 2002) reflects the toll fungal pathogens alone can take on plant growth and fecundity in agricultural systems. Some of the most notable examples of these impacts arise when pathogens are introduced into novel biological environments; invasive and emergent pathogens continue to frustrate the best efforts of resource managers, conservation biologists, and plant protection agencies (Weste and Marks 1987, Daughtrey *et al.* 1996, Goodell *et al.* 2000, Gordon *et al.* 2001, McDonald and Hoff 2001, Wingfield *et al.* 2001, Gilbert 2002, Rizzo and Garbelotto 2003, Parker and Gilbert 2004). Concurrent with efforts to reduce the impacts of unwanted diseases, there is broad interest among researchers, agriculturalists, and land managers to harness the destructive potential of plant pathogens to control weedy plants (Hasan and Ayres 1990, Charudattan and Dinoor 2000).

M.W. Cadotte et al. (eds.), Conceptual ecology and invasion biology, 289–305.

In natural ecosystems, pathogens have great potential to influence the dynamics and composition of plant populations and communities through density-dependent and coevolutionary dynamics (see reviews in Dinoor and Eshed 1984, Burdon 1987, Jarosz and Davelos 1995, Alexander *et al.* 1996, Alexander and Holt 1998, Gilbert 2002, 2004). In many plant communities, plant pathogens may prevent competitive exclusion and thus help maintain species diversity (Gillett 1962, Packer and Clay 2000, Gilbert 2002, Wright 2002, Gilbert 2005). Collective insights from research on epidemic diseases, biological control, and the evolutionary ecology of diseases in natural ecosystems provide a robust basis for identifying when pathogens are likely to be important in regulating plant populations, and their implications for understanding biological invasions.

Here we draw broadly from a diverse literature to place the regulation of plant populations by pathogens into the context of two key, complementary theories about the role of pathogens in biological invasions: Biotic Resistance (Maron and Vila 2001) and Escape from Natural Enemies (Keane and Crawley 2002) (Table 1). We then consider the practical implications for using introduced pathogens for classical biological control of introduced invasive plants (Table 1), illustrated with a detailed case study of the control of *Chondrilla* by rust fungi.

PLANT DISEASES AND POPULATION REGULATION

Pathogens reduce the fitness of individual plants by killing them, reducing growth, impeding competitive ability, or by rotting fruits or seeds (see recent review in Gilbert 2002). The greater probability of pathogen spread between closely spaced host plants means that most fungal plant diseases show density-dependent development. In addition, densely spaced plants may create microclimates that encourage pathogen growth, and hosts stressed by competition may be more susceptible to disease (Burdon and Chilvers 1982, Gilbert 2002). The combination of strong impacts on individual host plants and density-dependent disease development suggests that pathogens should be powerful regulators of plant populations. Nevertheless, aside from epidemics caused by introduced pathogens, there are remarkably few empirical studies showing that plant diseases are responsible for regulating plant population dynamics in natural ecosystems (Gilbert 2002). In large part this absence reflects the difficulties of isolating disease impacts from other factors and the strong focus on diseases with economic importance. But physiological or evolutionary responses by the host may also counteract the regulatory actions of pathogens in natural ecosystems. In particular, plants that survive after disease has killed or stunted competing conspecific neighbors may show a compensatory response that offsets numerical losses from disease at the population level (Friess and Maillet 1996, Alexander and Holt 1998, Alexander and Mihail 2000). In addition, maternally-transmitted induced resistance can generate cross-generational effects that ameliorate the numerical impacts of disease in succeeding generations (Agrawal *et al.* 1999). Finally, the genetics

Table 1 Implications of different features of plant-pathogen interactions for three areas of invasion biology: the two theories of biotic resistance and escape from natural enemies, and the field of classical biological control.

Features of plant-pathogen interactions	Biotic resistance	Escape from natural enemies	Biological control
Impact of pathogen on host	Pre-adapted pathogens in new range are highly virulent on naïve hosts.	Pathogens in native range limit host density or distribution. Introduced plants leave virulent pathogens behind.	To be effective control agents, pathogens must be important in population regulation in native range.
Interactions with environment (disease triangle)	Local pathogens are adapted to local environment.	Plants may escape natural enemies if introduced into an environment not conducive to disease development.	Introduced biocontrol pathogens may fail if poorly adapted to local environment.
Host range of pathogen	Generalist pathogens are more likely to acquire newly introduced host species.	Assumes specialist pathogens play a unique role in regulating plant populations.	Only specialist pathogens can be utilized as control agents.
Natural history/ life history of pathogen	Density-dependent disease development may change impact on invading hosts as invasion proceeds.	Pathogens with resting structures or alternate life history strategies most likely to accompany introduced plants; such pathogens may have greater virulence.	Pathogens with resting structures should show greater success.
Rapid evolutionary changes	Pathogens may acquire introduced host through evolutionary host shift. Virulence may increase quickly when host becomes common.	Host may lose defenses to original pathogens after introduction.	Host may develop resistance to introduced biocontrol agent.

of plant-pathogen interactions can be highly dynamic, with large changes in pathogen virulence or host resistance evolving in a few generations (e.g., Burdon and Thompson 1995, Bishop *et al.* 2000). Such rapid evolutionary changes are expected to strongly influence the impact of pathogens on host numerical dynamics (Alexander *et al.* 1996).

BIOTIC RESISTANCE AND ESCAPE FROM NATURAL ENEMIES

For disease to develop, virulent pathogens, susceptible host plants, and suitable environmental conditions must converge (commonly called the Disease Triangle). Changes in any of these components can reduce or increase how much diseases affect plant population dynamics. If plants are introduced to an environment where virulent pathogens are not present or where environmental conditions do not favor disease development, the plant population may be released from previous regulation by pathogens. Similarly, plants introduced to a new locale with novel, virulent pathogens or where environmental conditions favor disease development may experience unprecedented population regulation by pathogens. Such changes in the prevalence and severity of diseases under different conditions have long been thought to play an important role in the process of biological invasions in two hypothetical ways:

1. In natural habitats, native pathogens colonize and are highly virulent on naïve, introduced plant species and prevent population growth (*Biotic Resistance*).
2. Pathogens were important in regulating the plant population in its native range, but are not in the introduced range (*Escape from Natural Enemies*).

Biotic resistance

The great majority of plant introductions do not result in invasions (Mack 1995, Williamson 1996). Rather, many introduced species either fail to thrive altogether or are restricted to human cultivation, unable to build self-sustaining populations in wild plant communities or even in disturbed rangeland communities. Research attempting to identify plant traits that confer invasiveness (Reichard and Hamilton 1996, Rejmánek and Richardson 1996) shows that our predictive ability is imperfect, and it seems to be more difficult to predict which introductions will fail than which will succeed (Reichard and Hamilton 1996). Biotic resistance is one possible explanation for why some introductions fail when they "should" succeed; native pests and pathogens colonize exotic plants and eliminate them before they can establish a viable population (Elton 1958, Simberloff 1986, Mack 1996).

Pathogens will contribute to biotic resistance only when three conditions are met. First, the pathogens involved must not be narrow host specialists. In plant communities with high host diversity (and corresponding low host density) generalist plant pathogens should dominate (Gilbert *et al.* 2002, Gilbert 2005), which could contribute to the high invasion resistance of hyper-diverse ecosystems like tropical rain forests (Fine 2002). Second, because densities of the introduced host will usually be low immediately following introduction, pathogens involved in biotic resistance would not be those for which transmission or host switching was strongly density-dependent. Third, the pathogen must exact a high fitness cost on the host. Combinations of generalization and high virulence are not unusual in pathogens, especially for species that have long resting stages or saprophytic ability (e.g., *Phytophthora cinnamomi*, *Verticillium dahliae*, *Fusarium oxysporum*). Even pathogens that can attack many hosts have differential impacts on different host species. If a pathogen has a greater negative effect on the competitive ability of the non-native hosts than on the native hosts, then it will contribute to the competitive exclusion of the non-native (Keane and Crawley 2002).

Since even fairly specialized pathogens often infect many members of the same genus or family, it is likely that the number of pathogens competent to infect a novel host depends on whether it is phylogenetically related to native species already in the community. Therefore, biotic resistance should favor invasion by species with no close relatives. Despite this clear prediction, and the accepted practice of using phylogenetic relationships in quarantine and trade policy to target relatives of weedy plants or identify potential carrier hosts, surprisingly few studies of invasion even mention the phylogenetic structure of the invaded communities. In one rare attempt, Mack (1996) found that for 5 out of 6 regional floras, naturalized species were more common in genera with no native congeners than those with native congeners, consistent with the idea of biotic resistance. However, this study was not able to control for the effect of opportunity (whether plants without native congeners have a higher probability of successful invasion given their probability of introduction). Using a different approach, Duncan and Williams (2002) compiled a list of all plant species that have ever been introduced for cultivation in New Zealand. They found that introduced species in genera that already had resident natives were more likely, not less likely, to successfully naturalize. They suggest that species with native congeners may share characteristics that make them more fit in the introduced range, and this factor overwhelms the potential effect of local natural enemies.

There is a striking need for more studies to determine whether and when biotic resistance by native pathogens occurs. Because there is little information on where and when unintentional species introductions fail, assigning mechanisms to these failures has been nearly impossible. Horticulture, forestry, and agriculture each provide examples of endemic pathogens that have decimated introduced plant species so that growing these species is no longer economically viable (Mack 1996, Coutinho *et al.* 1998, Wingfield *et al.* 2001). However, extrapolating from agricultural or silvicultural examples to invasive introduced species requires caution, since the regeneration of hosts is controlled, preventing the host population from evolving resistance or tolerance to the pathogen. Additionally, simplified systems in agriculture and forestry may simply lack moderating effects of greater biotic complexity in less managed systems (e.g., hyperparasites of the pathogens).

Escape from natural enemies

Introduced plants that become invasive weeds are among our most challenging environmental problems (D'Antonio and Vitousek 1992, Office of Technology Assessment 1993, Vitousek *et al.* 1996, Parker *et al.* 1999, Mack *et al.* 2000), and escape from natural enemies provides a mechanism to explain the increased growth and density of introduced species (Darwin 1859, Crawley 1987, Blossey and Nötzold 1995, Tilman 1999, Maron and Vila 2001, Siemann and Rogers 2001, Keane and Crawley 2002). The "Natural Enemies Hypothesis" posits that introduced species leave behind their natural enemies — herbivores, seed predators, and pathogens — and are thereby released from a key regulating factor, leading to a dramatic increase in plant vigor, population growth, and/or competitive ability. This idea forms the justification for classical biological control, in which natural enemies are brought from the native range to control weedy invaders (Huffaker and Messenger 1976, DeBach and Rosen 1991). Yet despite the importance of the Natural Enemies Hypothesis, rigorous empirical tests are few (Maron and Vila 2001).

A number of studies have tested whether introduced plants tend to grow faster or larger and whether there is evidence of reduced allocation to herbivore defenses in the new range (Crawley 1987, Blossey and Nötzold 1995, Siemann and Rogers 2001) but see (Willis *et al.* 2000, Thebaud and Simberloff 2001, Vila *et al.* 2003). While several of these studies have suggested ecological release and evolution away from defense toward competitive ability, they lack direct evidence of involvement by natural enemies (Blossey and Nötzold 1995, Siemann and Rogers 2001). Other studies have tested the Natural Enemies Hypothesis by comparing the impact of natural enemies on exotic and native species within the introduced range. In a review of 13 studies, Keane and Crawley (2002) found some cases in which generalist herbivores showed higher impacts on native species than non-natives, and others characterized by the reverse pattern. There have been surprisingly few studies with pathogens; in one test Goergen and Daehler (2001) found that smut fungi caused greater reproductive loss in a native grass (*Heteropogon contortus*) than an introduced grass (*Pennisetum setaceum*).

Enemy removal experiments are an important but underutilized tool in comparing the fitness effect of pathogens or herbivores on exotic and native species within the introduced range (Keane and Crawley 2002). Blaney and Kotanen (2001) used a fungicide experiment to remove the effects of soilborne fungi and oomycetes on the survival of seeds of native vs. introduced plants in two habitats. They found no support for a release from fungal pathogens in the seed bank of exotic species compared with native species. More recently, Parker and Gilbert (unpublished data) found no difference in frequency of infection, leaf damage, fitness effects of foliar and damping-off pathogens, or pathogen diversity between sympatric suites of native and non-native clovers. Although fungal exclusion experiments in this system revealed significant impacts of pathogens, there was no difference in the response of native and non-native species.

Finally, other studies have taken the approach of comparing disease on a single host species in its native and invaded ranges. Wolfe (2002) surveyed for the anther smut fungus *Microbotryum violaceum* in 50 native populations and 36 introduced populations of *Silene latifolium*. He found significantly more populations infected, and at much higher infection frequencies, in the native range than in the introduced range. Mitchell and Power (2003) used published records of pathogen associations with 473 plant hosts in their native and introduced ranges. Plants were infected by 77% fewer pathogen species in their naturalized range. There was also an indication that species leaving behind proportionally more natural enemies were more likely to be categorized as noxious or invasive species by land managers and public agencies. Using an experimental approach, Beckstead and Parker (2003) directly measured the demographic effect of pathogens on an invader in the context of known information from the species' native range. *Ammophila arenaria* in its native Europe is limited to an early-successional role in shifting beach sands by soil-borne pathogens (Van der Putten *et al.* 1993). As an invader on the west coast of the U.S.A., *Ammophila* remains dominant for long periods of time. However, escape from natural enemies does not explain this contrast between its native and invasive ecological roles. By replicating experiments done in the native range, Beckstead and Parker (2003) found the negative effect of soilborne pathogens on early growth in the invaded range was at least as large or larger than their effect in the native range.

To predict whether an introduced plant is likely to benefit from escaping natural enemies, we need to understand the relative importance of host-specialist vs. host-generalist pathogens and pests in the invaded habitat (Maron and Vila 2001, Keane and Crawley 2002). If specialist pathogens predominate and host shifts are rare, native plants may be suppressed more than competing invasive species. On the other hand, if generalist pathogens dominate in a site and do not show a preference for native host species, one would not expect an introduced plant to experience release. Therefore, the wide range of results seen in the above empirical studies may be in part explained by the relative importance of specialist and generalist natural enemies. A greater understanding of the phylogenetic structure of pathogen host ranges, coupled with analysis of the phylogenetic structure of natural plant communities (Webb *et al.* 2002) may help predict the relative importance of specialist vs. generalist pathogens in different kinds of plant communities. For instance, in a high-diversity lowland tropical rainforest (300 + tree species), host generalists dominated the polypore fungal community; all of the more common fungal species were found on multiple families of host trees (Gilbert *et al.* 2002, Ferrer and Gilbert 2003). In contrast, in a nearby low diversity mangrove forest with only three tree species present (each from a different family), 88% of all polypore fungal collections belonged to just three fungal species, and each species was highly specialized on just one mangrove species

(Gilbert and Sousa 2002). Researchers are just now beginning to address the range of host specialization in different plant communities, and the wide range of outcomes suggests that many more studies will be needed before we can formulate predictive generalizations.

INTENTIONALLY INTRODUCED PATHOGENS FOR BIOLOGICAL CONTROL

A direct application of the Natural Enemies Hypothesis is deploying natural enemies from the native range of an invasive plant to control the invader population. Such classical biological control uses plants and pathogens with shared evolutionary histories but a new environmental context. There are several ways in which biological control interactions may be different from native pathogens attacking introduced hosts. First, unlike native pathogens, which are presumably adapted to the local climate, the introduced biological control agent experiences a novel environment, which could have a large impact on disease development. Case studies of failed biological control efforts provide us with many examples of the importance of the disease triangle (Morin *et al.* 1996). Second, both the host and pathogen are likely to be genetically depauperate. However, the pathogen will have been chosen specifically to be virulent on the invasive host, placing the host at a relative disadvantage for evolutionary responses. Third, only fairly host-specific pathogens are selected for biological control releases, meaning that pathogen numerical dynamics should always be closely linked to individual host density. In fact, biological control releases are an excellent opportunity to study factors influencing numerical dynamics. Not only should there be a tight connection between pathogen and host density, but initial conditions of the interaction are well known. That is, the host population is originally free of that pathogen and is usually at high density. A successful epidemic provides an opportunity to quantify both frequency-dependence of transmission of the pathogen and density-dependence as the host density declines.

Predicting the short-term and long-term success of particular biological control introductions is a matter of obvious practical importance. To make such predictions, we need to understand how the numerical dynamics of a host plant following introduction of its biocontrol agent depend on host density, disease incidence, genetic variation, and evolutionary changes in virulence or resistance. Surprisingly, while there are some cases for which we have good information on the dynamics of host numbers after release of a control agent (e.g., Hasan and Ayres 1990, Morris 1997), for many other releases the details of changes are not well documented. In particular, we should ask (i) is control more successful in genetically depauperate weeds? (ii) Do transmission rate and demographic impact of the pathogen attenuate as the host population declines? (iii) Do pathogen and host reach a stable equilibrium or are they dependent on metapopulation dynamics to persist in the landscape? Detailed information on numerical dynamics in biological control systems is scarce, but data are nearly nonexistent for long-term genetic changes in the host or pathogen. There is great, untapped potential for biological control introductions to be used to understand the factors that drive the ecological end evolutionary dynamics of the plant-pathogen interactions. In fact, the only biocontrol study we found that tracked changes in pathogen virulence or host resistance for a plant-pathogen system was for the rust *Puccinia chondrillina* on *Chondrilla juncea* (see case study below).

Evolutionary ecology of biocontrol with pathogens

The evolutionary dynamics of interactions between weeds and pathogen biocontrol agents have important implications for the long-term success of biological control programs. Generally, we need to know, (i) has virulence of the biological control pathogen changed over time, and has this increased or decreased the success of control? (ii) Has the host developed resistance over time? (iii) Has host specificity changed over time? Recent analysis of emerging diseases has suggested that ecological host shifts (i.e., having a preadapted ability to use a newly encountered host) may predominate as causes of novel epidemics, and that host shifts may only rarely be caused by mutations that allow colonization of a new host (Schrag and Wiener 1995). However, the difficulty of observing such genetic events may distort our perspective. Van Klinken and Edwards (2002) synthesized information on host range from 352 biological control programs using herbivores. They found that host shifts were more quantitative than qualitative, that is, while preference and efficiency on novel hosts evolved, there was no evidence of evolutionary changes in fundamental host-range. Such an analysis should be done for pathogens used as biological control agents. While pathogens are generally thought to offer the opportunity for high host specificity, some have argued that high host specificity may be correlated with evolutionary lability (Brooks and McLennan 1993, Secord and Kareiva 1996). Because they are unable to simply move from an unacceptable host to a more suitable one as an animal might, specialist pathogens may experience even stronger selection for host shifts than herbivores (Roy 2001). Knowing the frequency of evolutionary host shifts in pathogens and understanding the conditions under which they occur are critical to the process of risk assessment in biological control (Secord and Kareiva 1996).

General theory of host-pathogen interactions has played a large role in the choice of biological control agents in the past (McFadyen 1998). For example, it has long been thought that sexually reproducing weeds would be harder to control because their higher levels of genetic variation interfere with pathogen population growth (Burdon *et al.* 1981); however, a more recent analysis has disputed this assertion (Chaboudez and Sheppard 1995). Similarly, the belief that pathogen populations are locally adapted to their host genotypes has had a large influence on the process of selection of control agents, with genetic analysis playing an increasing role in the careful matching of agent genotypes with the populations of origin for the weed (e.g., Holden and Mahlberg 1996). However, the evidence for close local adaptation of pathogens to their host populations is mixed, and in fact resistance and gene-for-gene virulence should fluctuate in an asynchronous, frequency-dependent way (reviewed in Parker and Gilbert 2004). Biological control practitioners have clearly based their introduction strategies on theoretical considerations, but the simpler rules are now coming under question (McFadyen 1998). There is a need for clear predictions and modern empirical work testing those predictions to help inform the practice of biological control for the future.

An evolutionary ecology case study — *Chondrilla* and *Puccinia* on three continents

No system of classical biological control of a weed by a pathogen demonstrates the importance of interactions between environment, genetics, and numerical dynamics as well as that of *Chondrilla juncea*, rush skeletonweed (Asteraceae). Native to Eurasia, this species was introduced to the eastern United States in the late 1800s, to Australia in the early 1900s, and to the western United states in the 1930s (McVean 1966, Pryor 1967, Schirman and Robocker 1967, Supkoff *et al.* 1988). *Chondrilla* is a significant economic problem in wheat-growing regions (Panetta and Dodd 1995), and is also a widespread rangeland weed. Of three biocontrol agents that were introduced, *Puccinia chondrillina* was the most effective at reducing plant vigor (Supkoff *et al.* 1988), and within two decades of its introduction into Australia and California, *Chondrilla* densities were reduced to those typical in its native range (Wapshere *et al.* 1974, Cullen *et al.* 1982, Supkoff *et al.* 1988).

The extreme host specificity found in the *Chondrilla/Puccinia* system makes this example particularly interesting (Hasan 1972). *Chondrilla* is a triploid apomict, and thus reproduction is clonal. In Australia, three clonal types are present, each with a different leaf width; *P. chondrillina* causes disease only on the narrow-leaf type. The original infestation of *Chondrilla* was primarily this narrow-leaf type, but after the successful biocontrol of that clone, the intermediate-leaved clone has spread (Hanley and Groves 2002). Now there is concerted effort to introduce new strains of *Puccinia chondrillina* that are able to attack and control the other clones of *Chondrilla* (Hanley and Groves 2002).

In western North America, three different *Chondrilla* genotypes were found, distinguishable by their multi-locus isozyme phenotypes (Hasan *et al.* 1996); two genotypes are thought to have originated in Yugoslavia (Hasan and Delfosse 1995). The western US genotypes were largely resistant to the rust strain that controlled the narrow-leaf *Chondrilla* in Australia, so additional rust isolates were evaluated for use in biocontrol. Rusts collected from Yugoslavia, the putative site of origin of the US invaders, showed high virulence on some of the US genotypes, but other plants were little affected. In addition, rusts from other regions also showed high virulence. Genetic matching of hosts between the native and introduced range may yield well-adapted biocontrol pathogens when the target weed is genetically uniform, but this approach assumes high local adaptation of the pathogen to the host. As mentioned above, such local adaptation may not be generally found in natural populations.

Sexual reproduction in both the host and pathogen play a role in this story. Teliospores of *P. chondrillina* only germinate after cold winters (Adams and Line 1984), so sexual recombination of the pathogen occurs readily only in colder regions, such as parts of eastern continental Europe. In these areas of the native range, *Chondrilla* also shows higher clonal diversity and possibly diploid sexual populations (Chaboudez and Sheppard 1995). Interestingly, populations of *Chondrilla* with high vs. low clonal diversity in the native range appear to suffer similar degrees of rust infection (Chaboudez and Sheppard 1995). In northern North America, where cold winters allow for sexual recombination of the pathogen, the evolutionary dynamics of the host-pathogen interaction are potentially more complex (Hasan *et al.* 1996). In Australia, they are particularly concerned about the possibility of introduction of new, sexual types of *Chondrilla*, which could result in a situation where the host reproduces sexually but the pathogen can not (Chaboudez and Sheppard 1995).

This example shows the importance of the "disease triangle" interaction between host, pathogen, and environment in the dynamics of plant-pathogen interactions. The case of *Chondrilla/Puccinia chondrillina* is one of the few for which we have such detailed information about the host's and the pathogen's genetic makeup, and more importantly, about how the interaction has played out in a number of different regions. Pathogen biological control cases offer the potential to learn a great deal about the roles of genetics, numerical dynamics, evolutionary dynamics, and environmental factors in determining the long-term outcomes of host-pathogen interactions.

CONCLUSIONS

From the studies described above, it appears that pathogens are sometimes important in the regulation of natural plant populations, may constrain populations introduced to new regions, and may, by their absence, release introduced plants from an important source of regulation. We suggest two key directions that would most advance our understanding of the importance of pathogens in plant population regulation: an integration of numerical and evolutionary dynamics for both the pathogens and plants, and a greater breadth of studies to include more plant-pathogen systems.

Biological control of invasive weeds offers exciting opportunities to evaluate the importance of pathogens in numerical regulation of plant populations, and at the same time to follow genetic changes in plant and pathogen populations. Careful monitoring for changes in host and pathogen genotypes, along with numerical dynamics, should be integral to any introduction of pathogens for biological control of invasive weeds. Equally important, we must move beyond spinning narratives and making general predictions from a handful of examples. We have a good idea of the role pathogens *should* play in plant population regulation and biological invasions (Table 1); we now need to collect data from a diversity of systems to evaluate our predictions. How often and under what conditions are plant pathogens significant forces in regulating wild plant populations in their native ranges? How often do introduced plants fail to establish because they are attacked by local pathogens? How often is escape from pathogens a key to determining whether an introduced plant invades natural habitats? Through pathogen exclusion and addition experiments, common gardens, phylogenetic analysis of host ranges, analysis of rapid evolutionary changes in plant-pathogen interactions, and careful, creative natural history of plant diseases, we will illuminate the role of plant pathogens in biological invasions.

ACKNOWLEDGMENTS

We thank B. Ayala, K. Dlugosch, J. Hagen, C. Hays, B. Hardcastle, J. Hein, R. Hufft, S. Lambrecht, S. Langridge, M. Los Huertos, D. Plante, W. Satterthwaite, Y. Springer, and two anonymous reviewers for helpful discussions and critical comments on drafts of this manuscript. Preparation was supported in part by NSF grants DEB-9808501 to IMP, DEB-0096298 and DEB-0096398 to GSG, and USDA-NRI #2000-00891 to IMP.

REFERENCES

Adams, E. B. and R. F. Line. 1984. Biology of *Puccinia chondrillina* in Washington. Phytopathology **74**, 742-745.

Agrawal, A., C. Laforsch and R. Tollrian. 1999. Transgenerational induction of defences in animals and plants. Nature **401**, 61-63.

Alexander, H. M. and R. D. Holt. 1998. The interaction between plant competition and disease. Perspectives in Plant Ecology, Evolution, and Systematics **1**, 206-220.

Alexander, H. M. and J. D. Mihail. 2000. Seedling disease in an annual legume: consequences for seedling mortality, plant size, and population seed production. Oecologia **84**, 249-253.

Alexander, H. M., P. H. Thrall, J. Antonovics, A. M. Jarosz and P. V. Oudemans. 1996. Populations dynamics and genetics of plant disease: A case study of anther-smut disease. Ecology **77**, 990-996.

Anagnostakis, S. L. 1987. Chestnut blight the classical problem of an introduced pathogen. Mycologia **79**, 23-37.

Beckstead, J. and I. M. Parker. 2003. Invasiveness of *Ammophila arenaria*: Release from soilborne pathogens? Ecology **84**, 2824-2831.

Bishop, J. G., A. M. Dean and T. Mitchell-Olds. 2000. Rapid evolution in plant chitinases: molecular targets of selection in plant-pathogen coevolution. Proc. Natl. Acad. Sci. USA **97**, 5322-5327.

Blaney, C. S. and P. M. Kotanen. 2001. Effects of fungal pathogens on seeds of native and exotic plants: A test using congeneric pairs. J. Appl. Ecol. **38**, 1104-1113.

Blossey, B. and R. Nötzold. 1995. Evolution of increased competition ability in invasive nonindigenous plants: a hypothesis. Journal of Ecology **83**, 887-889.

Brooks, D. R. and D. A. McLennan. 1993. Parascript: parasites and the language of evolution. Smithsonian Institution Press, Washington.

Burdon, J. J. 1987. Diseases and Plant Population Biology. Cambridge University Press, Cambridge.

Burdon, J. J. and G. A. Chilvers. 1982. Host density as a factor in plant disease ecology. Annual Review of Phytopathology **20**, 143-166.

Burdon, J. J., R. H. Groves and J. M. Cullen. 1981. The impact of biological control on the distribution and abundance of *Chondrilla juncea* in south-eastern Australia. Journal of Applied Ecology **18**, 957-966.

Burdon, J. J. and J. N. Thompson. 1995. Changed patterns of resistance in a population of *Linum marginale* attacked by the rust pathogen *Melampsora lini*. Journal of Ecology **83**, 199-206.

Chaboudez, P. and A. W. Sheppard. 1995. Are particular weeds more amenable to biological control? — a re-analysis of mode of reproduction and life history. Pages 95-102 *in* E. S. Delfosse and R. R. Scott, editors. Biological Control of Weeds: Proc. VIIIth Int. Symp. Biol. Contr. Weeds. CSIRO Publications, Melbourne, Australia.

Charudattan, R. and A. Dinoor. 2000. Biological control of weeds using plant pathogens: Accomplishments and limitations. Crop Protection **19**, 691-695.

Coutinho, T. A., M. J. Wingfield, A. C. Alfenas and P. W. Crous. 1998. Eucalyptus rust: a disease with the potential for serious international implications. Plant Disease **82**, 819-825.

Crawley, M. J. 1987. What Makes a Community Invasible? Pages 429-454 *in* A. J. Gray, M. J. Crawley and P. J. Edwards, editor. British Ecological Society Symposium, Vol. 26. Colonization, Succession and Stability. Blackwell Scientific Publications, Oxford, England.

Cullen, J. M., R. H. Groves and J. F. Alex. 1982. The influence of *Aceria chondrillae* on the growth and reproductive capacity of *Chondrilla juncea*. Journal of Applied Ecology **19**, 529-538.

D'Antonio, C. M. and P. M. Vitousek. 1992. Biological invasions by exotic grasses, the grass/fire cycle and global change. Annual Review of Ecology and Systematics **23**, 63-87.

Darwin, C. 1859. On the origin of species by means of natural selection, or, The preservation of favoured races in the struggle for life. J. Murray, London.

Daughtrey, M. L., C. R. Hibben, K. O. Britton, M. T. Windham and S. C. Redlin. 1996. Dogwood Anthracnose: understanding a disease new to North America. Plant Disease **80**, 349-358.

DeBach, P. and D. Rosen. 1991. Biological control by natural enemies. Cambridge University Press, Cambridge.

Dinoor, A. and N. Eshed. 1984. The role and importance of pathogens in natural plant communities. Annual Review of Phytopathology **22**, 443-466.

Donaldson, D., T. Kiely and A. Grube. 2002. Pesticides Industry Sales and Usage. 1998 and 1999 Market Estimates. Office of Pestice Programs, U. S. Environmental Protection Agency, Washington, D. C.

Duncan, R. P. and P. A. Williams. 2002. Darwin's naturalization hypothesis challenged. Nature **417**, 608-609.

Elton, C. S. 1958. The Ecology of Invasions by Animals and Plants. Methuen, London.

Ferrer, A. and G. S. Gilbert. 2003. Effect of tree host species on fungal community composition in a tropical rain forest in panama. Diversity and Distributions **9**, 455-468.

Fine, P. V. A. 2002. The invasibility of tropical forests by exotic plants. Journal of Tropical Ecology **18**, 687-705.

Friess, N. and J. Maillet. 1996. Influence of cucumber mosaic virus infection on the intraspecific competitiv ability and fitness of purslane (*Portulaca oleracea*). New Phytologist **132**, 103-111.

Fry, W. E. and S. B. Goodwin. 1997. Resurgence of the Irish potato famine fungus: after 150 years, the late blight fungus is again menacing farmers. Bioscience **47**, 363-371.

Gilbert, G. S. 2002. Evolutionary ecology of plant diseases in natural ecosystems. Annual Review of Phytopathology **40**, 13-43.

Gilbert, G. S. 2005. The dimensions of plant disease in tropical forests. Pages 141-164 *in* D. R. F. P. Burslem, M. A. Piinard and S. Hartley, editors. Biotic Interactions in the Tropics. Cambridge Univ. Press, Cambridge.

Gilbert, G. S., A. Ferrer and J. Carranza. 2002. Polypore fungal diversity and host density in a moist tropical forest. Biodiversity and Conservation **11**, 947-957.

Gilbert, G. S. and W. P. Sousa. 2002. Host specialization among wood-decay polypore fungi in a Caribbean mangrove forest. Biotropica **34**, 396-404.

Gillett, J. B. 1962. Pest pressure, an underestimated factor in evolution. Systematics Association Publication Number **4**, 37-46.

Goergen, E. and C. Daehler. 2001. Inflorescence damage by insects and fungi in native pili grass (*Heteropogon contortus*) versus alien fountain grass (*Pennisetum setaceum*) in Hawai'i. Pacific Science **55**, 129-136.

Goodell, K., I. M. Parker and G. S. Gilbert. 2000. Biological impacts of species invasions: Implications for policy makers. Pages 87-117 *in* J. Caswell, editor. Incorporating Biological, Natural and Social Sciences in Sanitary and Phytosanitary Standards in International Trade. National Academy Press, Washington, D. C.

Gordon, T. R., A. J. Storer and D. L. Wood. 2001. The Pitch Canker epidemic in California. Plant Disease **85**, 1128-1139.

Hanley, M. E. and R. H. Groves. 2002. Effect of the rust fungus *Puccinia chondrillina* TU 788 on plant size and plant size variability in *Chondrilla juncea*. Weed Research **42**, 370-376.

Hasan, S. 1972. Specificity and host specialization of *Puccinia chondrillina*. Annals of Applied Biology **72**, 257-263.

Hasan, S. and P. G. Ayres. 1990. Tansley Review No. 23 The control of weeds through Fungi. Principles and prospects. New Phytologist **115**, 201-222.

Hasan, S., P. Chaboudez and C. Espiau. 1996. Isozyme patterns and susceptibility of North American forms of *Chondrilla juncea* to European strains of the rust fungus

Puccinia chondrillina. Pages 367-373 *in* E. S. Delfosse and R. R. Scott, editors. Biological control of weeds. CSIRO Publications, East Melbourne, Victoria, Australia.

Hasan, S. and E. S. Delfosse. 1995. Susceptibility of the Australian native, *Heliotropium crispatum*, to the rust fungus *Uromyces heliotropii* introduced to control common heliotrope, *Heliotropium europaeum*. Biocontrol Science and Technology **5**, 165-174.

Holden, A. N. G. and P. G. Mahlberg. 1996. Rusts for the biological control of leafy spurge (*Euphorbia esula*) in North America. Pages 419-424 *in* E. S. Delfosse and R. R. Scott, editors. Biological control of weeds. CSIRO Publications, East Melbourne, Victoria, Australia.

Huffaker, C. B. and P. S. Messenger, editors. 1976. Theory and practice of biological control. Academic Press, New York.

Jarosz, A. M. and A. L. Davelos. 1995. Effects of disease in wild plant-populations and the evolution of pathogen aggressiveness. New Phytologist **129**, 371-387.

Keane, R. M. and M. J. Crawley. 2002. Exotic plant invasions and the enemy release hypothesis. Trends in Ecology and Evolution **17**, 164-170.

Mack, R. N. 1995. Understanding the processes of weed invasions: the influence of environmental stochasticity. *in* C. H. Stirton, editor. Weeds in a Changing World. British Crop Protection Council, Brighton, UK.

Mack, R. N. 1996. Biotic barriers to plant naturalization. Pages 39-46 *in* V. C. Moran and J. H. Hoffmann, editors. Proceedings of the IX International Symposium on Biological Control of Weeds. University of Cape Town, Stellenbosch, South Africa.

Mack, R. N., D. Simberloff, W. M. Lonsdale, H. Evans, M. Clout and F. A. Bazzaz. 2000. Biotic invasions: Causes, epidemiology, global consequences, and control. Ecological Applications **10**, 689-710.

Maron, J. L. and M. Vila. 2001. When do herbivores affect plant invasion? Evidence for the natural enemies and biotic resistance hypotheses. Oikos **95**, 361-373.

McDonald, G. I. and R. J. Hoff. 2001. Blister rust: an introduced plague. Pages 193-220 *in* D. F. Tomback, S. F. Arno and R. E. Keane, editors. Whitebark Pine Communities. Ecology and Restoration. Island Press, Washington DC.

McFadyen, R. E. C. 1998. Biological control of weeds. Annual Review of Entomology **43**, 369-393.

McVean, D. N. 1966. Ecology of *Chondrilla juncea* L. in southeastern Australia. Journal of Ecology **54**, 345-365.

Mitchell, C. E. and A. G. Power. 2003. Release of invasive plants from fungal and viral pathogens. Nature **421**, 625-627.

Morin, L., A. BA and S. HE. 1996. Rust epidemics, climate and control of *Xanthium occidentale*. Pages 385-391 *in* V. C. Moran and J. H. Hoffmann, editors. Proc. 9th Int. Symp. Biol. Control Weeds. Univ. Cape Town Press., Stellenbosch.

Morris, M. J. 1997. Impact of the gall-forming rust fungus *Uromycladium tepperianum* on the invasive tree *Acacia saligna* in South Africa. Biological Control **10**, 75-82.

Office of Technology Assessment. 1993. Harmful Non-Indigenous Species in the United States. OTA-F-565, U. S. Congress.

Packer, A. and K. Clay. 2000. Soil pathogens and spatial patterns of seedling mortality in a temperate tree. Nature **404**, 278-281.

Panetta, F. D. and J. Dodd. 1995. *Chondrilla juncea*. Pages 67-84 *in* R. H. Groves, R. C. H. Shepherd, and R. G. Richardson, editors. The Biology of Australian Weeds, Vol. 1. Richardson, Melbourne, Australia.

Parker, I. M. and G. S. Gilbert. 2004. The evolutionary ecology of novel plant-pathogen interactions. Annual Review of Ecology, Evolution, and Systematics **35**, 675-700.

Parker, I. M., D. Simberloff, W. M. Lonsdale, K. Goodell, M. Wonham, P. M. Kareiva, M. H. Williamson, B. Von Holle, P. B. Moyle, J. E. Byers and L. Goldwasser. 1999. Impact: Toward a framework for understanding the ecological effects of invaders. Biological Invasions **1**, 3-19.

Pryor, M. R. 1967. Skeletonweed control. Pages 27-30 *in* 19th Annual California Weed Conference. California Weed Conference, Sacramento, California.

Reichard, S. H. and K. Hamilton. 1996. Predicting invasions of woody plants introduced into North America. Conservation Biology.

Rejmánek, M. and D. M. Richardson. 1996. What attributes make some plant species more invasive? Ecology **77**, 1655-1661.

Rizzo, D. M. and M. Garbelotto. 2003. Sudden oak death: endangering California and Oregon forest ecosystems. Frontiers in Ecology and the Environment **1**, 197-204.

Roy, B. A. 2001. Patterns of association between crucifers and their flower-mimic pathogens: Host jumps are more common than coevolution or cospeciation. Evolution **55**, 41-53.

Schirman, R. and W. C. Robocker. 1967. Rush skeletonweed—threat to dryland agriculture. Weeds **15**, 310-312.

Schrag, S. J. and P. Wiener. 1995. Emerging infectious disease: what are the relative roles of ecology and evolution? Trends in Ecology & Evolution **10**, 319-324.

Secord, D. and P. Kareiva. 1996. Perils and pitfalls in the host specificity paradigm. Bioscience **46**, 448-453.

Siemann, E. and W. E. Rogers. 2001. Genetic differences in growth of an invasive tree species. Ecology Letters **4**, 514-518.

Simberloff, D. 1986. Introduced insects: a biogeographic and systematic perspective. Pages 3-26 *in* H. A. Mooney and J. A. Drake, editors. Ecology of Biological Invasions of North American and Hawaii. Springer-Verlag, New York.

Supkoff, D. M., D. B. Joley and J. J. Marois. 1988. Effect of introduced biological control organisms on the density of *Chondrilla juncea* in California USA. Journal of Applied Ecology **25**, 1089-1096.

Thebaud, C. and D. Simberloff. 2001. Are plants really larger in their introduced ranges? American Naturalist **157**, 231-236.

Tilman, D. 1999. The ecological consequences of changes in biodiversity: a search for general principles. Ecology **80**, 1455-1474.

Van der Putten, W. H., C. Van Dijk and B. A. M. Peters. 1993. Plant-specific soil-borne diseases contribute to succession in foredune vegetation. Nature **362**, 53-56.

van Klinken, R. D. and O. R. Edwards. 2002. Is host-specificity of weed biological control agents likely to evolve rapidly following establishment? Ecology Letters **5**, 590-596.

Vila, M., A. Gomez and J. L. Maron. 2003. Are alien plants more competitive than their native conspecifics? A test using *Hypericum perforatum* L. Oecologia **137**, 211-215.

Vitousek, P. M., C. M. D'Antonio, L. L. Loope and R. Westbrooks. 1996. Biological invasions as global environmental change. American Scientist **84**, 468-478.

Wapshere, A. J., S. Hasan, W. K. Wahba and L. Caresche. 1974. The ecology of *Chondrilla juncea* in the western Mediterranean. Journal of Applied Ecology **11**, 783-799.

Webb, C. O., D. D. Ackerly, M. A. McPeek and M. J. Donoghue. 2002. Phylogenies and community ecology. Annual Review of Ecology and Systematics **33**, 475-505.

Weste, G. and G. C. Marks. 1987. The biology of *Phytophthora cinnamomi* in Australasian forests. Annual Review of Phytopathology **25**, 207-229.

Williamson, M. H. 1996. Biological Invasions. Chapman Hall, New York.

Willis, A. J., J. Memmott and R. I. Forrester. 2000. Is there evidence for the post-invasion evolution of increased size among invasive plant species? Ecology Letters **3**, 275-283.

Wingfield, M. J., B. Slippers, J. Roux and B. D. Wingfield. 2001. Worldwide movement of exotic forest fungi, especially in the tropics and the southern hemisphere. Bioscience **51**, 134-140.

Wolfe, L. M. 2002. Why alien invaders succeed: Support for the escape-from-enemy hypothesis. American Naturalist **160**, 705-711.

Wright, S. J. 2002. Plant diversity in tropical forests: a review of mechanisms of species coexistence. Oecologia **130**, 1-14.

Chapter fourteen

Exploring the relationship between niche breadth and invasion success

D. P. Vázquez

INTRODUCTION

The ultimate goal of invasions biology should be to generate a body of general, predictive theory. Ideally, such theory would help determine whether a species with a particular set of traits will be able to invade a given system. In its almost fifty years, the discipline has seen many attempts to provide such predictability through a focus on the traits of the potential invaders and those of the invaded community. Regrettably, most attempts have been so far unsuccessful.

One attempt at generalization that has been often discussed in the literature is the idea that niche breadth is positively correlated with invasion success, so that species with broad niches ("generalists") are more likely to invade than species with narrower niches ("specialists"), because they are more likely to find the necessary resources or environmental conditions; throughout this paper, I will refer to this notion as the "niche breadth–invasion success hypothesis." As with many other ideas in ecology, it is difficult to track the historical development of this hypothesis. Probably one of the first to include some measure of niche breadth in thinking about invasion success was H. G. Baker. Baker (1965, 1974) defined the traits of what he called the "ideal weed" — species prone to become

M.W. Cadotte et al. (eds.), Conceptual ecology and invasion biology, 307–322.

weedy and invade disturbed habitats. Among these traits, three were related to niche breadth, namely (i) that the species can produce seeds in a wide range of environmental circumstances and has high tolerance of (and often plasticity in face of) climatic and environmental variation; (ii) that the species has no special environmental requirements for germination (i.e., wide environmental niche); and (iii) that, when cross pollinated, pollination can be achieved by a nonspecialized flower visitor or by wind (i.e., wide pollination niche; note, however, that it is assuming wide niche breadth of pollinators, not of the plant, which brings up the issue of asymmetric specialization discussed below).

More recently, Rejmánek (2000) proposed ideas similar to those of Baker's in the broader context of plant invasions in general. He described what he thinks are ten major predictors of plant invasiveness, three of which are related to niche breadth: (i) "fitness homeostasis," the ability of an individual or population to maintain relatively constant fitness over a range of environments; fitness homeostasis will depend on phenotypic plasticity (homeostasis at the individual level) and population genetic polymorphism (homeostasis at the population level); (ii) size of "primary" (native) geographical ranges (see also Daehler and Strong 1993), which ultimately depends on population fitness homeostasis; and (iii) specificity of mutualisms, so that species that depend on non-specific mutualisms are more likely to overcome many abiotic and biotic barriers in new environments (see also Richardson *et al.* 2000).

Many others have discussed the idea that broader niches confer higher invasion success. For example, in a review on biotic homogenization McKinney and Lockwood (1999) concluded that extinction–resistant species and successful invaders are characterized by omnivory, rapid growth and dispersal or breeding in ephemeral habitats, whereas extinction-prone groups have a predominance of traits associated with specialization, slow reproduction and other traits not associated with opportunism. Similarly, Richardson *et al.* (2000) stated that "It is logical to predict that plants with specialized pollination syndromes are less likely to be pollinated in foreign environments than plants without specialized requirements" (p. 70). Likewise, Ricciardi and Rasmussen (1998) regard broad geographic, environmental and dietary ranges as general attributes of invasive aquatic species. Although these ideas are interesting and appealing, their evaluation is complicated by several conceptual and methodological problems. I will argue that progress towards a predictive theory of invasions will be difficult unless these issues are resolved.

Below I present a review of the literature, summarizing several studies that have evaluated the relationship between different aspects of niche breadth and invasion success. I then provide a critical examination of the mechanisms potentially involved in generating these results, stressing several conceptual issues that have been overlooked in previous discussions. My review is not exhaustive, and is taxonomically biased, focusing on plants, birds, and heterotrophic mutualists of plants. This bias reflects the fact that most studies evaluating this hypothesis have dealt with these groups. In summarizing the available literature

I use a vote-counting approach, because the number of studies available per taxon and measure of niche breadth was too small to allow the application of more quantitative meta-analytical techniques (see Arnqvist and Wooster 1995, Gurevitch *et al.* 2001).

The studies reviewed here are based on three different kinds of comparisons: some measure of niche breadth in native region of successful vs. failed invaders (SI vs. FI); niche breadth of invaders in their native vs. their introduced regions (NR vs. IR); and niche breadth of native vs. introduced species in the introduced region (NS vs. IS). It is important to note several similarities and differences about the three approaches. The first comparison (SI vs. FI) uses a binary measure of invasion success, asking whether success is explained by the niche breadth in the native range. In contrast, the second approach (NR vs. IR) uses a quantitative measure of success, such as the size of the introduced range, to determine whether invasion success is explained by niche breadth in the native range. The third approach (NS vs. IS) compares niche breadth of introduced species with native species in the introduced range (i.e., species native to the recipient community); here the question is whether the success of invaders (i.e., the fact that they established in a new environment) is explained by their broader niche compared to that of native species, in sharp contrast with the question asked in the previous two approaches. Thus, although the three types of comparisons are intended to address the question of whether niche breadth of potential invaders is related to their invasion success, they are not equivalent and are likely to provide different answers.

My review consists of three main parts: (i) the role of geographic, habitat and climatic range on plant and bird invasions; (ii) the role of mutualisms on plant and pollinator invasions; and (iii) the role of trophic breadth on bird invasions. I end with a section summarizing the results of the review and offering concluding remarks.

THE ROLE OF GEOGRAPHIC, CLIMATIC AND HABITAT RANGE

Patterns: studies on plants and birds

Studies that have evaluated the invasion success of plants and its relationship to geographic, climatic or habitat range are based on the three types of comparisons outlined above. Studies based on the first two kinds of comparisons (SI vs. FI and NR vs. IR) have found that invasion success is positively related to native geographic, climatic or habitat range, so that more successful invaders tend to have broader native ranges than less successful invaders (Table 1). In general, these results support the notion that the native range of plant species is a predictor of their success as invaders in new regions where they did not occur naturally (Daehler and Strong 1993, Rejmánek 2000).

Table 1 Studies evaluating the relationship between breadth of geographic, climatic or habitat range of plant species and invasion success.

Type of comparison*	Major taxon	Native region	Introduced region	Measure of success	Measure of niche breadth	Relat. with niche breadth[§]	Ref.
SI vs. FI	Higher plants	Central Europe	Mendoza and Buenos Aires provinces, Argentina	Establishment in new region	No. native habitat types	Positive	1
	Angiosperms	Europe	New Brunswick, Canada	Establishment in new region	Size of native range	Positive	4
NR vs. IR	Compositae, Gramineae	Eurasia	North America	Size of introduced range	No. native habitat types	Positive	2
	Angiosperms	South Africa	Australia	Weed status	No. native habitat types	Positive	3
	Angiosperms	South Africa	Australia	Weed status	No. climatic zones	Positive	3
	Bromus	Mediterranean regions	Mediterranean regions	No. climatic zones in native range	No. Mediterranean regions occupied	Positive	5
NS vs. IS	Higher plants	World/Hawaii	Hawaii	No. of habitat types occupied	No. habitat types	Negative	6

References: 1, Prinzing *et al.* (2002); 2, Rejmánek (1995); 3, Scott and Panetta (1993); 4, Goodwin *et al.* (1999); 5, Roy *et al.* (1991); 6, Kitayama and Mueller-Dombois (1995).

* Comparisons included: SI vs. FI: native range of successfully established species vs. unsuccessfully established species; NR vs. IR: native range vs. introduced range of introduced species; NS vs. IS: range of native vs. introduced species in the introduced region.

§ Indicates statistical significance of test and direction of effect: Positive, significant effect in the direction expected by the specialization–niche breadth hypothesis; Negative, significant effect in the direction opposite to the expected by the specialization–niche breadth hypothesis.

In contrast, the one study comparing breadth of habitat use by native and introduced species (NS vs. IS; Table 1) found a pattern opposite to the prediction of the niche breadth–invasion success hypothesis: introduced species were found in fewer habitat types than native species. This result is difficult to interpret, however, mainly because the number of habitats in which a species is found will depend on its dispersal ability and the time elapsed after its arrival to the environment, which may obviously bias the results towards native species, given their longer presence in the study area.

Studies evaluating the relationship between geographic, climatic or habitat range and invasion success in birds (Table 2) all used the first approach outlined above; that is, they compare the size of the native range between successful and unsuccessful invaders (SI vs. FI). These studies have used either size of native geographic range, number of native habitat types, or climatic zones occupied in the native region as measures of niche breadth.

In agreement with studies on plants discussed above, studies of bird invasions in Hawaii, Australia and worldwide found a positive relationship between invasion success and size of native geographic range used in the native range (Table 2). However, two studies, one in four oceanic islands and the other in New Zealand, failed to find evidence for such an effect (Table 2). The two studies using the number of habitat types occupied in the native range found that successfully invading species or families had significantly greater native habitat ranges than species or families that failed to invade (Table 2).

In summary, it appears that breadth of native geographic, climatic and habitat range is related to the invasion success of plant and bird species. Although there are exceptions to these patterns, they appear to have some degree of generality.

Mechanisms

Why does breadth of native geographic, habitat or climate range affect invasion success? Species with broad ranges may be more successful invaders than species with narrower ranges for several reasons. As discussed in the introduction, species' niches may determine their distribution, so that species with broader niches tend to have broader distributions and are so more likely to invade new environments (see also Brown 1984).

However, other factors not related to niche breadth may also affect species distributions. Under these scenarios, widespread species would have higher invasion success regardless of their niche breadth. First, species specialized on widespread resources, or species that are able to live under environmental characteristics that span large areas, may have wide distributions in spite of having a narrow tolerance to environmental conditions or of being able to use a narrow range of resources (Hanski *et al.* 1993, Gaston *et al.* 1997). This mechanism is particularly important for species that are able to exploit human-dominated ecosystems, which are themselves very widespread (Sol *et al.* 2002). Second, high dispersal ability can also lead to wide geographic ranges, at least in some groups, whereas limited dispersal can result in geographic ranges much narrower

Table 2 Studies evaluating the relationship between breadth of geographic, habitat range, trophic or behavioral breadth of bird species and invasion success.

Measure of niche breadth	Taxonomic level of analysis	Introduced region	Relationship with niche breadth[§]	Ref.
Size of native range	Species	Hawaii	Positive	1
		Australia	Positive	2
		New Zealand	NS	3
		Four oceanic islands	NS	4
		World	Positive	5
	Families[†]	World	Positive	6
No. native habitat types	Species	World	Positive	7
	Families[†]	World	Positive	6
Diet breadth	Species	New Zealand	NS[a]	3
		Nine oceanic islands	Positive[b]	8
	Families[†]	World	Positive[c]	6
Behavioral flexibility	Species	World	Positive	9, 10

References: 1, Moulton and Pimm (1986); 2, Duncan *et al.* (2001); 3, Veltman *et al.* (1996); 4, Lockwood *et al.* (1999); 5, Blackburn and Duncan (2001); 6, Cassey (2002); 7, Brooks (2001); 8, McLain *et al.* (1999); 9, Sol and Lefebvre (2000); 10, Sol *et al.* (2002).

Measures of diet breadth used in studies: [a]Diet breadth measured as carnivore or herbivore (specialist) and omnivore (generalist). [b]Diet breadth measured as frugivore or granivore (specialist) and omnivore (eating fruits and seeds; generalist); a fourth category was included for species that fed on neither fruits or nectar. [c]Diet breadth measured as number of seven major food types consumed.

§ Indicates statistical significance of test and direction of effect: Positive, significant effect in the direction expected by the specialization–niche breadth hypothesis; NS, statistically nonsignificant result.

† Cassey (2002) also did analyses at species level. However, his species-level multivariate analyses are mostly descriptive, and are difficult to compare with other studies listed in the table and were therefore not included.

than would be predicted based on a species' niche (Gaston 2003). Thus, species with high dispersal abilities may tend to have both wide native ranges and high chances of colonizing and spreading in new environments. Third, species with wider native ranges may be more likely to be transported to new areas by humans through passive sampling (Prinzing *et al.* 2002), particularly considering that widespread species also tend to be locally abundant (Gaston *et al.* 1997). Cassey *et al.* (2004) have recently shown that propagule pressure of introduced birds explains invasion success worldwide; these authors also found that propagule pressure is positively related to geographic and trophic breadth, and so these measures of niche breadth are not independent of propagule pressure and could be confounded with it. Understanding the role of the habitat and climatic components of niche breadth in determining invasion success will require teasing apart these factors.

THE ROLE OF SPECIALIZATION IN MUTUALISTIC INTERACTIONS

Patterns

Several studies have evaluated the relationship between specialization in mutualistic interactions and invasion success of plants and/or animal mutualists. Richardson *et al.* (2000) have reviewed the role of mutualisms in facilitating plant invasions. Their review suggests that plants with extremely specialized mutualistic interactions (including interactions with pollinators, seed dispersers, mycorrhizal fungi and nitrogen-fixing bacteria) may be prevented from establishing and spreading in a new environment. However, the rarity of extreme specialization in mutualistic interactions may greatly minimize the potential role of mutualisms in limiting plant invasion success.

One of the examples of how extremely specialized mutualistic interactions may prevent the establishment and spread of introduced plant populations discussed by Richardson *et al.* (2000) is the invasion ecology of figs (*Ficus*, Moraceae). Of the sixty species of *Ficus* introduced to Florida, USA, only three have become invasive, and they have done so only after the accidental introduction of their specific wasp (see Richardson *et al.* 2000 and references therein). However, Richardson *et al.* also discuss the case of *F. lutea* in South Africa, where one isolated female individual of this nonindigenous species has been pollinated by wasps that were not its normal pollinator. Although in this case the absence of male individuals prevented reproduction (hybrids with native *Ficus* did not produce viable seedlings), spread might have been possible if male individuals of *F. lutea* had also been introduced. As a side note, Richardson *et al.* (2000) point out that several pine species (*Pinus* spp.) native to North America and the Mediterranean Basin that are wind-dispersed in their native ranges are dispersed by cockatoos in Australia. Although birds destroy most of the seeds, some survive to establish isolated foci in habitats (eucalypt forests) that they would otherwise not have been able to invade. Thus, even apparent extreme specialization on a particular pollen vector (fig wasp) or seed dispersal agent (wind) may not preclude survival and successful reproduction in a new environment.

Albeit intriguing, the extreme specialization observed for *Ficus* is not representative of the majority of pollination interactions (Waser *et al.* 1996), and most species within a community fall somewhere along a gradient that spans from extreme specialization to extreme generalization (Bascompte *et al.* 2003, Jordano *et al.* 2003, Vázquez and Aizen 2003, 2004a). A similar situation occurs in seed dispersal mutualisms (Bascompte *et al.* 2003). The question is whether species located towards the specialized extreme of the continuum are less likely to invade successfully than more generalized species.

In a recent review of introduced bees and their ecological impacts, Goulson (2003) points out that most introduced bee species appear quite generalized. Goulson argues that among bees, which depend on floral food resources throughout their entire life cycles, dietary generalization seems to be a precondition for becoming successful invaders. However, this interpretation should be taken with caution. All species of bees known to have been introduced outside their native ranges (the honey bee, five bumblebees, ten megachillids and one halictid) have been purposefully introduced by humans to enhance crop pollination, usually of several species. Given this goal, dietary generalization of introduced bees is probably a desirable trait, and it is therefore difficult to tease apart the effect of human selection from the effect of dietary specialization on invasion success.

Another way of evaluating the niche breadth–invasion success hypothesis as it applies to mutualistic interactions is by comparing the degree of specialization in mutualistic interactions between native and introduced species in a given area. The expectation under the niche breadth–invasion success hypothesis is that interactions are less specific for introduced than for native species (because exotic species with more specific interactions may have been "filtered out" during the establishment process). I have found studies in four systems attempting to answer this question by comparing the degree of specialization of plant and/or pollinators between native and introduced species (i.e., a NS vs. IS comparison; see above). In addition, I have found comparable data for a fifth system, for which I calculated degree of specialization in a way comparable to the other studies. These studies are summarized in Table 3. None of the studies support the niche breadth–invasion success hypothesis, and in some cases the observed pattern was contrary to the expectation (i.e., introduced species had more specific interactions). Thus, the few available studies tell us that interaction specificity of introduced plants and pollinators does not influence their invasion success.

Mechanisms

From the studies reviewed above evaluating the role of mutualisms in invasions it appears that only extremely specialized mutualisms have the potential to influence invasion success, and in general specificity of mutualistic interactions does

Table 3 Studies that have compared degree of specialization of plants or pollinators between native and introduced species.

		Mean number of mutualist species			
Introduced region	Major taxon	Native	Exotic	Result of statistical test[§]	Ref.
Illinois, USA	Angiosperms	34.0	24.0	Negative	1
Nahuel Huapi, Argentina	Angiosperms	14.8	17.6	NS	2
Galápagos Islands	Angiosperms	2.2	1.3	Negative	3
Ile aux Aigrettes, Mauritius	Angiosperms	3.9	3.4	NS	4
Flores Island, Azores (Azores)	Angiosperms	3.3	2.3	NS	4
Nahuel Huapi, Argentina	Hymenoptera (rare species excluded)	6.3	11.3	NS	2
Nahuel Huapi, Argentina	Hymenoptera (Apidae only)	9.0	11.5	NS	2
Flores Island, Azores (Azores)	Pollinators	3.4	1.9	NS	4
Ile aux Aigrettes, Mauritius	Pollinators	4.1	3.8	NS	4

References: 1, Memmott and Waser (2002); 2, Morales and Aizen (2002); 3, McMullen (1993); 4, Olesen *et al.* (2002).

[§] Indicates statistical significance of test and direction of effect: Negative, significant effect in the direction opposite to the expected by the specialization–niche breadth hypothesis; NS, statistically nonsignificant result.

not seem to influence invasion success significantly. These results contrast with those obtained in studies evaluating the relationship between geographic, habitat or climatic range and invasion success reviewed in the previous section. How can we explain these contrasting results? Why does narrow climatic or geographic range appear to translate into low likelihood of invasion, whereas narrow range of mutualistic interactions does not? The answer to this question may require some conceptual refinements in our thinking about the relationship between niche breadth and invasion success as it applies to mutualism.

First, we need to distinguish between the fundamental and the realized niche (Hutchinson 1957). In the context of species interactions, fundamental specialization refers to the potential interactions that would lead to positive fitness for a given species, under any possible ecological circumstances (van Klinken and Edwards 2002, Vázquez and Aizen 2004b, Vázquez 2005). Thus, fundamental specialization will ultimately depend on the genetic background of a species, whereas realized specialization refers to the actual specialization attained under a particular ecological context. For example, most fig species are fundamentally specialized on one or a few species of pollinating wasps; they fail to reproduce if they are moved to different environments, unless their specialized pollinators are introduced as well. Conversely, many other plant species that are pollinated by a single pollinator in a particular habitat are often pollinated by a different species in other habitats where they are introduced (Richardson *et al.* 2000). Regrettably, the best we can do in the field is to measure the realized niche, but if we want to predict invasions what we need to know is the fundamental niche. In other words, knowing the niche breadth of a species in a particular environment does not tell us what the realized niche will be in another environment. Although studying the climatic component of the fundamental niche may be possible by, for example, experimentally growing a species under a wide range of environmental conditions, doing so in the context of species interactions may be simply unfeasible, because it would require exposing the species to all possible combinations of interaction partners that it could encounter in any given environment. Even many species selected as biocontrol agents, which are usually selected to be specialists, exhibit unexpected host shifts in their introduced ranges (Simberloff and Stiling 1996, Henneman and Memmott 2001, van Klinken and Edwards 2002, Pearson and Callaway 2003), and these shifts are usually not the result of changes in the fundamental niche through rapid evolution, but an expression of a different realized niche under the new environment (van Klinken and Edwards 2002).

Second, as I pointed out above, species that use widespread resources, live under widespread environmental conditions or use widespread habitats are likely to be widespread themselves. This idea can be easily extended to species interactions to argue that species that specialize on widespread mutualists may themselves be widespread. An emerging pattern in plant-animal mutualistic interactions is the existence of highly asymmetric specialization, whereby specialists tend to specialize on generalists (Bascompte *et al.* 2003, Vázquez and Aizen 2004a). Given that abundant, frequently interacting species tend to be more generalized than rare species (Dupont *et al.* 2003, Vázquez and Aizen 2003, 2004a), and that locally abundant species tend to be geographically widespread (Gaston *et al.* 1997), species specialized on widespread mutualists will have a greater chance of finding their mutualists in their introduced region. These general features of mutualistic interactions may substantially increase the likelihood of successful invasion by species that are relatively specialized on their mutualists.

THE ROLE OF TROPHIC BREADTH

Patterns

Another aspect of the niche that can potentially affect invasion success is dietary breadth, and it is frequently listed as one potential characteristic of successful invaders (Ehrlich 1986, Ricciardi and Rasmussen 1998, McKinney and Lockwood 1999). I discuss here a few studies that have addressed this issue in the context of bird invasions.

To my knowledge, three studies have evaluated the relationship between invasion success and diet breadth (Table 2). These studies compared trophic breadth in the native region of successful vs. failed invaders (SI vs. FI). McLain *et al.* (1999) defined specialists as species feeding on either fruits or seeds, and generalists as species feeding on both; they found that trophic generalists were more likely to be successful invaders of oceanic islands than trophic specialists. Veltman *et al.* (1996) defined herbivorous or carnivorous species as dietary specialists, and omnivores (species feeding on both plant and animal material) as generalists; they failed to find a statistical effect of size of native trophic range on the success of birds introduced in New Zealand. Finally, Cassey (2002) defined dietary breadth as the number of food types consumed out of seven possible categories. He found a weak but significant effect of dietary breadth explaining the invasion success of bird families.

Mechanisms

From this limited number studies and from their disparate results, it is difficult to make generalizations about the importance of trophic breadth for invasion success. Furthermore, a major limitation of the studies discussed above is that the measure of specialization used is probably too coarse and may therefore be a poor surrogate of true trophic specialization. For example, in the study by McLain *et al.* (1999), only fruits and seeds are considered to characterize diet breadth, although the authors state that "virtually all introduced species [in these islands] eat some insects and often other invertebrates" (p. 552). Thus, it is unclear whether what the authors measured actually represents diet breadth, or rather whether it represents something else related to eating fruits or seeds. In fact, the quantification of degree of specialization is probably one of the main conceptual and methodological hurdles in the study of species interactions (Martinez *et al.* 1999, Novotný *et al.* 2002, Vázquez and Aizen 2004b). Many ideas in ecology, evolution and biogeography depend on assumptions about the degree of specialization of species interactions. For example, because the bulk of the Earth's species are believed to be tropical herbivorous insects, estimates of the number of species on Earth depend critically on assumptions about their degree of specialization. The realization that tropical herbivorous insects were less specialized than previously thought led to revising the estimated number of global species from 30 million (Erwin 1982) to approximately 5 million (Ødegaard 2000, Novotný *et al.* 2002).

To circumvent the problem of the measurement of trophic specialization in birds, Sol and Lefebvre (2000) and Sol *et al.* (2002) proposed the use of "behavioral flexibility" as a surrogate of trophic breadth. These authors defined behavioral flexibility as "the frequency of new and unusual feeding behaviors reported in the short note section of ornithology journals" (Sol *et al.* 2002, p. 495). They argue that a taxonomic group in which large numbers of new feeding behaviors are observed is likely to change its foraging techniques or diet frequently, to eat a surprisingly large range of foods and to use handling behaviors and novel situations in a way that strikes ornithologists by its complexity and flexibility. They further found that behavioral flexibility of bird species in their native range is related to their success as invaders worldwide. This approach is a promising alternative to estimating trophic niche breadth from feeding records.

Regardless of the problems of estimation of trophic specialization, two issues discussed above in the context of geographic range and mutualisms also apply here. First, species specializing on widespread resources, particularly those that exploit human-dominated ecosystems, may be likely to invade in spite of being specialized. Second, even if accurate, estimates using feeding records provide a measure of the realized niche but not of the fundamental niche, which would be necessary predict invasion success in a new environment with a different set of food resources.

CONCLUDING REMARKS

My review of the literature suggests several generalizations about the role of niche breadth as a determinant of invasion success. In particular, native habitat, geographic or climatic range of plants and birds appear to be good predictors invasion success, whereas specificity of mutualistic interactions and trophic breadth appear to have a much weaker predictive power. However, even when results do fit the predictions of the niche breadth–invasion success hypothesis, it is usually not possible to know whether niche breadth is involved in generating the observed pattern. Future studies aiming at predicting the success of invaders should tease apart the relative contributions of niche breadth from other confounding factors. More broadly, I hope my review can serve as an example of how conceptual elaboration may be useful to improve our understanding in invasions biology.

ACKNOWLEDGMENTS

I thank the editors for the invitation to contribute to this volume. Sean McMahon, Ingrid Parker, David Richardson, Eric Seabloom, Daniel Sol and Montserrat Vilà made useful suggestions on the manuscript. I was supported by a Postdoctoral Fellowship at the National Center for Ecological Analysis and Synthesis (funded by NSF Grant #DEB-0072909, the University of California, and the Santa Barbara campus).

REFERENCES

Arnqvist, G. and D. Wooster. 1995. Meta-analysis: synthesizing research findings in ecology and evolution. Trends in Ecology and Evolution **10**, 236-240.

Baker, H. G. 1965. Characteristics and modes of origin of weeds. Pages 147-169 *in* H. G. Baker and C. L. Stebbins, editors. The Genetics of Colonizing Species. Academic Press, New York.

Baker, H. G. 1974. The evolution of weeds. Annual Review of Ecology and Systematics **5**, 1-24.

Bascompte, J., P. Jordano, C. J. Melián and J. M. Olesen. 2003. The nested assembly of plant-animal mutualistic networks. Proceedings of the National Academy of Sciences (USA) **100**, 9383-9387.

Blackburn, T. M. and R. P. Duncan. 2001. Determinants of establishment success in introduced birds. Nature **414**, 195-197.

Brooks, T. M. 2001. Are unsuccessful avian invaders rarer in their native range than successful invaders? Pages 125-155 *in* J. L. Lockwood and M. M. McKinney, editors. Biotic Homogenisation: the loss of diversity through invasion and extinction. Kluwer Academic Press, Dordrecht, The Netherlands.

Brown, J. H. 1984. On the relationship between abundance and distribution of species. American Naturalist **124**, 255-279.

Cassey, P. 2002. Life history and ecology influences establishment success of introduced land birds. Biological Journal of the Linnean Society **76**, 465-480.

Daehler, C. C. and D. R. Strong. 1993. Prediction and biological invasions. Trends in Ecology and Evolution **8**, 380-380.

Duncan, R. P., M. Bomford, D. M. Forsyth and L. Conibear. 2001. High predictability in introduction outcomes and the geographical range size of introduced Australian birds: a role for climate. Journal of Animal Ecology **70**, 621-632.

Dupont, Y. L., D. M. Hansen and J. M. Olesen. 2003. Structure of a plant-flower-visitor network in the high-altitude sub-alpine desert of Tenerife, Canary Islands. Ecography **26**, 301-310.

Ehrlich, P. R. 1986. Which animal will invade? Pages 79-95 *in* H. A. Mooney and J. A. Drake, editors. Ecology of Biological Invasions of North America and Hawaii. Springer-Verlag, New York.

Erwin, T. L. 1982. Tropical forests: their richness in coleoptera and other arthropod species. Coleopterists Bulletin **36**, 74-75.

Gaston, K. J. 2003. The Structure and Dynamics of Geographic Ranges. Oxford University Press, Oxford and New York.

Gaston, K. J., T. M. Blackburn and J. H. Lawton. 1997. Interspecific abundance–range size relationships: an appraisal of mechanisms. Journal of Animal Ecology **66**, 579-601.

Goodwin, B. J., A. J. McAllister and L. Fahrig. 1999. Predicting invasiveness of plant species based on biological information. Conservation Biology **13**, 422-426.

Goulson, D. 2003. Effects of introduced bees on native ecosystems. Annual Review of Ecology, Evolution, and Systematics **34**, 1-26.

Gurevitch, J., P. S. Curtis and M. H. Jones. 2001. Meta-analysis in ecology. Advances in Ecological Research **32**, 199-247.

Hanski, I., J. Kouki and A. Halkka. 1993. Three explanations of the positive relationship between distribution and abundance of species. Pages 108-116 *in* R. E. Ricklefs and D. Schluter, editors. Species diversity in ecological communities: historical and geographical perspectives. University of Chicago Press, Chicago, IL.

Henneman, M. L. and J. Memmott. 2001. Infiltration of a Hawaiian community by introduced biological control agents. Science **293**, 1314-1316.

Hutchinson, G. E. 1957. Concluding remarks. Cold Spring Harbor Symposia on Quantitative Biology **22**, 415-427.

Jordano, P., J. Bascompte and J. M. Olesen. 2003. Invariant properties in coevolutionary networks of plant-animal interactions. Ecology Letters **6**, 69-81.

Kitayama, K. and D. Mueller-Dombois. 1995. Biological invasion on an oceanic island mountain: Do alien plant species have wider ecological ranges than native species? Journal of Vegetation Science **6**, 667-674.

Lockwood, J. L., M. P. Moulton and K. L. Balent. 1999. Introduced avifaunas as natural experiments in community assembly. Pages 108-129 *in* E. Weiher and P. Keddy, editors. Ecological Assembly Rules: Perspectives, Advances, Retreats. Cambridge University Press, Cambridge.

Martinez, N. D., B. A. Hawkins, H. A. Dawah and B. P. Feifarek. 1999. Effects of sampling effort on characterization of food-web structure. Ecology **80**, 1044-1055.

McKinney, M. L. and J. L. Lockwood. 1999. Biotic homogenization: a few winners replacing many losers in the next mass extinction. Trends in Ecology and Evolution **14**, 450-453.

McLain, D. K., M. P. Moulton and J. G. Sanderson. 1999. Sexual selection and extinction: The fate of plumage-dimorphic and plumage-monomorphic birds introduced onto islands. Evolutionary Ecology Research **1**, 549-565.

McMullen, C. K. 1993. Flower-visiting insects of the Galápagos Islands. The Pan-Pacific Entomologist **69**, 95-106.

Memmott, J. and N. M. Waser. 2002. Integration of alien plants into a native flower-pollinator visitation web. Proceedings of the Royal Society of London Series B **269**, 2395-2399.

Morales, C. L. and M. A. Aizen. 2002. Does invasion of exotic plants promote invasion of exotic flower visitors? A case study from the temperate forests of the southern Andes. Biological Invasions **4**, 87-100.

Moulton, M. P. and S. L. Pimm. 1986. Species introductions to Hawaii. Pages 231-249 *in* H. A. Mooney and J. A. Drake, editors. Ecology of Biological Invasions of North America and Hawaii. Springer-Verlag, New York.

Novotný, V., Y. Basset, S. E. Miller, G. D. Weiblen, B. Bremer, L. Cizek and P. Drozd. 2002. Low host specificity of herbivorous insects in a tropical forest. Nature **416**, 841-844.

Ødegaard, F. 2000. How many species of arthropods? Erwin's estimate revised. Biological Journal of the Linnean Society **71**, 583-597.

Olesen, J. M., L. I. Eskildsen and S. Venkatasamy. 2002. Invasion of pollination networks on oceanic islands: importance of invader complexes and endemic super generalists. Diversity and Distributions **8**, 181-192.

Pearson, D. E. and R. M. Callaway. 2003. Indirect effects of host-specific biological control agents. Trends in Ecology & Evolution **18**, 456-461.

Prinzing, A., W. Durka, S. Klotz and R. Brandl. 2002. Which species become aliens? Evolutionary Ecology Research **4**, 385-405.

Rejmánek, M. 1995. What makes a species invasive? Pages 3-13 *in* P. Pyšek, editor. Plant invasions: general aspects and special problems. Academic Publishing, Amesterdam.

Rejmánek, M. 2000. Invasive plants: approaches and predictions. Austral Ecology **25**, 497-506.

Ricciardi, A. and J. B. Rasmussen. 1998. Predicting the identity and impact of future biological invaders: a priority for aquatic resource management. Canadian Journal of Fisheries and Aquatic Sciences **55**, 1759-1765.

Richardson, D. M., N. Allsopp, C. M. D'Antonio, S. J. Milton and M. Rejmánek. 2000. Plant invasions — the role of mutualisms. Biological Reviews **75**, 65-93.

Roy, J., M. L. Navas and L. Sonié. 1991. Invasion by annual grasses: a case study challenging the homoclime approach to invasions. Pages 207-224 *in* R. H. Groves and F. di Castri, editors. Biogeography of Mediterranean Invasions. Cambridge University Press, Cambridge.

Scott, J. K. and F. D. Panetta. 1993. Predicting the Australian weed status of southern African plants. Journal of Biogeography **20**, 87-93.

Simberloff, D. and P. D. Stiling. 1996. How risky in biological control? Ecology **77**, 1965-1974.

Sol, D. and L. Lefebvre. 2000. Behavioural flexibility predicts invasion success in birds introduced to New Zealand. Oikos **90**, 599-605.

Sol, D., S. Timmermans and L. Lefebvre. 2002. Behavioural flexibility and invasion success in birds. Animal Behaviour **63**, 495-502.

van Klinken, R. D. and O. R. Edwards. 2002. Is host-specificity of weed biological control agents likely to evolve rapidly following establishment? Ecology Letters **5**, 590-596.

Vázquez, D. P. and M. A. Aizen. 2003. Null model analyses of specialization in plant-pollinator interactions. Ecology **84**, 2493-2501.

Vázquez, D. P. and M. A. Aizen. 2004a. Asymmetric specialization: a pervasive feature of plant-pollinator interactions. Ecology **85**, 1251-1257.

Vázquez, D. P. and M. A. Aizen. 2004b. Community-wide patterns of specialization in plant-pollinator interactions revealed by null-models. Pages in press *in* J. Ollerton, editor. Specialization and generalization in plant-pollinator interactions. University of Chicago Press, Chicago.

Vázquez, D. P. 2005. Revisitando el nicho hutchinsoniano. Ecología Austral, in press.

Veltman, C. J., S. Nee and M. J. Crawley. 1996. Correlates of introduction success in exotic New Zealand birds. American Naturalist **147**, 542-557.

Waser, N. M., L. Chittka, M. V. Price, N. M. Williams and J. Ollerton. 1996. Generalization in pollination systems, and why it matters. Ecology **77**, 1043-1060.

Chapter fifteen

Interactions between invasive plants and soil ecosystems: positive feedbacks and their potential to persist

A. S. Thorpe and R. M. Callaway

INTRODUCTION

Plants indirectly affect their neighbors in many ways, but one of the most important is by altering the biotic, physical, and chemical characteristics of soils (Hobbie 1992; Angers and Caron 1998; Berendse 1998; Binkley and Giardina 1998; Northrup *et al.* 1998; Schlesinger and Pilmanis 1998; Van Breemen 1993; Wardle *et al.* 1998; Chen and Stark 2000; Eaton and Farrell 2004). These general effects have been understood for decades, but only recently have experiments demonstrated that complex interactions between plants and soil microbial communities can have strong effects on plant populations (Bever *et al.* 1997; Clay and Van der Putten 1999; Packer and Clay 2000), interactions among plant species (West 1996), and the organization of plant communities (Grime *et al.* 1987; Van der Putten *et al.* 1993; Bever 1994; Van der Putten 1997; van der Heijden *et al.* 1998; Hooper *et al.* 2000; Klironomos 2002). Soil communities alter competitive outcomes among plants through their pathogenic effects (Van der Putten and

M.W. Cadotte et al. (eds.), Conceptual ecology and invasion biology, 323–341.

Peters 1997), by favoring obligate mycorrhizal species over non-mycorrhizal or facultative mycorrhizal species (Hetrick *et al.* 1989; Hartnett *et al.* 1993), and by transferring resources and fixed carbon between species (Chiarello *et al.* 1982; Francis and Read 1984; Grime *et al.* 1987; Moora and Zobel 1996; Watkins *et al.* 1996; Simard *et al.* 1997; Marler *et al.* 1999; but see Robinson and Fitter 1999). By altering the biotic and abiotic characteristics of soils, plants can drive positive or negative feedbacks (Box 1), and these feedbacks can profoundly affect plant populations and communities.

Box 1

Positive feedbacks occur when plant species accumulate microbes that have beneficial effects on the plants that cultivate them, such as mycorrhizal fungi and nitrogen fixers. Positive feedbacks promote species dominance and are thought to lead to a loss of local community diversity (Bever *et al.* 1997, Bever 2002). Negative feedbacks occur when plant species accumulate pathogenic microbes in their rhizospheres and these interactions outweigh the benefits received from mutualistic interactions. Negative feedbacks create conditions that are increasingly hostile to the plants that cultivate the pathogens and are thought to promote community diversity (Van der Putten *et al.* 1993, Bever 1994, Klironomos 2002).

Jim Bever (1994; Bever *et al.* 1997) proposed the feedback model, in which a plant or population of plants alters the composition of the soil community. This change then feeds back to affect the growth and survival of the plant individual or population. This model describes how two different types of feedback could be established; positive feedback, which should occur when the negative effects of soil pathogens are outweighed by the beneficial effects of mutualistic mycorrhizae, and negative feedback, which is expected to occur either when the effects of pathogens outweigh the effects of mycorrhizae or when the soil community enhances the growth of competing plants more than that of the associated plant (Bever 1994; Bever *et al.* 1997). By favoring the local replacement of species, negative feedback is expected to maintain species diversity, whereas positive feedback is expected to lead to species dominance and a decrease in local species diversity (Bever 1994; Bever *et al.* 1997; Watkinson 1998). A growing body of research has demonstrated that plant-soil feedbacks can have important ecological consequences (Bever *et al.* 1996; Wardle and Nicholson 1996; Westover *et al.* 1997; Klironomos 2002). For example, in a study comparing feedback and relative abundance of 61 co-existing old field species from southern Ontario, Canada, Klironomos (2002) found a strong positive relationship between a plant's feedback with the soil community and its relative abundance (Fig. 1).

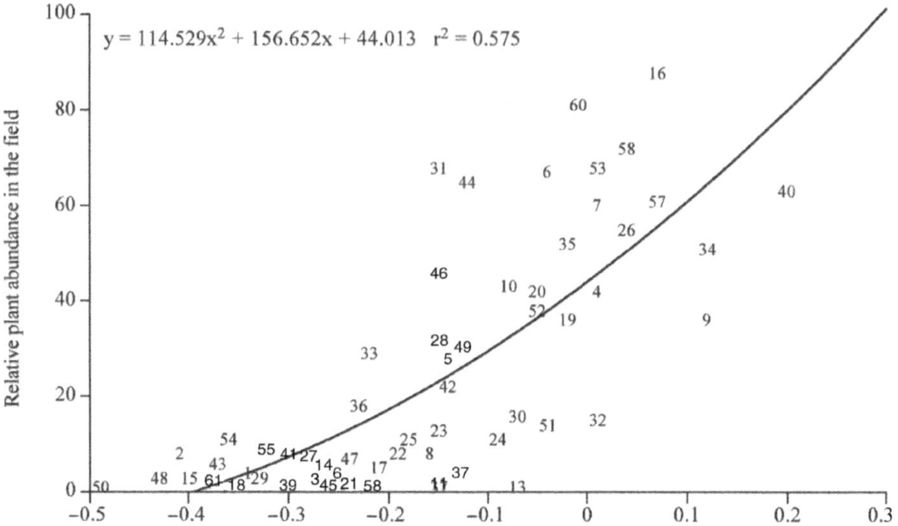

Fig. 1 The relationship between relative plant abundance in an old-field site and soil feedback response. Squares represent means. $y = 114.529x^2 + 156.652x + 44.013$, $r^2 = 0.575$, P = 0:0001. Numbers represent different plant species. 1, *Carex garberi*; 2, *Carex aurea*; 3, *Carex granularis*; 4, *Daucus carota*; 5, *Agrostis gigantea*; 6, *Solidago graminifolia*; 7, *Solidago nemoralis*; 8, *Aster simplex*; 9, *Aster vimineus*; 10, *Aster novaeangliae*; 11, *Circium vulgare*; 12, *Chenopodium ambrosioides*; 13, *Oenothera biennis*; 14, *Carex flava*; 15, *Juncus dudleyi*; 16, *Solidago canadensis*; 17, *Linaria vulgaris*; 18, *Cichorium intybus*; 19, *Circium arvense*; 20, *Solidago rugosa*; 21, *Geum aleppicum*; 22, *Satureja vulgaris*; 23, *Potentilla recta*; 24, *Coronilla varia*; 25, *Asclepias syriaca*; 26, *Achillea millefolium*; 27, *Apocynum cannabinum*; 28, *Hypericum perforatum*; 29, *Agrostis scabra*; 30, *Phleum pratense*; 31, *Poa compressa*; 32, *Echium vulgare*; 33, *Centaurea jacea*; 34, *Rudbeckia serotina*; 35, *Poa pratensis*; 36, *Dactylis glomerata*; 37, *Cerastium vulgatum*; 38, *Galium palustre*; 39, *Oenothera perennis*; 40, *Prunella vulgaris*; 41, *Trifolium pratense*; 42, *Convolvulus arvensis*; 43, *Silene cucubalus*; 44, *Erigeron strigosus*; 45, *Asparagus officinalis*; 46, *Hieracium auranticum*; 47, *Erigeron philadelphicus*; 48, *Veronica officinalis*; 49, *Plantago lanceolata*; 50, *Galium mollugo*; 51, *Hieracium pilosella*; 52, *Vicia cracca*; 53, *Hieracium pratense*; 54, *Medicago lupulina*; 55, *Ranunculus acris*; 56, *Taraxacum officinale*; 57, *Fragaria virginiana*; 58, *Chrysanthemum leucanthemum*; 59, *Tragopogon pratensis*; 60, *Bromus inermis*; 61, *Panicum lanuginosum*. Reprinted from Klironomos 2002.

Plant species that were found in low abundance in the field consistently displayed negative feedback interactions, whereas plants with high abundance either had low negative or positive feedback interactions (Klironomos 2002).

Plant invasions provide an exceptional opportunity for understanding how plants affect soils and drive feedback processes. Numerous studies have documented the effects of invasive species on the composition of soil biota (Belnap and Phillips 2001; Kourtev *et al.* 2002) and soil microbial function (Kourtev *et al.* 2002; Ehrenfeld 2003). However, much less is known about the role of feedbacks in the success of exotic plants. Feedbacks may also affect evolution — potentially most apparent in the rapid evolutionary changes that can accompany exotic invasion (Rice and Emory 2003).

If plant-soil feedbacks drive variation in fitness differences among individuals, then ultimately feedback interactions may be under selective pressure (Van der Putten 1997; Van Breeman and Finzi 1998). Based on theory developed by Bever *et al.* (1997) and Klironomos (2002), we hypothesize that positive feedbacks in general are likely to lead the plant community to shift towards a monoculture of the invasive species. However, we propose that the longevity of positive feedback interactions in evolutionary time may depend on whether the invasive plant is having its strongest effects on the pathogenic or the nutrient-cycling components of the soil ecosystem (Fig. 2). Microbes have a short generation time and thus can respond to evolutionary pressures quickly. Therefore pathogens may respond relatively rapidly to the invasion of non-native plants. This may ultimately lead to greater coexistence among species, where, although the invasive may still be present, it will not be the over-whelming community dominant. In contrast, feedbacks between invasive plants and nutrient cycles may be much more likely to persist. When the biotic component of the soil nutrient cycles is involved, it is expected that there would be a shift in the microbial community to populations that are better adapted to the new nutrient status of the ecosystem (Fig. 2; Atlas and Bartha 1998; Schimel and Bennett. 2004). In the case of alterations to abiotic components of the soil ecosystem, there is no direct selective pressure to stimulate adaptation (Fig. 2).

Here we discuss interactions between invasive species and the soil ecosystem (pathogenic and nutrient-cycling components), the mechanisms for these interactions, and evidence that these interactions have very different effects on the survival of invasive versus native species.

INVASIVE PLANTS AND SOIL PATHOGENS

One of the leading hypotheses for the remarkable success of some exotic species is that they have escaped the specialist enemies that control them in their native ranges (Keane and Crawley 2002). Embedded within this hypothesis is the idea that if microbial pathogens limit the growth of invasive plants in their native range, there will be negative feedbacks between the soil microbial community and the plant in the native range due to the accumulation of species-specific soil pathogens (Klironomos 2002; Mitchell and Power 2003; Callaway *et al.* 2004). In contrast, positive feedbacks may occur in the invaded range where exotic

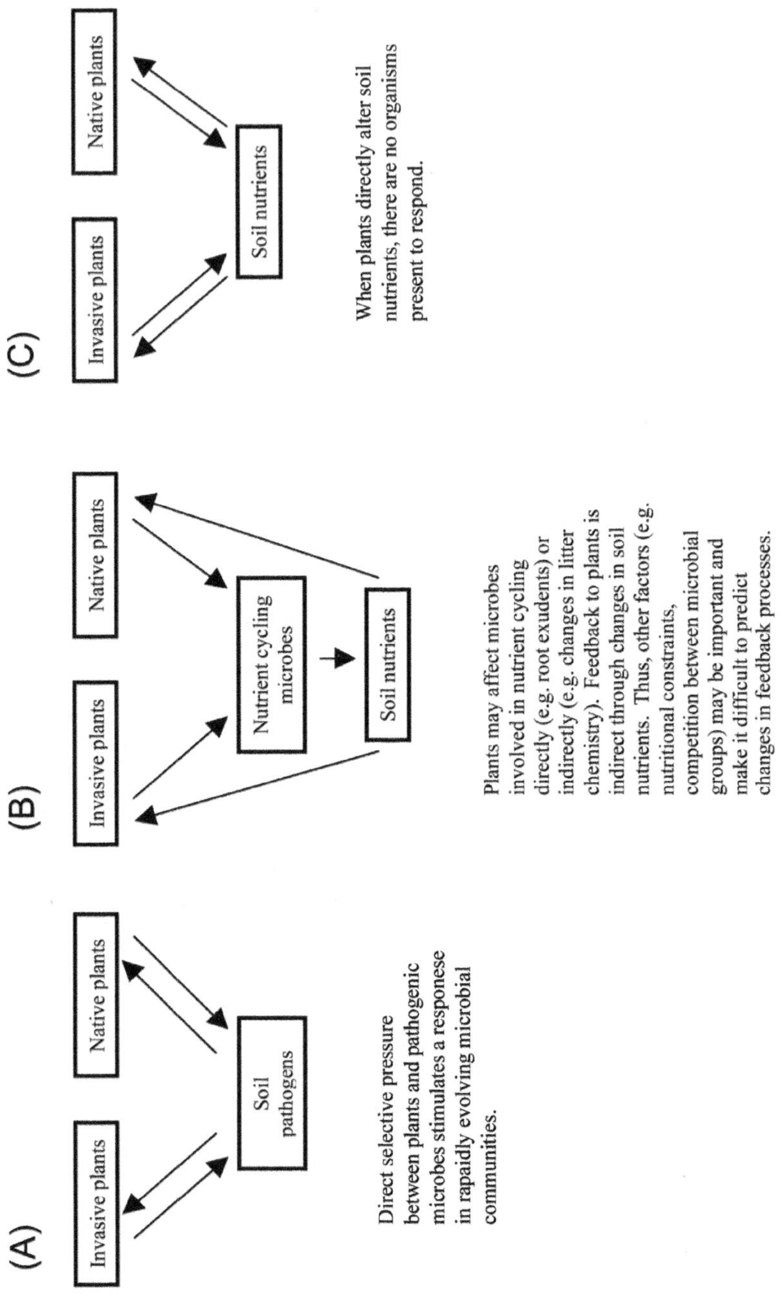
(A)
Invasive plants
Native plants
Soil pathogens
Direct selective pressure between plants and pathogenic microbes stimulates a responese in rapaidly evolving microbial communities.
(B)
Invasive plants
Native plants
Nutrient cycling microbes
Soil nutrients
Plants may affect microbes involved in nutrient cycling directly (e.g. root exudents) or indirectly (e.g. changes in litter chemistry). Feedback to plants is indirect through changes in soil nutrients. Thus, other factors (e.g. nutritional constraints, competition between microbial groups) may be important and make it difficult to predict changes in feedback processes.
(C)
Invasive plants
Native plants
Soil nutrients
When plants directly alter soil nutrients, there are no organisms present to respond.

Fig. 2

species are largely free from species-specific soil pathogens but can still interact with less host-specific mutualists such as mycorrhizal fungi and bacteria that drive nutrient cycles.

In this section, first we will describe the evidence that invasive species have escaped pressure from soil pathogens and the potential feedback effects on the fitness of invasive species. Second, we will discuss the implications of this type of feedback for ecosystem stability.

Do invasive species benefit from escaping soil pathogens?

In a review of 473 species that were naturalized in the United States, Mitchell and Power (2003) found that 84% fewer fungi and 24% fewer virus species infected the plant species in their invaded ranges compared to their native ranges. Furthermore, they reported that species that experienced greater release from microbial pathogens were more invasive (Mitchell and Power 2003). However, the relationships were relatively weak, particularly for invaders of natural areas, soil pathogens were not distinguished from other pathogens, and the presence or abundance of pathogens does not necessarily correlate with the strength of their effects.

More recently, experiments using soils from native and invaded ranges have suggested that some invasive species have escaped from soil pathogens. If invasive species suffer from the effects of soil pathogens in their native soils, then sterilization of native soils should result in an increase in the growth of the invasive. In contrast, in invaded soils, the invasive should be relatively free from pathogens but may benefit from less host-specific mutualistic microbes. If so, sterilization of invaded soils should result in a neutral to negative effect on the invasive species. These interactions have been explored for *Prunus serotina* (black cherry), *Centaurea maculosa* (spotted knapweed) and two *Acer* (maple) species.

Reinhart *et al.* (2003) compared the effects of soil microbes on the growth of *Prunus serotina* in both its native and invaded ranges. In its native, North American range, the soil microbial community occurring near *P. serotina* strongly inhibited the establishment of neighboring conspecifics and reduced seedling performance in the greenhouse. In contrast, in its non-native European range, *P. serotina* readily establishes in close proximity to conspecifics, and soil microbial communities enhance the growth of seedlings. Previous research in the native range of *P. serotina* demonstrated that soil-borne *Pythium* species (Oomycota) inhibit the survival, growth, and abundance of *P. serotina* (Packer and Clay 2000, 2002). Although the genus *Pythium* is found around the world, genotypes are often host-specific (Deacon and Donaldson 1993; Mills and Bever 1998). Thus, in the native range, *P. serotina* experiences negative plant-soil feedback interactions, likely due to the negative effects of *Pythium*. In contrast, in the invaded region, *P. serotina* experiences positive feedbacks due to escape from its main natural enemy (Reinhart *et al.* 2003).

Centaurea maculosa is one of western North America's worst invasive weeds. In several experiments, Callaway *et al.* (2004) have compared the effects of soil microbes from the native range in Europe to the effects of soil microbes collected from invasive populations in the northwestern United States. European soil biota had much stronger inhibitory effects on *C. maculosa* than North American soil biota. Sterilization of European soils caused, on average, a 166% increase in the total biomass of *C. maculosa*, suggesting a release from pathogenic microbes. In contrast, sterilizing invaded North American soils led at most to a slight increase in total biomass of 24%. For most North American soils, sterilization led to a decrease in growth of 20-30%, suggesting that *C. maculosa* had benefited from mutualistic soil microbes. These results support Mitchell and Power's (2003) conclusion that invasive species should suffer much higher fungal and viral infection in their home ranges compared to invaded ranges. They also suggest that in some cases, mutualisms may be more beneficial in non-native ranges because the negative effect of natural enemies do not attenuate the positive effect of mutualists.

Mutualists have also been found to play an important role in the plant-soil feedback interactions of two *Acer* species (Reinhart and Callaway, 2004). In the field, distances between *Acer* conspecifics were 56-77% less in their invaded ranges than in their native ranges. In a greenhouse experiment, the effect of soil microbial communities also differed between native and invaded ranges. Relative to sterilized controls, soil associated with both conspecifics and heterospecifics from the native range decreased the total biomass of *Acer* seedlings by 35% suggesting inhibition by pathogenic microbes. In the invaded range, soil associated with conspecifics decreased the biomass of *Acer* seedlings by an even greater magnitude, 112%. However, soil associated with heterospecifics in the non-native ranges increased biomass of *Acer* seedlings by 13%. Thus, while *Acer*s accumulate pathogens in their invaded range, the surrounding soil is relatively free from inhibitory microbes, potentially enhancing invasion by these trees.

Thus, there is evidence that, not only do invasive species escape the negative effects of soil pathogens in their invaded ranges, but that potentially due to the effects of mutualists, feedback effects in invaded ranges are often positive. Next, we will explore the potential for these feedback effects to affect community stability.

Will escape from negative feedbacks from soil pathogens persist?

The experiments described above indicate that invasive species are likely to experience positive feedback in their invaded habitat because they escape specialist soil pathogens at home but can utilize generalist mutualists where they invade. What remains to be determined is how this feedback affects community dynamics. Plants participating in strong positive feedbacks with soil biota are more likely to become community dominants than those that do not. The most complete study of these interactions was done by Klironomos (2002), who explored feedback interactions among plant species and soil microbial communities in grasslands in eastern North America. In experiments using only the mycorrhizal fraction of the microbial community, he found that the origin of the filtrate (from soils in which the same species or a different species had previously been grown) did not alter the response (either positive or neutral) to mycorrhizal fungi. In contrast, in experiments using only the pathogenic/saprobic fractions, the rare native species experienced negative feedbacks when the fractions were from soils that had previously grown the same species. However, the origin of the pathogenic/saprobic fraction had no effect on the growth of invasive species. Overall, relatively rare native species consistently exhibited negative feedback interactions with the soil microbial community (a relative decrease in growth on 'home' soil in which conspecifics had previously been grown), whereas invasive species consistently exhibited positive feedback interactions with the soil community. Similarly Agrawal *et al.* (2005) found that introduced plants were subject to half the negative soil feedback as congeneric species. How long, in terms of evolutionary time scales, such positive interactions will be maintained remains an unknown.

The basic nature of microbes suggests that they will be able to respond relatively quickly to pressures exerted by invasion by exotic plants. As discussed above, soil-borne pathogens can be relatively host specific (Neergaard 1977; Kirkpatrick and Bazzaz 1979; Agarwal and Sinclair 1997; Mills and Bever 1998). However, many soil-borne pathogens are generalists (Dix and Webster 1995). For example, Blaney and Kotanen (2001) found that seed germination of 15 congeneric pairs of invasive and native plant species from western Ontario displayed a similar positive response to the application of fungicide, suggesting non-species specificity of fungal seed pathogens in their system. Furthermore, microbes have a short generation time and thus can respond to evolutionary pressures within a short timeframe. Thus, pathogenic microbes may be able to rapidly switch to a new invasive host. *Sclerotinia sclerotiorum*, a fungus native to intermountain prairies invaded by *C. maculosa*, has been found to damage *C. maculosa* when applied to the rhizospheres at high concentrations (Jacobs *et al.* 1996, Ridenour and Callaway 2003). As described above, Reinhart and Callaway (2004) found that while the soil community associated with other tree species had a positive effect on the growth of invasive *Acers*, soils associated with conspecifics had a negative effect. This suggests that the pathogenic soil microbial community may have been able to adapt to this new host, and accumulation of soil pathogens eventually suppresses the offspring of *Acer* recruits.

In summary, some invasive species appear to have escaped pressure from soil pathogens and thus benefit from positive feedback interactions with the soil biota where they invade. Although not yet explicitly addressed in the literature, microbial communities may change over time and thus break down positive plant-soil microbial feedbacks. If this occurs, the abundance of the invasive species should decrease, and the community should move to a point where negative feedback interactions restrict the invasive's dominance. This was originally suggested by

Klironomos (2002) who observed that plant-specific pathogen loads are maximized under high population densities, particularly monocultures such as those created by some invasive species, and will eventually result in negative feedback on abundant plants. Next we consider how feedbacks driven by microbes in nutrient cycles might respond differently over evolutionary time than feedbacks driven by soil pathogens.

INVASIVE PLANTS AND SOIL NUTRIENT CYCLING

In general, plant-soil feedbacks are thought to be determined by the direct effects of pathogens and mutualists (Bever 1994; Mills and Bever 1998; Packer and Clay 2000; Bever 2002; Klironomos 2002), but other components of the soil ecosystem may participate in feedbacks. In particular, individual plant characteristics, such as phenology, nutrient uptake, litter-fall, tissue chemical composition, and association with symbiotic microbes, can have significant effects on soil nutrient cycles (Hobbie 1992; Angers and Caron 1998; Berendse 1998; Binkley and Giardina 1998; Northrup *et al.* 1998; Schlesinger and Pilmanis 1998; Van Breemen 1993; Wardle *et al.* 1998; Chen and Stark 2000; Eaton and Farrell 2004) which may, in turn, alter the growth and survival of the species that drive these effects. Because they are novel, may have different biochemical constituents (Bais *et al.* 2003, Vivanco *et al.* 2004), and are often dominant components of plant communities, invasive plants can have unusually strong effects on soil nutrient cycles (Vitousek 1986; Vitousek *et al.* 1987; Vitousek 1990; D'Antonio and Vitousek 1992; Ehrenfeld *et al.* 2001; Ehrenfeld and Scott 2001; Ehrenfeld 2003).

In the previous section we showed that many invasive species exhibit positive feedbacks after escaping soil pathogens, and then speculated that these feedbacks may eventually become neutral or negative as generalist pathogens switch to the host or specialists adapt. In contrast to this scenario in which invasive dominance may fade, positive feedbacks between invasive plants and soil nutrient cycles may lead to much longer time periods of invasive dominance. Nutritional constraints may lead to significant shifts in microbial communities, resulting in long-term changes in nutrient pools and cycling rates. These changes may also occur due to the introduction of novel plant-microbe interactions, such as symbiotic nitrogen-fixation. Similarly, the direct effects of invasive plants on soil nutrients may be particularly long-lived because there is no mediation by another organism with the potential to evolve.

In this section, we briefly review the mechanisms by which invasive species may alter soil nutrient cycles and illustrate the potential for long-lived positive feedback interactions by describing the interactions of *Bromus tectorum* (cheatgrass, downy brome) and *Myrica faya* (fire tree) with soil nutrient cycles in invaded communities in the western United States and Hawaii.

The effects of invasive species on nutrient cycles

There are many mechanisms by which invasive species may alter soil nutrient cycles (see review by Ehrenfeld 2003). Through changes in litter production and quality, invasive plants may increase (Ehrenfeld *et al.* 2001; Mack *et al.* 2001) or decrease (Saggar *et al.* 1999; Ehrenfeld *et al.* 2001; Evans *et al.* 2001) microbially-mediated decomposition and/or mineralization rates. For example, litter of *Microstegium vimineum*, an exotic C4 grass that has invaded Eastern deciduous forests, has a higher C:N ratio, decomposes slower, and immobilizes more N than litter from uninvaded forests (Ehrenfeld *et al.* 2001).

Invasive species may also alter the input of nitrogen by nitrogen-fixing bacteria. Nearly 10% of the invasive species listed by the U.S. Department of Agriculture are in the Fabaceae family (Ehrenfeld 2003), and changes in ecosystem nitrogen availability due to association of invasive plants with symbiotic nitrogen-fixing bacteria have been documented in several ecosystems (Versfeld and van Wilgren 1986; Vitousek *et al.* 1987; Stock *et al.* 1995; Yelenik *et al.* 2004). Furthermore, changes in litter quality from non-nitrogen fixing invaders may alter the abundance and activity of non-symbiotic nitrogen-fixing bacteria, as found in Hawaiian forests invaded by African grasses (Ley and D'Antonio 1998).

Invasive species may affect soil nutrient cycles through the production of secondary chemicals. Roots of *Centaurea maculosa* exude the polyphenol, (±)-catechin. (+)-Catechin displays strong antimicrobial properties for at least some groups of bacteria (Bais *et al.* 2002, 2003) and appears to affect at least some aspects of the soil nitrogen cycle (A. Thorpe, unpublished data). Furthermore, by chelating metal-phosphorus complexes, catechin may increase phosphorus availability in phosphorus-limited soils (Thorpe *et al.* in press; Stevenson and Cole 1999). An allelochemical produced by *Centaurea diffusa*, 8-hydroxyquinoline, may also alter nutrient cycling through antimicrobial (Vivanco *et al.* 2004) and chelation (The Merck Index 1996) properties. The dry mass of leaves of *Melaleuca* spp. (paperbark), which has invaded large areas of the coastal southeast United States, particularly the Everglades, is up to 7% monoterpenes (Boon and Johnstone 1997). These compounds inhibit microbial colonization and decomposition of leaf litter in both the native and invaded ranges of *Melaleuca* spp. (Boone and Johnstone 1997). It has also been suggested that allelopathic chemicals released by some invasive species may alter nitrogen-fixation in neighboring plants (Wardle *et al.* 1994, 1995). Many other invasive species produce chemicals with antimicrobial activity (Rice 1964; Ehrenfeld 2003), however, the role of these chemicals in the plants' invasive success is generally unknown.

In sum, there is good evidence that by introducing a novel characteristic (e.g. a higher C:N ratio, association with nitrogen-fixing bacteria, or exudation of an anti-microbial chemical), invasive species can alter soil nutrient cycles in invaded communities. Although explicit studies of the ramifications of such alteration of nutrient cycles are rare, these effects may ultimately feedback to the plants that cause them and affect the organization of plant communities. Two different species, *Bromus tectorum* and *Myrica faya* provide excellent examples of (1) how invasive plants may affect the soil, and (2) how soil changes affect the survival of invasive and native species. These studies also illustrate how positive feedbacks between invasive plants and soil nutrient cycles may persist in an ecosystem.

The effects of *Bromus tectorum* on soil nutrient cycles

Bromus tectorum is an annual (occasionally biennial) Eurasian grass that has invaded over 40.5 million ha in the Intermountain West of North America (Ypsilantis 2003). The effects of *B. tectorum* on nutrient cycles differ in fire-prone and non-fire-prone systems.

Bromus tectorum tends to germinate and complete its life cycle earlier than most native species in the systems it invades, and its dead, dry stems create an unusually large fuel load in the summer (Harris 1967; Mack 1981; Upadhyaya *et al.* 1986; Ypsilantis 2003). In fire-prone sagebrush-grassland ecosystems, fire recurrence intervals decrease from 20-100 years to 3-5 years (Mack 1981; Upadhyaya *et al.* 1986; Ypsilantis 2003). Since *B. tectorum* germinates earlier and grows faster than most native species (Harris 1967; Mack 1981; Upadhyaya *et al.* 1986; Ypsilantis 2003), this invader appears to take better advantage of the post-fire flush of nitrogen than native species (Lowe *et al.* 2003). Early nitrogen uptake by *B. tectorum* reduces total soil nitrogen and creates higher soil carbon to nitrogen ratios than native vegetation (Blank *et al.* 1994; Halvorson *et al.* 1997). *Bromus tectorum* may also limit nitrogen availability by shading biological soil crusts that fix nitrogen (Ypsilantis 2003).

In ecosystems that lack fire, there are very different interactions between *B. tectorum* and the soil ecosystem. Grassland communities in Utah invaded by *B. tectorum* have higher levels of exchangeable potassium and ratios of potassium or phosphorus to calcium carbonate and magnesium or iron oxides than uninvaded soils (Belnap and Phillips 2001; Belnap *et al.* 2003). It is unknown whether these nutrient differences are due to *B. tectorum* invasion or if *B. tectorum* preferentially invades sites with these characteristics, however it is clear that *B.tectorum* can dramatically alter phosphorus cycling in invaded soils. Although there is no net change in total soil phosphorus pools, *B. tectorum* appears to access forms of P that are recalcitrant and unavailable to natives, which increases levels of labile phosphorus (R.L. Sanford, *personal communication*).

By altering the biotic and abiotic components of nutrient cycles, *B. tectorum* alters nutrient availability in ways that ultimately feedback to increase its own survival relative to native species (Belnap and Phillips 2001; Evans *et al.* 2001). By altering fundamental ecosystem characteristics, these effects may significantly alter plant community structure and dynamics.

The effects of *Myrica faya* on soil nutrient cycles

Invasion of Hawaiian ecosystems by the nitrogen fixing tree *Myrica faya* (Gerrish and Mueller-Dombois 1980; Vitousek 1986; Vitousek and Walker 1989; Aplet 1990; Hughes *et al.* 1991) provides an example of how ecosystem scale changes may result in "invasional meltdown" (Box 2; Simberloff and van Holle 1999). The volcanic soils of Hawaii are usually nitrogen-limited (Vitousek 1986; Vitousek and Walker 1989). There are no native nitrogen-fixing plants that colonize early successional habitats in these systems (Vitousek 1986; Vitousek and Walker 1989), and invasion by *M. faya* introduces a novel ecosystem process that results in substantial increases in soil nitrogen (Vitousek 1986; Vitousek and Walker 1989). Most successful invasions in Hawaii have occurred on soils that are relatively fertile (Gerrish and Mueller-Dombois 1980) and the nitrogen fertilization that results from invasion by *M. faya* may facilitate invasion by other species with higher nitrogen requirements. For example, another invasive tree, *Psidium cattleianum* (strawberry guava) grew much larger when grown in soil from under *M. faya* than from soil collected under the native *Metrosideros polymorpha* (`Ohi`a lehua; Hughes *et al.* 1991). *Myrica faya* also enhances populations of exotic earthworms, which increase nitrogen burial and further alter nutrient cycling (Aplet 1990). Thus, invasion by *M. faya* results in positive feedbacks that not only enhance the invasion of this species, but promote invasion by other invasive species and lead to fundamental ecosystem changes.

Box 2

In 1999, Simberloff and van Holle introduced the concept of "invasional meltdown". During this process, invasion by one exotic species promotes invasion by other exotic species. These species interact synergistically, causing ecosystem scale changes (such as more frequent fire cycles or more rapid nitrogen cycling through increased nitrogen fixation) that ultimately result in the collapse of native ecosystems.

Will positive feedbacks between invasive plants and soil nutrient cycles persist?

As described above, since invasive plants often differ from native species in characteristics such as phenology, nutrient uptake, litter-fall, tissue chemical composition, and association with symbiotic microbes, they can have significant effects on soil nutrient cycles. Similar effects are often seen during succession (Berendse 1998; Schimel and Bennett 2004). We speculate that, in contrast to our predictions for evolutionary change for interactions between invasive plants and soil pathogens, positive feedbacks between invasive plants and soil nutrient cycles are more likely to result in long term shifts in community composition that resemble succession. These changes may provide examples of invasional meltdown (Simberloff and Von Holle 1999).

We know of no examples where, during succession, characteristics of the soil ecosystem revert back to those resembling the initial characteristics after disturbance. What little is known about the biotic components of the soil ecosystem during succession indicates that change in the microbial community is towards populations that are better adapted to the new nutrient conditions of the later successional stages (Schimel *et al.* 1998; Schimel and Bennett 2004). These successional shifts appear to be ecological in nature, not evolutionary, and driven by shifts in the species composition of communities over time. In contrast, in the case of alterations to abiotic components of the soil ecosystem, there is no direct selective pressure to stimulate adaptation. Ultimately, these types of changes require that the soil and plant communities may have to respond adaptively to new nutrient conditions.

When invader-soil ecosystem feedbacks occur through effects on nutrient cycles, we suggest that long term effects will be similar to those that occur during succession, the development of a new community composed of species that are better adapted to the specific nutrient conditions of the soil (e.g. Berendse 1998). This has been observed in systems invaded by both native (Maron and Jefferies 1999) and invasive (Hughes *et al.* 1991) nitrogen-fixing species. In California, Maron and Connors (1996) found that nitrogen-rich patches that were left by death of *Lupinus arboreous* (bush lupine) were invaded by exotic annual grasses. Similarly, as described in the preceding section, soils that have been enriched by the exotic nitrogen-fixer *M. faya* are more prone to invasion by other exotic plants, including *Psidium cattleianum* (strawberry guava; Hughes *et al.* 1991). In addition, although only a few native species have been found to "monopolize" sites by creating positive feedbacks (van der Putten 1997), invasive species may be particularly capable of this, especially if they introduce a novel ecosystem process to which native species are not adapted. For example, although *B. tectorum* first became established in fire-prone communities in the Great Basin over 100 years ago (Mack 1981), no other species, either native or invasive, have since been able to establish in substantial numbers. These examples show how ecosystem changes involving exotic species can result in unpredictable successional trajectories.

CONCLUSION

Exotic plant invasions can often involve positive feedbacks between the invader and the native soil community. Positive feedbacks have the potential to lead to long-lasting dominance in communities. However, these feedbacks can be caused by a number of different reasons, and different drivers of positive feedbacks may result in different predictions of how long invaders may dominate a community.

Typically, positive feedbacks are thought to be caused by the absence of soil pathogens but the presence of soil mutualists in invaded soils. Such pathogen-driven positive feedbacks may result in long term invasive dominance, but evolutionary changes in native pathogens might ultimately lead to suppression of the invader and a return to native coexistence. Positive feedbacks can also be caused by the effects of invasive plants on the soil biota that drive nutrient cycles or on the abiotic components of the nutrient cycles themselves. We hypothesize that positive feedbacks caused by the effects of the invader on the soil biota involved in nutrient cycles are likely to lead to community shifts resembling succession and perhaps invasional meltdown. Dominance by invaders that drive positive feedbacks through abiotic components of nutrient cycles may persist for much longer than dominance occurring through positive feedbacks involving the microbial community due to the lack of mediation by an organism with the potential to adapt. However, understanding potential evolutionary changes among invaders and soil ecosystems will help us to accurately predict the long-term effects of biological invasions.

ACKNOWLEDGEMENTS

The authors thank the NSF, USDA-NRI, and the DOD for financial support.

REFERENCES

Aplet, G. H. 1990. Alternation of earthworm community biomass by the alien *Myrica faya* in Hawaii. Oecologia **82**, 411-416.

Agarwal, V. K. and J. B. Sinclair. 1997. Principles of Seed Pathology, Second edition. Lewis Publishers, Boca Raton.

Agrawal, A. A., P. M. Kotanen, C. E. Mitchell, A. G. Power, W. Godsoe and J. Klironomos. Enemy Release? 2005. An experiment with congeneric plant pairs and diverse above- and below-ground enemies. Ecology **86**, 2979-2989.

Angers, D. A. and J. Caron. 1998. Plant-induced changes in soil structure: processes and feedbacks. Biogeochemistry **42**, 55-72.

Atlas, R. M. and R. Bartha. 1998. Microbial Ecology: Fundamentals and Applications. Benjamin Cummings, New York.

Bais, H. P., R. Vepachedu, S. Gilroy, R. M. Callaway and J. M. Vivanco. 2003. Allelopathy and exotic plants: from genes to invasion. Science **301**, 1377-1380.

Bais, H. P,, T. S. Walker, F. R. Stermitz, R. A. Hufbauer and J. M. Vivanco. 2002. Enantiomeric-dependent phytotoxic and antimicrobial activity of (±-catechin. A rhizosecreted racemic mixture from spotted knapweed. Plant Physiology **128**, 1173-1179.

Belnap, J. and S. L. Phillips. 2001. Soil biota in an ungrazed grassland: response to annual grass (*Bromus tectorum*) invasion. Ecological Applications **11**, 1260-1275.

Belnap, J., S. K. Sherrod, M. E. Miller. 2003. Effects of soil amendments on germination and emergence of downy brome (*Bromus tectorum*) and Hilaria jamesii. Weed Science. **51**, 371-378.

Berendse, F. 1998. Effects of dominant plant species on soils during succession in nutrient-poor ecosystems. Biogeochemistry **42**, 73-88.

Bever, J. D. 1994. Feedback between plants and their soil communities in an old field community. Ecology **75**, 1965-1977.

Bever, J. D., J. B. Morton, J. Antonovics and P. A. Schultz. 1996. Host-dependent sporulation and species diversity of arbusuclar myorrhizal fungi in a mown grassland. Journal of Ecology **84**, 71-82.

Bever, J. D., K. M. Westover and J. Antonovics. 1997. Incorporating the soil community into plant population dynamics: the utility of the feedback approach. Journal of Ecology **85**, 561-573.

Bever, J. D. 2002. Negative feedback within a mutualism: host-specific growth of mycorrhizal fungi reduces plant benefit. Proceedings Royal Society Lund B **269**, 2595-2601.

Binkley, D. and C. Giardina. 1998. Why do tree species affect soils? The warp and woof of tree-soil interactions. Biogeochemistry **42**, 89-106.

Blaney, C. S. and P. M. Kotanen. 2001. Effects of fungal pathogens on seeds of native and exotic plants: a test using congeneric pairs. Journal of Applied Ecology **38**, 1104-1113.

Blank, R. R., F. Allen and J. A. Young. 1994. Extractable anions in soils following wildfire in a sagebrush-grass community. Soil Science Society of America Journal **58**, 564-570.

Boon, P. I. and L. Johnstone. 1997. Organic matter decay in coastal wetlands: an inhibitory role for essential oil from *Melaleuca alternifolia* leaves? Archiv fur hydrobiolgie **138**, 428-449.

Callaway R. M., G. C. Thelen, S. Barth, P. W. Ramsey and J. E. Gannon. 2004. Soil fungi alter interactions between North American plant species and the exotic invader *Centaurea maculosa* in the field. Ecology.

Callaway, R. M., G. C. Thelen, A. Rodriguez and W. E. Holben. 2004. Release from inhibitory soil biota in Europe may promote exotic plant invasion in North America. Nature. **427**, 731-733.

Chen, J. and J. M. Stark. 2000. Plant species effects and carbon and nitrogen cycling in a sagebrush-crested wheatgrass soil. Soil Biology and Biochemistry **32**, 47-57.

Chiarello, N., J. C. Hickman and H. A. Mooney. 1982. Endomycorrhizal role in interspecific transfer of phosphorus in a community of annual plants. Science **217**, 941-943.

Clay, K. and W. H. Van der Putten. 1999. Pathogens and plant life-histories. Pages 275-301 in T. O. Vuorisalo and P. K. Mutikainen, editors. Life History in Plants. Kluwer Academic Publishers, Dordrecht.

D'Antonio, C. M and P. M. Vitousek. 1992. Biological invasion by exotic grasses, the grass/fire cycle and global change. Annual Review of Ecology and Systematics **23**, 63-87.

Deacon, J. W. and Donaldson, S. P. 1993. Molecular recognition in the homing responses of zoosporic fungi, with special reference to *Pythium* and *Phytophthora*. Mycological Research **97**, 1153-1171.

Dix, N. and J. Webster. 1995. Fungal Ecology. Chapman and Hall, London, United Kingdom.

Eaton, W. D. and R. E. Farrell. 2004. Catabolic and genetic microbial indices, and levels of nitrate, ammonium and organic carbon in soil from the black locust (*Robinia pseudo-acacia*) and tulip poplar (*Liriodendron tulipifera*) trees in a Pennsylvania forest. Biological Fertility of Soils. **39**, 209-214.

Ehrenfeld, J. G. 2003. Effects of exotic plant invasions on soil nutrient cycling processes. Ecosystems **6**, 503-523.

Ehrenfeld, J. G., P. Kourtev and W. Huang. 2001. Changes in soil functions following invasions of exotic understory plants in deciduous forests. Ecological Applications **11**, 1287-1300.

Ehrenfeld, J. G. and N. Scott. 2001. Invasive Species and the Soil: Effects on Organisms and Ecosystem Processes. Ecological Applications **11**, 1259-1260.

Evans, R. D., R. Rimer, L. Sperry and J. Belnap. 2001. Exotic plant invasion alters nitogen dynamics in an arid grassland. Ecological Applications **11**, 1301-1310.

Francis, R. and D. J. Read. 1984. Direct transfer of carbon between plants connected by vesicular-arbuscular mycorrhizal mycelium. Nature **307**, 53-56.

Gerrish, G. and D. Mueller-Dombios. 1980. Behavior of native and non-native plants in two tropical rainforests on Oahu, Hawaiian Islands. Phytocoenologia **8**, 237-295.

Grime, J. P., J. M. L. Mackey, S. H. Hllier and D. J. Read. 1987. Floristic diversity in a model system using experimental microcosms. Nature **328**, 420-422.

Halvorson, J. J., H. Bolton and J. L. Smith. 1997. The pattern of soil variables related to *Artemisia tridentata* in a burned shrub-steppe site. Soil Science Society of America Journal **62**, 287-294.

Harris, G. A. 1967. Some competitive relationships between *Agropryon spicatum* and *Bromus tectorum*. Ecological Monographs **37**, 89-111.

Hartnett, D. C., B. A. D. Hetrick, G. W. T. Wilson and D. J. Gibson. 1993. Mycorrhizal influence on intra- and interspecific neighbour interactions among co-occurring prairie grasses Journal of Ecology **81**, 787-795.

Hetrick, B. A. D., G. W. T. Wilson and D. C. Hartnett. 1989. Relationship between mycorrhizal dependence and competitive ability of two tallgrass prairie grasses. Canadian Journal of Botany **67**, 2608-2615.

Hobbie, S. E. 1992. Effects of plant species on nutrient cycling. Trends in Ecology and Evolution **7**, 336-339.

Hooper, D. U., D. E. Bignell and V. K. Brown. 2000. Interactions between aboveground and belowground biodiversity in terrestrial ecosystems: patterns, mechanisms, and feedbacks. BioScience **50**, 1049-1061.

Hughes, R., P. M. Vitousek and T. Tunison. 1991. Alien grass invasion and fire in the seasonal submontane zone of Hawai'i. Ecology **72**, 743-746.

Jacobs, J. S., R. L. Sheley and B. D. Maxwell. 1996. Effect of *Sclerotina sclerotiorum* on the interference between bluebunch wheatgrass (*Agropyron spicatum*) and spotted knapweed (*Cenatuarea maculosa*). Weed Technology **10**, 13-21.

Keane, R. M. and M. J. Crawley. 2002. Exotic plant invasions and the enemy release hypothesis. Trends in Ecology and Evolution **17**, 164-170.

Kirkpatrick, B. L., and F. A. Bazzaz. 1979. Influence of certain fungi on seed germination and seedling survival of four colonizing annuals. Journal of Applied Ecology **16**, 515-527.

Klironomos, J. 2002. Feedback with soil biota contributes to plant rarity and invasiveness in communities. Nature **417**, 67-70.

Kourtev, P. S., J. G. Ehrenfeld and W. Huang. 2002. Exotic species alter microbial structure and function in the soil. Ecology **85**, 3152-3166.

Ley, R. E. and C. M. D'Antonio. 1998. Exotic grass invasion alters potential rates of N fixation in Hawaiian woodlands. Oecologia **113**, 179-187.

Lowe, P. N., W. K. Lauenroth and I. C. Burke. 2003. Effects of nitrogen availability on competition between *Bromus tectorum* and *Bouteloua gracilis*. Plant Ecology **167**, 247-254.

Mack, R. N. 1981. Invasion of *Bromus tectorum* L. into western North America: an ecological chronicle. Agro-Ecosystems **7**, 145-165.

Mack, M. C., C. M. D'Antonio and R. E. Ley. 2001. Alteration of ecosystem nitrogen dynamics by exotic plants: a case study of C_4 grasses in Hawaii. Ecological Applications **11**, 1323-1335.

Marler, M. J., C. A. Zabinski and R. M. Callaway. 1999. Mycorrhizae indirectly enhance competitive effects of an invasive forb on a native bunchgrass. Ecology **80**, 1180-1186.

Maron, J. L., and P. G. Connors. 1996. A native nitrogen-fixing shrub facilitates weed invasion. Oecologia **105**, 302-312.

Maron, J. L. and R. L. Jefferies. 1999. Bush lupine mortality, altered resource availability, and alternative vegetation states. Ecology **80**, 443-454.

Merck Index. 1996. 8-Hydroxyquinoline. Merck and Co., Inc., Whitehouse, NJ. Page 4890.

Mills, K. E. and J. D. Bever. 1998. Maintenance of diversity within plant communities: soil pathogens as agents of negative feedback. Ecology **79**, 1595-1601.

Mitchell, C. G. and A. G. Power. 2003. Release of invasive plants from fungal and viral pathogens. Nature **421**, 625-627.

Moora, M. and M. Zobel. 1996. Effect of arbuscular mycorrhiza on inter- and intraspecific competition of two grassland species. Oecologia **108**, 79-84.

Neergaard, P. 1997. Seed Pathology. John Wiley, New York, New York, U.S.A. 839 p.

Northup, R. R., R. A. Dahlgren and J. G. McColl. 1998. Polyphenols as regulators of plant-litter-soil interactions in northern California's pygmy forest: A positive feedback? Biogeochemistry **42**, 189-220.

Packer, A. and K. Clay. 2000. Soil pathogens and spatial patterns of seedling mortality in a temperate tree. Nature **440**, 278-281.

Packer, A. and K. Clay. 2002. Soil pathogens and *Prunus serotina* seedlings and sapling growth near conspecific trees. Ecology **84**, 108-199.

Reinhart, K. O., A. Packer, W. H. Van der Putten and K. Clay. 2003. Plant–soil biota interactions and spatial distribution of black cherry in its native and invasive ranges. Ecology Letters **6**, 1046-1050.

Reinhart, K. O. and R. M. Callaway. 2004. Regional effects of soil biota facilitate exotic *Acer* invasions in Europe and North America. Ecological Applications **14**, 1737-1745.

Rice, E. L. 1964. Inhibition of nitrogen-fixing and nitrifying bacteria by seed plants. Ecology **45**, 824-837.

Rice, K. J. and N. C. Emory. 2003. Managing microevolution: restoration in the face of global change. Frontiers in Ecology **1**, 469-478.

Ridenour, W. L. and Callaway. R. M. 2003. Root herbivores, pathogenic fungi and competition between *Centaurea maculosa* and *Festuca idahoensis*. Plant and Soil **169**, 161-170.

Robinson, D. and A. Fitter. 1999. The magnitude and control of carbon transfer between plants linked by a common mycorrhizal network. Journal of Experimental Botany **50**, 9-13.

Saggar, S., P. D. McIntosh, C. B. Hedley and H. Knicker. 1999. Changes in soil microbal biomass, metabolic quotient, and organic matter turnover under *Hieracium* (H. *pilosella* L.). Biological Fertility of Soils **30**, 232-238.

Schimel, J. P. and J. Bennett. 2004. Nitrogen mineralization: challenges of a changing paradigm. Ecology **85**, 591-602.

Schimel, J. P., R. G. Cates and R. Ruess. 1998. The role of balsam popular secondary chemicals in controlling soil nutrient dynamics through succession in the Alaskan taiga. Biogeochemistry **42**, 221-234.

Schlesinger, W. H. and A. M. Pilmanis. 1998. Plant-soil interactions in deserts. Biogeochemistry **42**, 169-187.

Simard, S. W., D. A. Perry and M. D. Jones. 1997. Net transfer of carbon between ectomycorrhizal tree species in the field. Nature **388**, 579-582.

Simberloff, D. and B. Von Holle. 1999. Positive interactions of nonindigenous species: invasional meltdown? Biological invasions **1**, 21-32.

Stevenson, F. J. and M. A. Cole. 1999. Cycles of soil: carbon, nitrogen, phosphorus, sulfur, micronutrients. John Wiley and Sons, Inc., New York NY.

Stock, W. D., K. T. Wienand and A. C. Baker. 1995. Impacts of invading N_2-fixing *Acacia* species on patterns of nutrient cycling in two Cape ecosystems: evidence from soil incubation studies and ^{15}N natural abundance values. Oecologia **101**, 375-382.

Thorpe, A. S., V. Archer and T. H. Deluca. In Press. The invasive forb, *Centaurea maculosa*, increases phosphorus availability in Montana grasslands. Journal of Applied Soil Ecology.

Upadhyaya, J. K., R. Turkington and D. McIluride. 1986. The biology of Canadian weeds. 75. *Bromus tectorum* L. Canadian Journal of Plant Science **66**, 689-709.

Van Breemen, N. 1993. Soils as biotic constructs favouring net primary productivity. Geoderma **57**, 270-275.

Van Breemen, N. and A. C. Finzi. 1998. Plant-soil interactions: ecological aspects and evolutionary implications. Biogeochemistry **42**, 1-19.

van der Heijden, M. G. A. J. N. Klironomos, M. Ursic, P. Moutoglis, R. Streitwolf-Engel, T. Boller, A. Wiemken and I. R. Sanders. 1998. Mycorrhizal fungal diversity determines plant biodiversity, ecosystem variability and productivity. Nature **396**, 69-72.

Van der Putten, W. H. and B. A. M. Peters. 1997. How soil-borne pathogens may affect plant competition. Ecology **78**, 1785-1795.

Van der Putten, W. H., C. Van Dijk and B. A. M. Peters. 1993. Plant-specific soil-borne diseases contribute to succession in foredune vegetation. Nature **362**, 53-56.

Van der Putten, W. H. 1997. Plant - soil feedback as a selective force. Trends in Ecology and Evolution **12**, 169-170.

Versfeld, D. B. and B. S. van Wilgren. 1986. Impact of woody aliens on ecosystem properties. Pages 239-246 in I. A. W. Macdonald, F. J. Kruger and A. A. Ferrar, edictors. The ecology and management of biological invasions in southern Africa. Oxford University Press, Cape Town, South Africa.

Vitousek, P. M. 1986. Biological invasions and ecosystem properties: can species make a difference? Pages 163-178 in G. A. Mooney and J. A. Drake, editors. Ecology of Biological Invasions of North America and Hawaii. Springer-Verlag, New York, NY.

Vitousek, P. M. 1990. Biological invasions and ecosystem processes: towards an integration of population biology and ecosystem studies. Oikos **57**, 7-13.

Vitousek, P. M. and L. R. Walker. 1989. Biological invasion by *Myrica faya* in Hawaii : plant demography, nitrogen fixation, ecosystem effects. Ecological Monographs **59**, 247-65.

Vitousek, P. M., L. R. Walker, L. D. Whiteaker, D. Mueller-Dombois and P. A. Matson. 1987. Biological invasion by *Myrica faya* alters ecosystem development in Hawaii. Science **238**, 802-804.

Vivanco, J. M., H. P. Bais, F. R. Stermitz, G. C. Thelen and R. M. Callaway. 2004. Biogeographical variation in community response to root allelechemistry: novel weapons and exotic invasion. Ecology Letters **7**, 285-292.

Yelenik, S. G., W. D. Stock and D. M. Richardson. 2004. Ecosystem level impacts of invasive *Acacia saligna* in the South African fynbos. Restoration Ecology **12**, 44-51.

Ypsilantis, W. 2003. Risk of cheatgrass invasion after fire in selected sagebrush community types. US Bureau of Land Management.

Wardle, D. A., K. S. Nicholson and A. Rahman. 1995. Ecologicla effects of the invasive weed species *Senecio jacobaea* L. (ragwort) in a New Zealand pasture. Agriculture, Ecosystems, and Environment **56**, 19-28.

Wardle, D. A. and K. S. Nicholson. 1996. Synergistic effects of grassland plant species on soil microbial biomass and activity: implications for the ecosystem level effects of enriched plant diversity. Functional Ecology **10**, 410-416.

Wardle, D. A., K. S. Nicholson, M. Ahmed and A. Rahman. 1994. Interference effects of the invasive plant *Carduus nutans* L. against the nitrogen fixation ability of *Trifolium repens*/*L.* Plant and Soil **163**, 287-297.

Wardle, D. A., M. Nilsson, C. Gallet and O. Zackisson. 1998. An ecosystem-level perspective of allelopathy. Biological Review **73**, 305-319.

Watkins, N. K., A. H. Fitter, J. D. Graves. 1996. Carbon transfer between C_3 and C_4 plants linked by a common mycorrhizal network, quantified using stable carbon isotopes. Soil Biology and Biochemistry **28**, 471-477.

Watkinson A. R. 1998. The role of the soil community in plant population dynamics. Trends in Ecology and Evolution **13**, 171-2.

West, H. M. 1996. Influence of arbuscular mycorrhizal infection on competition between *Holcus lanatus* and *Dactylis glomerata*. Journal of Ecology **84**, 429-438.

Westover, K. M., A. C. Kennedy and S. E. Kelley. 1997. Patterns of rhizosphere microbial community structure associated with co-occurring plant species. Journal of Ecology **85**, 863-873.

Chapter sixteen

Invasion biology as a community process: messages from microbial microcosms

P. H. Warren, R. Law and A. J. Weatherby

INTRODUCTION

Invasions are widely regarded as a significant current problem in population, ecosystem and habitat management, with consequences including global homogenization of floras and faunas, extirpation of native species and interference with ecosystem functioning (Drake *et al.* 1989, Williamson 1996, Mack *et al.* 2000, Pimentel *et al.* 2000, Sakai *et al.* 2001). Such invasions occur, and have effects, at large scales, in real ecosystems. So why consider investigating such phenomena in artificial microcosm systems in the laboratory? We believe that there are several compelling reasons why laboratory microcosm systems have a valuable role in the study of biological invasions; these centre on the issues of manipulation, time scale and replication.

First, applied ecology seeks a predictive understanding of invasions: what determines which species will make successful invaders? Are some types of communities invasion prone? How will an invader impact on a community? There are various theoretical expectations, but testing these in the field is largely dependent on extracting statistical signals from documentation of the 'natural experiments' that real invasions provide. Opportunities for controlled experimentation are

M.W. Cadotte et al. (eds.), Conceptual ecology and invasion biology, 343–367.

limited and it is notoriously difficult to document systematically invasion attempts that fail (Simberloff 1989) — the necessary 'control' for evaluation of the determinants of invasion success. Microcosms, mesocosms and related techniques, such as enclosure experiments, provide the possibility of manipulating potential influences on invasion success, albeit at the expense of realism in complexity and scaling.

Second, invasions take time. The population dynamics of an invading species, its geographic spread, and the adjustment of the receiving communities are all typically expressed over tens to hundreds of generations, usually taking many years. However desirable, the exploration of such effects in invasions currently happening in the field may be impractical. Microcosm systems utilizing microorganisms, allow 'long term' effects and outcomes, in the range of tens to hundreds of generations, to be documented (Drake *et al.* 1996, Lawler 1998).

Third, as the wealth of case studies in the invasions literature testifies, invasions are rather individualistic events — a particular invader, invading a particular system, or region, under particular circumstances. This makes the appraisal of general patterns and causes difficult. In particular, if we are interested in prediction, or at least predictability, then it becomes important to know the probability of a particular outcome for what is, after all, a process likely to be strongly influenced by stochastic events. In experimental systems, replication of communities and invasion events makes it possible to assess the variation in the outcome of invasions, and the source of that variation.

A beaker in the laboratory is far from a field experiment in a natural ecosystem. Spatial and temporal scales, the types of organisms involved and the simplicity of the communities mean that microcosms cannot simply be taken as analogues of large natural ecosystems; the same is true for other sorts of microcosm and mesocosm experiments (Petersen and Hastings 2001). A microbial microcosm is better thought of as a small system in its own right and with its own peculiarities, rather than as a large system writ small. In a sense, deciding the extent to which results from microcosms can be applied to other types of system is equivalent to deciding the extent to which results from any one type of community can be applied to other disparate systems, for instance the extent to which results from phytoplankton communities can be applied to forests. Microcosms of protists certainly tell us something about the behaviour of microbial communities, but leave open open the question of whether such behaviour generalizes; the latter is also true for experiments on forests and phytoplankton.

Microcosms cannot, of course, replace large scale field experiments on specific ecoystems. Only through such experiments can some of the complexity and specific features of the system of interest be captured. However, large-scale field experiments are often concerned with testing general principles such as the roles of species richness, or species type (e.g., Wardle 2001, Hector *et al.* 2001), and the effects of fluctuation or heterogeneity in the abiotic enviroment (e.g., Davis and Pelsor 2001). Questions about how to generalize to other kinds of system apply as much in these field experiments as in microcosms. Many important ecological principles can be tested in microcosms as well as in field experiments (e.g., Petchey *et al.* 1999), and we take the view that micrososm experiments should complement these other approaches, rather than being alternatives to them.

Arguably therefore, laboratory microcosms have a useful role to play in investigating ecological principles of invasion, just as they have provided valuable investigative tools in other areas of community ecology (Lawler 1998). In fact, many of the processes hypothesized to be determinants of invasion success or impact are just those of 'natural' community change and regulation (population growth, species interactions, dispersal, trophic position, assembly, etc.) (Diamond and Case 1986, Shea and Chesson 2002). It is in the study of these general processes, rather than in the study of factors more specific to the biogeography of invasion (for example the loss of coevolved natural enemies (Torchin *et al.* 2003, Mitchell and Power 2003)) that laboratory microcosms are most likely to generate useful insights.

Here we examine various components of invasion biology and some of the insights laboratory microcosms have given us into these. We focus on microcosms of heterotrophic microbes, which we know best from our own work, and which have a long heritage in diverse areas of community ecology (e.g., see review in Lawler 1998). Our reason for this is that such systems exploit all the advantages of microcosms listed above, but it does mean that we do not cover other types of experimental invasions, such as those of plants into communities in pots, or similar experimental systems (e.g. Dukes 2001, Stampe and Daehler 2003, Dick *et al.* 2002), which are sometimes termed microcosms, valuable though such experiments are for investigating the details and mechanisms of particular invasions.

AN INVASION EXPERIMENT

In developing the ideas below, we draw substantially on information from microcosm experiments on invasibility which we conducted as part of a study of community assembly (Weatherby *et al.* 1998; Law *et al.* 2000; Warren *et al.* 2003). There are a number of results that emerge from this work which we feel are both pertinent to questions about invasions in general, and also illustrative of some of the areas in which microcosms systems can contribute uniquely to understanding invasion processes. Full details of the experimental methods can be found in the papers above, but a brief outline of the experiments is given here, for convenience.

We worked with a species pool of six heterotrophic protists (bacteriovores: *Tetrahymena pyriformis, Colpidium striatum, Paramecium caudatum*; an omnivore: *Blepharisma japonicum*; predators: *Euplotes patella, Amoeba proteus*) plus a mixed bacterial assemblage, in microcosms containing 70 mL of liquid medium. In an initial experiment (Weatherby *et al.* 1998), all possible combinations of the species were tested for their ability to coexist; of the 2^6 possible combinations of species, ten formed persistent communities (on a time scale of tens to hundreds of generations). We then took each of these persistent communities and challenged them experimentally with each non-resident species. Six replicates of each combination of community and introduced species were used starting from a pre-established resident community (set up for 6 weeks) and challenging it with an inoculum of the new species consisting of thirty individuals. The initial rate of increase of the invader, its long term persistence (up to 12 weeks after invasion), and the changes in community composition, were all recorded.

FACTORS AFFECTING ESTABLISHMENT OF INVADERS

Establishment as a probabilistic event

Chance plays an important part in whether a new species gets established locally (Crawley 1989). Even if a new species has vital rates that enable it to increase on the average, random events can catch up with it before it is able to get started. Such events include demographic stochasticity, environmental stochasticity, and the exact state of the resident community at the time when the new species is introduced.

It therefore helps to think of establishment as a probabilistic event. Theory from stochastic birth-death processes provides a quantitative framework on which to build ideas, at least so far as demographic stochasticity is concerned (Bailey 1964: 92-94, Goel and Richter-Dyn 1974:79). This theory considers a population in which individuals have a constant probability per unit time of giving birth b and of dying d, with an assumption that $b > d$. If the population starts with an inoculum of n independent individuals, it has a probability

$$P(n, N) = \frac{1 - (d/b)^n}{1 - (d/b)^N}$$

of reaching size N ($> n$) before it goes to extinction. For a species introduced into a well-mixed community close to steady state and at an abundance much lower than the resident species, b and d may be taken as approximately constant during the early stages of population growth, because most interactions are with the resident species. N can be set at a value above the level at which demographic stochasticity is important, to give an expression for the probability that a new species becomes successfully established

$$P(n) = 1 - (d/b)^n = 1 - (1 - r/b)^n \qquad (1)$$

where $r = b - d$ is the initial per capita rate of increase of the new species and $b > d$. ($P(n) = 0$, if $b < d$). Plotting this function shows, as one would expect, the species is more likely to become established the greater the inoculum size

n and the greater the per capita rate of increase r relative to the birth rate b, or equivalently the smaller the death rate d relative to the birth rate b (Fig. 1). (see Chapter 5 by Freckleton *et al.* for a broader discussion of the role of stochasticity in invasions.)

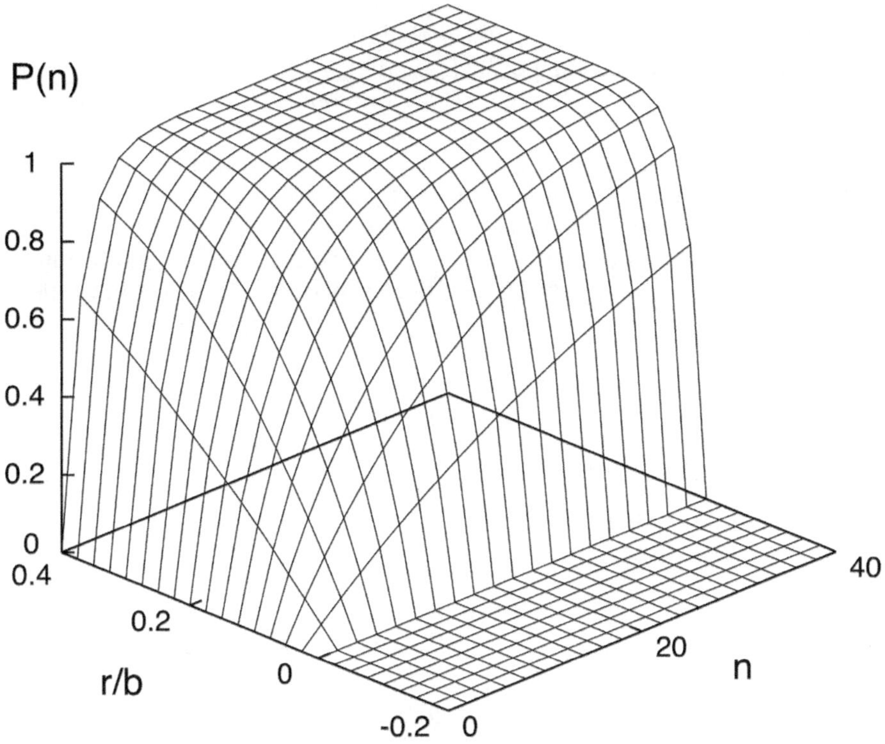

Fig. 1 Probability $P(n)$ that an introduced, asexual species reaches a size above a level at which demographic stochasticity is important starting with n propagules and having constant per capita birth b and death d rates with $r = b - d$. The surface is constructed from Equation 1.

Although the role that chance plays in establishment of new species is rarely a subject of direct experimentation, we found it played an important part in the replicate invasion trials we carried out during our work on community assembly of protists (Law *et al.* 2000). In this work, we took a species to have become established if it reached a threshold of 40 individuals at some stage during the course of the experiment. However, in about 25% of the resident-introduction treatments, the outcome remained uncertain in the sense that the introduced species became established in some replicates while disappearing in others (Fig. 2). Demographic stochasticity most likely played a part in this, although this does not preclude involvement of other factors as well. For instance, species that remain at low population sizes for long periods of time are also vulnerable to the continuing effects of environmental stochasticity, as in two of our microcosms where neither extinction nor the threshold for establishment had been reached by the time the experiment ended 12 weeks after inoculation. Also, the exact abundance of the resident species at the time of introduction, which we could not control for in our invasion trials, could influence the outcome. Such effects of abundance are suggested by the fact that the outcome of invasions was never in doubt in our microcosms lacking resident protists, but was uncertain in about a third of the treatments containing resident protists. Drake (1991), in microcosms of algae and grazing microcrustacea, suggested a possibly similar effect, where small variations in initial demography of introduced grazers among replicates caused marked variation in grazer population success in some communities, but not in others.

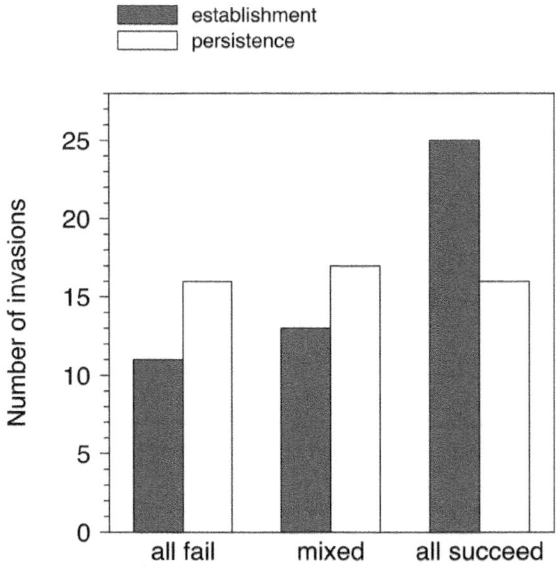

Fig. 2 Uncertainty in the outcome of resident-introduction treatments in a protist invasion experiment described in Law *et al.* (2000). Treatments are shown in three categories according to whether the introduced species failed to establish/persist in all replicates, succeeded in establishing/persisting in all replicates, or successfully invaded some, but failed in others (mixed). Bars show the number of treatments falling into each category. Shaded bars: establishment (reaching a threshold of 40 individuals at some point in the experiment); open bars: persistence (still present in the community after 12 weeks). Most treatments had 6 replicates though a few only had 4, or 5.

Propagule pressure and the probability of establishment

There is a general understanding that establishment of a new species is more likely the more individuals introduced at a time (Williamson 1996, Kola and Lodge 2001). It is intuitive, for instance, that demographic stochasticity should have its greatest effect on establishment of a species when just a few individuals are introduced because a sequence of relatively few deaths can take the new population to extinction. Although rather little is known about propagule pressure, this may well contribute in important ways to patterns of invasions observed in nature (Williamson 1996:46 *et seq.*; Lonsdale 1999). Among the most systematic investigations have been the release of biological control agents (e.g., Hopper and Roush 1993), where there is particular concern to ensure the control agents become successfully established (Shea and Possingham 2000). Equation 1 above provides a clear quantitative prediction about the effect of inoculum size n on the probability of successful establishment.

The quantitative dependence of invasion success on propagule pressure has received little systematic experimental investigation. Well-mixed microcosms, containing species with short generation times, would be a good point at which to begin examining this relationship.

Initial rate of increase and the probablity of establishment

One would expect that the larger the per capita rate of increase of a rare introduced species, the greater the likelihood that it becomes established. Apart from human pathogens (e.g., Anderson and May 1991:70, 87), rather little information is available on this initial rate of increase. Yet, it is intuitive that a high rate of increase should allow early escape from the region in which demographic stochasticity could drive the introduced species to extinction. Equation 1 makes this intuition precise, showing that what actually matters is the ratio of the death rate to birth rate, the ratio leading to a dimensionless expression independent of the time scale on which the dynamics are operating.

Microcosms, because of their potential for high levels of replication, provide scope for studying initial rates of increase and the probability of establishment. Our studies of protist microcosms (Law *et al.* 2000) illustrate the potential to escape from stochastic extinction through rapid population growth (Fig. 3). The graphs show the growth of *Blepharisma* populations, each starting with an inoculum of 30 individuals, introduced into four different types of microcosm: those containing *Paramecium*, *Colpidium* or *Tetrahymena*, and those with no other protists. When faced with *Paramecium*, most *Blepharisma* populations declined rapidly; in fact, two were extinct by day six. With *Colpidum*, *Blepharisma* teetered along at small population sizes; during this time, there would be a continuing risk of extinction. In the presence of *Tetrahymena*, *Blepharisma* populations grew much faster and soon got beyond a size at which extinction through demographic

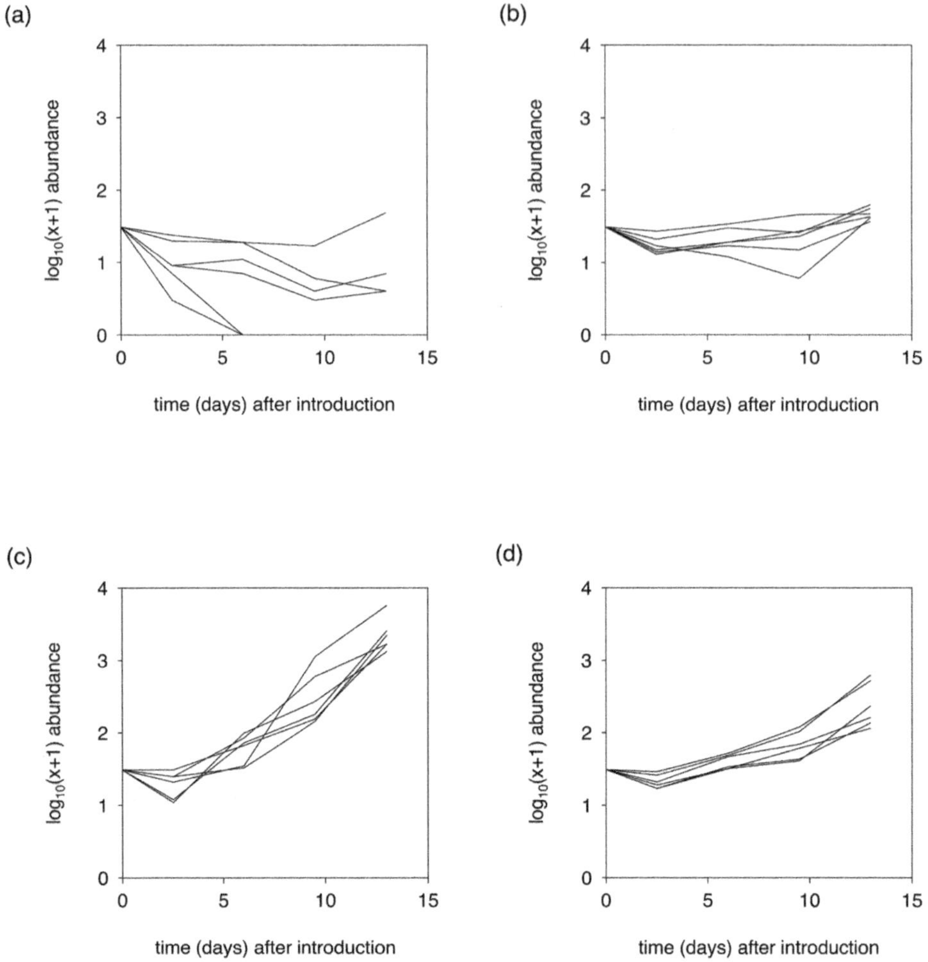

Fig. 3 Time series of the early growth of populations of *Blepharisma* introduced into established populations of (a) *Paramecium*, (b) *Colpidium*, (c) *Tetrahymena*, (d) no other protists, taken from an experiment described in Law *et al.* (2000). In each case six independent replicate introductions of 30 individuals are shown.

stochasticity could be an issue. The behaviour of *Blepharisma* in the absence of any protists was similar, although the growth rate was somewhat smaller.

Invasion biology would be relatively straightforward if ecologists could point to biological properties of species that make their initial rates of increase either positive or negative, though success in doing this has generally been mixed (Williamson

1996, Mack 1996, Kolar and Lodge 2001). The results on *Blepharisma* above are also not encouraging in this regard because success was clearly contingent on properties of the resident community. Nonetheless, our studies did show some large differences among introduced species in the initial rates of increase (Law *et al.* 2000). It was striking that the bacteriovore *Paramecium* was constitutively a strong invader, whatever resident community it encountered. There are evidently features of *Paramecium* that predispose it to grow especially well in microcosms, even when it seems likely to be in competition for resources with other species.

Resident community and the probability of establishment

Although certain species might be constitutively good or bad invaders, in general it is more realistic to envisage invasion as a joint property of an introduced species and the species already resident in the community, contingent on the interactions between the new species and the residents. The per capita birth and death rates in Equation 1 need to be thought of as being determined as much by the resident community at the time of introduction as by properties of the introduced species.

The importance of resident and introduced species as co-determinants of invasions was clear in our experimental introductions of protists into different communities (Law *et al.* 2000). We used analysis of variance to partition variation in initial rate of increase between introduced species, resident community, and introduction × community interaction. As one might expect, the analysis showed a strong introduction × community interaction. The variation in initial growth rates of *Blepharisma* (Fig. 3, and see discussion above), illustrates the effect of differences in resident species. *Blepharisma*, as an omnivore, is able to eat small protists and bacteria. In a community with *Paramecium*, another large protist with which it can only compete for bacteria, *Blepharisma* usually declined rapidly (Fig. 3a). When with *Colpidium*, which is somewhat smaller and also feeds on bacteria, *Blepharisma* was just about able to hold its own (Fig. 3b). In the presence of *Tetrahymena*, a small bacteriovore which *Blepharisma* can eat as well as bacteria, *Blepharisma* populations grew rapidly (Fig. 3c), faster in fact than when all competition for bacteria was removed (Fig. 3d).

Although invasion may depend on interactions among species, could communities have general properties that render them more or less liable to invasion? One long-standing prediction is that communities with a large number of species should be less readily invaded than those with few species (Elton 1958:147, Levine and D'Antonio 1999); this is on the grounds that the available niche space is more fully occupied. Among the few microcosm experiments that have investigated the relationship between species richness and invasibility are the studies by McGrady-Steed *et al.* (1997) and Robinson and Dickerson (1984). The former study found that the protist *Euplotes* was able to invade some (not all) microcosms of low species richness, and able to invade no communities of high richness. The latter, using three different species of protists as invaders, appeared to suggest that invasion success (measured as presence of the invader after three weeks) was not strongly influenced by community diversity, though a later reanalysis (Levine and D'Antonio 1999) found evidence for a negative effect of diversity on success. However, experiments of this kind have the property that richer communities contain more different species (Huston 1997) and are therefore more likely to contain particular species that discourage (or, potentially, promote) invasion by particular new species, making it difficult to attribute invasion outcomes to effects of diversity *per se* (Wardle 2001). Hodgson *et al.* (2002), looking at the ability of particular strains of the bacterium *Pseudomonas fluorescens* to increase when rare, in the presence of other strains, found an effect of diversity on this measure of invasibility, but that it was much weakened by statistical removal of this sampling effect.

Numerical studies of community models suggest that communities become increasingly resistant to invasion as time goes on as a result of the turnover of species (Post and Pimm 1983, Drake 1990b, Law and Morton 1996). Increasing invasion resistance can apply, even if the resident community is not accumulating species, and suggests that it should be relatively hard for new species to invade communities with long histories of assembly. Also one might expect communities constructed from large species pools to be less readily invaded because, with more combinations of species tested during assembly, a finer degree of tuning to invasion-resistant states becomes possible. To date we know of only one microcosm study of the effect of community age on invasibility (Long *et al.* unpublished ms); in this case, older communities were more resistant to two out of three species introduced than the younger communities. However, the older communities were created by allowing a longer period of time to elapse from a fixed initial set of species, so invasion resistance here may be better interpreted as a consequence of being further along a transient of community dynamics rather than as a consequence of turnover of species.

Environmental factors and the probability of establishment

The external environment of the community is a further important determinant of whether a new species becomes established. If resistance to invasion is a consequence of intense interactions between new species and residents, then factors that disrupt those interactions, might be expected to promote invasion. In a related vein, if there are environmental conditions that enhance growth or persistence of species in a community generally, and hence diversity, then such conditions may also increase the likelihood of success of an introduced species — when the sun shines, it shines on the just and the unjust alike (Levine 2000, Shea and Chesson 2002, Byers and Noonburg 2003). Both these possibilities argue for a significant role of environmental variation, either over time, or between communities, in determining the probability of establishment of a new species.

The environmental factor that has received most attention in this respect is disturbance, the essential conclusion being that, at least on some timescales, disturbance enhances invasion, presumably by reducing the intensity of biotic interactions (Crawley 1986, Rejmanek, 1989, Hobbs 1989, Thompson *et al.* 2001, Sher and Hyatt 1999, Davis *et al.* 2000, Shea and Chesson 2002). Although microcosms would seem an obvious system in which to examine this effect systematically, we know of only one such experiment (Long *et al.*, unpublished ms), a study using heterotrophic protists with disturbance generated by density-independent mortality of all resident species. In this case, disturbance affected the introduced species *Blepharisma*, *Paramecium* and *Colpidium* in quite different ways; the main effects of disturbance were to enhance the abundance of *Blepharisma*, to leave *Paramecium* unchanged, and to reduce the abundance of *Colpidium*.

A second microcosm study (Jiang and Morin 2004), with protists and rotifers, provides one of the first experimental tests of the effect of another environmental factor, energy availability, on invasion. In this case, increased energy availability promoted initial population growth of two different introduced species, supporting the possibility, previously speculated about theoretically (Shea and Chesson 2002, Byers and Noonburg 2003), that successful establishment and resident diversity might be positively correlated through the action of a common factor. However, Fukami and Morin (2003) also show, using microcosm systems, that the energy-diversity relationships in communities assembled by sequential invasions are affected by the order in which invasions occur, suggesting that the relationship between energy and invasion success may not always be a simple one.

CHANGES TO COMMUNITIES FOLLOWING ESTABLISHMENT OF INVADERS

Following successful establishment, an invading species typically builds up to a substantial population size, taking the community beyond the point at which Equation 1 can possibly be a good approximation. In effect, the invading species becomes abundant enough to change the birth and death rates of other species in the community, potentially causing change in their densities. The changing density of other species feeds back to the invader, potentially causing changes in its own birth and death rates. At this stage, there is no alternative to dealing with the full dynamics of the community augmented by the invader, and all the complexities that follow from this.

Despite the great complexities of community dynamics, there are some qualitative questions about the invader and the community it enters that can be addressed, and again we draw on examples from our experiments with protists to do this.

Do invaders persist in the long term?

A priori, there is no reason to expect species that become established in the short term to persist in the long term. It is quite possible for an invader to change the community in some way which is deleterious to itself, for instance by driving its prey to extinction. However, in analyses of communities with Lotka-Volterra dynamics, species able to increase from initial low numbers were also present in

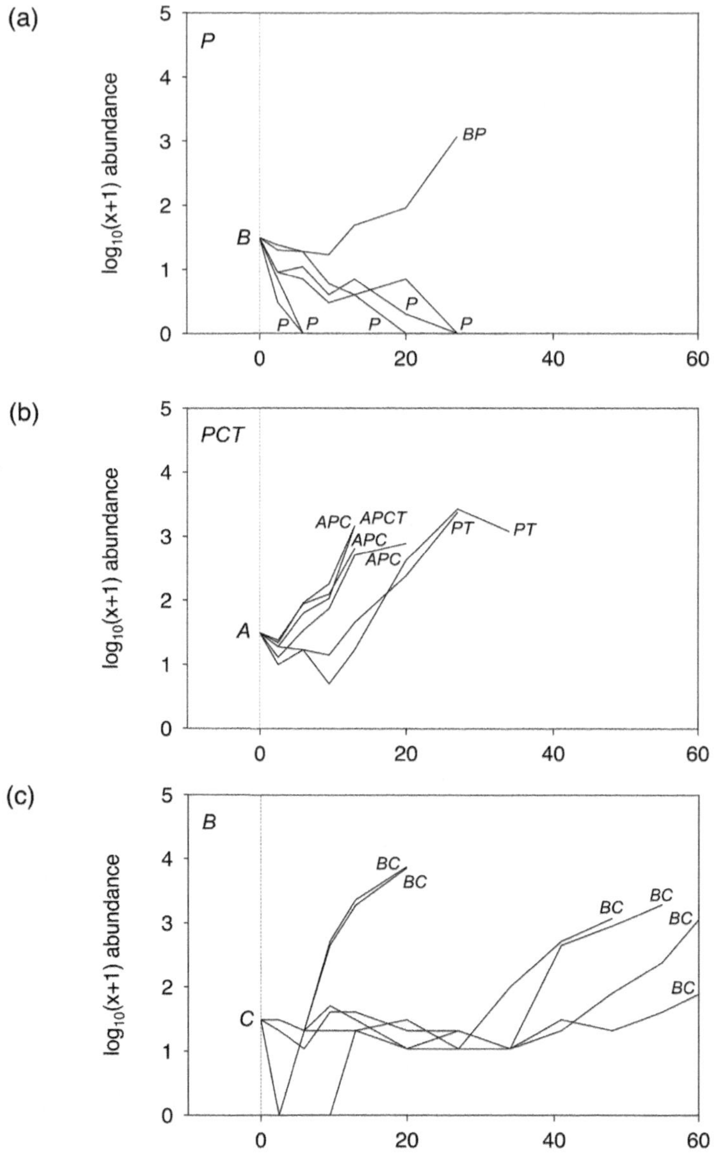

Fig. 4 Examples of initial growth of introduced protists and the communities that persisted in the long term (12 weeks) from experiments described in Law *et al.* (2000) and Warren *et al.* (2003). Lines show the abundances of the introduced species for a period determined by the population either showing consistent positive growth (and achieving numbers greater than the initial inoculum), going extinct, or, if neither condition was satisfied, to an arbitrary end point (60 days). The letter codes used represent the species: →

the long term (Law and Morton 1996). These studies used an asymptotic criterion called permanence to work out the species composition of new communities, thereby jumping over any effects that transient dynamics might have on persistence of species. Real communities have to pass along the transients and it is, of course, possible for these transients to bring the invader to a population size low enough for extinction by demographic stochasticity to become an issue again.

Questions of long-term persistence of invaders can be readily addressed in microcosms, where the component organisms have short generation times. Here we look at two aspects of this: first, the variation in long term persistence among replicate introductions, and second the correspondence between initial establishment and long term persistence of different combinations of residents and introduced species.

In our experiments, the long-term fate of species introduced to communities was far from uniform across the replicates within resident-introduction treatments. In about two-thirds of the forty-nine resident-introduction treatments, all invasions either failed or succeeded; in the remainder, the results were mixed (Fig. 2). In other words, although the communities contained the same species, were run under the same conditions, and were inoculated at the same time with the same number of individuals, there was still a substantial element of indeterminacy about the eventual fate of the introduced species. As already noted (Section 3.1), one explanation is the success or failure of initial establishment, as illustrated by the invasion of *Blepharisma* into systems containing *Paramecium* (e.g., Fig. 3a, Fig. 4a)— an interesting case in which the invader established in just one replicate, but in that system then persisted in the long term. However, it is notable that initial establishment and long term persistence were by no means equivalent (Fig. 2). This is illustrated by *Amoeba* introduced into a community of *Paramecium*, *Colpidium* and *Tetrahymena* (Fig. 4b): it always established itself at the start, but did not persist in all replicates in the long run. In passing, it is also interesting to note that even where long term persistence of the invader occurs in all replicates, there could be considerable variation in the timing and pattern of establishment (Fig. 4c), again making prediction from initial establishment to final outcome a difficult task.

In our experimental invasions the mismatch between the consistency of establishment and persistence, in individual replicate introductions (Fig. 2), was largely

---

A: *Amoeba*, B: *Blepharisma*, C: *Colpidium*, E: *Euplotes*, P: *Paramecium*, T: *Tetrahymena*. The letters in the top left hand corner of the graph indicate the resident community before introduction, the vertical dashed line shows the point at which introductions took place, and the letter at the start of the trajectories indicates the identity of the introduced species. The letter, or combination of letters associated with each line show the community composition of that microcosm at the end of the experiment (12 weeks after introduction). (N.B. population trajectory lines are only drawn to the point at which one of the above criteria was met, while the final species composition shown was that recorded after 12 weeks.)

driven by particular species (Fig. 5). In most invasions, species that showed clear initial success, were also present in the final community, but *Euplotes*, despite being a successful invader in some instances, never persisted in the long term, and *Amoeba*, though more persistent, often showed the same pattern (Fig. 5). Such species provide a counter example to the theoretical predictions outlined above (Law and Morton 1996) which suggest that successful invaders should become persistent members of the invaded system (even if they cause other changes). Interestingly, these species were the two obligate predators in our species pool and it may be that the discrepancy was related to trophic position. Demographic stochasticity is a possible cause of the vulnerability of introduced predators because, for reasons of either energy availability or oscillatory population dynamics, their population sizes could have been small from time to time.

There is clearly variation in the relationship between initial establishment and long term persistence at the level of individual replicates, which has obvious implications for predicting the outcome of particular instances of an introduction.

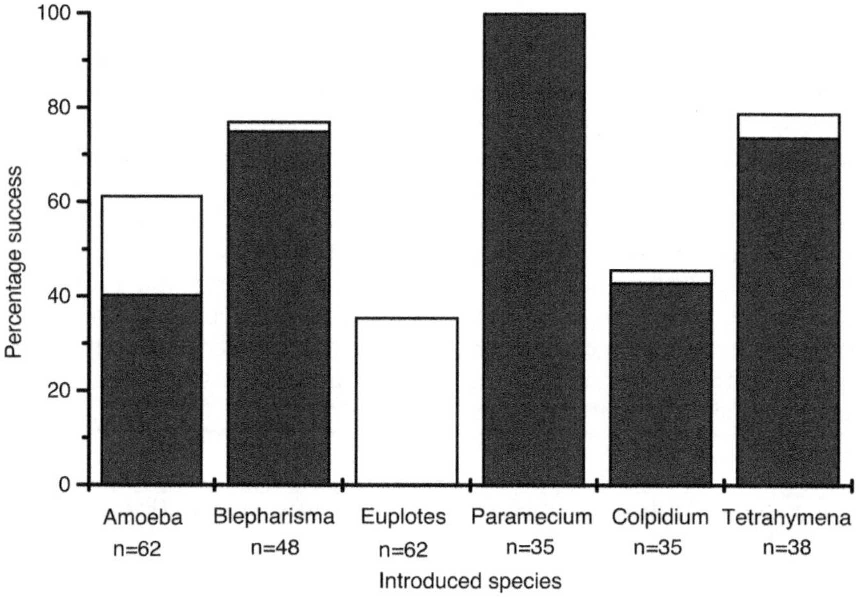

Fig. 5 Mismatch in establishment and long-term persistence of protists introduced into microcosms (from Weatherby 2000). Bars show the percentage of microcosms in which the introduced species became established (open bars), and persisted in the long term (18 weeks) (shaded bars). Data are given for all individual microcosms across all resident communities into which the species was introduced, so values take into account the mixed outcomes shown in Fig. 2; *n* is the number of microcosms into which each species was introduced (usually 6 replicates of each resident community, but in some cases only 5, or 4).

However, we can also look at the broader picture, and focus on the resident-introduction combinations in which the majority of replicates had the same outcome (both in terms of establishment and composition of the final community). Taking just these more consistent outcomes (a total of thirty five resident-introduction combinations), in the majority of cases the introduced species either established and persisted (15), or failed both to establish and to persist (15). There were relatively few cases in which species established and then failed to persist (5) and none in which species failed to establish but yet persisted. (This last category may seem a truism, but it is not so, because it would be possible for an invader to persist without showing net population growth, therefore failing to meet the criterion for establishment). Thus the message from these microcosm studies is that initial establishment of an introduced species within a community (which could feasibly be measured even in the field) is, a fair, but not perfect, predictor of its fate in the long term.

Do the invaders have an impact on the residents?

We have so far focused on the fate of the invader, but it is often not the invader as such which is of primary concern, but rather the impact of the invader on the communities it invades (Parker *et al.* 1999). Generalizations about these impacts are bound to be hard to make, because of the complex nonlinear couplings that characterize community dynamics. It is entirely possible for the impact to be negligible, or for other species to be driven to extinction. An invader may target particular species (e.g., Dutch Elm disease in the UK (Williamson 1996)), or have effects that ramify widely through the community (e.g., whole ecosystem effects of *Myrica faya* on volcanic sites in Hawaii (Vitousek and Walker 1989); the demise of kelp beds following killer whale arrival on the Pacific coast of N. America (Estes *et al.* 1998)). Microcosm communities are especially amenable to systematic investigation of impacts of invaders because we can take them apart in an orderly replicated way, to see what effect each invading species eventually has on the species composition of each community, and how this relates to the invader's own ability to persist. Various scenarios are possible: (1) the invader persists, and the resident community is changed; (2) the invader persists and the resident community is unchanged; (3) the invader fails to persist, but the resident community is changed; (4) the invader fails and the resident community is unchanged.

Situation (1) is the classic 'problem' invasion, an invader that becomes a part of the community and changes it, usually by generating one or more extinctions in the residents: the invasion of the community comprising *Blepharisma* and *Colpidium* by *Paramecium* (Fig. 6a) provides an example of this from our microcosm experiments. These are also the best documented invasion case studies in natural systems (e.g., Elton 1958, Drake *et al.* 1989) because invaders that persist are most readily documented and studied, and their impacts are of concern.

In (2) we have simple augmentation of the community by an invader. Of course the colonizing species may alter densities of other species, but not to a sufficient extent to cause species losses. Invasion of *Blepharisma* systems by *Paramecium* provides an example: although the resident persisted (Fig. 6b), it was about one fifth as abundant in the invaded systems as in the uninvaded controls at the end of the experiment. There are examples from natural systems of invaders which appear to add themselves to systems but cause little impact (e.g., the fulmar *Fulmaris glacialis*, an arctic seabird; Williamson 1996: 5 *et seq.*). The frequency of non-impactive invasions however remains a controversial issue (Parker *et al.* 1999, for example see: Simberloff 1981, Herbold and Moyle 1986, Pimm 1989) because it can be difficult to establish impacts and test for causality if there are no suitable uninvaded systems for comparison.

Scenario (3) requires that a species fails to persist, but has sufficiently strong effects on the resident community while it is present to cause changes in species composition. Such species can be termed 'catalysts'. In our experiments, catalysts were not common, but two species did play this role in some situations: principally *Euplotes* and also *Amoeba* in occasional replicates. For example, *Euplotes* established itself in a community comprising *Paramecium, Colpidium* and *Tetrahymena*, caused the latter two species to go extinct in most systems, and then went to extinction itself (Fig. 6c). In this example *Euplotes* did establish (increase in population size after introduction), but there were also cases (e.g., *Euplotes* invading the community of *Paramecium* and *Tetrahymena*) in which the invader did not consistently increase above its initial numbers, but nonetheless caused an extinction of one of the residents (*Tetrahymena*), before itself going extinct. There is some theoretical evidence of such effects in models of invasions into competitive systems (Case 1995). Some examples of situations in natural systems seem to show this behaviour (e.g., Flux 1994) but the apparent rarity of such observations (Williamson 1996) is consistent with the relative rarity of these effects in our experimental systems.

Finally we have scenario (4). In this case, the failure of the introduced species to persist, coupled with the lack of change in the resident community (e.g., Fig. 6d) suggests that, effectively, there is nothing we would recognize as an invasion at all. In the majority of introductions with this outcome this was very much the case, with the introduced species simply declining to extinction after introduction. However there were also a few cases in which introduced species showed successful initial establishment, while nonetheless eventually going extinct and leaving the resident set unchanged; for example, in most introductions of *Euplotes* into populations of *Tetrahymena, Euplotes* achieved populations of several thousands of individuals before disappearing. Evidently an introduced species can achieve a significant presence during some part of an ultimately unsuccessful invasion.

Because our microcosm experiments contained all possible permutations of resident community and introduced species, we can use these results to look at the frequencies of these various outcomes, and the circumstances under which they occur. Taking just the thirty-six resident-introduction treatments in which the majority of replicates ended up with the same species composition in the long term, the resident community was altered in just nine of these (Warren *et al.* 2003).

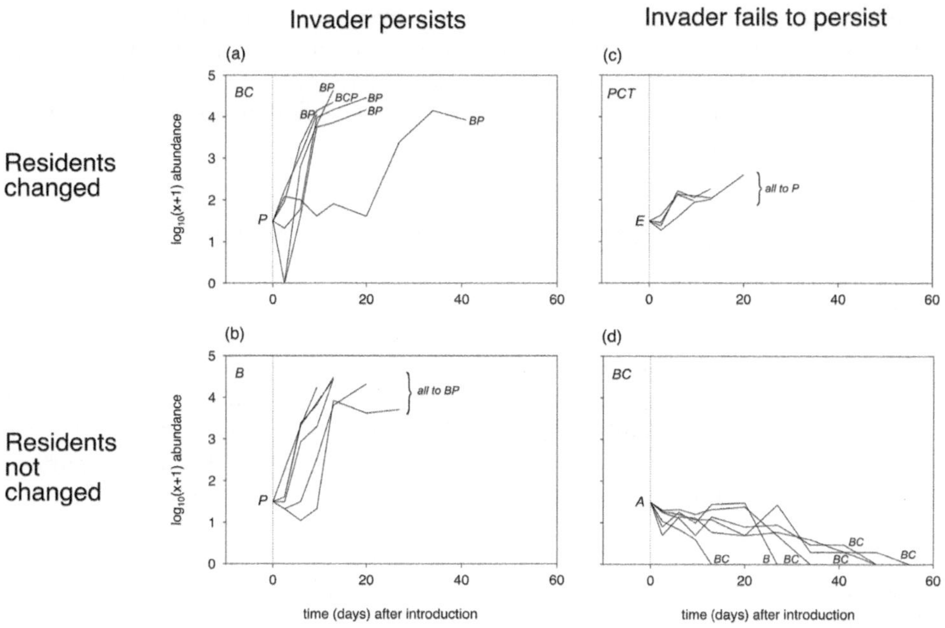

Fig. 6 Examples of initial growth of introduced protists and the communities that persisted in the long term, for combinations of persistence and non-persistence of the invader, and impact, or lack of impact, on the residents. Structure and notation as for Fig. 4.

As one might expect (e.g., Case 1990), changes to the resident community became more likely as the number of resident species increased, rising from about 11% in one-species communities to 27% in two-species communities, to 100% in three-species communities (though there are only two three species combinations). The effect of the introduced species was most often negative causing extinction of resident species (six cases), although there were three instances of the reverse effect, i.e., a resident species (*Colpidum* in two systems and *Paramecium* in one) for which persistence seemed to be enhanced by the presence of the invaders (*Amoeba, Paramecium* and *Euplotes*).

Putting these results together, we can break down the thirty six resident-introduction treatments into those in which the introduced species persisted and the resident community was changed (5), the introduced species persisted but the resident community was not changed (10), the introduced species did not persist, but the resident community was changed (4), and the introduced species did not persist and the residents were unchanged (17). Clearly there is no significant association between the likelihood of establishment of introduced species and the likelihood of the resident community being altered (Fisher's Exact Test $p = 0.44$). If this sort of pattern turns out to have any sort of generality, it lends weight to the view that knowing a species is likely to become established and persist in a system may provide no useful guide to whether it is going to change the existing community in other ways.

INVASIONS AND COMMUNITY ASSEMBLY

Although our main focus here has been on the components of individual invasions, the arrival of a new species, and any changes in community composition that accompany this, are just a single step in a longer-term process of community development through the gradual turnover of species. Such turnover, generally termed 'community assembly', is determined by factors acting at very different scales, from biogeographic constraints on the species pool, through to patterns of niche differentiation among competitors in a community (Diamond and Case 1986, Drake 1990a, Gray *et al.* 1987, Weiher and Keddy 1999), but has the same underlying driver: the arrival, establishment, and sometimes persistence, of species from outside the system: i.e., invasion. Here, certainly, the links between community ecology and invasion biology are explicit: invasions drive community change. Of course in most of the 'invasions' effecting such changes the invader is not from a new region, dissociated from the species in the resident community, but nonetheless, the components of the invasion process we have discussed above apply equally to what might be termed 'local invasions' and, therefore, to community change in general.

Microcosm systems, for the same reasons that make them useful for studying invasions, have played a significant role in exploring community assembly. In particular microcosms of algae, protists and small metazoans have been used to examine the effects of non-simultaneous invasion, invasion order and invasion rate on the development of community structure (Dickerson and Robinson 1984, 1985, 1986, Robinson and Dickerson 1987, Robinson and Edgemon 1988, Drake 1991, Fukami and Morin 2003). The general conclusion from these studies is that contingencies in the invasion process can result in substantial differences in the structure of the final communities that are formed. Such studies are generally presented in the context of community assembly, and document the outcomes of different sequences or rates of species arrival, rather than the details of individual invasions which have been our focus here, but it is clear that the sorts of effects we see in individual invasions, in particular stochastic effects in low populations and the interdependence of success on invader and resident community identities, could be the drivers of such assembly sequence effects (e.g., Drake 1991). Drake's study even found evidence to suggest that invasibility of systems with identical species compositions may to be affected by differences in their assembly history. Microcosms are also being used to examine the effects of the continuous assembly and disassembly of communities in multiple habitat patches, where invasions are a necessary and natural part of the maintenance of populations and communities within a 'landscape' (Drake *et al.* 1993, Holyoak and Lawler 1996a,b, Warren 1996, Burkey 1997, Holt *et al.* 2002), and with the possible, though largely unexplored, opportunity such work provides for examining the spread of invaders through a community landscape.

Putting these components together allows us to construct a description of the community space (persistent communities) and the routes that community assembly can take through that space as a result of invasions into each community (Zimmerman *et al.* 2003, Warren *et al.* 2003), which we can term an assembly graph (e.g., Fig. 7). Assembly graphs can clarify features of community assembly that would otherwise remain obscure. For example, the graph shown in Fig. 7 has a cyclic end-state driven by the particular characteristics of *Euplotes*. *Euplotes* invades the community composed of *Paramecium* and *Tetrahymena*, eliminates the *Tetrahymena* and then itself goes to extinction, leaving a community of just *Paramecium* which can then be reinvaded by *Tetrahymena* to return to the original community. Another feature of the graph is the existence of communities, such as that comprising *Colpidium* and *Tetrahymena*, which persist if all species are introduced together but which cannot readily be created by sequential invasion from any other persistent community. The generality of such features remains to be established, but what is clear is that understanding the invasion process lies at the heart of understanding this fundamental aspect of community ecology.

MICROCOSMS AS SYSTEMS FOR STUDYING INVASION BIOLOGY

Remarkably little work on biological invasions has been done in microcosm systems, yet even from the few studies there are, a number of messages emerge. Work in microcosms highlights the probabilistic nature of the invasion process, the joint dependence of invasion success on both invader identity and resident community composition, the relationships between different phases of the invasion process (establishment, persistence, impact), and the importance of disturbance, energy, and species number.

This seems a respectable yield of information from a modest amount of experimental work. These results were made possible by the having the facility to manipulate resident and introduced species, by following the outcomes over tens to hundreds of generations, and by putting in place enough replication to partition variation within and between treatments. There is little likelihood of being able to carry out experiments at equivalent temporal and spatial scales in the field. Field studies have provided important insights into all these elements of invasions, but microcosms bring the facility to relate the long and short-term outcomes of the invasion process, and quantify the stochasticities involved. Of course, the scarcity of such studies in microcosms precludes much assessment of the generality of these results, as it would in any system. The difference is that, in microcosms, testing this generality is an entirely feasible prospect.

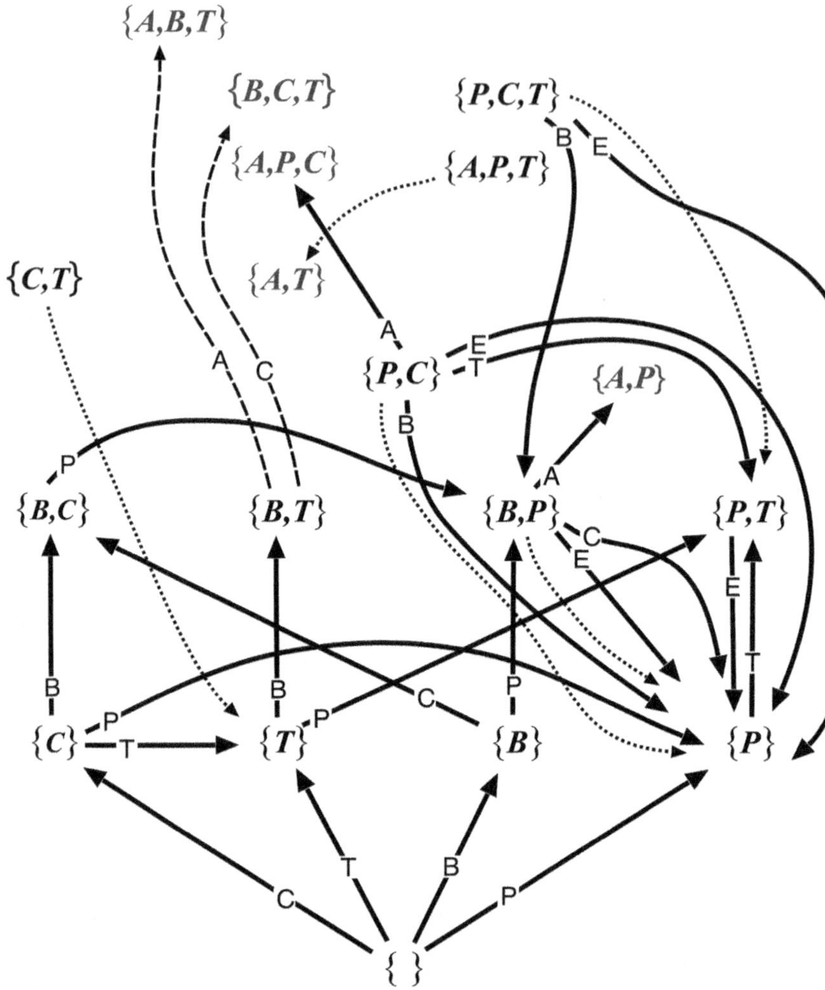

Fig. 7 The assembly graph for the six species protist system described in Warren *et al.* (2003). The letter codes (as for Fig. 5) in brackets represent the persistent communities, and the arrows represent transitions between those communities produced by introductions of the species represented by the letter code on the arrow. Arrows are based on the majority transitions observed in the experiments in each case, so that only a single outcome from any particular invasion is depicted (see Warren *et al.* 2003 for an alternative presentation that includes the variation in outcomes). Dotted arrows show where certain resident communities had a tendency to collapse eventually, even in the absence of invasions, and the dashed arrows are transitions that are based on information from only two replicates. The communities depicted in grey are those formed as a result of invasions, but not found in initial experiments to establish which sets of species could coexist, and hence for which no invasion trials were carried out, and so we have no information on their invasibility. (Figure reproduced by permission of the Ecological Society of America).

That such questions could be addressed is, though, perhaps not the issue. Those sceptical of the microcosm approach would question whether even if we had such studies we would be any the wiser about invasions in the field. The arguments about using microcosms in ecology are well rehearsed (e.g., Drake *et al.* 1996, Carpenter 1996, 1999, Drenner and Mazumder 1999, Lawler 1998, Jessup *et al.* 2004) but against the limitations we have to set the advantages outlined above. Microcosms do not allow us to make practical management decisions about specific invasions, but they do allow us to test our intuition, reasoning and theory, as to the basic ecological principles that underpin them (Drake *et al.* 1996, Lawler 1998). We take the view that microcosms can, and should, play a significant role in developing our understanding of invasion processes. We will understand invasions better armed with good experimental data from microcosms, in combination with theory, field experiments and observational data, than from any of these sources alone, and we will also end up knowing a great deal more about how communities work as a consequence.

ACKNOWLEDGEMENTS

We are grateful to the editors for the invitation to contribute to this volume. Our thanks go to Peter Morin, Lin Jiang and Zac Long, for kindly sending us manuscripts prior to publication, to Mark Williamson and Ken Thompson for helpful discussions about invasions, and to Jeremy Fox, Tad Fukami, Owen Petchey, David Wardle and Mark Williamson for thoughtful comments on the manuscript. The experimental work discussed in this chapter was funded by a grant from the UK Natural Environment Research Council.

REFERENCES

Anderson, R. M. and May, R. M. (1991) Infectious diseases of humans. Oxford University Press Oxford, UK.

Bailey, N. T. J. (1964) Elements of stochastic processes. Wiley, New York, U. S. A.

Burkey, T. V. (1997) Metapopulation extinction in fragmented landscapes: using bacteria and protozoa communities as model ecosystems. Am. Nat. **150**, 568-591.

Byers, J. E. and Noonburg, E. G. (2003) Scale dependent effects of biotic resistance to biological invasion. Ecology **84**, 1428-1433.

Carpenter, S. R. (1996) Microcosm experiments have limited relevance for community and ecosystem ecology. Ecology **77**, 677-680.

Carpenter, S. R. (1999) Microcosm experiments have limited relevance for community and ecosystem ecology: reply. Ecology **80**, 1085-1088.

Case, T. J. (1990) Invasion resistence arises in strongly interacting species-rich models of competition communities. Proc. Natn. Acad. Sci. U. S. A. **87**, 9610-9614.

Case, T. J. (1995) Surprising behavior from a familiar model and implications for competition theory. Am. Nat. **146**, 961-966.

Crawley, M. J. (1986) The population biology of invaders. Phil. Trans. Roy. Soc. Lond. B **314**, 711-731.

Crawley, M. J. (1989) Chance and timing in biological invasions. pp. 407-423 In: Drake, J. A., Mooney, H. A., di Castri, F., Groves, R. H., Kruger, F. J., Rejmanek, M. and Williamson, M. (eds.) Biological invasions: a global perspective. SCOPE 37. John Wiley and Sons, Chichester, UK.

Davis, M. A, Grime, J. P. and Thompson, K. (2000) Fluctuating resources in plant communities: a general theory of invasibility. J. Ecol. **88**, 528-534.

Davis, M. A. and Pelsor M. (2001) Experimental support for a resource-based mechanistic model of invasibility. Ecol. Lett. **4**, 421-428.

Diamond, J. and Case, T. J. (1986) Overview: introductions, extinctions, exterminations and invasions. pp. 65-79 In: Diamond, J. and Case, T. J. (eds.) Community Ecology. Harper and Row, New York, U. S. A.

Dick, J. T. A., Platvoet, D. and Kelly D. W. (2002) Predatory impact of the freshwater invader Dikerogammarus villosus (Crustacea: Amphipoda). Can. J. Fish. Aquat. Sci. **59**, 1078-1084.

Dickerson, J. E. and Robinson, J. D. (1984) The assembly of microscopic communities: patterns of species importance. Trans. Am. Microsc. Soc. **103**, 164-171.

Dickerson, J. E. and Robinson, J. V. (1985) Microcosms as islands: a test of the MacArthur-Wilson equilibrium theory. Ecology **66**, 966-980.

Dickerson, J. E. and Robinson, J. V. (1986) The controlled assembly of microcosmic communities: the selective extinction hypothesis. Oecologia **71**, 12-17.

Drake, J. A. (1990a) Communities as assembled structures: do rules govern pattern? Trends Ecol. Evol. **5**, 159-164.

Drake, J. A. (1990b) The mechanics of community assembly and succession. J. Theor. Biol. **147**, 213-234.

Drake, J. A. (1991) Community-assembly mechanics and the structure of an experimental species ensemble. Am. Nat. **137**, 1-26.

Drake, J. A., Flum, T. E., Witteman, G. J., Voskuil, T., Holyman, A. M., Creson, C., Kenny, D. A., Huxel, G. R., Larue, C. S. and Duncan, J. R. (1993) The construction and assembly of an ecological landscape. J. Anim. Ecol. **62**, 117-130.

Drake, J. A., Huxel, G. R. and Hewitt, C. L. (1996) Microcosms as models for generating and testing community theory. Ecology **77**, 670-677.

Drake, J. A., Mooney, H. A., di Casti, F., Groves, R. H., Kruger, F. J., Rejmánek, M. and Williamson, M. (ed.) (1989) Biological invasions: a global perspective. SCOPE 37. John Wiley and Sons Chichester, UK.

Drenner, R. W. and Mazumder, A. (1999) Microcosm experiments have limited relevance for community and ecosystem ecology. Ecology **80**, 1081-1085.

Dukes, J. S. (2001) Biodiversity and invasibility in grassland microcosms. Oecologia **126**, 563-568.

Elton, C. S. (1958) The ecology of invasions by animals and plants. Menthuen, London, UK.

Estes, J. A., Williams, M. T. and Doak, D. F. (1998) Killes whale predation on sea otters linking oceanic and nearshore ecosystems. Science **282**, 473-476.
Flux, J. E. (1994) World distribution. pp. 8-21 In: Thompson, H. V. and King, C. M. (eds.) The European rabbit: the history and biology of a successful invader. Oxford University Press, Oxford.
Fukami, T. and Morin, P. J. (2003) Productivity-biodiversity relationships depend on the history of community assembly. Nature **424**, 423-426.
Goel, N. S. and Richter-Dyn, N. (1974) Stochastic models in biology. Academic Press New York, U. S. A.
Gray, A. J. and Crawley, M. J. and Edwards, P. J. (ed.) (1987) Colonization, succession and stability. Blackwell Science, Oxford, UK.
Hector, A., Dobson, K., Minns, A., Bazeley-White, E. and Lawton, J. H. (2001) Community diversity and invasion resistance: an experimental test in a grassland ecosystem and a review of comparable studies. Ecol. Res. **16**, 819-831.
Herbold, B. and Moyle, P. B. (1986) Introduced species and vacant niches. Am. Nat. **128**, 751-760.
Hobbs, R. J. (1989) The nature and effects of disturbance relative to invasions. Ch. 17 pp. 389-405 In: Drake, J. A., Mooney, H. A., di Castri, F., Groves, R. H., Kruger, F. J., Rejmánek, M. and Williamson, M. (ed.) Biological invasions: a global perspective. SCOPE 37. John Wiley and Sons, Chichester, UK.
Hodgson, D. J., Rainey, P. B. and Buckling, A. (2002) Mechanisms linking diversity, productivity and invasibility in experimental bacterial communities. Proc. Roy. Soc. Lond. B **269**, 2277-2283.
Holt, A. R., Warren, P. H. and Gaston, K. J. (2002) The importance of biotic interactions in abundance-occupancy relationships. J. Anim. Ecol. **71**, 846-854.
Holyoak, M. and Lawler, S. P. (1996) Persistence of an extinction prone predator-prey interaction through metapopulation dynamics. Ecology **77**, 1867-1879.
Holyoak, M. and Lawler, S. P. (1996) The role of dispersal in predator-prey metapopulation dynamics. J. Anim. Ecol. **65**, 640-652.
Hopper, K. R. and Roush, R. T. (1993) Mate finding, dispersal, number released, and the success of biological control introductions. Ecol. Entomol. **18**, 321-323.
Huston, M. A. (1997) Hidden treatments in ecological experiments: re-evaluating the ecosystem function of biodiversity. Oecologia **110**, 449-460.
Jiang, L. and Morin, P. J. (2004) Productivity gradients cause positive diversity-invasibility relationships in microbial communities. Ecol. Lett. **7**, 1047-1057.
Jessup, C. M., Kassen, R., Fordel, S. E., Kerr, B., Buckling, A., Rainey, P. B. and Bohannan, B. J. M. (2004) Big questions, small worlds: microbial model systems in ecology. Trends Ecol. Evol. **19**, 189-197.
Kola, C. S. and Lodge, D. M. (2001) Progress in invasion biology: predicting invaders. Trends Ecol. Evol. **16**, 199-204.
Law, R. and Morton, R. D. (1996) Permanence and the assembly of ecological communities. Ecology **77**, 762-775.
Law, R., Weatherby, A. J. and Warren, P. H. (2000) On the invasibility of persistent protist communities. Oikos **88**, 319-326.

Lawler, S. P. (1998) Ecology in a bottle: using microcosms to test theory. pp. 236-253 In: Resetarits, W. J. and Bernado, J. (eds.) Experimental ecology: issues and perspectives. Oxford University Press, New York.

Levine, J. M. (2000) Species diversity and biological invasions: relating local processes to community patterns. Science **288**, 852-854.

Levine, J. M. and D'Antonio, C. M. (1999) Elton revisited: a review of evidence linking diversity and invasibility. Oikos **87**, 15-26.

Lonsdale, W. M. (1999) Global patterns of plant invasions and the concept of invasibility. Ecology **80**, 1522-1536.

Mack, R. N. (1996) Predicting the identity and fate of plant invaders: emergent and emerging approaches. Biol. Conserv. **78**, 107-121.

Mack, R. N., Simberloff, D., Lonsdale, M. W., Evans, H., Clout, M. and Bazazz, F. A. (2000) Biotic invasions: causes, epidemiology, global consequences, and control. Ecol. Applic. **10**, 689-710.

McGrady-Steed, J., Harris, P. M. and Morin, P. J. (1997) Biodiversity regulates ecosystem predictability. Nature **390**, 163-165.

Mitchell, C. E. and Power, A. G. (2003) Release of invasive plants from fungal and viral pathogens. Nature **421**, 625-627.

Parker, I. D., Simberloff, D., Lonsdale, W. M., Goodell, K., Wonham, M., Kareiva, P. M., Williamson, M. H., Von Holle, B., Moyle, P. B., Byers, J. E. and Goldwasser, L. (1999) Impact: toward a framework for understanding the ecological effects of invaders. Biol. Invasions **1**, 3-19.

Petchey, O. L., McPhearson, P. T., Casey, T. M. and Morin, P. J. (1999) Environmental warming alters food web structure and ecosystem function. Nature **402**, 69-72.

Petersen, J. E. and Hastings, A. (2001) Dimensional approaches to scaling experimental ecosystems: Designing mousetraps to catch elephants. Am. Nat. **157**, 324-333.

Pimentel, D., Lach, L., Zuniga, R. and Morrison, D. (2000) Environmental and economic costs of nonindigenous species in the United States. Bioscience. **50**, 53-65.

Pimm, S. L. (1989) Theories of predicting success and impact of introduced species. pp. 351-367 In: Drake, J. A., Mooney, H. A., di Casti, F., Groves, R. H., Kruger, F. J., Rejmánek, M. and Williamson, M (ed.) Biological invasions: a global perspective. SCOPE 37. John Wiley and Sons, Chichester, UK.

Post, W. M. and Pimm, S. L. (1983) Community assembly and food web stability. Math. Biosci. **64**, 169-192.

Rejmánek, M. (1989) Invasibility of plant communities. Ch. 16 pp. 369-388. In: Drake, J. A., Mooney, H. A., di Casti, F., Groves, R. H., Kruger, F. J., Rejmánek, M and Williamson, M. (ed.) Biological invasions: a global perspective. SCOPE 37. John Wiley and Sons, Chichester, UK.

Robinson, J. V. and Dickerson J. E. (1984) Testing the invulnerability of laboratory island communities to invasion. Oecologia **61**, 169-174.

Robinson, J. V. and Dickerson, J. E. (1987) Does invasion sequence affect community structure? Ecology **68**, 587-595.

Robinson, J. V. and Edgemon, M. A. (1988) An experimental evaluation of the effect of invasion history on community structure. Ecology **69**, 1410-1417.

Sakai, A. K., Allendorf, F. W., Holt, J. S., Lodge, D. M., Molofsky, J., With, K. A., Baughman, S., Cabin, R. J., Cohen, J. E., Ellstrand, N. C., McCauley, D. E., O'Neil, P., Parker, I. M., Thompson, J. N. and Weller, S. G. (2001) The population biology of invasive species. Ann. Rev. Ecol. Syst. **32**, 305-332.

Shea, K. and Chesson, P. (2002) Community ecology theory as a framework for biological invasions. Trends Ecol. Evol. **17**, 170-176.

Shea, K. and Possingham, H. P. (2000) Optimal release strategies for biological control agents: an application of stochastic dynamic programming to population management. J. Appl. Ecol. **37**, 77-86.

Sher, A. A. and Hyatt, L. A. (1999) The disturbed resource-flux invasion matrix: a new framework for patterns of plant invasion. Biol. Invasions **1**, 107-114.

Simberloff, D. (1981) Community effects of introduced species. pp. 53-83 In: Nitecki, M. H. (ed.) Biotic crises in ecological and evolutionary time. Academic Press, New York, USA.

Simberloff, D. (1989) Which insect introductions succeed and which fail? Ch. 3 pp. 61-75 In: Drake, J. A., Mooney, H. A., di Casti, F., Groves, R. H., Kruger, F. J., Rejmánek, M. and Williamson, M. (ed.) Biological invasions: a global perspective. SCOPE 37. John Wiley and Sons, Chichester.

Stampe, E. and Daehler, C. C. (2003) Mycorrhizal species identity affects plant community structure and invasion: a microcosm study. Oikos **100**, 362-372.

Thompson, K., Hodgson, J. G. and Grime, J. P. and Burke M. J. W. (2001) Plant traits and temporal scale: evidence from a 5-year invasion experiment using native species. J. Ecol. **89**, 1054-1060.

Torchin, M. E., Lafferty, K. D., Dobson, A. P., McKenzie, V. J. and Kuris, A. M. (2003) Introduced species and their missing parasites. Nature **421**, 628-630.

Vitousek, P. M. and Walker, L. R. (1989) Biological invasion by *Myrica faya* in Hawaii: plant demography, nitrogen fixation, ecosystem effects. Ecol. Monogr. **59**, 247-265.

Wardle, D. A. (2001) Experimental demonstration that plant diversity reduces invasibility: Evidence of a biological mechanism or a consequence of sampling effect? Oikos **95**, 161-170.

Warren, P. H. (1996) Dispersal and destruction in a multiple habitat system: and experimental approach using protist communities. Oikos **77**, 317-325.

Warren, P. H., Law, R. and Weatherby, A. J. (2003) Mapping the assembly of protist communities in microcosms. Ecology **84**, 1001-1011.

Weatherby, A. J. (2000) Species coexistence and community assembly in protist microcosms. unpublished PLD thesis. University of Sheffield, UK.

Weatherby, A. J., Warren, P. H. and Law, R. (1998) Coexistence and collapse: an experimental investigation of the persistent communities of a protist species pool. J. Anim. Ecol. **67**, 554-566.

Weiher, E. and Keddy, P. (eds.) (1999) Ecological assembly rules: perspectives, advances, retreats. Cambridge University Press, Cambridge, UK.

Williamson, M. (1996) Biological invasions. Chapman and Hall London, England.

Zimmermann, C. R., Fukami, T. and Drake, J. A. (2003) An experimentally-derived map of community assembly space. In: Minai, A. and Bar-Yam, Y. (eds.) Unifying themes in complex systems II: Proceedings of the second international conference on complex systems. New England Complex Systems Institute. Perseus Press.

Large-scale consequences and patterns of invasions

Chapter seventeen

Understanding invasions in patchy habitats through metapopulation theory

K. C. Harding, J. M. McNamara and R. D. Holt

INTRODUCTION

Many invasive species experience a world that is heterogeneous and patchy, and experience it at multiple scales. In this chapter we will show how metapopulation theory can be used to explore general mechanisms which determine the initial spread and long term establishment of invasive species in novel environments. Metapopulation ecology is a branch of ecology that is focused on how species persist in fragmented or 'subdivided' habitats (Hanski 1999, Hanski and Gaggiotti 2004). Although real landscapes are often highly complex, with admixtures of smooth gradients, fractal edges, and juxtaposed habitats varying in quality, for many species it is natural to construe their world as consisting of sets of suitable habitat patches, arrayed within a largely inhospitable matrix separating those patches. Some such species inhabit naturally patchy habitats, such as islands or ponds (Fig. 1). Some metapopulations appear in artificially constructed habitat networks, such as the remaining fragments of a rain forest surrounded by cattle pastures, or concrete reefs in marine systems. Patch networks can also be formed in more subtle ways, but still make functional metapopulations. The patchy distribution of a host species for a specialist herbivore, parasite, parasitoid, or

M.W. Cadotte et al. (eds.), Conceptual ecology and invasion biology, 371–403.

Fig. 1 Patchy world, ponds in the Northern Prairie farmland. Photo: Stewart, R.E. *et al.* (1971). From Northern Prairie Wildlife Research Center.

commensal (or indeed any vital and spatially delimited resource) can lead to metapopulation dynamics for the specialist. Biological nonlinearities can transform smooth environmental gradients into patchy distributions. For instance, strong Allee effects (positive density dependence at low numbers) can amplify patchiness in the environment. When dispersal is limited, permanent range limits can arise in patchy environments even without overall environmental gradients in the landscape (Keitt *et al.* 2001). Even without Allee effects, if the density of suitable patches declines along an environmental gradient, a species range limit may emerge, when there is no decline in the quality of suitable habitat patches (Holt and Keitt 2000).

A metapopulation is defined as a set of populations connected by dispersal. Dispersal becomes particularly vital to understanding regional population persistence when local populations can become extinct, and must be re-established via colonization. Classical metapopulation theory thus focuses on systems where there is a potential for recurrent extinction. In all metapopulations, long-term species persistence relies on the balance between the rate of extinction of local populations and the rate of new colonizations of empty habitat patches (Fig. 2).

Metapopulation theory is well suited for describing invasions and the establishment of new species into patchy habitats. There are three basic questions

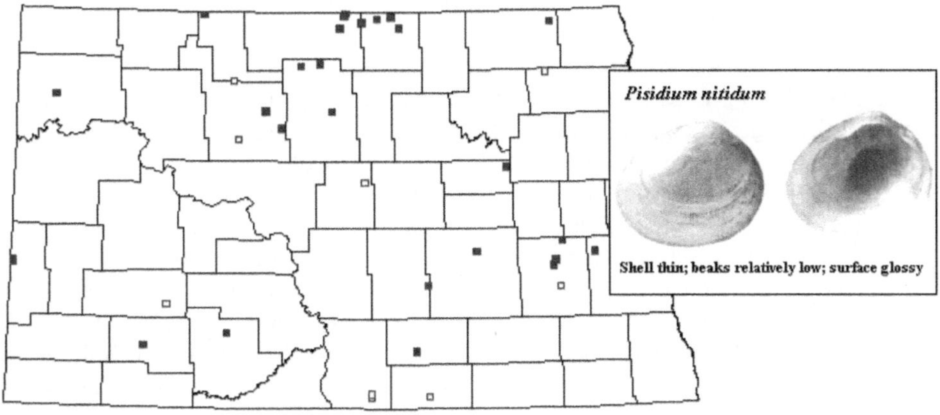

Fig. 2 Blue squares indicate ponds occupied by *Pisidium nitidum* in the year of sampling. White square are ponds where *P. nitidum* has gone extinct. Stewart and Kantrud (1971). From Northern Prairie Wildlife Research Center

one can ask about an invasion. First, can an invasion occur at all? Second, given that an invasion occurs, how rapidly does it occur? Finally, following invasion, what impacts does the invader have upon the prior resident community? Metapopulation theory can be used to address each of these basic issues.

We introduce a simple model structure that can be used to summarise and classify some different types of metapopulation invasion scenarios. We examine some special cases, starting with a case in which the invasive species does not compete with any other ecologically similar species. The dynamics of the invasion in this case is then governed by the biology of the invader only, and classic single species metapopulation theory applies. We then consider a general metapopulation model in which the invading species interacts with a native species. We define criteria for when initial invasion occurs and for when long-term coexistence is possible. We also look at some interactions that are specific cases of this general model: a) the invader completely excludes the native species, b) the invader is dependent on the presence of a native species, and, c) there is an intermediate level of competition. We also touch on how spatial heterogeneity can influence invasion. Our overview highlights some general insights about the population biology of invasion which emerge from metapopulation theory.

NO COMPETITION WITH NATIVE SPECIES — CLASSIC METAPOPULATION BIOLOGY APPLIES

Levins metapopulation

Some invasions occur without any obvious competition from native species. For instance, when the marsh-loving muskrat was introduced into central Europe, its numbers grew exponentially (Skellam 1951). In that environment, there were no other obvious aquatic mammals which could interfere with the initial establishment and subsequent growth of this species. In these cases single species metapopulation theory can be applied to investigate basic properties of the invasion (e.g., for the muskrat, its required habitat typically occurs in discrete patches — marshes — separated by a relatively hostile matrix of terrestrial habitats). Moreover, if resident species are relatively fixed in abundance (e.g., the invader initially has little impact), then the residents can be viewed as fixed habitat factors, either hampering or assisting invasion by the focal species. We shall distinguish between two important stages in the invasion process: a. the initial phase of potential successful invasion, or failure, when the invader occurs in relatively few patches, and b. the long term establishment of an invasive species.

First, let us define the model framework. A simple model that captures much of the special metapopulation level effects can be built on the foundation of Richard Levins' (1969) seminal paper on metapopulation theory. He regarded the environment as composed of T patches of equal quality (where T is large). He modelled the dynamics of a particular species within this landscape, assuming that each patch is either occupied by members of the species, or empty. Occupied patches send out migrants that settle on patches at random. Each occupied patch sends out migrants that are successful in finding another patch at rate m. We assume for the moment that we can ignore the spatial arrangement of patches, and that the dispersers from all occupied patches merge into a regional migrant pool. From this pool, they settle uniformly across all patches, both occupied, and empty. If there are N occupied patches, the total pool size is mN, and so the rate of arrival of immigrants on each empty (or occupied) patch is simply $\alpha = mN/T$. Since there are $T - N$ empty patches the rate of colonization of new patches is $K\alpha\,(T - N)$, where K is the probability that a single empty patch is colonized (i.e. converted from empty to occupied by a resident population of the species) by an immigrant. Note that if $K < 1$, this means that some colonizations fail. Levins also assumed that each occupied patch becomes extinct (that is changes from occupied to empty) at con stant rate e. These simple assumptions can later be generalized, but are very useful as a first step for developing an intuitive feeling for metapopulation dynamics (Fig. 3). With these assumptions the rate of change in the number of occupied patches is

$$\frac{dN}{dt} = K\alpha\,(T - N) - eN. \qquad (1)$$

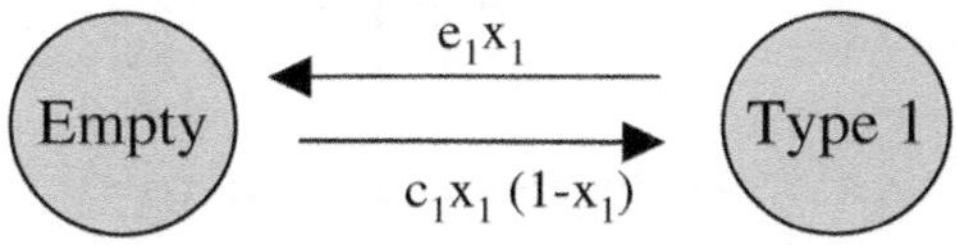

Fig. 3 The classic Levins (1969) metapopulation model for a single species. A given patch can be in one of two states, empty or occupied (by species 1). The rate of change is given by the colonization and extinction rates and the proportion x_1 of occupied patches. (c_1 corresponds to K_α in eqns. 2 and 3).

Let $x = N/T$ be the proportion of patches that are occupied. Then the above equation can be rewritten in terms of x as

$$\frac{dx}{dt} = K\alpha\,(1 - x) - ex, \tag{2}$$

where

$$\alpha = mx. \tag{3}$$

It can be seen from these equations that there is an equilibrium (where the rates of extinction and colonization are equal) at which a certain proportion $x^* = 1 - e/Km$ of patches are occupied (Levins 1969). The equilibrium is stable since the fraction of occupied patches will always tend towards this equilibrium. (To see that the equilibrium is stable note that if we write equation (2) as $dx/dt = f(x)$ then $f'(x^*) < 0$.)

Even this very simple model can be used to illuminate the conditions causing successful versus failed invasions. To see how an invader will do when it is introduced to an empty patch network, we focus on very low x values (this is usually a reasonable assumption regarding the initial conditions of natural or inadvertent invasions, but it may not always be appropriate for deliberate, large-scale introductions). From equations (2) and (3)

$$\frac{1}{x}\frac{dx}{dt} = Km - e \text{ for small } x. \tag{4}$$

Thus the species can invade if $Km > e$. If invasion is possible the invading species will increase its presence until the equilibrium size x^* is reached. (see Box 1 for a list of model parameters.)

Box 1

Levins model defenitions

T = the number of habitat patches available
N = the number of habitat patches occupied by the study species
m = rate of migration from an occupied patch
α = the rate of arrival of immigrants to each of the T patches
K = the probability that a single empty patch is colonized (i.e., converted from empty to occupied by a resident population) by a single immigrant
e = rate at which an occupied patch goes extinct
(Overall) rate of colonization = $K\alpha(T-N)$
(Overall) rate of extinction = eN
x^* = equilibrium patch occupancy (where rate of colonization and rate of extinction are equal)

Habitat loss model

U = the number of unsuitable patches
h = fraction of the landscape that is habitable for the study species

Generalized model

$E(\alpha)$ = single patch extinction rate as a function of immigration
$C(\alpha)$ = single patch colonization rate as a function of immigration
Rescue effect = immigration into a patch lowers patch extinction risk
Anti-rescue effect = immigration into a patch increases patch extinction risk
Allee effect = colonization success is a non-linear function of immigration, accelerating with increasing immigration rate
Anti-Allee effect = non linear colonization success, decelerating with increasing immigration rate

Model with two distinct habitats

h_i = the fraction of a landscape that is covered by type i habitat patches
p_i = the fraction of the landscape that is both of habitat type i and occupied
e_i = extinction rate of type i patches
c_{ij} = colonization onto empty patches of type i, due to migrants generated by occupied patches of type j
λ_i = rate of increase in patch occupancy of species when it is rare, and only patch type i is present.

Box 1

Two species metapopulation model

N_1 = the number of patches occupied by species 1
N_B = the number of patches occupied by both species
Patch states = Empty, Type 1, Type 2, Type B
m_1 = migration rate of species 1 migrants from a Type 1 patch

$\tilde{m}_1$ = migration rate of species 1 migrants from a Type B patch
$C_1(\alpha_1)$ = colonization rate of an empty patch by species 1 immigrants

$\tilde{C}_1(\alpha_1)$ = colonization rate of a Type 2 patch by species 1 immigrants
$E_1(\alpha_1)$ = extinction rate of a Type 1 patch
$\tilde{E}_1(\alpha_1)$ = extinction rate of species 1 in a Type B patch

Biological conclusions from Levins model

Failed invasions can arise either from processes that affect extinction, or from the details of the process of colonization. There are many reasons extinction rates may be high in a novel environment. If within each patch the intrinsic growth rate (r) of the invader is negative (i.e., $r < 0$, so births < deaths), then extinction is inevitable (unless there is very rapid evolution, see Gomulkiewicz and Holt 1995). This could arise because of unfavorable abiotic conditions, or because superior competitors or effective predators are present. But even if $r > 0$, if a patch has a very low local carrying capacity, it may suffer high rates of extinction due to demographic stochasticity; if it has a low r, it may also not be able to recover rapidly from catastrophes, and so randomly walk to extinction. Environments with frequent catastrophes are likely to generate high extinction rates for invasive species, so all else being equal, are not as likely to be invaded.

But equal weight should be given to constraints on the colonization process as potential causes of failed invasions. In the terms of the basic model, if m is reduced, so is the likelihood of invasion. The quantity m combines two things: a per capita rate of emigration from an occupied patch, and the number of individuals present there and available to emigrate. Species that are sedentary, or philopatric, are likely to have a low m. Such species should tend to be poor invaders. For species with equal per capita rates of emigration, those species with the highest abundance within occupied patches can generate a high abundance of migrants, and so experience the highest colonization rate into empty patches. Large m should thus reflect both high fecundity and high survival of emigrants across the matrix habitat separating habitable patches. The probability of successful colonization (K) also varies greatly among species. For instance, a high K can be expected in species where the migrants have high fecundity, so that one or a few immigrants can establish a successful population (by quickly producing a large number of offspring, thereby avoiding initial demographic stochasticity), and when individuals are long-lived. Birth-death process models (e.g., as in MacArthur and Wilson 1967) reveal that in comparing species with a given intrinsic growth rate (r), the probability of extinction is lower for species with lower absolute death rates. Conversely, species which are at risk of frequent stochastic mortality events are likely to have low intrinsic colonization propensity, and so have a low K. Immigrants are likely to be unfamiliar with the details of local resources, refuges, and mortality risks, and if learning new conditions is slow or unreliable, immigrants may be vulnerable to rapid elimination in novel environments.

One mathematically slight but biologically significant generalization of the Levins' model suggested by Lande (1987) and Nee and May (1992) is to recognize that in a landscape, not all empty patches are suitable. If we assume that out of T patches, U are unsuitable, then $h = 1 - U/T$ is the fraction of the landscape that is potentially occupiable by the species. Working back through the same steps as before, we end up with $dx/dt = K\alpha(h\text{-}x)\text{-}ex$. The invasion criterion now becomes $Kmh > e$. As noted for a somewhat different version of the model in Holt and Keitt (2000), this simple formulation of invasion encompasses a great deal of biology. In addition to the mechanisms mentioned in the previous paragraphs, another reason why an invasion could fail is that insufficient suitable habitat is available. Even if each suitable patch could sustain a population on its own over a reasonable time scale, if there are too few of them, colonization may not be able to match even a low rate of extinction. Holt and Keitt (2000) show how these ideas permit one to consider in a unified fashion many different hypotheses that have been put forth regarding the factors causing distributional range limits.

The theoretically best invader would have low extinction risk and high m. The theoretical 'best' is likely to be unfeasible, because there are biological constraints, such as tradeoffs between life history traits (e.g., reproduction and longevity in animals, and between development of vegetative propagules and seed production in plants).

Dynamic Levins type metapopulation

The Levins model clarifies many essential features of metapopulation dynamics, but it rests on quite simple assumptions about how migration influences both colonization and extinction (Hanski 1999). The basic Levins framework has been extended to several specific cases in different studies, including for example a linear rescue effect (Hanski 1982, Gotelli and Kelly 1993), and a specific nonlinear Allee effect (Hanski 1994). The Levins model was recently generalized to allow for these and other dynamic influences from migration on both colonization and extinction rates (compare Figs 3 and 4) (Harding and McNamara 2002). The generality is obtained by expressing both colonization and extinction rates as functions of immigration rate, as we now describe.

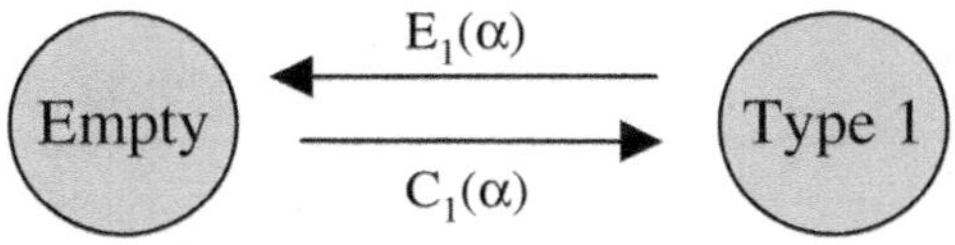

Fig. 4 An extension of Levins' model (Fig. 3). Here the rate of colonization and extinction are not fixed constants, but can be any function of immigration rate (α). This model can therefore capture several types of single species dynamics, such as Allee effects, rescue effects and anti-rescue effects (Harding and McNamara 2002).

It is useful to think of metapopulation processes at two scales. The smallest scale is the patch. The subpopulation inhabiting a patch has a given risk of local extinction $E(\alpha)$, that may be influenced by the rate of immigration α to the patch. Empty patches are colonized at rate $C(\alpha)$, that very likely depends on α. Given these 'local rates of change' the overall rate of change in the proportion ($x = N/T$) of patches that are occupied in the whole metapopulation is

$$\frac{dx}{dt} = C(\alpha)\,(1 - x) - E(\alpha)x \tag{5}$$

(Harding and McNamara 2002). (Here we again assume that all patches are potentially occupiable.) The Levins model is then a special case of this model with extinction function $E(\alpha) = e$ and colonization function $C(\alpha) = K\alpha$. As described below, a variety of biological processes can be mimicked phenomenologically with this extended framework.

Dynamic extinction functions

The original Levins extinction function has a constant value, e, that is independent of immigration rate. In a metapopulation it can be important to allow the extinction rate of a patch to depend on immigration into that patch. Immigrants can for example lower patch extinction risk by providing a demographic contribution to declining populations or by introducing new good genes (e.g., reducing inbreeding depression (Sacceri *et al.* 1998)); this is the rescue effect (Brown and Kodric-Brown 1977, Hanski 1982). Following a catastrophe, immigrants can speed the recovery of the population, allowing it to more rapidly rebound from dangerously low densities. Depending on the underlying processes leading to a rescue effect, the extinction function can have different shapes (Fig. 5).

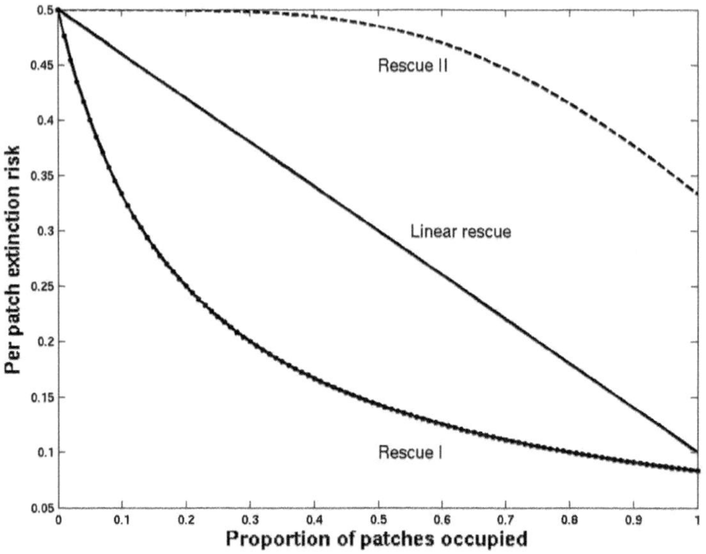

Fig. 5 Illustration of rescue effect functions. The rate of per patch extinction can sometimes decrease as immigration rate increases. Three different types of response to increased immigration are illustrated.

However, in other circumstances immigration can lead to increased extinction risk — we term this pattern an anti-rescue effect. This can for example be due to immigrants carrying parasites or diseases, or because immigrants carry different genes, breaking up locally adapted beneficial gene combinations (Rolán-Alvarez *et al.* 1997). In predator-prey systems, increasing immigration can sometimes destabilize local interactions, making excursions to low densities and hence local extinctions more likely (Schöps 1999, Holt 2002; see review in Hoopes *et al.* 2005). With all these mechanisms, increasing immigration rate over some range of values is likely to boost local extinction rates.

Dynamic colonization functions

In Levins' model the colonization rate per patch increases linearly with the immigration rate, and hence increases linearly with the number of occupied patches.

However, colonization might take other forms as well. It is common for a species to have problems in initial establishment in a patch when the number of immigrants is low, due to demographic skews (lack of adults or males for example), difficulty in finding mates, or inbreeding. The phenomena of disproportionately low colonization success when the immigration rate is low is an example of an **Allee effect** (Allee 1931, Hanski 1994).

The opposite, when colonization by a single migrant is more successful when there are few other migrants is termed an **anti-Allee effect** (Harding and McNamara 2002). It can occur if migrants do not settle at random but avoid conspecifics (e.g., territoriality), so that migrants actively seek unoccupied patches. This leads to a very rapid colonization initially, but after a while most patches are occupied and the effect should become less important. It is also conceivable that if resources are exploited which have low rates of renewal, an increase in the number of immigrants could lead to rapid overexploitation, resource depletion, and thus a reduced probability of successful establishment. Both these mechanisms can lead to an anti-Allee effect at the metapopulation level.

Impacts by resident generalist predators upon colonization by an invasive prey species are likely to shift between anti-Allee effects and Allee effects, depending on the quantitative details of the predator's functional response, and the number of immigrants. If a resident predator has a type III (sigmoidal) functional response, then over a range of low prey numbers, an increase in prey density due to an increasing number of migrants leads to an increase in the per capita mortality rate inflicted on the prey (e.g., because the predator becomes increasingly aware of the novel prey with increasing exposure to it, and so is more likely to cause prey extinction with increasing numbers of colonists). This can be viewed as an anti-Allee effect (which could broadly describe any kind of negative density dependence operating at low densities which can negatively affect colonization success). But typically at higher prey densities, most predators have saturating functional responses (e.g., due to limited handling time or gut capacity), so the per capita mortality rate experienced by prey declines with increasing prey density. In this range, an increasing number of migrants may enhance colonization, by diluting the impact of predation. So whether or not a colonising species experiences an Allee, or an anti-Allee effect, due to resident predators will depend upon both the predator's functional response, and the number of migrants.

Who is a good invader in the single species system?

To analyse whether a species can invade into a patchy landscape, we suppose that the species has been introduced in a small number of patches. We can then ask whether the number of occupied patches tends to grow or decline. In terms of the generalized metapopulation model described above we are concerned with whether $dx/dt > 0$ when x is small. This depends on the behaviour of the colonization and extinction functions for small x, and hence small α. Suppose that

$C(\alpha) \approx K\alpha$ for small α, (6)

and

$E(\alpha) \approx e$ *for small* α. (7)

Thus for small numbers of occupied patches the metapopulation behaves as described by Levins' model with parameters K and e, although it may behave very differently when the number of occupied patches increases. With this assumption, approximation (4) is still valid, so that the species can invade if $Km > e$.

When does an invasive species get established in a single species metapopulation?

Figure 6 a-f illustrate the extinction rate (white surface) and colonization rate (black 'hills') for single species metapopulations. It shows how different combinations of $C(\alpha)$ and $E(\alpha)$ influence the equilibria (where the two surfaces meet). Depending on the functions $C(\alpha)$ and $E(\alpha)$ there may exist at least one non-trivial equilibria in patch occupancy ($x^* > 0$) such that total colonization and extinction rates balance when a proportion x^* of patches are occupied. In some cases there may be more than one such equilibrium (i.e., for a given m the two rates are equal at two or three different x, see Fig. 6 b,c,e,f) (Harding and McNamara 2002). The exact shapes of $C(\alpha)$ and $E(\alpha)$ are important also for the extinction risk of the entire metapopulation, at low patch numbers (Fig; 7). The occurrence of multiple equilibria has earlier been studied in more complex metapopulation models (Hanski and Gyllenberg 1993, Gyllenberg *et al.* 1997). If at least one equilibrium exists then there is at least one stable equilibrium. Here by 'stable' we mean that if x is initially close to x^* then the proportion of occupied patches will tend to x^* over time. In Levins' model there is a unique equilibrium (that is stable) if $Km > e$ (Fig. 6 a) and there is no equilibrium if $Km \leq e$. Thus in this particular model a stable equilibrium exists if and only if the species can invade!

For other shapes of $C(\alpha)$ and $E(\alpha)$ it is also true that if a species can invade, then the proportion of occupied patches will certainly increase and tend to a stable equilibrium. In general, however, the converse need not hold, and the existence of an equilibrium is not equivalent to the ability to invade. To see this, suppose that there is an Allee effect. Thus the patch colonization function $C(\alpha)$ is an accelerating function of α for small α. Let the patch extinction function be as in Levins model; i.e., $E(\alpha) = e$ (a constant). Then for suitable choice of e the total extinction rate exceeds the total colonization rate for small x, but colonization exceeds extinction if x is somewhat larger (Fig. 6e). Thus the species cannot invade; however, if the proportion of patches occupied were artificially increased so as to exceed some critical level, then the proportion would continue to increase and tend to a stable equilibrium.

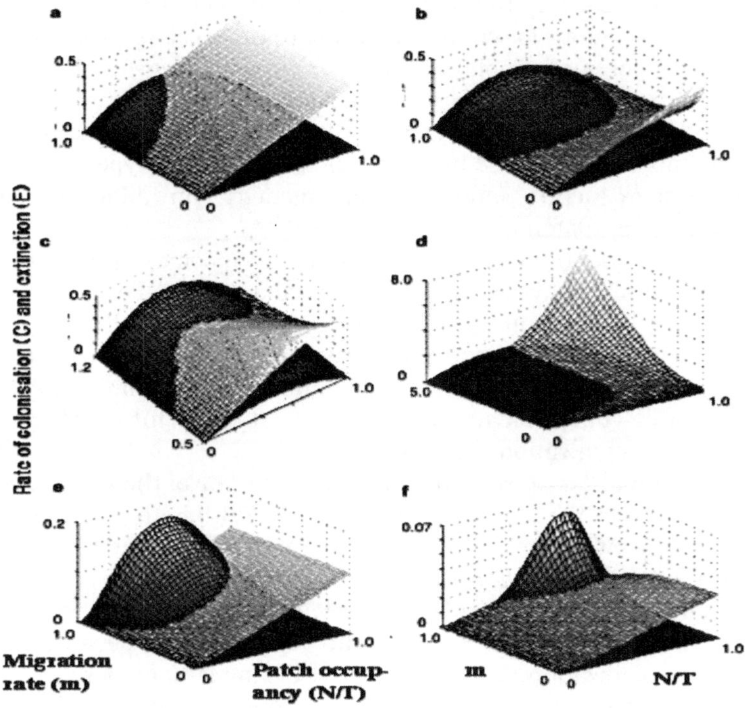

Fig. 6 A range of metapopulations with different dynamics of the model in Fig. 4. White surfaces are total extinction risks and black hills total colonization rates. a. An ordinary Levins metapopulation. b. A rescue effect makes extinction risk decline with higher migration rates. c. A different kind of rescue effect. d. Anti rescue effect leads to higher extinction risk with higher migration rates. e. The colonization function is bell-shaped; an Allee effect prevails. f. An example with both a rescue and Allee effect.

Heterogeneous landscapes

Holt (1997) considered a generalization of the standard Levins' metapopulation model to a landscape with two distinct habitats. A fraction h_i of the landscape is comprised of patches of type *i*. The fraction of habitat patches that are type *i* and occupied by the species is denoted by p_i. Necessarily, we have $p_i \leq h_i$, and $h_1 + h_2 \leq 1$. (If in the latter case, the inequality holds ($h_1 + h_2 < 1$), some patches in the landscape are unsuitable for occupancy.) Using the standard Levins' notation, e_i denotes extinction on occupied patches of type *i*, and c_{ij} is colonization onto empty patches of type *i*, due to migrants generated by occupied patches of type *j*. The equation for dynamics of p_1 is $dp_1/dt = (c_{11}\,p_1 + c_{12}\,p_2)\,(h_1 - p_1) - e_1\,p_1$ (a similar equation describes dynamics in the second habitat type).

The total occupancy by the species in the landscape is $p = p_1 + p_2$. If we sum the dynamical equations for change in occupancy, we get an expression for dp/dt. In general, this does not simplify to a form in just p. However, if all the colonization rates c_{ij} are equal to a constant, c, and both patches have the same extinction rates, the original Levins model emerges. In other words, arbitrary distinctions among habitats can simply be ignored. However, if patch type influences either colonization or extinction, landscape heterogeneity can influence metapopulation dynamics (e.g., patch types differ in K or m).

All the factors mentioned above which influence colonization and extinction rates can potentially show variation among habitat types. For instance, an individual born in a habitat patch of type *i* before emigrating might become acclimatized to the abiotic conditions in its natal site (e.g., as in physiological adaptation to thermal conditions), and so either preferentially seek out similar habitats when it disperses, or be vulnerable to high mortality if it settles into a different habitat, leading to failed colonization.

If only patch type *i* is present, the initial growth rate of the species, when rare, is $\lambda_i = c_{ii} h_i - e_i$ (as in the standard Levins model). With both patch types present, the asymptotic growth rate of the invading species turns out to be $\lambda = (1/2)\,(\lambda_1 + \lambda_2 + \sqrt{[(\lambda_1 - \lambda_2)^2 + 4\, c_{12}\, c_{21}\, h_1\, h_2]})$ (Holt 1997).

Manipulating this expression leads to several simple conclusions. First, a species which is a generalist may be able to invade and persist because it can enjoy cross-habitat colonization, whereas otherwise similar habitat specialists would go extinct (i.e., experience failed invasions). Second, sometimes utilizing a second habitat may be crucial in permitting a species to persist in a landscape, even if there is no colonization among patches of this second type (condition 5 in Holt 1997). This can happen if all colonization from and of the second habitat type is into and from habitats of the first habitat type (e.g., because patches of the second habitat type are widely spaced), but the second habitat type has a lower extinction rate than does the primary habitat type. This can be viewed as a kind of 'spatial storage effect'. It would be interesting in future work to extend this two-habitat model using the generalized approach of Harding and McNamara (2002). Colonization among some habitats may permit rescue effects, even while colonization among others resembles the anti-rescue pattern. In such cases, the expected dynamics of invasion would be quite sensitive to the relative proportions of the landscape occupied by the distinct habitat types.

INTERACTIONS WITH A NATIVE SPECIES

An invasive species can interact with a native metapopulation in the same patch network in various ways, with net effects on the likelihood of invasion ranging from positive to negative. For instance, when species compete for the same limiting resource within patches, competitive exclusion is likely. Metapopulation dynamics can nonetheless permit invasion by an inferior species, if it is superior at colonization of empty patches (Levins and Culver 1971, Tilman 1988). This requires a tradeoff between competition and colonization, allowing the inferior competitor to exploit more effectively the empty habitat patches left by the superior competitor when it suffers extinctions (Amarsekare 2003).

Metapopulation dynamics can also lead to exclusion that would otherwise not occur (Holt 1997). For instance, one species can delimit a second species indirectly, by hosting a pathogen that is more harmful to the second species (Holt and Lawton 1993). An example is the gradual eradication of the native noble cray fish (*Astacus astacus*) in Sweden which is being replaced by the North American signal cray fish (*Pacifastacus leniusculus*), mediated by a shared infectious disease. The invasive species carries a fungus (*Aphanomyces astaci*) to which it itself appears to be immune, whereas the native species is severely impacted by the parasite (Bangyeekhun 2002). In a metapopulation, such indirect exclusion may occur even if the two species occupy distinct habitat patches; in this example, if fungal spores disperse widely the invasive species could provide a landscape 'reservoir' for the pathogen, which can eliminate the native species even in patches where the invasive species itself never occurs.

There can also be facilitative interactions between invasive and resident species. In the literature on succession, there are many examples of early colonizers facilitating invasion by later colonists (e.g., nitrogen-fixers may need to colonize prior to other species) (Connell and Slatyer 1977). In this case, one species alters the abiotic environment so as to enhance colonization or reduce extinction for a second species. Moreover, one species may require another species as a resource. Invasions by specialist predators, parasites or herbivores will almost always depend upon the presence of their required prey or hosts.

A flexible model which allows for many kinds of interactions

We will now look at invasions in which each of two species has interlinked metapopulation dynamics, extending a phenomenological metapopulation model of Hanski (1999) and others (Fig. 8) to include different types of colonization and extinction functions, and differential migration rates. We use this model to classify different types of interactions between an invasive species and a native species.

We consider a metapopulation with T identical patches. Each patch can be in one of four states: Empty, Type 1 when species 1 is present, Type 2 when species 2 is present, or Type B, when both are present (Fig. 9).

Migration

Each Type 1 patch sends out (successful) species 1 migrants at rate m_1. Each Type 2 patch sends out (successful) species 2 migrants at rate m_2. Each Type B patch sends out (successful) species 1 migrants at rate $\tilde{m}_1$ and sends out (successful) species 2 migrants at rate $\tilde{m}_2$. Note that the difference between m_1 and $\tilde{m}_1$ measures how much species 2 can suppress or enhance the migration rate of

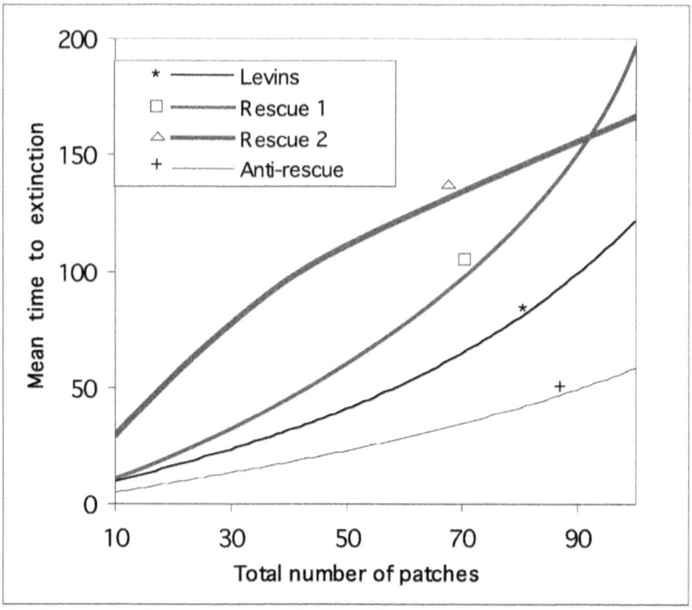

Fig. 7 The risk of extinction at low patch numbers depend on the shape of the *C* and *E* functions (see Fig. 6) From Harding and McNamara 2002.

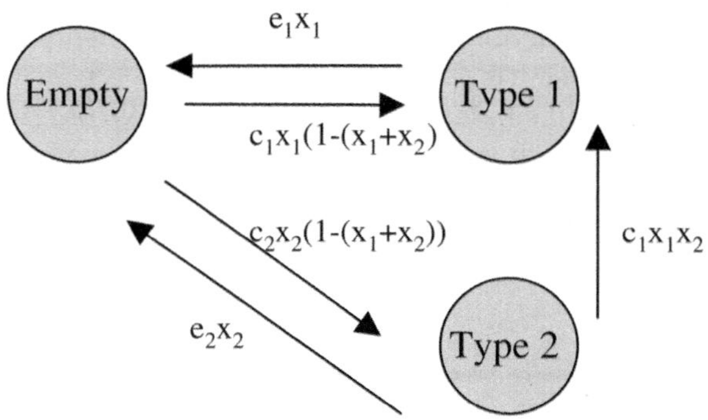

Fig. 8 A metapopulation model for two species. Species 1 is competitively superior and excludes species 2 from patches. Species 2 can persist if it has a higher colonization to extinction ratio than species 1 (Nee and May 1992).

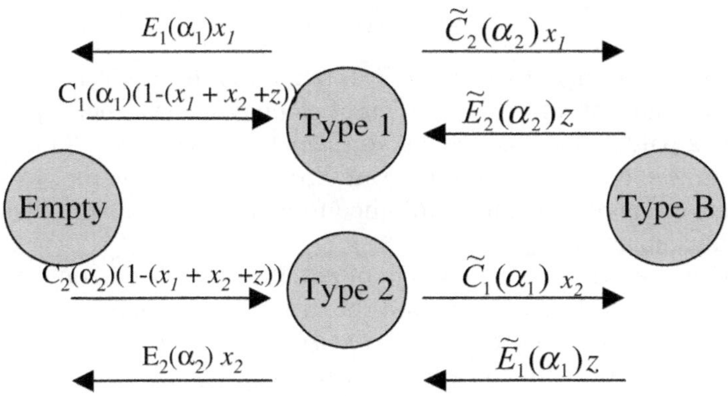

Fig. 9 A metapopulation model specifically designed to study invasive species (eq. 9-12). A patch can be in one of four states: empty, occupied by species 1, or species 2 or occupied by both species (B). This model allows us to explore many sorts of influences that invasive species can have on a native species. The extinction (E) and colonization rates (C) are functions of immigration (α).

species 1. Similarly for species 2, the difference in migration parameters reflects interspecific impacts on the rate at which emigrants are emitted from jointly occupied patches. These differences could occur due to changes in density within patches, or due to changes in individual behavior in the face of the other species (e.g., a prey species may spend more time in hiding in the face of a predator, and thus be less likely to emigrate from patches containing both predators and prey).

Initially, we assume that migrants settle at random. Suppose that the number of Type 1 and Type B patches are N_1 and N_B respectively. Then species 1 migrants arrive at each patch in the environment at rate

$$\alpha_1 = (N_1 m_1 + N_B \tilde{m}_1)/T. \tag{8}$$

The rate of arrival of species 2 migrants, α_2, is similarly defined.

Colonization

Each empty patch is colonized by species 1 migrants (and hence becomes a Type 1 patch) at rate $C_1(\alpha_1)$. Empty patches are colonized by species 2 migrants (and become Type 2) at the rate $C_2(\alpha_2)$. Each Type 2 patch is colonized by species 1 migrants (and hence becomes a Type B patch) at rate $\tilde{C}_1(\alpha_1)$. Similarly each Type 1 patch is colonized by species 2 migrants and is converted to a Type B patch at the rate $\tilde{C}_2(\alpha_2)$. Note that we never have an empty patch simultaneously colonized by both species 1 and 2.

Extinction

A Type 1 patch changes to an empty patch at rate $E_1(\alpha_1)$. A Type 2 patch changes to an empty patch at rate $E_2(\alpha_2)$. A Type B patch (which has both species) changes to a Type 2 patch when species 1 goes extinct, at rate $\tilde{E}_1(\alpha_1)$. Similarly a Type B patch changes to a Type 1 patch at rate $\tilde{E}_2(\alpha_2)$. We assume that in a patch with both species, there are not simultaneous extinctions, taking Type B patches directly back to an empty state.

In this model, the rate of change of patch state depends on migration rates from both the two single species and the mixed species patches. The two species can thus influence each other by altered extinction and colonization rates, both of which can involve altered migration rates. (see Fig. 9.)

Possible interactions between species

Now consider interactions between the two species (Fig. 9). If species 1 competes with species 2, then species 2 may be adversely affected by the presence of species 1 in various ways. It may be more difficult for species 2 to colonise a patch that is already occupied by species 1 than to colonise an empty patch; i.e., $\tilde{C}_2(\alpha_2) < C_2(\alpha_2)$. If species 1 can colonise a patch currently occupied by species 2, so as to produce a patch with both species, then this may increase the probability that species 2 goes extinct on the patch; i.e., $\tilde{E}_2(\alpha_2) > E_2(\alpha_2)$. The presence of species 1 may also reduce the rate at which species 2 sends out migrants from the patch; i.e., $\tilde{m}_2 < m_2$. This will reduce the immigration rate α_2, which will tend to reduce the rate at which patches are colonized by species 2.

If $\tilde{m}_2 < m_2$, there is a possible beneficial effect to species 2 for species 1 to also be present in a patch. Given a strong anti-rescue effect, (i.e., increased immigration increases extinction rate) a reduction in the immigration rate α_2 may *reduce* the rate at which species 2 dies out on patches. Whether competition from species 1 is beneficial or not may depend on the number of patches already occupied by species 2. For example, if the anti-rescue effect only becomes marked at high immigration rates, competition may be deleterious to species 2 when it is rare but beneficial when it is common. In a sense, given metapopulation dynamics and an anti-rescue effect, a competitor (as measured by a reduction in the size of the migrant pool generated by patches) may dynamically be a mutualist.

If instead of being a competitor, species 2 is a predator that benefits from the presence of prey species 1, we would have $\tilde{C}_2(\alpha_2) > C_2(\alpha_2)$, $\tilde{E}_2(\alpha_2) < E_2(\alpha_2)$ and $\tilde{m}_2 > m_2$. In the extreme case where species 2 is a specialist consumer (e.g., a parasite or parasitoid) that is totally dependent on the presence of species 1: $C_2(\alpha_2) = 0$, $E_2(\alpha_2) = \infty$ and $m_2 = 0$, and the model reduces to a three patch system similar to epidemiological models. The direction of the inequalities for the growth of species 1 depend on the biology of the interaction. In a typical predator-prey interaction, the inequalities are all reversed for species 1.

In some cases, however, as noted by Holt (1997, 2002) a predator may instead reduce the rate of extinction of its prey; for instance the presence of the predator may prevent the prey species from overexploiting its own resource base. Although this is an interesting possibility (e.g., generating alternative landscape states), below we focus instead on more classical predator-prey systems where the predator is harmful for prey persistence in a patch.

Dynamics of the two species metapopulation

We now analyse the rates at which patch occupancy in the two species metapopulation changes over time. Denote the fraction of patches with species 1 only by x_1, the fraction with species 2 only by x_2 and the fraction with both species present by z. The fraction of empty patches is thus $1 - (x_1 + x_2 + z)$, and the rates of change of patch occupancy are

$$\frac{dx_1}{dt} = C_1(\alpha_1)\,(1 - (x_1 + x_2 + z)) - E_1(\alpha_1)x_1 - \tilde{C}_2(\alpha_2)x_1 + \tilde{E}_2(\alpha_2)z \qquad (9)$$

$$\frac{dx_2}{dt} = C_2(\alpha_2)\,(1 - (x_1 + x_2 + z)) - E_2(\alpha_2)x_2 - \tilde{C}_1(\alpha_1)x_2 + \tilde{E}_1(\alpha_1)z \qquad (10)$$

$$\frac{dz}{dt} = x_1\tilde{C}_2(\alpha_2) + x_2\tilde{C}_1(\alpha_1) - z(\tilde{E}_1(\alpha_1) + \tilde{E}_2(\alpha_2)) \qquad (11)$$

Here the immigration rates can be expressed in terms of x_1, x_2 and z as

$$\alpha_1 = x_1 m_1 + z\tilde{m}_1 \text{ and } \alpha_2 = x_2 m_2 + z\tilde{m}_2 \qquad (12)$$

Criteria for invasion

Initial invasion

We now describe the criteria for the initial phase of an invasion. We assume that the native species (species 1) currently occupies a proportion x_1 of all patches, where this proportion is a stable equilibrium for the metapopulation with just species 1 present. We then introduce the invader (species 2) into a few patches, and ask whether the frequency of occurrence of the rare invading species increases or declines. Since the invading species is rare, x_2 and z are small. Consequently the rate at which species 2 migrants arrive at each patch, α_2, is also small. Thus all that matters for the initial invasion are the limiting values (for small α_2) of the colonization and extinction functions for species 2. By analogy with

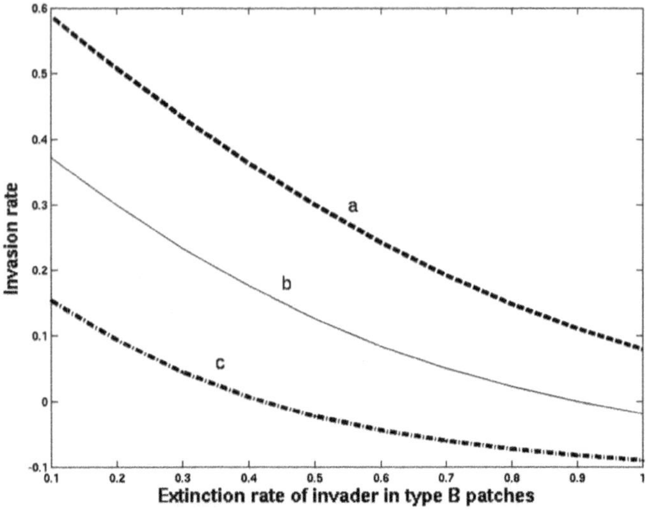

Fig. 10 One example of how initial invasion rate is influenced by $\tilde{e}_2$, the extinction risk of the invader in type B patches (Model structure as in Fig. 9). The three curves represent three different $\tilde{m}_2$, migration rates from type B patches, a: 0.8, b: 0.5, c: 0.2. The figure is based on equations 15-18 and parameter values are $K_2 = 1$, $x_1 = 0.7$, $e_2 = 0.2$, $\tilde{c}_1 = 0.1$, $\tilde{K}_2 = 1$, $\tilde{e}_1 = 0.05$, $m_2 = 0.5$.

equations (6) and (7) we assume that $C(\alpha_2) \approx K_2\alpha_2$, $\tilde{C}(\alpha_2)x_2 \approx \tilde{K}_2\alpha_2$, $E(\alpha_2) \approx e_2$ and $\tilde{E}(\alpha_2) \approx \tilde{e}_2$ for small α_2. During the initial phase of invasion by species 2, the immigration rate of species 1 migrants can be taken as constant and equal to its value at the equilibrium occupancy x_1 of species 1. We thus also assume that the rate at which a Type 2 patch is converted to a Type B patch, $\tilde{c}_1 \equiv \tilde{C}_1(m_1x_1)$, is constant. Similarly the rate at which a Type B patch converts to Type 2, $\tilde{e}_1 \equiv E_1(m_1x_1)$, is constant. We now substitute these expressions into equations (10) and (11), express α_2 in terms of x_2 and z (eq. 12), and since x_2 and z are small ignore terms involving x_2^2, x_2z and z^2. The rate of change in proportion of invaded patches early in the invasion process are then given by the following linearised equations

$$\frac{dx_2}{dt} = (K_2(1 - x_1)m_2 - (e_2 + \tilde{c}_1))x_2 + (K_2(1 - x_1)\tilde{m}_2 + \tilde{e}_1)z$$

$$\frac{dz}{dt} = (\tilde{K}_2x_1m_2 + \tilde{c}_1)x_2 + (\tilde{K}_2x_1\tilde{m}_2 - (\tilde{e}_1 + \tilde{e}_2))z$$

These equations can be written as

$$\frac{dx_2}{dt} = Ix_2 + Jz \tag{13}$$

$$\frac{dz}{dt} = \tilde{J}x_2 + \tilde{I}z \tag{14}$$

where

$$I = K_2(1 - x_1)\, m_2 - (e_2 + \tilde{c}_1) \tag{15}$$

$$\tilde{I} = \tilde{K}_2 x_1 \tilde{m}_2 - (\tilde{e}_1 + \tilde{e}_2) \tag{16}$$

$$J = K_2(1 - x_1)\, \tilde{m}_2 + \tilde{e}_1 \tag{17}$$

$$\tilde{J} = \tilde{K}_2 x_1 m_2 + \tilde{c}_1 \tag{18}$$

The quantity I is the net rate at which a Type 2 patch leads to the production of Type 2 patches: I is the rate of colonization of empty patches by migrants from the Type 2 patch (i.e., $K_2(1 - x_1)\, m_2$) minus the rate of disappearance of the focal Type 2 patch as a result of extinction or colonization by species 1 (i.e., $e_2 + \tilde{c}_1$) (see eq. (15)). The term J can be interpreted as the rate of "crossing over"; the rate at which a Type B patch contributes to the formation of Type 2 patches. There are two ways this contribution is made. The term $K_2(1 - x_1)\, \tilde{m}_2$ is the rate at which migrants from the Type B patch convert empty patches into Type 2 patches by colonization, and $\tilde{e}_1$ is the rate at which the focal Type B patch itself changes to a Type 2 patch as a result of the extinction of the species 1 occupants (see eqn. (17)).

Similarly, $\tilde{I}$ is the net rate at which a Type B patch leads to the production of Type B patches. The quantity $\tilde{K}_2 x_1 \tilde{m}_2$ is the rate at which migrants from the Type B patch colonise Type 1 patches, and so convert them to Type B, and $\tilde{e}_1 + \tilde{e}_2$ is the rate at which the Type B patch reverts to a patch with just a single species present (eq. 16). The cross term $\tilde{J}$ is the net rate at which a single Type 2 patch contributes to the formation of Type B patches. Again there are two elements to this contribution. Migrants from the Type 2 patch colonise Type 1 patches, and the patch may itself be colonized by species 1 (see eq. 18).

Let $w = x_2 + z$ be the total proportion of patches in which species 2 is present. To analyse whether species 2 can initially invade we look at how this quantity changes over time. The linearised equations (13) and (14) imply that, whatever the initial values of x_2 and z, the proportion x_2/w will settle down to an equilibrium value. As this equilibrium is approached

$$\frac{dw}{dt} \rightarrow \lambda\,,$$

where the rate of initial invasion λ is the largest eigenvalue of the matrix $A = \begin{vmatrix} I & J \\ \tilde{J} & \tilde{I} \end{vmatrix}$.

Thus species 2 can invade the metapopulation if $\lambda > 0$, while if $\lambda < 0$ the species goes extinct in the metapopulation. Figure 10 illustrates how invasion rate can be influenced by the extinction risk of the invader in patches of Type B ($\tilde{e}_2$) for one combination of parameter values. In this case increasing extinction risk in Type B patches is very costly for the invader and at too high $\tilde{e}_2$ invasion even becomes impossible (Fig. 10).

Below we shall describe the early invasion process for two special cases, before considering the general criteria for invasion.

An 'invisible' invader

Suppose that the invader, species 2, cannot compete for patches with species 1. Specifically species 2 is unable to colonise patches already occupied by species 1, and species 1 immediately wipes out species 2 when species 1 colonises a Type 2 patch. Then there exist no patches containing both species so $z = 0$, so that equations (13) and (14) reduce to the single equation

$$\frac{dx_2}{dt} = Ix_2 . \qquad (19)$$

The rate of growth of the number of patches containing species 2 is just I, so invasion is possible if $I > 0$. If x_1 approaches zero, this expression reduces to the condition for invasion by a single species discussed earlier. As x_1 increases, from (15), there is a value < 1 at which the invader is excluded. In the simple Levins model, for invasion to be possible requires that the invader have a higher equilibrium patch occupancy (were it alone) than does the resident species (see below).

A 'parasite type' invader

Now suppose that species 2 is totally reliant on the presence of the native species 1. For example, species 2 might be a parasite or a specialist predator of species 1 (e.g., parasitic wasps are often obligate specialists on particular host species, Nouhuys and Hanski 2002), or an obligate commensal. Specifically assume that species 2 is unable to colonise empty patches. Furthermore, on patches where both species are present (a Type B patch) if species 1 dies out then species 2 immediately dies out as well, so that the patch becomes empty. Then $x_2 = 0$, and the rate of growth in the number of patches containing species 2 is just $\tilde{I}$ since equations (13) and (14) reduce to the single equation

$$\frac{dz}{dt} = \tilde{I}z . \qquad (20)$$

Thus in this special case invasion is possible if $\tilde{I} > 0$. From (16), we see that an increase in the occupancy by the resident species 1 tends to facilitate invasion by species 2.

General criteria for invasion

We now consider the general case. It is easy to show that if $I > 0$ then $\lambda > 0$, so invasion is guaranteed. This is not surprising; equation (19) shows that even when competition by species 1 is at its strongest, species 2 can invade by just colonising empty patches. A *fortiori*, if $I > 0$, species 2 can invade whatever the form of the interaction between the two species. Similarly, $\tilde{I} > 0$ implies $\lambda > 0$; if the invader were able to spread when restricted to patches occupied by species 1, then it is certainly a successful invader, regardless of its performance by itself when confronted with empty patches.

Thus, either $I > 0$ or $\tilde{I} > 0$ is sufficient to ensure that species 2 can invade. However, when $I < 0$ and $\tilde{I} < 0$ it may still be possible for species 2 to invade because Type B and Type 2 patches tend to enhance each others formation. In other words migrants from Type 2 patches may colonise Type 1 patches and so convert them to Type B, and Type 2 migrants from Type B patches may colonise empty patches and so convert them to Type 2. It is easy to show that when $I < 0$ and $\tilde{I} < 0$ species 2 can invade ($\lambda > 0$) provided that $J\tilde{J} > I\tilde{I}$.

Invasion in the long run

We now look at a special case, illustrating how the metapopulation system can develop in the long run, after the initial invasion. We are especially interested in characterizing the conditions for long-term coexistence of the invader and the native species.

Consider the example when two species can only colonise empty patches, so that there are no patches with both species present. Who ever comes first wins the patch (a so called 'lottery model'). Thus the presence of one species reduces the number of patches that are available for the other species to occupy. (Note, however, that so far as a species is concerned it is not simply as if the metapopulation as a whole had fewer patches present. This is because migrants of the species still settle on patches at random, so those migrants that settle on a patch currently occupied by the other species are lost. Thus there is a dilution of migrants that settle on empty patches.) When the two species locally exclude one another the equations governing metapopulation dynamics (which are now just equations (9) and (10)) reduce to

$$\frac{dx_1}{dt} = C_1(\alpha_1)\,(1 - (x_1 + x_2)) - E_1(\alpha_1)x_1 \qquad (21)$$

$$\frac{dx_2}{dt} = C_2(\alpha_2)\,(1-(x_1+x_2)) - E_2(\alpha_2)x_2 \qquad (22)$$

For simplicity of exposition we assume that when only species 1 is present there is a unique equilibrium level of patch occupancy, which we denote by x^*_1. Similarly when species 2 is present alone it has a unique equilibrium level of patch occupancy x^*_2.

To analyse the ability of each species to invade when the other is present we assume that $C_1(\alpha_1) \approx K_1\alpha_1$ and $E_1(\alpha_1) \approx e_1$ for small α_1, and $C_2(\alpha_2) \approx K_2\alpha_2$ and $E_2(\alpha_2) \approx e_2$ for small α_2; note that these functions may deviate from this form when α_1 and α_2 are not small. Motivated by Levins' model we define

$$\hat{x}_1 = 1 - \frac{e_1}{K_1 m_1} \qquad \text{and} \qquad \hat{x}_2 = 1 - \frac{e_2}{K_2 m_2} \qquad (23)$$

These are the equilibrium levels of single species occupancy when the above approximations for small α_1 and α_2 hold exactly for all α_1 and α_2. When x_1 is small we have $C_1(\alpha_1)\,(1-(x_1+x_2)) - E_1(\alpha_1)x_1 \approx K_1 m_1(\hat{x}_1 + x_2)x_1$. Thus by equation (21)

$$\frac{dx_1}{dt} > 0 \Leftrightarrow \hat{x}_1 > x_2 \qquad \text{when } x_1 \text{ is small.} \qquad (24)$$

That is the initial rate of increase in patch occupancy by species 1 is positive if the patch occupancy it initially seems to aim towards in the absence of species 2 ($\hat{x}_1$) is higher than the current patch occupancy by the other species (x_2). Thus if species 2 is at its equilibrium of patch occupancy x^*_2, species 1 can invade if

$$\hat{x}_1 > x^*_2 \qquad (25)$$

Similarly if species 1 is at its equlibrium of patch occupancy x^*_1, then species 2 can invade if

$$\hat{x}_2 > x^*_1 \qquad (26)$$

If both expressions (25) and (26) hold, one will see robust coexistence, in that each species can increase when it is rare and the other species is at its respective equilibrium (see further below).

The dynamics of the lottery model with Levins functions

In analysing the possibility for coexistance it is of utmost importance to consider the shape of the extinction and colonization functions. The lottery model outlined above has been shown to exclude coexistence (Hanski 1999; see also Shurin *et al.* 2004). However, this conclusion is based on assuming Levins' static functions. When each of the single species metapopulations are as in Levins' basic model we have $C_1(\alpha_1) = K_1\alpha_1$ and $E_1(\alpha_1) = e_1$ for all α_1, and $C_2(\alpha_2) = K_2\alpha_2$ and $E_2(\alpha_2) = e_2$ for all α_2. In this special case $\hat{x}_1 = x^*_1$ and $\hat{x}_2 = x^*_2$. (I.e. the equilibrium patch occupancy is equal to what is expected by the rate of initial invasion.) Thus if species 2 is at its equilibrium species 1 can invade if and only if $x^*_1 > x^*_2$. Similarly, if species 1 is at its equilibrium occupancy then species 2 can invade if and only if $x^*_2 > x^*_1$. Thus the species with the highest equilibrium patch occupancy when alone can invade the other, and it is never possible for both species to be able to invade each other. Figure 11 shows the metapopulation dynamics in the whole of the* $(x_2 x^*_1)$ plane. As can be seen, the metapopulation dynamics leads to the inevitable extinction of the species with the smallest equilibrium patch occupancy on its own. Thus, with these assumptions, coexistence is not possible, and the species with the largest equilibrium occupancy excludes the other (Hanski 1999 p. 118).

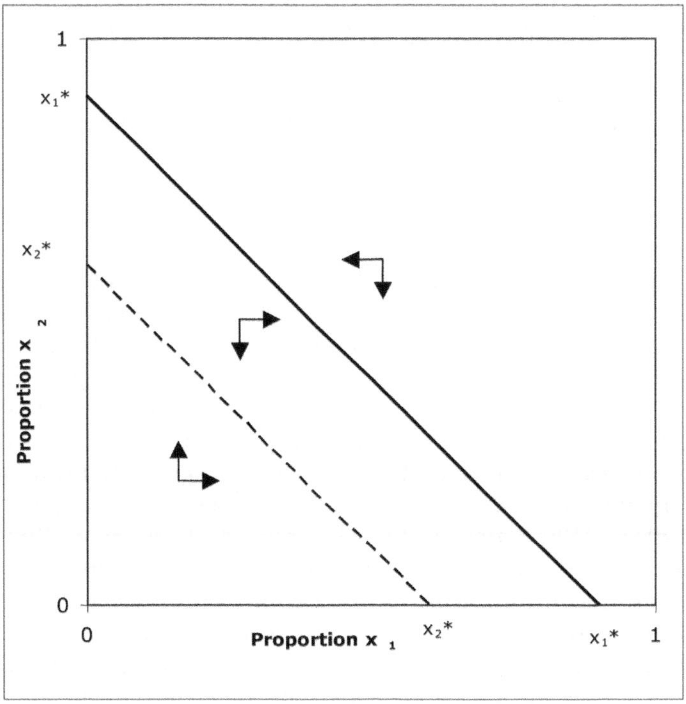

Fig. 11 In a lottery model parameterised with Levins' static functions coexistence of species 1 (x-axis) and species 2 (y-axis) is not possible. The proportion of patches occupied by species 1 will continue to increase until species 2 goes extinct.

The lottery model with general functions

To analyse metapopulation dynamics in the general case when the colonization and extinction rate can have many different shapes we consider how the signs of $\frac{dx_1}{dt}$ and $\frac{dx_2}{dt}$ vary in the (x_1, x_2) plane. When species 2 is absent ($x_2 = 0$) patch occupancy of species 1 increases with time when occupancy is below the equilibrium occupancy x^*_1; i.e.,

$$\frac{dx_1}{dt} > 0 \Leftrightarrow x_1 < x^*_1 \qquad \text{when } x_2 = 0 \tag{27}$$

From criteria (27) and (24) the curve

$\frac{dx_1}{dt} = 0$ in the (x_1, x_2) plane joins the point $(x^*_1, 0)$ to the point $(0, \hat{x}_1)$.

Similarly the curve $\frac{dx_2}{dt} = 0$ joins the point $(\hat{x}_2, 0)$to the point $(0, x^*_2)$.

These curves are illustrated in Fig. 12. The figure also shows the resulting metapopulation dynamics. From criteria (25) and (26), and as the figure illustrates, if $\hat{x}_1 > x^*_2$ and $\hat{x}_2 > x^*_1$ then each species can invade the other at the latter's equilibrium patch occupancy. The metapopulation dynamics then leads to stable co-existence of the two species. The conditions for mutual invasion can be satisfied when the initial rate of increase in patch numbers is higher than would be expected from its equilibrium. This phenomenon is likely to arise if extinction rate is initially lower or colonization initially higher compared to the rate at higher patch occupancies. This type of disproportionally high initial rate of increase in patch occupancy will be found in populations with anti-rescue effects and/or anti-Allee effects. When such effects are present, the initial rate of increase of each single species metapopulation can be high without a correspondingly high patch occupancy at equilibrium. This reduced equilibrium size reduces competition, allowing co-existence, even in the lottery model (see Fig. 12).

The dynamics and stability of a two-species system depend on the exact shape of the colonization and extinction functions. We have illustrated this for the simple lottery model; more complex forms of interactions among species require further analysis.

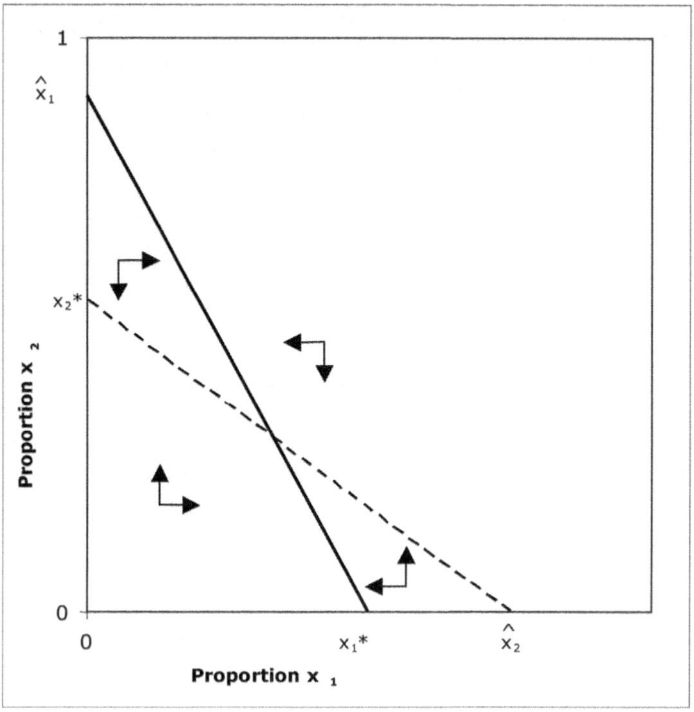

Fig. 12 A lottery model which allows different shapes of the *C* and *E* functions can have stable coexistence of two species. The initial rate of invasion must be disproportionally high, which can be the case for example when there is an anti-rescue effect.

EXPLICIT SPATIAL STRUCTURE AND FINITE NUMBER OF PATCHES

The framework described above assumes the number of patches to be very large. However, if the number of patches is less than about 100 (Hanski 1999, p. 56), single stochastic events of colonization and extinction become important. The risk that all patches will go extinct simultaneously (metapopulation extinction) depends on the combination of metapopulation size and the shape of the per patch extinction function (Fig. 7). A consideration of the effects of finite patch numbers is inevitable when considering the consequences of explicit space and landscape structure.

A very active area of research at present is focused on making metapopulation theory spatially explicit. The basic idea is that in a metapopulation with a finite number of patches each patch has its own specific characteristic and landscape position, which in turn influence its probability of extinction and likelihood of colonization. Metapopulation dynamics of the entire ensemble reflects both heterogeneity among patches (e.g., in area, which can influence extinction) and the specific landscape structure and pattern of connectivity of the patches. The size and composition of local populations also influence invasion speed (Neubert and Caswell 2000). Ovaskainen and Hanski (2004) provide an excellent overview of recent theory in this area, and here we simply sketch some of the basic issues, with an emphasis upon implications for invasion biology. Metapopulation models with a spatially explicit structure provide important linkages with landscape ecology, and with applied issues such as population viability analyses.

One biological assumption that is built into the generalized Levins model we have considered above is that all empty patches are accessible to migrants emanating from all occupied patches. In general, dispersal is spatially limited, and so this assumption will be violated. This may be particularly important to consider in the early stages of an invasion, when a colonising species has established a beachhead on just one or a few sites. From these initial sites, with limited dispersal there may be a small, finite number of empty sites available for colonization. Mathematically, the dynamics can be described as a Markov chain (in discrete time) or Markov process (in continuous time). Extinction can arise because of an analogue of demographic stochasticity at the metapopulation scale. Even if initial invasion is successful, the rate of increase (and rate of spatial spread) of the colonising species can be damped, because many colonists emerging from the center of the invasion simply move to patches that are already occupied. In reaction-diffusion style models of invasive species, heterogeneity in dispersal rates seems to exert a particularly strong effect on the overall rate of invasion of species (Shigesada and Kawasaki 1997). In spatially-explicit landscape models, there can be critical thresholds in habitat connectivity, below which species are likely to go extinct, even though a substantial number of patches exist which are perfectly suitable for occupancy (With 2002). We might call this the 'Tantalus effect' in metapopulation ecology (after the Greek legend of Tantalus, who was tormented by an eternity of hunger and thirst, with luscious grapes and clean water visible, but just barely out of reach...).

In general, spatially explicit metapopulation models seem to imply that localised dispersal makes invasion more difficult. However, were one to place the generalized Levins model discussed above into a spatially explicit landscape, a richer array of outcomes might be observed. For instance, with Allee effects, given uniform dispersal over a large number of patches from an initial propagule, too few individuals might enter any given patch to ensure successful colonization. If instead dispersal is localised, most individuals might enter the same, relatively few patches, thus increasing the effective migrant density in those patches. This could facilitate colonization. What one might observe in this case is a successful nucleus of initial colonization, followed by accretionary growth as neighboring patches get successfully colonized. Alternatively, if there are gaps in availability of suitable habitats, Allee effects could lead to constraints on the ability of an invasive species to expand much beyond the sites of its original colonization (Keitt *et al.* 2001).

One complication that arises in spatially-explicit metapopulations is that if there is heterogeneity among patches (e.g., in attributes that influence colonization), such heterogeneity is likely to exhibit spatial autocorrelation. This can lead to population synchrony which can strongly influence the probability of regional extinction (i.e., a failed invasion) (Caswell and Cohen 1995, Engen *et al.* 2002). In general, for population persistence the size of the metapopulation should be substantially greater than the scale at which such population synchrony is observed. Working out the implications of spatial and temporal autocorrelation for metapopulation dynamics is an important area for future work.

EMPIRICAL STUDIES AND THE ROLE OF THEORY

In attempts to 'test' predictions from theoretical models, empirical metapopulation biologists almost without exception will reach the conclusion that their study system does not fit the assumptions of the model. Or, they might express it the other way around — the model does not fit their observations. But there are many uses of theoretical models, and not all models are meant to be tested. One of their main functions is as tools for thought. Such models can be used to expose the logic of processes assumed to be operating in particular systems, showing clearly how predictions depend on assumptions. Theoretical models can be used as a starting point to generate questions about empirical systems, and as new biological knowledge about the driving forces of that system are discovered, new simplistic general models can be developed, or new more detailed models can be elaborated. Occasionally a model might predict the behaviour of a system (Box 2), although stochasticity due to weather and the chaotic dynamics of nature will mostly hide such correspondence. However, models are useful if correctly handled; incorporating mechanisms that we think are important can reveal the potential behaviour of a system. Just as in learning about the simple building bricks of hydrology and gas kinetics, metereologists also learn more about the nature of currents and winds and even climate change, and yet are unable to predict the weather at a certain spot more than a few days ahead, so the mismatch between a theoretical model and empirical systems in ecology rarely provides sensible 'tests' of the theory itself.

Metapopulation theory in the generalized form presented here is a philosophical tool that highlights how species can persist despite the fact every single subpopulation faces a substantial extinction risk. It does not attempt to describe the detailed behaviour of a given real system at a given time, but captures essentials of forces that apply to real systems.

For example, Gutiérrez *et al.* (2001) studied the spatial distribution of four lepidopteran butterfly species. Because the four species all had the same favourite host plant (*Lotus corniculatus*) the authors claimed this must be a system where the predictions from multi-species models would be realised. However, it turned out that environmental factors other than the favourite host plant determined the distribution of the four species (for example; alternative host plants, differences between coastal or inland sites, and vulnerability to grazing). One of the species was present in the entire landscape and might not act as a metapopulation at all. The study claims to have proven that the assumptions of phenomenological multi-species metapopulation models are unrealistic. But these models can only provide interesting insights into systems where several species actually do compete for the same patches, and where colonization-extinction dynamics and interspecific interactions are in fact major forces at work. These assumptions need to be assessed, and it is not clear that the empirical system of Gutierrez *et al.* (2001) provides a good match to the metapopulation assumption. We suggest that on *a priori* grounds it is reasonable to believe that there are multi-species assemblages which quite faithfully match the assumptions of metapopulation theory (e.g., aquatic predators and prey in ponds occupy patches which may dry and suffer extinctions, parasites in the same host can interact and all go extinct when the host dies, etc.) and where multi-species metapopulation models are likely to produce some interesting insights.

The old Popperian view that hypothesis must be confirmed or falsified in critical tests has led to much confusion at the interface between theoretical and empirical metapopulation biology. Theoretical models can not say more than 'given exactly these conditions we will get this or that outcome; they can not be validated or falsified in systems with other conditions and other prevailing mechanisms. It would be like telling the metereologists that they should be skeptical about the proposition 'warm air is less dense than cold air', simply because they cannot accurately predict today's rainfall in a specific location, such as Ullapool.

Box 2

Biological control experiment in a snail metapopulation living in ponds in Guadeloupe

Biomphalaria glabrata is a native snail species that is the intermediate host for human intestinal schistosomes (a trematode infection). A Venezuelan snail *Marisa cornuarietis* was introduced in 1987. The invader destroyed local populations of water lilies that are important habitats for the native *B. glabrata*. A strong decline in patch occupancy in the *B. glabrata* metapopulation followed in the ponds where the competitor was introduced. Interestingly, nearby control ponds (without the introduced competitor) also had declining patch occupancy (Pointier *et al.* 2004). Although the authors state that environmental circumstances might have contributed to the decline of *B. glabrata* in the un-invaded control ponds, they point out that the observed pattern matches that predicted by simple theoretical metapopulations models — lower patch occupancy leads to lowered overall colonization rate, and the whole metapopulation network can be affected by the removal of a few patches.

CONCLUSION

We have summarised some aspects of how generalized metapopulation theory is relevant for understanding invasions into patchy habitats. When species survive in a landscape as a metapopulation it is critical that colonization success is higher than the extinction rate of subpopulations. In conservation the worry is often that an invader competes with a native species. We have refined a two species metapopulation model to capture different kinds of interactions, including competition. There is a full range of behaviours that emerge, depending on the type of colonization and extinction function of the invader, and the type of influence of the native species. The existence of Allee and anti-Allee effects opens up a rich variety of possible dynamical outcomes. The invasive species becomes extinct if its initial success in colonzing empty patches or resisting extinction in occupied patches at low patch number is too low; we have suggested a number of particular mechanisms that can lead to such exclusion. If the initial growth rate is positive, it can become established and coexist sustainably with the native species or force the latter to go extinct. If the invasive species competes for resources this can reduce patch occupancy for the native species and if the native species is vulnerable, and has an unstable equilibrium beyond which there is no return, then extinction of the native species can be sudden and difficult to reverse. We predict metapopulation theory will become significant more refined and applied to study invasive species problems.

ACKNOWLEDGEMENTS

KCH was funded by the Swedish Research Counil (Vetenskapsrådet) and RDH by the University of Florida Foundation.

REFERENCES

Allee, W. C. 1931. Animal aggregations: a study in general sociology. Chicago University Press, Chicago.

Amarsekare, P. 2003 Competitive coexistence in spatially structured environments: a synthesis. Ecology Letters **6**, 1109-1122.

Bangyeekhun, E. 2002. Parasites on crayfish. Characterisation of their pathogenesis host interactions and diversity. Acta Universitatis Upsaliensis. Comphrehensive Summaries of Uppsala Dissertations from the Faculty of Science and Technology 737. 46 pages.

Brown, J. H. and A. Kodric-Brown. 1977. Turnover rates in insular biogeography: effect of immigration on extinction. Ecology **58**, 445-449.

Caswell, H. and J. Cohen. 1995. Red, white and blue: environmental variance spectra and coexistence in metapopulations. Journal of Theoretical Biology **176**, 301-316.

Connell, J. H. and R. O. Slatyer. 1977. Mechanisms of succession in natural communities and their role in community stability and organization. American Naturalist **122**, 661-696.

Engen, S., R. Lande and B. -E. Saether. 2002. The spatial scale of population fluctuations and quasi-extinction risk. American Naturalist **160**, 439-451.

Gomulkiewicz, R. and R. D. Holt. 1995. When does evolution by natural selection prevent extinction? Evolution **49**, 201-207.

Gotelli, N. J. and W. G. Kelly. 1993. A general model of metapopulation dynamics. Oikos **68**, 36-44.

Gutiérrez, D., J. L. León-Cortés, R. Menédez, R. J. Wilson, M. J. R. Cowley and C. D. Thomas. 2001. Metapopulations of four lepidopteran herbivores on a single host plant, Lotus corniculatus. **82**, 1371-1286.

Gyllenberg, M., I. Hanski and A. Hastings. 1997. Structured metapopulation models. Pages 93-122 in Hanski, I. and M. E. Gilpin, eds. Metapopulation Biology. Academic Press, London.

Harding, K. C. and J. M. McNamara. 2002. A unifying framework for metapopulation dynamics. American Naturalist **160**, 173-185.

Hanski, I. 1982. Dynamics of regional distribution: the core and satellite species hypothesis. Oikos **38**, 210-221.

Hanski, I. 1994. A practical model of metapopulation dynamics. Journal of Animal Ecology **63**, 151-162.

Hanski, I. 1999. Metapopulation Ecology. Oxford University Press, Oxford.

Hanski, I. and M. Gyllenberg. 1993. Two general metapopulation models and the core-satellite species hypothesis. American Naturalist **142**, 17-41.

Holt, D. R. 1997. From metapopulation dynamics to community structure. Pages 149-164 in I. Hanski and M. E. Gilpin, eds. Metapopulation Biology. Academic Press, San Diego.

Holt, R. D. and J. H. Lawton. 1993. Apparent competition and enemy-free space in insect host-parasitoid communities. American Naturalist **142**, 623-645.

Holt, R. D. and T. Keitt. 2000. Alternative causes for range limits: a metapopulations perspective. Ecology Letters **3**, 41-47.

Holt, R. D. 2002. Food webs in space: On the interplay of dynamic instability and spatial processes. Ecological Research **17**, 261-273.

Hoopes, M. F., R. D. Holt and M. Holyoak. 2005. The effects of spatial processes on two-species interactions. Page: 35-67. In Metacommunities: Spatial Dynamics and Ecological Communities, eds. M. Holyoak, M. A. Leibold and R. D. Holt. U. of Chicago Press, Chicago.

Keitt, T. H., M. A. Lewis and R. D. Holt. 2001. Allee effects invasion pinning and species borders. American Naturalist **157**, 203-216.

Lande, R. 1987. Extinction thresholds in demographic models of territorial populations. American Naturalist **130**, 624-635.

Levins, R. 1969. Some demographic and genetic consequences of environmental heterogeneity for biological control. Bulletin of the Entomological Society of America **15**, 237-240.

Levins, R. and D. Culver. 1971. Regional coexistence of species and competition between rare species. Proceedings of the Natl. Academy of Sciences. USA, **68**, 1246-1248.

MacArthur, R. H. and E. O. Wilson 1967. The theory of island biogeography. Princeton University Press, Princeton.

Neubert, M. and H. Caswell. 2000. Demography and dispersal: calculations and sensitivity analysis of invasion speed for structured populations. Ecology **81**, 1613-1628.

Nee, S. and R. M. May, 1992. Dynamics of metapopulations: habitat destruction and competitive coexistence. Journal of Animal Ecology **61**, 37-40.

Nouhuys, S. V. and I. Hanski. 2002. Colonisation rates and distances of a host butterfly and two specific parasitoids in a fragmented landscape. **71**, 639-650.

Ovaskainen, O. and I. Hanski. 2004. Metapopulation dynamics in highly fragmented landscapes. Pages 73-104 in I. Hanski and O. E. Gaggiotti, editors. Ecology, Genetics and Evolution of Metapopulations, Elsevier/Academic, Burlington.

Pointier, J. P. and P. David. 2004. Biological control of Biomphalaria glabrata, the intermediate host of schistosomes, by Marisa cornuarietis in ponds of Guadeloupe: long term impact on the local snail fauna and aquatic flora. Biological Control **29**, 81-89.

Rolán-Alvarez, E., K. Johannesson and J. Erlandsson. 1997. The maintenance of a cline in the marine snail Littorina saxatilis: the role of home site advantage and hybrid fitness. Evolution **51**, 1838-1847.

Sacceri, I., M. Kuussaari, M. Kankare, P. Vikman, W. Fortelius and I. Hanski. 1998. Inbreeding and extinction in a butterfly metapopulation. Nature **392**, 491-494.

Schöps, K. 1999. Effect of deforestation on the spatial dynamics of an endangered weevil species. Page 45 in Habitat Loss Helsinki, 7-12 Sept. Hakapaino Oy, Helsinki.

Shigesada, N. and K. Kawasaki. 1997. Biological Invasions: Theory and Practice. Oxford University Press, Oxford.

Shurin, J. B., P. Amaresekare, J. M. Chase, R. D. Holt, M. F. Hoopes, R. Law and M. A. Leibold. 2004. Alternative stable states and regional community structure. Journal of Theoretical Biology **227**, 359-368.

Skellam, J. G. 1951. Random dispersal in theoretical populations. Biometrika **38**, 196:218.

Tilman, D. 1988. Plant strategies and the dynamics and structure of plant communities. Princeton University Press, Princeton, NJ.

With, K. A. 2002. Using percolation theory to assess landscape connectivity and effects of habitat fagmentatin. Pages 105-130 in K. J. Gutzwiller, editor. Applying Landscape Ecology in Biological Conservation. Springer-Verlag, New York.

Chapter eighteen

Competition and the assembly of introduced bird communities

R. P. Duncan and D. M. Forsyth

INTRODUCTION

A central goal of ecology is to understand what determines the number and identity of species in ecological communities. Of the many species that could potentially occupy an area, why do only a particular subset of species actually co-occur and what determines the identity of those species? And to what extent are these patterns the product of deterministic processes rather than stochastic events?

Much interest has focused on interspecific competition as a process shaping co-occurrence patterns (Strong *et al.* 1984, Diamond and Case 1986): some species capable of joining a local community may be excluded by the presence of competitors. If competition is sufficiently strong and pervasive enough to structure ecological communities then certain 'assembly rules' should govern how communities are put together (Diamond 1975, Weiher and Keddy 1999). In particular, competition should be more intense, and competitive exclusion more likely, among species of similar size and morphology that compete for similar resources. In these circumstances we expect co-occurring species to be morphologically different from each other and to exhibit a pattern of 'morphological overdispersion' (Pimm 1991).

M.W. Cadotte et al. (eds.), Conceptual ecology and invasion biology, 405–421.

Many studies have attempted to evaluate the importance of interspecific competition in shaping co-occurrence patterns by testing for morphological overdispersion (e.g., Strong *et al.* 1979, Grant and Abbot 1980, Bowers and Brown 1982, Diamond and Case 1986, Dayan and Simberloff 1994, Kingston *et al.* 2000). Such tests have generally involved comparing the morphological patterning of species in a local community with that of a larger pool of species potentially capable of joining that community. If species in the local community exhibited stronger morphological overdispersion than expected under a process of random assembly from this larger pool then it was concluded that the local community was structured by competition. Critical to the outcome of such tests is deciding which species to include in the larger pool of species that could potentially join the local community (Colwell and Winkler 1984, Gotelli and Graves 1996). Ideally, all species known to have been present at some stage during the process of assembling that community would be included. Mistakenly including species that were never part of the process (e.g., because they never reached the site) can bias the outcome of tests for overdispersion (Colwell and Winkler 1984). Unfortunately, in most situations we do not know how local communities were assembled and which species were (or were not) present during the process. Consequently, evidence for competitive exclusion based on tests for morphological overdispersion is often subject to uncertainties associated with choosing the appropriate species pool.

Since the 17th century there has been a substantial increase in the movements of people around the globe and an associated increase in the numbers of species transported to and released in new environments (Williamson 1996, Mack *et al.* 2000, Duncan *et al.* 2003). At some locations there are sufficiently detailed records that we can reconstruct the history of these introductions. In particular, there are excellent records documenting the bird species introduced to locations such as the New Zealand and Hawaiian islands, including data on the species introduced, dates of introduction, whether they established wild populations or not, and measures of effort such as the numbers of individuals or the numbers of release events of each species introduced (Moulton and Pimm 1983, Veltman *et al.* 1996). These records provide a remarkable opportunity to test hypotheses about the factors structuring ecological communities (Pimm 1991, Lockwood *et al.* 1999) because they provide a near complete list of the species present during the assembly process, thus reducing the uncertainty associated with choosing the appropriate species pool. Furthermore, because bird introductions to many places (including Hawaii and New Zealand) occurred decades to centuries ago, there should have been sufficient time for the communities to reach compositional equilibrium: species that are going to be excluded by competition should have already been so.

Because of these advantages, records of bird introductions have been used extensively to test for patterns consistent with the outcome of competition. In addition to tests for morphological overdispersion, studies have tested for 'priority effects' (Moulton 1993): birds introduced at a later date in the assembly process should be more likely to fail because they have to compete with a greater diversity of already established species at higher abundance. Using these approaches, several studies have identified patterns consistent with competition and argued that it is an important process structuring introduced bird assemblages (Moulton and Pimm 1983, Moulton 1985, Moulton and Pimm 1986a, b, 1987, Pimm 1991, Moulton and Lockwood 1992, Lockwood *et al.* 1993, Moulton 1993, Lockwood and Moulton 1994, Brooke *et al.* 1995, Moulton and Sanderson 1997, Lockwood *et al.* 1999, Moulton *et al.* 2001).

Nevertheless, the results of some studies have been equivocal (Simberloff 1992, Moulton *et al.* 1996, Duncan 1997) and several objections have been raised regarding the analysis and interpretation of these data (Simberloff and Boecklen 1991, Duncan and Blackburn 2002). In particular, patterns consistent with competition can be caused by or confounded with variation in other factors. Specifically, Duncan (1997) showed that, while passerine birds introduced to New Zealand at a later date were more likely to fail (a priority effect consistent with the outcome of competition), this result was confounded with variation in introduction effort. Birds introduced at a later date were also released in smaller numbers and so were more likely to fail for that reason. Similarly, Moulton *et al.* (2001) found significant morphological overdispersion among gamebirds successfully introduced to New Zealand, but Duncan and Blackburn (2002) showed that this pattern could not have been due to competition because the distribution of most species did not overlap in space or time.

Our aim in this chapter is to try and resolve these difficulties and to gain greater insight into how historical events, including the timing and effort put into introductions, can interact with a deterministic process, competition, to affect the outcome of introductions and hence the composition and structure of introduced species assemblages. We investigate, using a simple mathematical model, how three factors (introduction effort, the abundance of a competitor, and the strength of competition) interact to determine the probability that an introduced species will establish in the presence of a competitor. We then examine how well the predictions of the model are supported by data using the historical record of passerine introductions to New Zealand.

A MATHEMATICAL MODEL OF ESTABLISHMENT

Since introductions typically involve the release of few individuals, stochastic processes can play a critical role in determining whether they succeed or fail (Richter-Dyn and Goel 1972, Lande 1993, Grevstad 1999). We therefore used a simple two-species stochastic birth-death model (SBD, Renshaw 1991) to explore how interactions between the number of individuals released, the abundance of a previously established competitor and per-capita strength of interspecific competition affect the probability of a species establishing.

Let N_1 be the number of individuals of species 1 released at a location and let N_2 be the abundance of species 2, a potential competitor, already present at that location. We assume that both species have identical birth and death rates and that interspecific competition is symmetric; we are primarily interested in how differences in the initial population sizes of the two species and the per-capita strength of competition affects the probability that the introduced species 1 establishes.

Assuming that isolated populations grow subject to logistic growth and that density-dependence (both within and between-species) affects only the death rate, changes in the size of populations of the two species can be described by a pair of differential equations:

$$\frac{dN_1}{dt} = B(N_1) - D(N_1) = N_1 b - N_1(d + cN_1 + eN_2),$$

and

$$\frac{dN_2}{dt} = B(N_2) - D(N_2) = N_2 b - N_2(d + cN_2 + eN_1),$$

where $B(N_i) = N_i b$ is the birth rate of species i, with b the instantaneous birth rate, and $D(N_i) = N_i(d + cN_i + eN_j)$ is the death rate, with d the instantaneous death rate, c is a coefficient reflecting the per-capita strength of intraspecific competition, and e is a coefficient reflecting the per-capita strength of interspecific competition. If the two populations are at size N_1 and N_2 at a given time then the next event will be one of the following (Renshaw 1991):

1. A birth of species 1 (transition $N_1 \rightarrow N_1 + 1$) with probability $B(N_1)/R(N)$;
2. A death of species 1 (transition $N_1 \rightarrow N_1 - 1$) with probability $D(N_1)/R(N)$;
3. A birth of species 2 (transition $N_2 \rightarrow N_2 + 1$) with probability $B(N_2)/R(N)$;
4. A death of species 2 (transition $N_2 \rightarrow N_2 - 1$) with probability $D(N_2)/R(N)$,

where $R(N) = B(N_1) + D(N_1) + B(N_2) + D(N_2)$.

We implemented a SBD model by generating two independent random numbers, Y_1, Y_2 on the uniform unit interval [0,1]. If:

1. $Y_1 \leq B(N_1)/R(N)$ then the next event was a birth of species 1.
2. $Y_1 > B(N_1)/R(N)$ and $\leq [B(N_1) + D(N_1)]/R(N)$ then the next event was a death of species 1.
3. $Y_1 > [B(N_1) + D(N_1)]/R(N)$ and $\leq [B(N_1) + D(N_1) + B(N_2)]/R(N)$ then the next event was a birth of species 2.
4. $Y_1 > [B(N_1) + D(N_1) + B(N_2)]/R(N)$ then the next event was a death of species 2.

The time between events is an exponentially distributed random variable so that the inter-event time, t, was simulated as:

$$t = \frac{-[\log_e(Y_2)]}{R(N)},$$

and the time advanced by that amount (Renshaw 1991). This process was reiterated for 1000 steps. We defined species 1 as having established if it persisted to the end of the 1000 step simulation (for the parameter values we used, populations that persisted for 1000 steps were typically fluctuating around carrying capacity and were likely to persist for much longer).

There is strong theoretical (Lande 1993, Grevstad 1999) and empirical evidence (Inchausti and Halley 2003, Drake and Lodge 2004) that environmental stochasticity affects the persistence of species over and above the effects of demographic stochasticity alone. Hence, in addition to the demographic stochasticity modelled above, we also modelled environmental stochasticity by allowing the death rate of each species to vary independently between steps according to the following equation:

$$d_i = d + \text{abs}\,[N(0,\, u)],$$

where d_i is the instantaneous death rate of species i and $N(0,u)$ is a random normal deviate with mean 0 and variance u. This acts to model deviations away from some minimum instantaneous death rate, d, with larger values of u implying greater variation in death rate between steps and thus greater environmental stochasticity.

To investigate the general behaviour of the model, we held parameters constant at the following values in our simulations: environmental stochasticity $u = 0.4$, birth rate $b = 2$, death rate $d = 1$, and intrapecific competition coefficient $c = 0.02$. We set initial values of N_1 as 1, 2, 3, 4, 5, 10, 15, and 20, and N_2 as 5, 10, 20, and 40, and the interspecific competition coefficient e as 0, 0.01, 0.02, 0.03, 0.04, and 0.05. For each combination of parameter values for N_1, N_2 and e, we ran the SBD model 1000 times and calculated the proportion of those 1000 runs that resulted in species 1 establishing successfully. This proportion is an estimate of the probability of establishment under the set of conditions defined by the parameters.

The results of our simulations are summarised in Fig. 1. First, the probability of establishment increases as introduction effort (N_1, the initial abundance of species 1) increases. Second, when interspecific competition occurs (i.e., $e > 0$) the probability of establishment declines as both the per-capita competition coefficient, e, and the initial abundance of the competitor species, N_2, increases. Moreover, these three factors interact such that competition has little effect on the probability of establishment when a competitor species occurs at low initial abundance, regardless of the strength of competition. Here, stochastic events dominate and the number of individuals released is critical. When competitor species occur at higher initial abundance, the probability of establishment depends increasingly on the strength of competition.

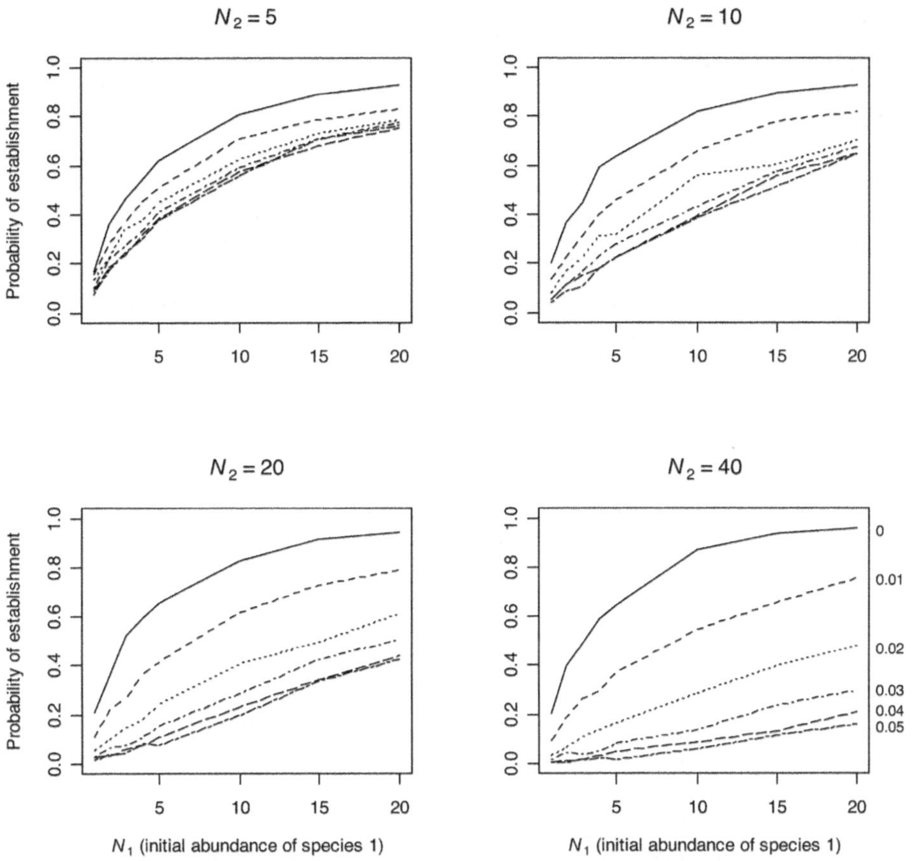

Fig. 1 Results of the stochastic birth-death model showing the probability of establishment as a function of the initial abundance of an introduced species (N_1), given a resident competitor at four different abundances (N_2, the four panels), and with the per-capita strength of interspecific competition taking six values from 0 to 0.05 (shown as different lines).

These simulation results highlight two predictions. First, the abundance of a competitor in addition to the per-capita strength of competition should be important in determining the outcome of an introduction. Hence, we would expect to observe both morphological overdispersion (resulting from the competitive exclusion of morphologically similar species) and priority effects (a higher chance of failure among later introductions because already resident competitors have had time to increase in abundance) in introduced assemblages structured by competition. Second, these factors should interact such that the influence of per-capita strength of competition depends on the relative abundances of the introduced and competitor species, and the importance of competition depends on introduction effort (Fig. 1). Thus, inter-specific competition is predicted to be of much less importance in affecting establishment when a species is introduced in small numbers (stochastic processes dominate) or when the resident competitor occurs at low abundance. The effects of competition should be most apparent when both the introduced and competitor species occur at high abundance.

CONFRONTING THE MODEL WITH DATA: THE ASSEMBLY OF INTRODUCED PASSERINE COMMUNITIES IN NEW ZEALAND

Data

We used historical data documenting the timing and numbers of 42 passerine species introduced to four acclimatisation districts in New Zealand (Auckland, Wellington, Canterbury, and Otago) to test these predictions. The data are from Duncan (1997) and list the passerine species introduced to each district, the year of introduction, whether species established a wild population or not, the year species went extinct if they failed to establish, and introduction effort measured as the number of separate release events. For each species introduced to each district (a total of 97 introductions) we used these data, and the morphological data described below, to estimate the strength of interspecific competition, the relative abundance of the closest competitor, and introduction effort.

We obtained morphological data by measuring museum specimens. In all cases, we measured specimens collected in the country where the species was most likely to have been introduced from and, where possible, we measured specimens collected around the time that introductions to New Zealand occurred (the late 1800s). The country of origin for most European species was United Kingdom while for North American species we measured specimens collected on the west coast of the United States, a likely source area for introductions to New Zealand. For *Lonchura punctulata*, which is native to Asia, we measured specimens collected in Australia because this species was introduced to New Zealand from there (Thomson 1922). All of the specimens we measured were housed in the British Museum of Natural History at Tring, UK.

We measured 20 individuals of each species, 10 male and 10 female, except for four species where we measured 12–19 individuals, with approximately equal numbers of males and females. We measured six morphological characters on each individual: the length of the culmen (from the base of the skull to the tip of the upper mandible), the width and depth of the beak at the anterior margin of the nares, the length of the tarsus, the length of the wing (from the wrist to the tip of the longest primary), and the length of the tail.

Following Ricklefs and Travis (1980) and Moulton and Pimm (1987), we summarised morphological differences among species using the first two axes derived from a principal components analysis (PCA) based on log-transformed values of all six morphological characters, with introductions to each district analysed separately (Fig. 2). For each district, we used the Euclidean distance between species in this morphological space to estimate how morphologically similar species are to each other, and used this as a measure of the strength of interspecific competition.

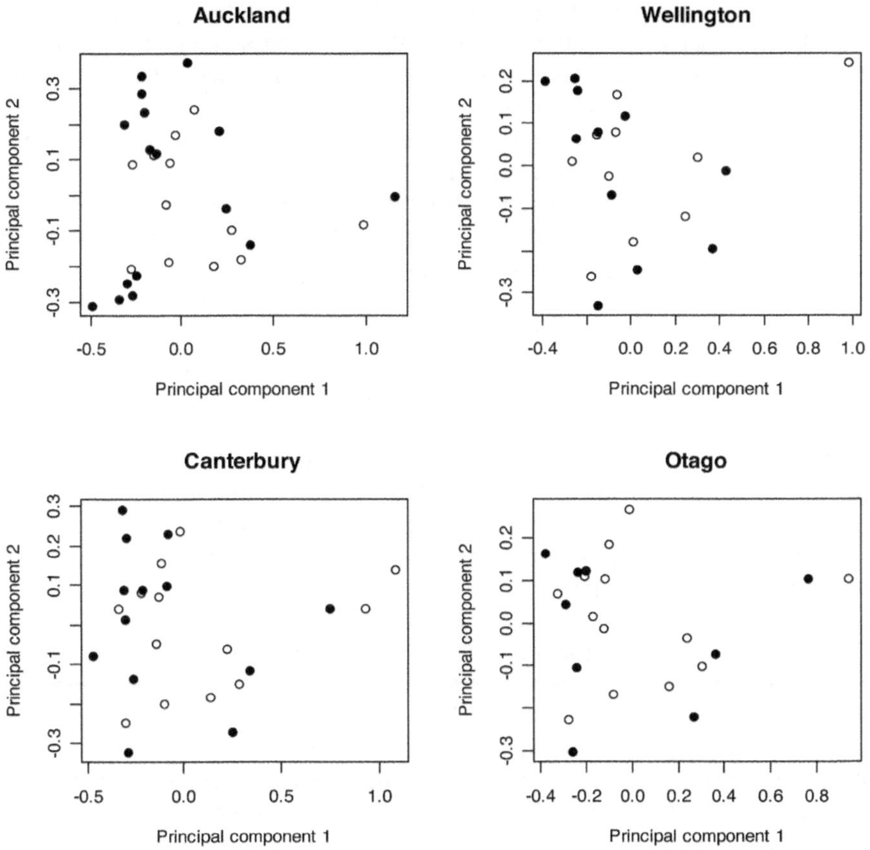

Fig. 2 Plots of the first two axes from a principal components analysis summarising morphological differences among the passerine species introduced to each of four acclimatisation districts in New Zealand (see text). Species that established are shown as open circles, species that failed are shown as closed circles.

Previous studies have used minimum spanning trees coupled with randomisation tests to search for overdispersion among successfully established species in morphological space (e.g., Moulton and Pimm 1987, Moulton and Lockwood 1992, Lockwood *et al.* 1993, Lockwood and Moulton 1994, Lockwood *et al.* 1999). Here we adopt a regression approach because we want to simultaneously consider the influence of several variables on establishment probability. For introductions to each district, we used the (log-transformed) distance between a species and its closest neighbour in morphological space as an estimate of the strength of interspecific competition; species with a nearest neighbour more similar in morphology are assumed to have faced stronger competition for the same resources. A species could have as its nearest neighbour only species that were present in the same district at the same time (determined from the year of introduction to a district and the year of extinction for species that failed). This avoids a problem, often present in the use of minimum spanning trees, whereby all species are included as potential competitors even if the distributions of some species fail to overlap in either space or time (Duncan and Blackburn 2002).

The level of competition that individuals of a species experience will be a function of their similarity to a competitor and the abundance of that competitor. No species abundance estimates are available for the period during which most introductions occurred in New Zealand (the late 1800s). However, Thomson (1922) observed that introduced passerines increased rapidly in abundance following establishment. Although legislation was originally passed protecting introduced birds, Thomson (1922) stated that "*within ten or fifteen years of the coming in of the birds, their numbers increased to such an extent, that the protection afforded them had to be taken away, and restrictive legislation imposed*". During this period of rapid population increase the number of years elapsed since introduction should provide a reasonable index of relative abundance, at least for those species that successfully established. We therefore used the time delay (in years) separating the introduction of a target species and its nearest competitor as an estimate of that competitor's relative abundance. Our assumption is that the longer the time delay, the longer a competitor's population will have had to increase, and the higher its abundance will be.

Model development and analysis

Our simulations predict how (i) number of release events (a measure of introduction effort), (ii) log-transformed distance to the nearest neighbour in morphological space (a measure of the per-capita strength of interspecific competition), and (iii) time delay between introductions (an estimate of the relative abundance of the closest competitor) should interact to generate patterns in establishment success (Fig. 1). To see how well the data fit these predictions we considered how well a series of candidate logistic regression models could explain variation in establishment probability for passerine introductions to New Zealand. Our approach is to fit a series of candidate models to the data, to identify the best-fitting model as a basis for inference, and then to compare the results from that best-fitting model with the simulation results.

Our response variable is binary: whether a species introduced to a district established or not. We considered 17 candidate logistic regression models that might explain variation in establishment success all of which included the number of release events, which we know is an important explanatory variable. Fourteen of the models included combinations of morphological distance, introduction delay and possible interactions between these variables that capture the range of possible outcomes (models 1-14 in Table 1). The remaining models (15-17) included combinations of morphological distance and year of first introduction. We considered models containing the variable year of first introduction because species introduced a long time after their closest competitor (i.e., a long introduction delay) will also have a later date of introduction in absolute terms. It has been noted that, in addition to facing greater competition, later introductions would have faced additional threats, including a greater diversity and abundance of introduced predators (Duncan *et al.* in press). We include year of first introduction as a surrogate for other factors that could have varied through time and affected establishment probability. We did not consider interactions between year of first introduction and other variables because we had no *a priori* reason to think they should be important.

We used the small sample version of Akaike's Information Criterion (AIC_c) to identify which of the 17 candidate models best fitted the data (Burnham and Anderson 2001, 2002). The best-fitting model has the smallest AIC_c value, and we ranked models from best to worst based on the difference between each model's AIC_c and the AIC_c of the best-fitting model (Δ_i values). We then calculated the Akaike weights (w_i, see Burnham and Anderson 2002), which provide a weight of evidence for each model given as an approximate probability that the best-fitting model is in fact the best out of the candidate set. If there is clearly a single best model, as judged by the weight of evidence, then that model can be used as a basis for inference. In our case, however, no single model provided a best fit to the data, with several candidate models providing a similarly good fit. Rather than arbitrarily selecting one of several well fitting models, we used information contained in all of these models as a basis for robust multi-model inference. To do this, we identified all candidate models for which the sum of the Akaike weights from largest to smallest was < 0.99. From this set of best-fitting models we calculated model-averaged parameter estimates, weighting the parameters in each model by the Akaike weight for that model and then summed the weighted estimates across all models (Burnham and Anderson 2002). We then based our inference on these model-averaged values.

Results

Candidate models that included the variable 'year of first introduction' instead of the variable 'introduction delay' received virtually no support (models 15-17 in

Table 2, weight of evidence, w_i, all ≤ 0.001), implying that variation in establishment success is better explained by the time delay between the introduction of a target species and its closest competitor, rather than the absolute date of introduction.

Table 1 The 17 candidate models we considered to explain variation in the probability that a passerine species introduced to an acclimatisation district in New Zealand established or not. The explanatory variables are: nrel = the number of release events in the district (a measure of introduction effort); dist = the logarithm of the distance in morphological space from the target species to its nearest neighbour (a measure of the per-capita strength of interspecific competition); delay = the time delay (in years) between introduction of the target species and introduction of its morphologically nearest neighbour (a measure of the relative abundance of the closest competitor); yi = the year of first introduction.

Model	Explanatory variables
1	nrel
2	nrel + dist
3	nrel + delay
4	nrel + dist + delay
5	nrel + dist + nrel × dist
6	nrel + delay + nrel × delay
7	nrel + dist + delay + nrel × dist
8	nrel + dist + delay + dist × delay
9	nrel + dist + delay + nrel × delay
10	nrel + dist + delay + nrel × dist + dist × delay
11	nrel + dist + delay + nrel × dist + nrel × delay
12	nrel + dist + delay + nrel × delay + dist × delay
13	nrel + dist + delay + nrel × dist + nrel × delay + dist × delay
14	nrel + dist + delay + nrel × dist + nrel × delay + dist × delay + nrel × dist × delay
15	nrel + yi
16	nrel + dist + yi
17	nrel + dist + yi + nrel × dist

Only one of the models with $w_i < 0.01$ did not include the three key explanatory variables, implying that all three variables were important in explaining patterns of establishment. However, no single model clearly provided a best-fit to the data, with the top five models having Δ_i values < 2, implying that each had "substantial" support (Burnham and Anderson 2002). Probability of establishment is plotted as a function of the three explanatory variables, based on model-averaged parameter estimates, in Fig. 3. The probability of establishment was positively related to introduction effort, negatively with increasing strength of competition, and negatively with the time delay between introductions.

Table 2 Comparison of the 17 candidate models explaining variation in the probability that a passerine species introduced to an acclimatisation district in New Zealand established or not. Model number corresponds to those in Table 1. *K* is the number of parameters in the model, Δi is the difference between the Akaike Information Criterion (AIC_c) for each model and that of the best-fitting model (model 4), and w_i are the Akaike weights.

Model	log-likelihood	*K*	AIC_c	Δ_i	w_i	$\sum w_i$
4	-41.87	4	92.175	0	0.228	0.228
9	-41.251	5	93.161	0.986	0.139	0.368
3	-43.553	3	93.363	1.188	0.126	0.494
8	-41.45	5	93.559	1.384	0.114	0.608
7	-41.644	5	93.947	1.772	0.094	0.702
6	-42.937	4	94.308	2.133	0.079	0.781
12	-40.791	6	94.515	2.34	0.071	0.852
11	-41.072	6	95.078	2.903	0.053	0.905
10	-41.104	6	95.14	2.965	0.052	0.957
13	-40.454	7	96.167	3.992	0.031	0.988
14	-40.39	8	98.417	6.242	0.01	0.998
16	-47.953	4	104.341	12.165	0.001	0.999
15	-49.051	3	104.359	12.184	0.001	0.999
1	-50.842	2	105.812	13.637	0	0.999
2	-49.804	3	105.867	13.691	0	1
17	-47.689	5	106.038	13.863	0	1
5	-49.452	4	107.339	15.164	0	1

Discussion

The relationships between establishment probability and the three explanatory variables revealed by the data for passerine introductions to New Zealand (Fig. 3) agree remarkably well with the form of the relationships predicted by our simple stochastic birth-death model incorporating interspecific competition (Fig. 1). In both the data and the model, the probability of establishment increased with greater introduction effort. Moreover, for a given level of introduction effort establishment probability declined when introduced species faced a competitor

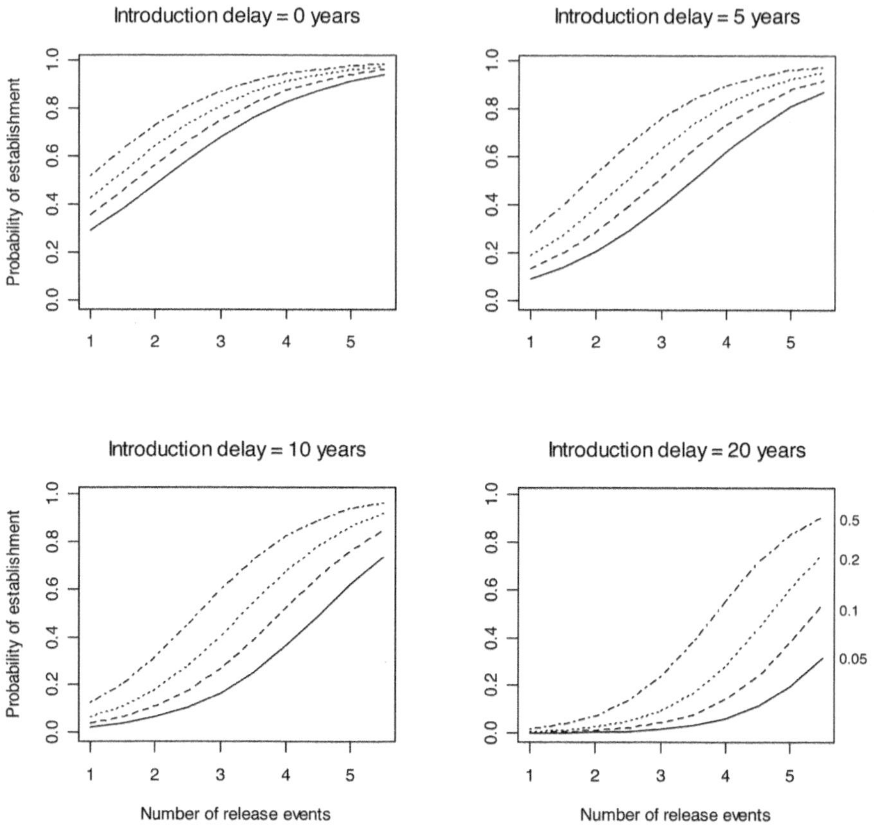

Fig. 3 The relationship between probability of establishment and number of release events for four values of introduction delay (the number of years after introduction of its closest competitor that a target species is introduced) and for four values of morphological distance (0.05, 0.1, 0.2, 0.5; the distance in morphological space [Fig. 2] from a species to its nearest neighbour, shown as different lines), for passerine birds introduced to four acclimatisation districts in New Zealand.

that was more morphologically similar to them, and declined further when that competitor had been introduced earlier than the target species, allowing time for the competitor to have attained higher abundance. This agreement between the relationships predicted by the model and those observed in the data suggests that our simple stochastic birth-death model captures important features underpinning establishment success in this system. In particular, our results suggest that interspecific competition affected establishment probability for passerine species introduced to New Zealand, and that this assemblage has therefore been structured by competition.

Patterns consistent with the outcome of competition have been observed in bird introduction data when competition could not have been a cause of those patterns (Moulton *et al.* 2001, Duncan and Blackburn 2002), and patterns consistent with competition have been shown to be confounded with other factors (Simberloff and Boecklen 1991, Duncan 1997). We emphasise that we have attempted to control for these effects; first, by considering only situations in which species overlapped in space and time, and could therefore have competed; second, by incorporating the confounding effects of introduction effort into our model and analysis; and third, by showing that including introduction delay (a competition effect) provides a better fit to the data than including absolute date of introduction (a surrogate for other variables that might have changed through time).

While historical records of bird introductions provide a unique opportunity to investigate the role of competition (and other factors) in structuring species assemblages (Pimm 1991, Lockwood *et al.* 1999), we stress the need to account for such potentially confounding factors. Moreover, our results highlight that the importance of competition, relative to other factors, in determining establishment success can vary depending on circumstances. Competition should be of less importance in structuring a group of species all introduced at the same time and in low numbers because stochastic events will largely determine which of these species establish and which fail (Fig. 1).

These results have important implications for studies that assess the importance of competition in determining invasion success, regardless of the habitat or taxa considered. This is because the supply of invading propagules and the abundances of resident competitors invariably differ widely between locations and studies, and interactions between these factors influence the role that competition plays. Such differences may contribute to variation in the relative importance of competition that is observed among experimental invasion studies (Levine *et al.* 2004).

The central message of our study is that both stochastic events (in this case mediated by introduction effort) and deterministic processes (in this case the outcome of interspecific competition) are important in community assembly, and that the timing and sequence of introductions (i.e., history) matters due to priority effects. Our results suggest that New Zealand would have ended up with a different assemblage of introduced passerines had the effort put into introducing species been redistributed, or had the sequence and timing of introductions been reshuffled. The finding that historical contingency is important in community assembly is not new and has been demonstrated repeatedly using mathematical models and in laboratory systems (e.g., Gilpin and Case 1976, Drake *et al.* 1993, Fukami and Morin 2003). Insights into the assembly process provided by historical records of bird introductions hint that such effects could be just as pervasive in real communities.

ACKNOWLEDGEMENTS

Thanks to Robert Prys-Jones for access to the bird collections at Tring and the New Zealand Marsden Fund for funding data collection. We thank Tim Blackburn, Phill Cassey, Julie Lockwood and an anonymous reviewer for helpful discussion and comments.

REFERENCES

Bowers, M. A. and J. H. Brown. 1982. Body size and coexistence in desert rodents: chance or community structure? Ecology **63**, 391-400.

Brooke, R. K., J. L. Lockwood and M. P. Moulton. 1995. Patterns of success in passeriform bird introductions on Saint Helena. Oecologia **103**, 337-342.

Burnham, K. P. and D. R. Anderson. 2001. Kullback-Liebler information as a basis for strong inference in ecological studies. Wildlife Research **28**, 111-119.

Burnham, K. P. and D. R. Anderson. 2002. Model selection and multimodel inference, 2nd edition. Springer-Verlag, New York.

Colwell, R. K. and D. W. Winkler. 1984. A null model for null models in biogeography. Pages 344-359 *in* D. R. Strong, D. Simberloff, L. G. Abele, and A. B. Thistle, editors. Ecological communities: conceptual issues and the evidence. Princeton University Press, Princeton, NJ.

Dayan, T. and D. Simberloff. 1994. Mophological relationships among coexisting heteromyids: an incisive dental character. American Naturalist **143**, 462-477.

Diamond, J. and T. J. Case. 1986. Community ecology. Harper and Row, New York.

Diamond, J. M. 1975. Assembly of species communities. *in* M. L. Cody and J. M. Diamond, editors. Ecology and evolution of communities. Harvard University Press, Cambridge.

Drake, J. A., T. E. Flum, G. J. Witteman, T. Voskuil, A. M. Hoylman, C. Creson, D. A. Kenny, G. R. Huxel, C. S. Larue and J. R. Duncan. 1993. The construction and assembly of an ecological landscape. Journal of Animal Ecology **62**, 117-130.

Drake, J. M. and D. M. Lodge. 2004. Effects of environmental variation on extinction and establishment. Ecology Letters **7**, 26-30.

Duncan, R. P. 1997. The role of competition and introduction effort in the success of passeriform birds introduced to New Zealand. American Naturalist **149**, 903-915.

Duncan, R. P. and T. M. Blackburn. 2002. Morphological overdispersion in game birds (Aves: Galliformes) successfully introduced to New Zealand was not casued by interspecific competition. Evolutionary Ecology Research **4**, 551-561.

Duncan, R. P., T. M. Blackburn and P. Cassey. in press. Factors affecting the release, establishment and spread of introduced birds in New Zealand. *in* W. G. Lee, editor. Biological invasions in New Zealand. Springer-Verlag, Berlin.

Duncan, R. P., T. M. Blackburn and D. Sol. 2003. The ecology of bird introductions. Annual Review of Ecology, Evolution and Systematics **34**, 71-98.

Fukami, T. and P. J. Morin. 2003. Productivity-biodiversity relationships depend on the history of community assembly. Nature **424**, 423-426.

Gilpin, M. E. and T. J. Case. 1976. Multiple domains of attraction in competition-communities. Nature **261**, 40-42.

Gotelli, N. J. and G. R. Graves. 1996. Null models in ecology. Smithsonian Institute Press, Washington.

Grant, P. R. and I. Abbot. 1980. Interspecific competition, island biogeography and null hypotheses. Evolution **34**, 332-341.

Grevstad, F. S. 1999. Factors influencing the chance of population establishment: implications for release strategies in biocontrol. Ecological Applications **9**, 1439-1447.

Inchausti, P. and J. Halley. 2003. On the relation between temporal variability and persistence time in animal populations. Journal of Animal Ecology **72**, 899-908.

Kingston, T., G. Jones, A. Zubaid and T. H. Kunz. 2000. Resource partitioning in rhinolophoid bats revisited. Oecologia **124**, 332-342.

Lande, R. 1993. Risks of population extinction from demographic and environmental stochasticity and random catastrophes. American Naturalist **142**, 911-927.

Levine, J. M., P. B. Adler and S. G. Yelenik. 2004. A meta-analysis of biotic resistance to exotic plant invasions. Ecology Letters **7**, 975-989.

Lockwood, J. L. and M. P. Moulton. 1994. Ecomorphological pattern in Bermuda birds: the influence of competition and implications for nature preserves. Evolutionary Ecology **8**, 53-60.

Lockwood, J. L., M. P. Moulton and S. K. Anderson. 1993. Morphological assortment and the assembly of communities of introduced passeriforms on oceanic islands: Tahiti versus Oahu. American Naturalist **141**, 398-408.

Lockwood, J. L., M. P. Moulton and K. L. Balent. 1999. Introduced avifaunas as natural experiments in community assembly. Pages 108-129 *in* E. Weiher and P. A. Keddy, editors. Ecological assembly rules. Cambridge University Press, Cambridge.

Mack, R. N., D. Simberloff, W. M. Lonsdale, H. Evans, M. Clout and F. A. Bazzaz. 2000. Biotic invasions: causes, epidemiology, global consequences, and control. Ecological Applications **10**, 689-710.

Moulton, M. P. 1985. Morphological similarity and coexistence of congeners: an experimental test with introduced Hawaiian birds. Oikos **44**, 301-305.

Moulton, M. P. 1993. The all-or-none pattern in introduced Hawaiian passeriforms: the role of competition sustained. American Naturalist **141**, 105-119.

Moulton, M. P. and J. L. Lockwood. 1992. Morphological dispersion of introduced Hawaiian finches: evidence for competition and a Narcissus effect. Evolutionary Ecology **6**, 45-55.

Moulton, M. P. and S. L. Pimm. 1983. The introduced Hawaiian avifauna: biogeographical evidence for competition. American Naturalist **121**, 669-690.

Moulton, M. P. and S. L. Pimm. 1986a. The extent of competition in shaping an introduced avifauna. Pages 80-97 *in* J. Diamond and T. J. Case, editors. Community ecology. Harper & Row, New York.

Moulton, M. P. and S. L. Pimm. 1986b. Species introductions to Hawaii. Pages 231-249 *in* H. A. Mooney and J. A. Drake, editors. Ecology of biological invasions of North America and Hawaii. Springer, New York.

Moulton, M. P. and S. L. Pimm. 1987. Morphological assortment in introduced Hawaiian passerines. Evolutionary Ecology **1**, 113-124.
Moulton, M. P. and J. G. Sanderson. 1997. Predicting the fates of passeriform introductions on oceanic islands. Conservation Biology **11**, 552-558.
Moulton, M. P., J. G. Sanderson and R. F. Labisky. 2001. Patterns of success in game bird (Aves: Galliformes) introductions to the Hawaiian Islands and New Zealand. Evolutionary Ecology Research **3**, 507-519.
Moulton, M. P., J. G. Sanderson and D. Simberloff. 1996. Passeriform introductions to the Mascarenes (Indian Ocean): an assessment of the role of competition. Écologie **27**, 143-152.
Pimm, S. L. 1991. The balance of nature? Chicago University Press, Chicago.
Renshaw, E. 1991. Modeling biological populations in time and space. Cambridge University Press, Cambridge.
Richter-Dyn, N. and N. S. Goel. 1972. On the extinction of a colonizing species. Theoretical Population Biology **3**, 406-433.
Ricklefs, R. E. and J. Travis. 1980. A morphological approach to the study of avian community organization. Auk **97**, 321-338.
Simberloff, D. 1992. Extinction, survival, and effects of birds introduced to the Mascarenes. Acta Oecologica **13**, 663-678.
Simberloff, D. and W. Boecklen. 1991. Patterns of extinction in the introduced Hawaiian avifauna: a reexamination of the role of competition. American Naturalist **138**, 300-327.
Strong, D. R., D. Simberloff, L. G. Abele and A. B. Thistle. 1984. Ecological communities: conceptual issues and the evidence. Princeton University Press, Princeton.
Strong, D. R., L. A. Szyska and D. Simberloff. 1979. Tests of community-wide character displacement against null hypotheses. Evolution **33**, 897-913.
Thomson, G. M. 1922. The naturalisation of animals and plants in New Zealand. Cambridge University Press, Cambridge.
Veltman, C. J., S. Nee and M. J. Crawley. 1996. Correlates of introduction success in exotic New Zealand birds. American Naturalist **147**, 542-557.
Weiher, E. and P. A. Keddy. 1999. Ecological assembly rules. Cambridge University Press, Cambridge.
Williamson, M. 1996. Biological Invasions. Chapman & Hall, London.

Chapter nineteen

Room for one more? Evidence for invasibility and saturation in ecological communities

S. A. Smith and J. B. Shurin

INTRODUCTION

Identifying the contributions of local and regional processes in shaping biological communities is key to understanding global patterns of species richness (Cornell 1999, Gaston 2000). Ecologists have typically fallen into two traditions to explain variation in diversity, one emphasizing local processes such as competition, predation, mutualism and the interaction between organisms and their environment, and the other focused on speciation, extinction and dispersal over broad regional scales (Terborgh and Faaborg 1980; Cornell 1985, 1993, 1999; Cornell and Lawton 1992; Ricklefs 1987, 2004). The local school of thought posits that species richness in communities is determined by processes that influence demographic rates within habitat patches (Chase and Leibold 2003). The regional approach argues that dispersal, speciation, extinction and the history of community assembly are of primary importance in shaping biological communities (Ricklefs 1987, 2004, Caswell and Cohen 1993, Gaston 2000,

M.W. Cadotte et al. (eds.), Conceptual ecology and invasion biology, 423–447.

Gaston 2003, Mora *et al.* 2003, Fukami 2004, Smith *et al.* 2004, Smith and Bermingham submitted). The critical distinction between the two approaches hinges on whether local species richness is saturated, or whether communities have the intrinsic capacity to support more species than they actually contain. If the supply of species through colonization or speciation is much greater than the number that can coexist due to local ecological constraints, then communities are said to be saturated and under strong local control (Cornell and Lawton 1992). Alternatively, if speciation and dispersal provide fewer species than sites can support, then regional control over communities is dominant and local richness is unsaturated. The question of whether communities are near or far from saturation is critical to our understanding of patterns of the distribution and diversity of organisms.

The concept of saturation is of fundamental interest to both community ecology and invasions biology because it has profound implications for the invasibility and post-invasion dynamics of local communities. Invasibility and saturation are related but distinct aspects of communities. For instance, there are two potential outcomes of species introductions in saturated communities. First, such communities might resist invasion by species introduced via natural or anthropogenic dispersal. In this case, both the composition and species richness of local communities are in a state of equilibrium and change very little with the introduction of novel species. Second, local communities might be both invasible *and* saturated with species. New species could be added to local communities via natural colonization or human introduction, however, these species displace local residents, maintaining a constant number of species (Brown *et al.* 1997, 2001). Therefore, the number of species, but not their identities, would remain constant over time. Alternatively, in unsaturated communities, dispersal events often result in successful invasion without causing the extirpation of resident species. Net species richness increases with enhanced dispersal because the addition of new species exceeds the loss of residents. A final possibility is that very high dispersal rates result in "super-saturation" as communities support a large number of sink populations that are excluded locally in the absence of dispersal (Shmida and Ellner 1984, Mouquet and Loreau 2003). Patterns of community invasibility and changes in species richness with increased immigration rates among communities, or the length of evolutionary history, can therefore be used to evaluate the degrees of saturation and invasibility of communities.

This chapter assesses three lines of evidence for saturation and invasibility in biological communities. First, we examine the outcomes of faunal exchanges resulting from recent anthropogenic modification of the landscape and past geologic events to evaluate the effect of enhanced immigration between regions on species richness. Second, we examine recent trends in the numbers of anthropogenic species invasions and extinctions observed over broad spatial scales. The global explosion in exotic species invasions suggests that contemporary communities are certainly invasible (Lodge 1993, Moyle and Light 1996, Vitousek *et al.* 1997, Cohen and Carlton 1998, Hobbs and Mooney 1998, Ricciardi 2001).

Whether biological invasions are accompanied by differing numbers of extinctions can be useful for evaluating whether local diversity in these communities is near or below saturation (Sax and Gaines 2003). We compare the number of successful invasions versus extirpations in order to evaluate how species richness changes when humans introduce species. Third, we examine experimental species introductions to evaluate the invasibility of local communities and compare degrees of saturation among different groups of organisms. Intentional introductions allow us to assess the success rate of colonization attempts, an important aspect of saturation as uncontrolled introductions do not tell us how many introduced species fail to colonize.

Finally, we compare the empirical evidence for saturation and invasibility with the predictions of four major theoretical perspectives on broad-scale community structure. Models of patch-dynamics, ecological drift (e.g., neutral models), species-sorting along environmental gradients and mass-effect systems make divergent predictions regarding community invasibility and saturation. Faunal exchanges, patterns of human-mediated invasions and extinctions, and experimental introductions may be useful for distinguishing the utility of the different models for describing real communities. We synthesize data and theory pertaining to the saturation of local communities and discuss avenues for future research.

BIOTIC INTERCHANGES: INVASIBLE AND UNSATURATED COMMUNITIES???

Biotic interchanges occur when barriers separating distinct biological communities break down, enabling the reciprocal exchange of resident species. Removal of dispersal barriers increases the likelihood of dispersal between previously isolated communities, permitting strong inferences regarding the importance of dispersal *per se* in determining patterns of the distribution and diversity of organisms. Here, we use evidence from contemporary and historic biotic interchanges to assess whether communities are saturated with species by documenting changes in species richness following removal of dispersal barriers. Moreover, because the regional pool of species capable of participating in cross-barrier exchanges is often known, it is possible to evaluate the success rate of colonizations in both directions.

Contemporary biotic interchanges

Contemporary human-mediated biotic interchanges provide a powerful test for saturation of biological communities over moderate to large geographic scales because the outcome can be observed directly without relying on the fossil record. Biotic interchanges following the construction of both the Panama and Suez Canals are well documented (Por 1978, 1989, Smith *et al.* 2004) and provide insights into the question of community saturation. In the case of the Panama Canal, the recent geologic origin of the Isthmus (3 million years old) and its young freshwater fish fauna (Bermingham and Martin 1998, Perdices *et al.* 2002) suggest that regional processes might be important in limiting species richness. Mesoamerican freshwater fish assemblages contain an order of magnitude fewer species than their putative source populations in northwestern South America (Smith and Bermingham unpublished data), suggesting that speciation and immigration may be more important in limiting local diversity in this system. Consequently, communities of freshwater fishes in the Panama Canal region might be both susceptible to invasion and unsaturated because the past supply of species may be lower than the number of species that the rivers could support. Alternatively, species richness might remain below the speciation-extinction equilibrium (Losos and Schluter 2000) because the process of speciation may be so slow that few communities ever approach evolutionary equilibrium (Losos and Schluter 2000, Ricklefs and Bermingham 2001).

Completion of the Panama Canal in 1914 created a freshwater corridor between the Rio Chagres on the Caribbean slope and the Rio Grande on the Pacific slope of the Isthmus of Panama, enabling the freshwater fishes of previously isolated drainage basins to intermingle. Three fish species colonized the Rio Chagres from the Rio Grande, and five species colonized the Rio Grande from the Rio Chagres. Colonists represented 75% and 38.5% of the total source pool of obligate freshwater fish species in the Rio Chagres and Rio Grande that were not already present in both basins, suggesting that freshwater fish communities in the Rio Chagres and Rio Grande were susceptible to invasion (Table 1). No cases of local extirpation were documented; species richness following dispersal events increased by 10% in the Rio Chagres, and 22% in the Rio Grande (Smith *et al.* 2004). In addition, colonists became common in newly invaded communities, achieving a level of distribution and abundance comparable with or greater than that in their source communities (Smith *et al.* 2004).

The biotic interchange that occurred following completion of the Suez Canal in 1869, which linked the Red Sea and the eastern Mediterranean Sea, is qualitatively similar to that which occurred following the construction the Panama Canal. Hundreds of Red Sea marine species colonized the Mediterranean via the Suez Canal, including 90 mollusks and 44 fishes (Por 1989). Lessepsian migrants, species that colonized the Mediterranean from the Red Sea, constitute approximately 10% of the total species richness of the Levant Basin (Por 1978, 1989). Many fewer colonizations of Mediterranean species have been recorded in the Red Sea. Fifty-three marine species have been reported to have participated in the faunal exchange (Por 1978). Large differences in the number of successful invasions suggest potential differences in the invasibility of the two communities, however, passive transport by unidirectional currents partially explains the observed asymmetry of invasion through the Suez Canal.

Invasions by species migrating through the Suez Canal have continued for more than 130 years since the canal's construction, whereas no extinctions have been attributed to invasions following faunal mixing (Por 1978, 1989). Rather, a bathymetric readjustment of resident species has occurred whereby resident

Table 1 The obligate freshwater fish assemblages of the Rio Grande and Rio Chagres drainage basins sampled by both Meek and Hildebrand (1916) and Smith *et al.* 2004. Colonists are denoted in bold-italic text and 'new' refers to drainages colonized following completion of the Panama Canal. The species in Table 1 are organized by their salinity tolerance, and then by alphabetical order of family, genus and species names. A nested ANOVA analysis of species log transformed abundances in adjacent watersheds was performed in order to determine if species were more successful in their native or newly colonized drainages. * denotes values significant at p = 0.05 and † denotes values significant at p = 0.01, and the location of the symbol indicates whether the species was more successful in its native or recently invaded drainage.

Primary and secondary freshwater fishes	Rio Grande	Rio Chagres
Astyanax aeneus	+	new
Astyanax ruberrimus	+	+
Brycon chagrensis	new	+
Brycon petrosus	new	+
Bryconamericus emperador	+	+
Compsura gorgonae	+	+
Gephyrocharax atricaudata	+	+
Hyphessobrycon panamensis	–	+
Pseudocheirodon affinis	–	+
Roeboides occidentalis	+*	new
Roeboides guatemalensis	new	+*
Cyphocharax magdalenae	+	new†
Hoplias microlepis	+	+
Piabucina panamensis	+	+
Ancistrus chagrensis	new	+*
Chaetostoma fischeri	–	+
Hypostomus panamensis	+	+
Rineloricaria uracantha	–	+
Pimelodella chagresi	+	+
Rhamdia quelen	+	+
Brachyhypopomus occidentalis	+	+
Rivulus brunneus	–	+
Aequidens coeruleopunctatus	+	+
Archocentrus panamensis	–	+
Geophagus crassilabris	–	+
Vieja maculicauda	new	+
Brachyrhaphis cascajalensis	–	+
Brachyrhaphis episcopi	+	+
Neoheterandria tridentiger	+	+
Poecilia gillii	+	+
Poeciliopsis elongata	+	–

species maintain their dominance in deeper, cooler water (Por 1978). Invaders also clearly play an important role in the community dynamics of invaded regions, both in terms of total contribution to the species richness and abundance (Ben-Tuvia 1985). Fisheries data suggest that a large portion of new immigrants (~45%) belong to the common fishes of the Levant basin. It is important to note that the Suez Canal data was collected at very large spatial scales, therefore more subtle effects of invasions may not be apparent. Biotic interchange following completion of the Suez Canal provides strong evidence that reciprocal interchange increased regional species richness without resulting in the regional extinction of the resident fauna.

Pre-historic biotic interchanges

Biotic interchanges have occurred frequently over the last 25 My, resulting from climatic changes and rearrangement of continental landmasses that break down barriers to dispersal. Vermeij (1991a) reviewed the outcome of biotic interchanges that occurred during the Neogene. One pattern that clearly emerges is that communities are invasible, however, invading species in most cases only make up a small portion of the total regional species pool in the donor biota (Vermeij 1991a). For example, between 23-46% of the shallow-water molluscan species in the cool-temperate North Pacific, and ~1% of the regional species pool of the North Atlantic participated in the trans-Arctic interchange (Vermeij 1991b). Similar patterns arise in other instances of biotic interchanges (Vermeij 1991a). The low apparent invasion rates for some pre-historic biotic interchanges could, however, be the result of the large spatial scale at which the data has been recorded. Perhaps many species in the donor species pool are not adapted to conditions in the recipient biogeographic region. The richness of the native species pool may therefore overestimate the true number of species capable of participating in the biotic interchange in question. Alternatively, the native biota may repel invasion by a large portion of the potential invaders. In either case, the fossil record suggests that communities involved in faunal exchanges added species, but that many likely invasion attempts also failed.

The central question pertaining to community saturation is whether biotic interchanges result in extirpation of resident species. The concept that the arrival of invaders from a more sophisticated biota might lead to the extinction of resident species via competitive displacement has a long history (Wallace 1876, Simpson 1950, 1980, Webb and Marshall 1982, Webb 1985, Marshall and Cifelli 1990). Patterns of extinction resulting from competitive displacement would present strong evidence for the importance of local ecological control of community membership. The exchange of mammals across the Isthmus of Panama during the Great American Biotic Interchange has been cited as an example where invading northern species drove southern resident species to extinction via competitive displacement (Marshall 1981, Bakker 1983). However, whether competitive displacement resulted in differential rates of extinction between

North and South American mammals has been contested (Lessa *et al.* 1997). Moreover, there are many instances of invasions that did not result in the extinction of resident species. No increases in extinction rates occurred following biotic interchange in the marine realm regarding transequitorial, trans-Arctic and trans-Pacific interchanges (Vermeij 1989a,b, Vermeij 1991a,b, Linberg 1991), or following the invasion of African and North American mammals into Asia during the Miocene and Pliocene (Barry *et al.* 1991, Flynn *et al.* 1991). Invasion events during biotic interchanges most often enhance the species richness of recipient communities (Barry *et al.* 1991, Flynn *et al.* 1991, Vermeij 1991a). These patterns bolster those of the contemporary interchange literature in that invasions rarely cause the global extinction of resident species and often result in increases in species richness at the regional scale. This suggests that biological communities do not resist all invasion attempts and are under-saturated with species. The outcomes of biotic interchanges emphasize the importance of dispersal in setting an upper limit to the number of species coexisting in a given area.

In this section, we argue that the outcome of contemporary and historic biotic interchanges provide strong evidence that communities are both invasible and unsaturated. However, the above examples lack information on the effects of invasion at multiple spatial scales. It is therefore impossible to determine how changes in richness at continental or oceanic scales are reflected at smaller local scales. Increases in large-scale richness might be accompanied by decreases at local scales if native species are much reduced in range or abundance. In addition, beta diversity or regional distinctiveness has clearly declined in many cases as a result of biotic homogenization (McKinney and Lockwood 1999, Rahel 2000, 2002) The spatial grain and extent of observation strongly influences our perception of community saturation.

CONTEMPORARY BIOLOGICAL INVASIONS: COMPARING INVASION AND EXTINCTION EVENTS

The explosion in the number of human-mediated invasions clearly attests that contemporary biological communities are invasible (Lodge 1993, Moyle and Light 1996, Vitousek *et al.* 1997, Cohen and Carlton 1998, Hobbs and Mooney 1998, Ricciardi 2001). Studying community dynamics following invasions permits inferences about the processes regulating species richness. Here we examine evidence from a number of contemporary invasions and extinctions in different ecosystems and address their significance for the invasibility and saturation of ecological communities.

Regional scale patterns

The literature on contemporary anthropogenic invasions indicates that, at broad regional scales (e.g., islands or archipelagos), invasions exceed extinctions, resulting in net increases of species richness (Case 1996, Hobbs and Mooney 1998, Sax *et al.* 2002, Sax and Gaines 2003, Table 2). Increases in species richness varied between continents and islands, as well as among taxa exhibiting differing degrees of dispersal limitation. Larger increases in species richness were encountered on islands as opposed to continents, especially in the case of plants where an astounding 107% increase in species richness was documented on oceanic islands (Sax *et al.* 2002). Comparing the increase in species richness among taxa indicates that the most dispersal limited taxa experienced the largest increases in species richness following human-mediated invasion events. For example, increases in the species richness of freshwater fishes and plants far exceeded those of birds. Most birds can disperse over much larger spatial scales than freshwater fishes which depend on direct connections between water bodies, and flowering plants, whose seeds are passively dispersed by wind or animals (Sax *et al.* 2002). It is therefore not surprising that dispersal limits the number of fish and plant species found at the regional scale to a greater extent than in birds. Contemporary invasions exceed extinctions, supporting the evidence from faunal exchanges that communities are unsaturated.

Local scale patterns

Strong species interactions at the local scale may not manifest themselves as saturation at broad regional scales. For instance, Byers and Noonburg (2003) analyzed models with strong local interactions and spatial heterogeneity in resources. They found that high diversity at the regional scale due to broad heterogeneity led to high invasibility, even though local interactions generated strong biotic resistance at smaller scales. Studies of changes in species richness following invasion events at the local scale may therefore provide more powerful tests for community saturation. Gido and Brown (1999) recorded changes in the species richness of freshwater fishes in 125 drainage basins in North America and found that invasions exceeded extinctions in 112 of 125 cases. Mean species richness of the 125 drainage basins increased by 21.4%, while richness of freshwater fishes in California increased 206% (Table 2, Marchetti *et al.* 2002), suggesting that local communities were not resistant to invasion or saturated with species. In addition, competition with invaders has rarely been implicated as a cause of native species extinctions (Davis 2003). Rather, predation pressure (Kaufman 1992, Fritts and Rodda 1998) and habitat change (e.g., Hawaiian Islands) are responsible for many of the impacts of invaders on natives. These results concur with those of contemporary and past biotic interchanges where increasing the size of the available regional species pool resulted in net increases in species richness (see section 2). Alternatively, drainage basins may contain multiple local populations and correspond more to the regional than local scale so that the spatial extent is still too large for saturation to become evident. Nevertheless, patterns of invasions and extinctions at smaller scales (e.g., drainages or islands) provide no more evidence for saturation than those over broader regions.

Table 2 Patterns of changes in species richness following invasion events. Data compiled from (1)-Hobbs and Mooney (1998), (2)-Sax, Brown and Gaines (2002), (3)-Case (1996), (4)-Gido and Brown (1999), (5)-Marchetti *et al.* 2001 and (6)-Duncan and Lockwood (2001).

Taxa	Location	Mean Net Increase in species richness	# of samples	% of total present SR that is introduced	Spatial scale
Plants (1)	California	15.698	1	13.969	State of California
Plants (1)	Western Australia	8.166	1	7.9189	Western Australia
Plants (2)	Oceanic Islands	107.428	11	51.555	Islands
Mammals (1)	California	10	1	9.604	State of California
Mammals (1)	Western Australia	4.255	1	10.759	Western Australia
Birds (1)	California	6.415	1	6.690	State of California
Birds (1)	Western Australia	1.960	1	1.923	Western Australia
Land Birds (2)	Oceanic Islands	−4.210	19	54.344	Islands
Birds (3)	Islands	12.165	48	12.532	Islands
Freshwater fishes (1)	California	56.060	1	40	State of California
Freshwater fishes (1)	Western Australia	15	1	13.043	Western Australia
Freshwater fishes (4)	USA	21.430	125	14.186	Drainage basin
Freshwater fishes (5)	California	205.732	43	42.439	Drainage basin
Freshwater fishes (6)	Tennessee	9.157	6	8.590	Ecoregions
Mussels (6)	Tennessee	−6.185	7	25.495	Ecoregions
Reptiles (1)	California	3.947	1	3.797	State of California
Amphibians (1)	California	4.444	1	4.255	State of California
Butterflies (1)	California	−1.581	1	0.393	State of California

ASSESSING METHODOLOGICAL LIMITATIONS AND CONSTRAINTS

There are several important caveats to consider when inferring the mechanisms structuring communities from patterns of invasion and extinction at local and regional scales. For example, extinctions may be more difficult to detect than invasions. Few extinctions may be recorded as species that are not observed cannot be assumed absent, whereas invasions are unambiguously invasions. In addition, changes in species richness may ignore other large effects on range or abundance of native species following species invasions. That is, many formerly abundant, widespread species may be reduced to few small remnant populations by interactions with invaders without becoming extinct. Such changes in distribution and abundance are not captured by presence/absence data.

Extinctions caused by invaders may also proceed slowly and occur long after the initial colonization (Ricklefs 2004), suggesting that there may be a time lag between invasion and extinction events. However, invasions following the elimination of dispersal barriers do not result in measurably greater extinction rates over evolutionary time scales (Barry *et al.* 1991, Flynn *et al.* 1991, Linberg *et al.* 1991). Nevertheless, examples of invasions that lead to mass extinction events are well documented (e.g., Nile perch in Lake Victoria, Kaufman 1992; brown tree snake in Guam, Fritts and Rodda 1998; and zebra mussels in the Great Lakes, Nalepa and Schloeser 1993, Vanderploeg *et al.* 2002).

Elevated rates of species introductions coupled with large correlated losses of diversity have led to the assumption that exotic species introductions are one of the leading causes of species extinctions (but see Gurevitch and Padilla 2004). The observation that exotic species are the third most prevalent threat to endangered species throughout the world (Sala *et al.* 2000) suggests that many extinctions resulting from biotic exchange are still to come. However, it has often been difficult to identify the role of exotic invaders in causing species extinctions because most imperiled species face more than one threat simultaneously (Gurevitch and Padilla 2004). Experiments that do not suffer from the confounding effects of multiple aspects of human alteration of the environment are required to assess the relative importance of invasions as a cause of extinction (Ricciardi 2004).

Finally, invasive species have been documented to preferentially occupy human dominated landscapes whereas native species have a greater affinity for native, undisturbed habitat (Diamond and Veitech 1981, Simberloff 1992, Smallwood 1994, Case 1996, Rahel 2002, Sax *et al.* 2002, Duncan *et al.* 2003). Many invaders may not interact with resident species because of habitat partitioning between native and introduced species, therefore complicating the use of biological invasions as tests of community saturation. In addition, the high apparent invasibility of communities may be as much a product of anthropogenic disturbances that weaken biotic resistance as human-aided dispersal (Seabloom *et al.* 2003). Nevertheless, net increases of species richness at local and regional scales following invasion events suggest that many communities are invasible and unsaturated. This conclusion is supported by patterns of past and contemporary faunal interchange. In the next section, we review experimental studies of community saturation and compare their results with those of patterns from unmanipulated systems.

EXPERIMENTALLY STAGED INVASIONS

Macroecological patterns suggest that communities are both unsaturated and invasible, that biotic resistance to invasion in communities is weak relative to dispersal limitation, and that native species are rarely excluded from communities by interactions with invaders (Stohlgren *et al.* 1999, Sax and Gaines 2003). This conclusion has important practical implications as it suggests that managing dispersal vectors is more effective for containing invaders than relying on native communities to prevent their spread. However, the conclusion that species interactions and biotic resistance are unimportant for shaping communities or preventing invasions may be incorrect. First, species interactions likely occur at scales much smaller than the broad regions (e.g., island chains or watersheds) that are the units of observation for biogeographical studies. A new colonist likely interacts only with other species within a fairly small neighborhood, not the entire biota of the region. In addition, while invasions are often dramatic and obvious, failed invasion attempts mostly pass unobserved. We do not know whether the regional differences in diversity that remain following wholesale species introductions are due to environmental constraints, species interactions, or lack of dispersal opportunities for some species. Finally, extinction of species may be slow relative to invasion (Ricklefs 2004), and some species may be greatly reduced in abundance or geographic range by invaders without being completely extirpated. Regional diversity loss due to invaders may therefore be a weak indicator of the importance of biotic exclusion. Local interactions may be difficult to detect in macroecological patterns, but still be critical for constraining invasions.

The final approach to assaying community saturation and invasibility that we discuss is through experimental species introductions. Field manipulations of invasion offer a number of insights that cannot be obtained from observational data. First, the rates of success and failure can be accurately characterized as the pool of introduced species is known. Second, aspects of the invader pool (e.g., propagule load or timing) and recipient community (e.g., diversity, composition, productivity or disturbance) can be directly manipulated. These allow assignment of the causes of invasion success or failure, and interpretation of the traits of species or communities that influence invasiveness and invasibility. Field experiments are also generally conducted at much smaller local scales such as the neighborhood of plant communities, and may therefore be more likely to reveal the role of species interactions.

Efforts to experimentally test community saturation are increasingly common. Assaying invasibility at the community level involves introducing potentially invasive species, usually ones that are present regionally but absent locally. The majority of field studies of community saturation are seed additions in herbaceous plant communities (Table 3). Turnbull *et al.* (2000) reviewed the literature on seed addition experiments and concluded that approximately 50% of all plant species show evidence for seed limitation. A sessile growth form may render terrestrial plants especially vulnerable to dispersal limitation as propagules have only one opportunity to find suitable habitat patches. Small scale variability in light or soil conditions may constrain appropriate sites for germination to a small fraction of the available habitat. A limited number of studies have been performed in other communities, including pond zooplankton (Shurin 2000) and laboratory studies with protozoans (Fox *et al.* 2000).

Table 3 A summary of experimental studies of community saturation and invasibility. Percent introduction success is measured as the proportion of introduced species observed in the experimental system and includes successful germination. "Increased local richness" indicates whether local richness was greater in the introduction treatment than the control. "Richness-invasibility" indicates whether the success of invaders declined at higher native species richness, and "disturbance treatment" indicates whether perturbations to the native community affected invasibility. A dash indicates that the treatment was not performed or the hypothesis was not tested.

Study	System	% introduction success	Increased local richness?	Richness-invasibility?	Disturbance treatment?
Burke and Grime 1996	grassland	59.3	–	–	Y
Tilman 1997	grassland	54.8	Y	negative	–
Shurin 2000	zooplankton	3.3	N	negative	Y
Zobel *et al.* 2000	grassland	60.0	Y	none	Y
Foster 2001	grassland	19.4	Y	–	Y
Levine 2001	sedges	42.4	Y	negative	Y
Foster *et al.* 2002	grassland	28.5	–	positive	–
Fargione *et al.* 2003	grassland	11.1	Y	negative	–
Foster and Tilman 2003	grassland	27.1	Y	–	–
Mouquet *et al.* 2004	grassland	62.5	Y	–	–

Table 3 summarizes experiments testing saturation and reveals a number of patterns. First, every experimental species addition found that some species are absent from local communities due to dispersal limitation. The introduction success rate into intact recipient communities ranged from 3.3% for pond zooplankton to much higher (11.1-62.5%) in herbaceous plant communities. The observation that zooplankton communities appear closer to saturation than plants may reflect either biological or methodological differences between the studies. Zooplankton may be more effective dispersers than plants, or may be better able to find suitable microsites within patches (e.g., ponds) because they are motile. Alternatively, the criterion for successful invasion in the plant studies was typically germination, not successful reproduction or positive population growth. Many of the seedlings that germinated may not have eventually established, therefore the experiments may overestimate invasion success. However, a number of studies lasted for several years and found that introduced plants persisted. The results of the experiments support large-scale patterns indicating that contemporary biological communities are invasible.

Experiments also indicate that plant communities are often unsaturated in that introductions (simulating increased connectivity to the larger regional species pool) enhanced local diversity. Seven out of eight studies found higher species richness in the introduction treatments than in the controls (all of the plant experiments that contrasted diversity in the seed addition treatment with the control, Table 3). By contrast, Shurin's (2000) pond zooplankton experiments found no difference in diversity between invasion treatments and the control, indicating that pond zooplankton communities may be closer to regional saturation than terrestrial plants. Experimental introductions indicate that many communities are unsaturated with species at both local (neighborhood) and regional scales.

However, a number of patterns from the experimental literature contrast with the interpretation of macro-scale patterns as indicating weak biotic resistance and a dominance of dispersal limitation, relative to species interactions, in structuring communities (e.g., Stohlgren *et al.* 1999, Sax and Gaines 2003). First, a number of studies found declining invasion success (measured either as the percent occurrence of invasive species or their overall abundance) at higher local richness of resident species (Table 3). This result suggests that although communities are unsaturated in that they can support more species, local richness approaches saturation at high levels in that it becomes increasingly difficult for additional species to invade. Increasing biotic resistance most likely results from greater intensity of local interactions at high diversity. The increased biotic resistance at high diversity could be the result of either stronger competitive interactions among established species or by the inhibitory effects of established species on recruitment (Tilman 2004).

Second, a number of experiments employed disturbance treatments that weakened interactions between the resident community and invaders. Comparing invasibility in disturbed vs. intact native communities illustrates which species are excluded from the recipient community due to interactions with natives versus inability to tolerate local abiotic conditions. Studies employing disturbance treatments (either through removal of native species or enhanced productivity by nutrient additions) found that invasibility was dramatically increased by altering the biotic environment. For instance, Shurin (2000) found that reducing the abundance of native species allowed four times as many invaders to establish, and to obtain sixteen times greater total abundance. These results indicate that, although communities may be invasible and unsaturated due to dispersal limitation, local interactions may still be strong enough to exclude a large fraction of potential invaders, and that many more species may be excluded than succeed in invading intact communities. This contrast illustrates the importance of field experiments. Analysis of patterns of diversity following faunal exchange and anthropogenic introductions may be interpreted to indicate that biotic resistance seldom, if ever, repels invaders (Stohlgren *et al.* 1999, Sax and Gaines 2003). This conclusion emphasizes successful invasions over failures, however the latter may be more common and often occur due to local interactions.

METACOMMUNITY PREDICTIONS

The global explosion in the incidence of exotic species reflects the results of human activities that increase rates of migration among previously isolated biogeographic regions. Within regions, however, humans may either increase habitat connectivity by acting as dispersal vectors or decrease it through habitat fragmentation. The effects of modifying rates of dispersal within regions and immigration from other regions are quite distinct. Here, we focus on the effects of increased movement rates among regions, putting aside the question of how human-induced environmental change affects dispersal within regions. Metacommunity models offer different theoretical predictions regarding the invasibility and saturation of local communities and the outcome of biological invasions. Comparing responses of diversity to changes in immigration rates with predicted outcomes may be useful for evaluating different perspectives on broad-scale community structure.

Theoretical and empirical work on metacommunities falls under four paradigms which we call **patch-dynamics**, **neutral**, **species-sorting** and **mass-effects** (Leibold *et al.* 2004, Chase *et al.* 2005). The four approaches differ in their assumptions about relative rates of dispersal, the intensity of interspecific interactions, and the degree of spatial heterogeneity among habitats (Fig. 1). Patch dynamics, the oldest perspective, is represented by island biogeography and metapopulation ecology. This approach considers habitats as divided into homogeneous patches among which species disperse. The simplest version of the patch-dynamics framework assumes that species are indifferent to one another's presence in patches, however such models have been modified to incorporate local interactions that lead to exclusion or facilitation (Levins and Culver 1971, Hastings 1980). Dispersal in the patch-dynamics framework is slow relative to demographic rates so that patches are either occupied or empty, but local dynamics are ignored. Neutral theory (Bell 2001, Hubbell 2001) represents a variation on patch dynamics where all individuals are identical in all relevant traits, local interactions are strong and lead to priority effects, and dispersal is spatially constrained. Species may coexist in patch-dynamics models at equilibrium by trade-offs in their ability to colonize and compete in patches (Levins and Culver 1971,

Hastings 1980), while coexistence in neutral models is determined by a balance between slow exclusion and the input of species, either from outside the region or through speciation (Bell 2001, Hubbell 2001).

The species-sorting approach posits that species disperse sufficiently rapidly that local environmental constraints limit distributions to a greater extent than the supply of colonists (Leibold *et al.* 2004). The species-sorting view can be considered classically niche-based in that species are selected by the local environment, and sites contain most or all species capable of coexisting under the particular local conditions. Species-sorting assumes sufficiently fast dispersal that local interactions play a dominant role over movement of species in constraining community membership. Finally, the mass-effects or source-sink perspective holds that dispersal can be so rapid that populations are maintained that would otherwise decline to extinction in the absence of input of individuals from other habitats (Shmida and Ellner 1984). Mass-effects models rely on heterogeneity in the environment that creates differences in fitness such that some habitats produce excess individuals that disperse to poor-quality patches where they have negative growth rates. Species-sorting holds that dispersal is fast relative to rates of population extinction, but not so rapid that the input of individuals is large relative to local reproduction. By contrast, mass-effects systems occur when dispersal is fast relative to local demographic rates. The differences among the four metacommunity approaches are illustrated in Fig. 1.

The four metacommunity perspectives make divergent predictions regarding the invasibility and saturation of communities and the predicted outcome of increased immigration rates among biogeographic provinces (Table 4, Chase *et al.* 2005). For example, patch-dynamics models without interactions predict that local communities are always invasible and unsaturated. Consequently, invasions following increased immigration rates should be commonplace. Patch-dynamic models with strong competitive exclusion predict that more species can always be added regionally as long as they show sufficiently strong competition-colonization tradeoffs (Tilman 1994). However, the necessary tradeoff becomes increasingly stringent as regional richness increases, therefore communities become increasingly resistant to invasion as more species are added to the region (Shurin and Allen 2001). Neutral models predict that species can always invade local communities regardless of local diversity and the rate of local dispersal because the demographic properties of individuals are equivalent. Consequently, resident species are unable to exclude invaders from local communities. Species-sorting models predict that dispersal is frequent enough to allow local assembly processes to reach their 'end-point' trajectories and that communities therefore resist further invasions. Source-sink models predict that communities become harder to invade with increasing dispersal beyond the maximum local richness point. Local richness plateaus under moderate levels of local dispersal (Mouquet and Loreau 2003) above which communities become increasingly difficult to invade because better regional competitors (dispersers) eliminate other species (Chase *et al.* 2005).

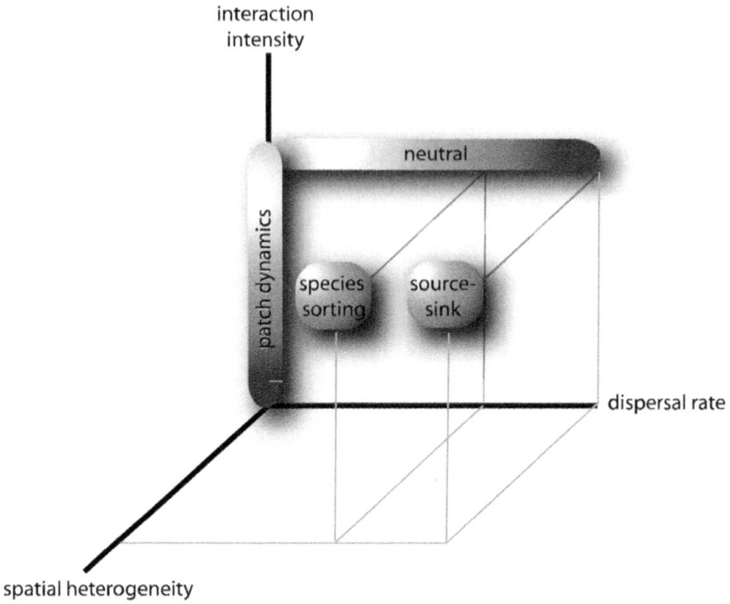

Fig. 1 The relationships among the four major metacommunity paradigms in terms of their assumptions about interaction intensity, spatial heterogeneity and dispersal. Patch dynamics models assume slow dispersal (relative to extinction probabilities), and either interactive (e.g., competition-colonization tradeoffs) or non-interactive communities. Species sorting assumes dispersal is fast relative to extinction, but slow relative to local demographic rates, high spatial heterogeneity and strong interactions. Neutral models assume a continuum of dispersal rates, strong local interactions with pre-emptive competition, and homogeneous patches. Source-sink models assume that dispersal is rapid relative to local dynamics, high spatial heterogeneity in fitness among patches, and strong local interactions.

Metacommunity models also differ in their predictions about community saturation. Both non-interactive patch dynamic models and neutral theory predict that local diversity is never saturated and new species can always be added to local communities. At high levels of local dispersal and species richness, however, neutral theory predicts that greater immigration will increase the rate at which community composition turns over in local communities due to faster extinction of resident species. If more species are added to local communities, the average population size of resident species must decrease under zero-sum dynamics, rendering rare species more susceptible to stochastic fluctuations in population sizes. In contrast, species-sorting, patch-dynamics with interactions and mass-effects models all predict saturation of diversity. In interactive patch-dynamics and mass-effects models, local communities approach saturation with increased dispersal because it becomes more difficult for new species to satisfy the necessary trade-off between dispersal ability and the potential to exclude other species (Shurin and Allen 2001). Moreover, in mass-effects models, source-sink dynamics breakdown at high dispersal rates and better regional competitors eliminate other species (Mouquet and Loreau 2003). In pure species-sorting models without dispersal limitation, exotic species are unlikely to be able to invade because the community is collectively well-adapted to local conditions. Local communities are therefore assumed to be saturated under the species sorting paradigm.

Table 4 Predictions of the 4 major metacommunity paradigms regarding the invasibility and saturation of ecological communities.

Paradigm	Invasible	Saturated
Patch-dynamics: non-interactive	Yes	No
Patch-dynamics: with strong competitive exclusion	Yes, but increasingly difficult with increasing diversity	Yes; increasingly difficult at high diversity to satisfy colonization/competition trade-off
Neutral	Yes	No, but community composition turns over more rapidly with increasing species richness
Species-sorting	No	Yes
Mass-effects	Yes, but increasingly difficult with increasing diversity and frequency of local dispersal	Yes, increasingly difficult at high diversity and dispersal to satisfy colonization/ competition trade-off

Patterns of invasibility and saturation from the biological invasions literature are useful for evaluating the ability of the four metacommunity paradigms to describe community structure and dynamics. The finding that most communities are invasible and unsaturated is consistent with the predictions of any of the models assuming dispersal limitation (patch-dynamics, neutral and mass-effects), and challenges the prediction of species sorting that communities resist invasion because they are saturated with species. The observation of many introductions and few extinctions is inconsistent with models where regional coexistence occurs through spatial refugia (interactive patch dynamics and mass-effects). These models predict that enhanced immigration can result in loss of diversity through greater exclusion by good competitors that are weak dispersers (Mouquet and Loreau 2003). The experimental result of declining invasion success with increasing local diversity, however, is inconsistent with neutral and non-interactive patch-dynamics models, but consistent with all the other metacommunity models. The discrepancy between patterns at local vs. regional scales could reflect the relative importance of different mechanisms operating at varying spatial scales. Species richness might be more important in determining invasion success at local scales, whereas other ecological factors such as propagule pressure or resource availability might govern patterns of community invasibility and saturation at community-wide scales (Levine 2000, Kennedy *et al.* 2002, Byers and Noonburg 2003). One explanation for this pattern is that changes in the number of available resources (as well as disturbance or propagule pressure) can cause invasion success to become positively correlated with diversity at large scales by decreasing mean interaction strength among competing species (Byers and Noonburg 2003).

Patterns in the biological invasions literature suggest that non-interactive patch-dynamics and neutral models successfully predict the dynamics of invasions and extinctions at large spatial scales, while failing to incorporate the mechanisms responsible for invasion resistance with increasing diversity at small spatial scales. Interactive patch-dynamics and mass-effects models adequately describe patterns at small and larger spatial scales. However, the prediction of declining diversity at the highest levels of dispersal was not born out. The invasibility and undersaturation of communities is not predicted by species-sorting models with global dispersal, or by interactive patch- dynamics and mass-effects models with extremely high rates of dispersal. However the decline in invasion success at higher diversity and the large effects of disturbance in weakening biotic resistance to invasion found in the experimental literature supports the species-sorting perspective.

CONCLUSION

Reconciling small-scale, mechanistic studies of the factors that allow or prevent invasion with broad-scale patterns of diversity gain and loss remains one of the most pressing challenges in community ecology and invasion biology. The under-saturation of ecological communities, evident from macro-scale patterns of invasion and extinction in faunal exchanges and contemporary invasions, has been interpreted as indicating that local interactions exert relatively weak influence over community structure (Stohlgren *et al.* 1999, Sax and Gaines 2003). However, even strong local competitive interactions can play out very slowly over broad, heterogeneous landscapes leading to patterns of invasibility and unsaturated communities at regional scales. Furthermore, experimental introductions provide evidence that local ecological interactions play a major role in determining the success or failure of invasions. The apparent unsaturation of communities does not, therefore, lend support to a strictly dispersal-limited, non-interactive view of ecological communities.

Dispersal and local interactions have long been treated as mutually exclusive alternative explanations for community structure. However, recent theoretical work has revealed a number of ways in which dispersal and local interactions can jointly regulate community structure (Shurin and Allen 2001, Byers and Noonburg 2003, Mouquet and Loreau 2003). The movement of individuals among patches gives coherence to local competitive interactions within a region and therefore, local community dynamics may play themselves out at regional spatial scales (Ricklefs 2004). Indeed, the metacommunity approach suggests that processes which have traditionally been considered local and regional in nature now need to be seen as spatially linked. Extending local competitive interactions to a broad region slows the time scale of competitive exclusion to that of species production (Ricklefs 2004), which has important implications for invasions biologists. Global extinctions may lag behind the wave of invasions so that many pending local extirpation events have not yet occurred.

Finally, the literature on invasions and extinctions reveals that the apparent patterns are strongly scale-dependent. Invasion may increase both local (e.g., Table 3) and regional diversity (Sax and Gaines 2003), however it is apparent that beta diversity or regional distinctiveness in composition is declining. The homogenization of the biosphere is occurring at ever increasing rates (McKinney and Lockwood 1999, Rahel 2000, 2002). Synergistic interactions among invaders leading to biological meltdowns (Simberloff and Von Holle 1999, Ricciardi 2001), and among other human-mediated threats to global biodiversity (Sala *et al.* 2000), will certainly increase the vulnerability of natural communities to the introduction of exotic species. The continuous and ever increasing rain of propagules of exotic species (MacIsaac *et al.* 2002, Grigorovich *et al.* 2003) continues to exceed the ability of natural communities to repel them. Despite documented evidence that local communities resist invasions (section 5), macro-scale patterns of biotic interchanges and biological invasions (sections 2 and 3) suggest that it is unlikely that local communities can be managed in ways that will prevent all invasions. Furthermore, increased frequency of human-mediated transport of exotic species coupled with the increased vulnerability of ecosystems to invasion will certainly increase the probability of introducing exotic species with novel ecological functions into naïve ecosystems. Such invasion events clearly have disastrous repercussions on the species richness as well as the functioning of local ecosystems (Kaufman 1992, Nalepa and Schloeser 1993, Fritts and Rodda 1998, Vanderploeg *et al.* 2002). In order to curtail the introduction of exotic species and prevent their sometimes disastrous consequences, our results suggest that managers and policy makers alike need to focus efforts on managing dispersal vectors and preventing the establishment of exotic species rather than relying on intact native communities to repel invaders.

ACKNOWLEDGEMENTS

The authors would like to thank Rod Docking, Joanna Mackenzie, R. Craig MacLean and Michelle Palmer for helpful comments on a previous draft of this manuscript. SAS was supported by an NSERC PGS B. An NSERC grant to JBS provided additional funding.

REFERENCES

Bakker, R. T. 1983. The deer flees, the wolf pursues: inconguencies in predator-prey evolution. Pages 350-382 in D. Futuyma, and M. Slatkin, editors. Coevolution. Sinauer, Sunderland.

Barry, J. C., M. E. Morgan, A. J. Winkler, L. J. Flynn, E. H. Linday, L. L. Jacobs and D. Pilbeam. 1991. Faunal interchange and Miocene terrestrial vertebrates of southern Asia. Paleobiology **17**, 231-245.

Bell, G. 2001. Neutral Macroecology. Science **293**, 2413-2418.

Ben-Tuvia, A. 1985. The impact of the Lessepsian Suez Canal Fish migration on the eastern Mediterranean ecosystem. Pages 367-375 in M. Moraitou-Apostolopoulou and V. Kiortsis, editors. Mediterranean marine ecosystems. Plenum Press, New York.

Bermingham, E. and A. P. Martin. 1998. Comparative mtDNA phylogeography of neotropical freshwater fishes: testing shared history to infer the evolutionary landscape of lower Central America. Molecular Ecology **7**, 499-517.

Brown, J. H., T. J. Valone and C. G. Curtin. 1997. Reorganization of an arid ecosystem in response to recent climate change. Proceedings of the National Academy of Science **94**, 9729-9733.

Brown, J. H., S. K. Morgan Ernest, J. M. Parody and J. P. Haskell. 2001. Regulation of diversity: maintenance of species richness in changing environments. Oecologia **126**, 321-332.

Burke, M. J. W. and J. P. Grime. 1996. An experimental study of plant community invasibility. Ecology **77**, 776-790.

Byers, J. E. and E. G. Noonburg. 2003. Scale dependent effects of biotic resistance to biological invasions. Ecology **84**, 1428-1433.

Case, T. J. 1996. Global patterns in the establishment and distribution of exotic birds. Biological Conservation **78**, 69-96.

Caswell, H. and J. E. Cohen. 1993. Local and regional regulation of species-area relations: a patch-occupancy model. Pages 99-107 *in* R. E. Ricklefs, and D. Schluter, editors. Species diversity in ecological communities: Historical and geographical perspectives. University of Chicago Press, Chicago.

Chase, J. M. and M. A. Leibold. 2003. Ecological niches: linking classical and contemporary approaches. University of Chicago Press, Chicago.

Chase, J., P. Amarasekare, K. Cottenie, A. Gonzalez, R. Holt, M. Holyoak, M. Hoopes, M. Leibold, M. Loreau, N. Mouquet, J. B. Shurin and D. Tilman. 2005. Competing theories for competitive metacommunities. Pages 335-356 in M. Holyoak, R. Holt and M. Leibold, editors. Metacommunities. In Press.

Cohen, A. N. and J. T. Carlton. 1998. Accelerating invasion rate in a highly invaded estuary. Science **279**, 555-558.

Cornell, H. V. 1985. Species assemblages of Cynipid Gall Wasps are not saturated. American Naturalist **126**, 565-569.

Cornell, H. V. 1993. Unsaturated patterns in species assemblages: the role of regional processes in settling local species richness. Pages 243-252 in R. E. Ricklefs and D.

Schluter, editors. Species diversity in ecological communities: historical and geographical perspectives. University of Chicago Press, Chicago.

Cornell, H. V. 1999. Unsaturation and regional influences on species richness in ecological communities: A review of the evidence. Ecoscience **6**, 303-315.

Cornell, H. V. and J. H. Lawton. 1992. Species interactions, local and regional processes, and limits to the richness of ecological communities: a theoretical perspective. Journal of Animal Ecology **61**, 1-12.

Davis, M. A. 2003. Biotic globalization: Does competition from introduced species threaten biodiversity? Bioscience **53**, 481-489.

Diamond, J. M. and C. R. Veitch. 1981. Extinctions and introductions in the New Zealand avifauna: cause and effect? Science **211**, 499-501.

Duncan, J. R. and J. L. Lockwood. 2001. Spatial homogenization of the aquatic fauna of Tennessee: extinction and invasion following land use change and habitat alteration. Pages 245-257 in J. L. Lockwood and M. L. McKinney, editors. Biotic homogenization. Kluwer Academic Press/Plenum Publishers, New York.

Duncan, R. P., T. M. Blackburn and D. Sol. 2003. The ecology of bird introductions. Annual Review of Ecology and Systematics **34**, 71-98.

Fargione, J., C. S. Brown and D. Tilman. 2003. Community assembly and invasion: An experimental test of neutral versus niche processes. Proceedings of the National Academy of Sciences of the United States of America **100**, 8916-8920. .

Foster, B. L. 2001. Constraints on colonization and species richness along a grassland productivity gradient: the role of propagule availability. Ecology Letters **4**, 530-535.

Foster, B. L., V. H. Smith, T. L. Dickson and T. Hildebrand. 2002. Invasibility and compositional stability in a grassland community: relationships to diversity and extrinsic factors. Oikos **99**, 300-307.

Foster, B. L. and D. Tilman. 2003. Seed limitation and the regulation of community structure in oak savanna grassland. Journal of Ecology **91**, 999-1007.

Fox, J. W., J. McGrady-Steed and O. L. Petchey. 2000. Testing for local species saturation with nonindependent regional species pools. Ecology Letters **3**, 198-206.

Fritts, T. H. and G. H. Rodda. 1998. The role of introduced species in the degradation of island ecosystems: A case history of Guam. Annual Review of Ecology and Systematics **29**, 113-140.

Flynn, L. J., R. H. Tedford and Q. Zhanxiang. 1991. Enrichment and stability in the Pliocene mammalian fauna of North China. Paleobiology **17**, 246-265.

Fukami, T. 2004. Community assembly along a species pool gradient: implications for multiple-scale patterns of diversity. Population Ecology **46**, 137-147.

Gaston, K. J. 2000. Global patterns in biodiversity. Nature **405**, 220-227.

Gaston, K. J. 2003. Ecology: the how and why of biodiversity. Nature **421**, 900-901.

Gido, K. B. and J. H. Brown. 1999. Invasion of North American drainages by alien fish species. Freshwater Biology **42**, 387-399.

Grigorovich, I. A., R. I. Colautti, E. L. Mills, K. Holeck, A. G. Ballert and H. J. MacIsaac. 2003. Ballast-mediated animal introductions in the Laurentian Great Lakes: retrospective and prospective analyses. Canadian Journal of Fisheries and Aquatic Science **60**, 740-756.

Gurevitch, J. and D. K. Padilla. 2004. Are invasive species a major cause of extinctions? Trends in Ecology and Evolution **19**, 470-474.

Hastings, A. 1980. Disturbance, coexistence, history and competition for space. Theoretical Population Biology **18**, 363-373.

Hobbs, R. J. and H. A. Mooney. 1998. Broadening the extinction debate: population deletions and additions in California and Western Australia. Conservation Biology **12**, 271-283.

Hubbell, S. P. 2001. The unified theory of biodiversity and biogeography. Princeton University Press, Princeton.

Kaufman, L. 1992. Catastrophic change in species-rich freshwater ecosystems: the lessons of Lake Victoria. BioScience **42**, 846-858.

Keit, T. H. and P. A. Marquet. 1996. The introduced Hawaiian avifauna reconsidered: Evidence for self-organization and criticality? Journal of Theoretical Biology **182**, 161-167.

Kennedy, T. A., S. Naeem, K. M. Howe, J. M. H. Knops, D. Tilman and P. Reich. 2002. Biodiversity as a barrier to ecological invasion. Nature **417**, 636-638.

Leibold, M. A., M. Holyoak, N. Mouquet, P. Amarasekare, J. M. Chase, M. F. Hoopes, R. D. Holt, J. B. Shurin, R. Law, D. Tilman, M. Loreau and A. Gonzalez. 2004. The metacommunity concept: a framework for multi-scale community ecology. Ecology Letters **7**, 601-613.

Lessa, E. P., B. V. Valkenburgh and R. A. Farina. 1997. Testing hypotheses of differential mammalian extinctions subsequent to the Great American Biotic Interchange. Palaeogeography, Palaeoclimatology, Palaeoecology **135**, 157-162.

Levine, J. M. 2000. Species diversity and biological invasions: Relating local process to community pattern. Science **288**, 852-854.

Levine, J. M. 2001. Local interactions, dispersal, and native and exotic plant diversity along a California stream. Oikos **95**, 397-408.

Levins, R. and D. Culver. 1971. Regional coexistence of species and competition between rare species. Proceedings of the National Academy of Science **68**, 1246-1248.

Linberg, D. R. 1991. Marine biotic interchange between the Northern and Southern hemispheres. Paleobiology **17**, 308-324.

Lodge, D. M. 1993. Biological invasions: Lessons for ecology. Trends in Ecology and Evolution **8**, 133-37.

Losos, J. B. and D. Schluter. 2000. Analysis of an evolutionary species-area relationship. Nature **408**, 847-850.

MacIsaac, H. J., T. C. Thomas, C. Robbins and M. A. Lewis. 2002. Modeling ships' ballast water as invasion threats to the Great Lakes. Canadian. Journal of Fisheries and Aquatic Science **59**, 1245-1256.

Marchetti, M. P., T. Light, J. Feliciano, T. Armstrong, Z. Hogan, J. Viers and P. B. Moyle. 2001. Homogenization of California's fish fauna through abiotic change. Pages 259-278 in J. L. Lockwood and M. L. McKinney, editors. Biotic Homogenization. Kluwer Academic Press/Plenum Publishers, New York.

Marshall, L. G. 1981. The Great American Biotic Interchange-an invasion influenced crisis for South American mammals. Pages 133-230 in M. H. Nitecki, editor. Biotic crises in ecological and evolutionary time. Academic Press, New York.

Marshall, L. G. and R. L. Cifelli. 1990. Analysis of changing diversity patterns in Cenozoic land mammal age fauna, South America. Palaeovertebrata **4**, 203-209.

McKinney, M. L. and J. L. Lockwood. 1999. Biotic homogenization: a few winners replacing many losers in the next mass extinction. Trends in Ecology & Evolution **14**, 450-453.

Mora, C., P. M. Chittaro, P. F. Sale, J. P. Kritzer and S. A. Ludsin. 2003. Patterns and processes in reef fish diversity. Nature **421**, 933-936.

Mouquet, N. and M. Loreau. 2003. Community patterns in source-sink metacommunities. American Naturalist **162**, 544-557.

Mouquet, N., P. Leadley, J. Meriguet and M. Loreau. 2004. Immigration and local competition in herbaceous plant communities: a three-year seed-sowing experiment. Oikos **104**, 77-90.

Moyle, P. B. and T. Light. 1996. Biological invasions of fresh water: empirical rules and assembly theory. Biological Conservation **78**, 149-161.

Nalepa, F. T. and D. W. Schloesser. 1993. Zebra mussels: Biology, impacts and control. Lewis Publishers, Boca Raton.

Perdices, A., E. Bermingham, A. Montilla and I. Doadrio. 2002. Evolutionary history of the genus Rhamdia (Teleostei: Pimelodidae) in Central America. Molecular Phylogenetics and Evolution **25**, 172-189.

Por, F. D. 1978. Lessepsian Migration: The influx of the Red Sea Biota into the Mediterranean by way of the Suez Canal. Verlag, Berlin.

Por, F. D. 1989. The legacy of Tethys-An aquatic biogeography of the Levant. Kluwer Academic Press, London.

Rahel, F. J. 2000. Homogenization of fish faunas across the United States. Science **288**, 854-856.

Rahel, F. J. 2002. Homogenization of freshwater faunas. Annual Review of Ecology and Systematics **33**, 291-315.

Ricciardi, A. 2001. Facilitative interactions among aquatic invaders: is an "invasional meltdown" occurring in the Great Lakes? Canadian Journal of Fisheries and Aquatic Science **58**, 2513-2525.

Ricciardi, A. 2004. Assessing species invasions as a cause of extinction. Trends in Ecology and Evolution **19**, 619.

Ricklefs, R. E. 1987. Community diversity: relative roles of local and regional processes. Science **235**, 167-171.

Ricklefs, R. E. 2004. A comprehensive framework for global patterns in biodiversity. Ecology Letters **7**, 1-15.

Ricklefs, R. E. and E. Bermingham. 2001. Nonequilibirum Diversity Dynamics of the Lesser Antillean Avifauna. Science **294**, 1522-1524.

Sala, O. E., F. S. Chapin, J. J. Armesto, E. Berlow, J. Bloomfield, R. Dirzo, E. Humber-Sanwald, L. F, Huenneke, R. B. Jackson, A. Kinzig, R. Leemans, D. M. Lodge, H. A. Mooney, M. Oesterheld, N. LeRoy Poff, M. T. Sykes, B. H. Walker, M. Walker and D. H. Wall. 2000. Global biodiversity scenarios for the year 2100. Science **287**, 1770-1774.

Sax, D. F., S. D. Gaines and J. H. Brown. 2002. Species invasions exceed extinctions on islands worldwide: A comparative study of plants and birds. American Naturalist **160**, 766-783.

Sax, D. F. and S. D. Gaines. 2003. Species diversity: from global decreases to local increases. Trends in Ecology and Evolution **18**, 561-566.

Seabloom, E. W., W. S. Harpole, O. J. Reichman and D. Tilman. 2003. Invasion, competitive dominance and resource use by exotic and native California grassland species. Proceedings of the National Academy of Sciences of the United States of America **100**, 13384-13389.

Shmida, A. and S. Ellner. 1984. Coexistence of plant species with similar niches. Vegetatio **58**, 29-55.

Shurin, J. B. 2000. Dispersal limitation, invasion resistance, and the structure of pond zooplankton communities. Ecology **81**, 3074-3086.

Shurin, J. B. and E. G. Allen. 2001. Effects of competition, predation, and dispersal on species richness at local and regional scales. American Naturalist **158**, 624-637.

Shurin, J. B. and D. S. Srivastava. 2005. New perspectives on local and regional diversity: beyond saturation. Pages 399-417 in M. Holyoak, M. Leibold and R. Holt, editors. Metacommunities: spatial dynamics and ecological communities. University of Chicago Press, Chicago.

Simberloff, D. 1992. Extinction, survival, and effects of birds introduced to the Mascarenes. Acta Oecologia **13**, 663-78.

Simberloff, D. and B. V. Holle. 1999. Positive interactions of nonindigenous species: invasional meltdown? Biological Invasions **1**, 21-32.

Simpson, G. G. 1950. History of the fauna of Latin America. American Journal of Science **38**, 361-389.

Simpson, G. G. 1980. Splendid isolation: the curious history of South American mammals. Yale University Press, New Haven.

Smallwood, K. S. 1994. Site invasibility by exotic birds and mammals. Biological Conservation **69**, 251-59.

Smith, S. A. and E. Bermingham. The biogeography of lower Mesoamerican freshwater fishes. Submitted to the Journal of Biogeography.

Smith, S. A., G. Bell and E. Bermingham. 2004. Cross-Cordillera exchange mediated by the Panama Canal increased the species richness of local freshwater fish assemblages. Proceedings of the Royal Society of London B **271**, 1889-1896.

Stohlgren, T. J., D. Binkley, G. W. Chong, M. A. Kalkhan, L. D. Schell, K. A. Bull, Y. Otsuki, G. Newman, M. Bashkin and Y. Son. 1999. Exotic plant species invade hotspots of native plant diversity. Ecological Monographs **69**, 25-46.

Terborgh, J. W. and J. Faaborg. 1980. Saturation and bird communities in the West Indies. American Naturalist **116**, 178-195.

Tilman, D. 1994. Competition and biodiversity in spatially structured habitats. Ecology **75**, 2-16.

Tilman, D. 1997. Community invasibility, recruitment limitation, and grassland biodiversity. Ecology **78**, 81-92.

Tilman, D. 2004. Niche tradeoffs, neutrality, and community structure: A stochastic theory of resource competition, invasion and community assembly. Proceedings of the National Academy of Sciences USA **101**, 10854-10861.

Turnbull, L. A., M. J. Crawley and M. Rees. 2000. Are plant populations seed-limited? A review of seed sowing experiments. Oikos 88, 225-238. Vanderploeg H. A., T. F. Nalepa, D. J. Jude, E. L. Mills, K. T. Holeck, J. R. Liebig, I. A.

Grigorovich and H. Ojaveer. 2002. Dispersal and emerging ecological impacts of Ponto-Caspian species in the Laurentian Great Lakes. Canadian Journal of Fisheries and Aquatic Sciences **59**, 1209-1228.

Vermeij, G. J. 1989a. Invasion and extinction: The last three million years of North Sean pelecypod history. Conservation Biology **3**, 274-281.

Vermeij, G. J. 1989b. Geographical restriction as a guide to the causes of extinction: the case of the cold northern oceans during the Neogene. Paleobiology **15**, 335-356.

Vermeij, G. J. 1991a. When biotas meet: Understanding biotic interchange. Science **253**, 1099-1104.

Vermeij, G. J. 1991b. Anatomy of an invasion: the trans-Arctic interchange. Paleobiology **17**, 281-307.

Vitousek, P. M., C. M. D'Antonio, L. D. Loope, M. Rejmanek and R. Westbrooks. 1997. Introduced species: a significant component of human-caused global change. New Zealand Journal of Ecology **21**, 1-16.

Wallace, A. R. 1876. The geographical distribution of animals. Hafner, New York.

Webb, S. D. and L. G. Marshall. 1982. Historical biogeography of recent South American land mammals. Pages 39-52 in M. A. Mares and H. H. Genoways, editors. Mammalian biology in South America. The University of Pittsburgh, Pittsburgh.

Webb, S. D. 1985. Late Cenezoic dispersals between the Americas. Pages 357-386 in F. G. Stehli and S. D. Webb, editors. The Great American Biotic Interchange. Plenum, New York.

Zobel, M., M. Otsus, J. Liira, M. Moora and T. Mols. 2000. Is small-scale species richness limited by seed availability or microsite availability? Ecology **81**, 3274-3282.

Chapter twenty

The biogeography of naturalized species and the species-area relationship: reciprocal insights to biogeography and invasion biology

D. F. Sax and S. D. Gaines

INTRODUCTION

"He who admits the doctrine of the creation of each separate species, will have to admit, that a sufficient number of the best adapted plants and animals have not been created on oceanic islands; for man has unintentionally stocked them from various sources far more fully and perfectly than has nature."

— Charles Darwin, 1859.

Darwin was not arguing that species invasion was a preferable condition. Indeed, he argued that introduced species were likely to drive many island endemics

M.W. Cadotte et al. (eds.), Conceptual ecology and invasion biology, 449–480.

ics to extinction. Instead, he was drawing on patterns of species introduction and naturalization to make inferences about the capacity of islands to hold many more species than had naturally colonized or evolved *in situ*. Darwin was clearly a broad thinker and in his quest to understand the natural world he frequently drew upon forms of evidence that might be viewed as "artificial" or "unnatural". For example, he drew heavily upon evidence from artificial breeding to better understand the process of natural selection. Therefore, it is not surprising that he drew upon evidence from naturalized species, i.e., introduced species that form self-sustaining populations, to better understand biogeographic principles of the natural world. Besides his comments quoted above, he also used naturalized species, namely their known rates of spread, to infer that native species would have had ample time to dramatically alter their geographic ranges during the last glacial period. These two observations of Darwin's are but a few of the many possible insights that studying naturalized species can provide for the fields of biogeography and ecology.

In this chapter, we will endeavor to accomplish several goals. First, we review work done to date (post Darwin) that explores insights that studying naturalized species can provide to understanding natural patterns and processes. Second, in keeping with the theme of this book, we will attempt to use these same patterns to better understand naturalized species themselves, i.e., to understand patterns and processes relevant to invasion biology. Finally, we will explore an understudied topic in the biogeography of naturalized species, namely the species-area relationships of naturalized species, focusing among other topics on the issue quoted above by Darwin, i.e., the capacity of islands to hold additional species.

A REVIEW OF BIOGEOGRAPHIC PATTERNS OF NATURALIZED SPECIES

The fields of biogeography and invasion biology share similar histories. Both fields have been the subject of study by many of the best-known ecologists, evolutionary biologists and naturalists of the 18th through 20th centuries. Notable contributions to Biogeography have been made by Linneaus, Buffon, Hooker, Darwin, Wallace, Darlington, Dobzhansky, Mayr, Grinnell, Elton, Lack, Wilson and MacArthur to name but a few — for a more complete list see Lomolino *et al.* (2004). Contributors to the field of invasion biology have been no less illustrious, and in fact are often many of the same individuals — for a treatment of this subject see Davis (this volume). Despite their long histories, both biogeography and invasion biology have only relatively recently garnered sufficient attention to warrant the establishment of journals and societies dedicated to these topics; for example, the *Journal of Biogeography* was established in 1974, while the journal *Biological Invasions* was established in 1999. It is perhaps not surprising then, that most research that integrates these two fields has been conducted in the past decade. While relatively recent, this body of work has great potential, both to inform our understanding of biogeographic patterns and processes, as well as to inform our understanding of species invasions. Here, we review several topics that have been particularly relevant to both these endeavors.

Latitudinal gradients in richness and range size

The latitudinal gradient in species richness from a peak in the tropics to a minima at the poles is one of the most general phenomena in biogeography (Brown and Lomolino 1998, Sax 2001). In fact, unlike most "general" patterns in biology, there are very few known exceptions. Perhaps the best known of these exceptions is that of benthic marine macroalgae, which generally peak in richness at mid-latitudes (e.g., Gaines and Lubchenco 1983, Luning 1990). Mid or high-latitude peaks in richness also occur for individual families and genera, such as ichneumonid wasps, penguins, and pines (Brown and Lomolino 1998, Stevens and Enquist 1998). In spite of these notable exceptions, however, all major groups of vertebrates, most invertebrate groups, vascular and non-vascular plants, and many other taxonomic groups show this latitudinal gradient (see review in Willig *et al.* 2003). European naturalists first prominently described this pattern during the Age of Exploration, for example Humboldt (1807, reprinted by Hawkins 2001) eloquently describes increases in plant richness as one approaches the tropics. The 'gradient' itself, as opposed simply to differences between the temperate zones and tropics, became more apparent and better characterized as more biogeographic data became available during the twentieth century (e.g., Fischer 1960, Simpson 1964; for a review of this topic see Brown and Sax 2004). Research describing this gradient continues to this day (e.g., Willig *et al.* 2003, Rodriguero and Gorla 2004).

A second gradient that is not as well characterized is that of variation in geographic range size across latitude. The first inkling of this pattern is present in the seminal work of Eduardo Rapoport (1982), who demonstrated that the geographic extent of mammalian subspecies tended to increase with increasing latitude. It was not until 1989, however, that George Stevens first characterized these patterns at the species level, showing that species tended on average to have large latitudinal extents at higher latitudes. Stevens (1989) christened this pattern Rapoport's rule and set off a flurry of research on this topic, greatly revitalizing studies of latitudinal gradients in biogeography. At present, the generality of the latitudinal gradient in species range size is much debated (e.g., Gaston *et al.* 1998), but there is broad and uncontested evidence that this pattern is robust within temperate latitudes of the North Hemisphere for many taxonomic groups, e.g., birds, mammals, crayfish, trees, etc. (Stevens 1989, France 1992, Letcher and Harvey 1994, Gaston *et al.* 1998).

What is entirely unclear, for both these gradients, is whether there is any universally applicable explanation for them. Over the years a number of hypotheses have been proposed to explain these gradients, particularly the gradient in species richness. Perhaps the most obvious of these explanations has been the observation that there is generally more available energy, heat, and water as one approaches the tropics, which could promote or maintain higher levels of biological diversity in terrestrial environments (e.g., Pianka 1966); note that while variation in available water is clearly not a limiting component of species richness in the ocean, that variation in energy and heat may be of direct importance in surface waters and of indirect importance in the deep-sea, where the latitudinal gradient in species richness also occurs (e.g., Willig *et al.* 2003). Other reasonable explanations for the latitudinal gradient in species richness, however, have also been proposed and many of these invoke temporally explicit explanations for these patterns. For example, it has been suggested that long-term stability of resource availability or conditions have allowed more species to accumulate (or be preserved) in the tropics (e.g., Pianka 1966). In terrestrial environments, for example, the effects of ice-ages and other changes in climate have been reduced in the tropics relative to high latitude regions, where glaciers have repeatedly cleared away most forms of life (Brown and Lomolino 1998). Other temporally explicit explanations suggest that higher speciation rates (at low latitudes) may be responsible for the richness gradient (e.g., Rohde 1992). For the latitudinal gradient in range size, similar explanations (or logical continuations of these explanations) have been employed; for example, glaciers or other forms of environmental perturbation may have differentially removed species with small geographic ranges from high latitudes (Brown 1995, Sax 2001). The gradient in range size has also been suggested to be a consequence of the gradient in richness, with the assumption that regions with high species richness would cause those species (presumably because of some sort of limiting biotic interactions, such as competition) to on average have smaller geographic ranges (Brown 1995).

Through the years the number of hypotheses proposed (particularly for the latitudinal gradient in species richness) has swelled. In 1966 Pianka characterized six principle hypotheses, by 1992, Rohde characterized more than 20 hypotheses, and today there are more than 30 hypotheses for these gradients described in the literature (Willig *et al.* 2003). The trouble, clearly, has been in reducing the number of explanations. Herein, however, lies one of the principal difficulties of many biogeographic pursuits, namely the difficulty in rejecting hypotheses when manipulative experiments are impossible (or nearly impossible) to pursue and the relative merits of individual hypotheses are instead measured against correlative data. Herein, also lies one of the principal advantages of studying naturalized species, as their introduction and subsequent naturalization provides a manipulation for a large uncontrolled 'experiment', one that would otherwise be impossible to perform at such large spatial and temporal scales. Similarly, the occurrence of these uncontrolled experiments in many regions and continents around the world provides the replication needed to draw general insights from the 'results' of these experiments.

The specific advantage of studying naturalized species, with respect to this issue, is that the great majority of these species have only been present in their naturalized ranges for a few hundred years at most. This means that any patterns in geographic gradients in richness or range size observed in naturalized species cannot be *directly* attributable to effects of glaciations, other historical processes that have differentially affected different latitudes, or to different evolutionary rates of accumulation over time (the potential *indirect* effects of these process will be discussed later). This means that if one can account for differences in patterns of introduction, we may be able to learn a great deal about what structures (or what doesn't structure) the geographic distributions of species by studying patterns of naturalized species.

To examine these issues, Sax (2001) examined the geographic distribution of naturalized species of several taxonomic groups on several continents: plants in Europe, fishes and birds in North America, and mammals in Europe, North and South America. This work shows that within continental areas of the tropics few species have invaded, and that those that have tend to be human commensals with very large geographic ranges (we shall return to the issue of few invasions within tropical mainland areas in a section below). At extra-tropical latitudes, however, where most naturalized species are distributed, naturalized species show geographic patterns of distribution that are qualitatively similar to those shown by native species, i.e., richness peaks at low latitudes and declines toward the poles, while average geographic range size is smallest at low latitudes and increases towards the poles (Figs. 1 and 2); for a discussion of the small idiosyncrasies in these patterns see Sax (2001). These latitudinal patterns do not appear to be driven exclusively by patterns of introduction effort per se, at least for birds, because the greatest number of introduction attempts for birds in North America have been at high temperate latitudes, i.e., above 35 degrees latitude, where human populations are largest, and not at low temperate latitudes, i.e., below 35 degrees, where the number of naturalized species is greatest. These patterns also do not appear to be an indirect consequence of patterns that existed in naturalized species' native ranges (see Sax 2001 for evidence and details). This suggests that latitudinal gradients in richness and range size can form relatively quickly (over the ecological time of a few hundred years since exotic species were introduced) and that temporally dependent explanations for these gradients (such as glaciation events and differential rates of speciation) may not be necessary to explain these patterns. This represents a potentially significant contribution to our understanding of these gradients by providing a means by which we can potentially reject certain hypotheses as general and necessary explanations.

In addition to learning about what structures biogeographic gradients, studying these patterns with naturalized species also provides insights into invasion biology. If species are more likely to become established at low temperate latitudes than high temperate latitudes, independent of deliberate attempts at introduction (Sax 2001), then regions may be differentially susceptible to invasion. The evidence presented, however, suggests that we should adopt a more complicated vision of "invasion susceptibility" than is generally used. This is because, despite more species invading low temperate latitudes, these non-native species tend to acquire small geographic ranges. In contrast, few species tend to invade high latitudes, but the ones that do tend to acquire large geographic ranges.

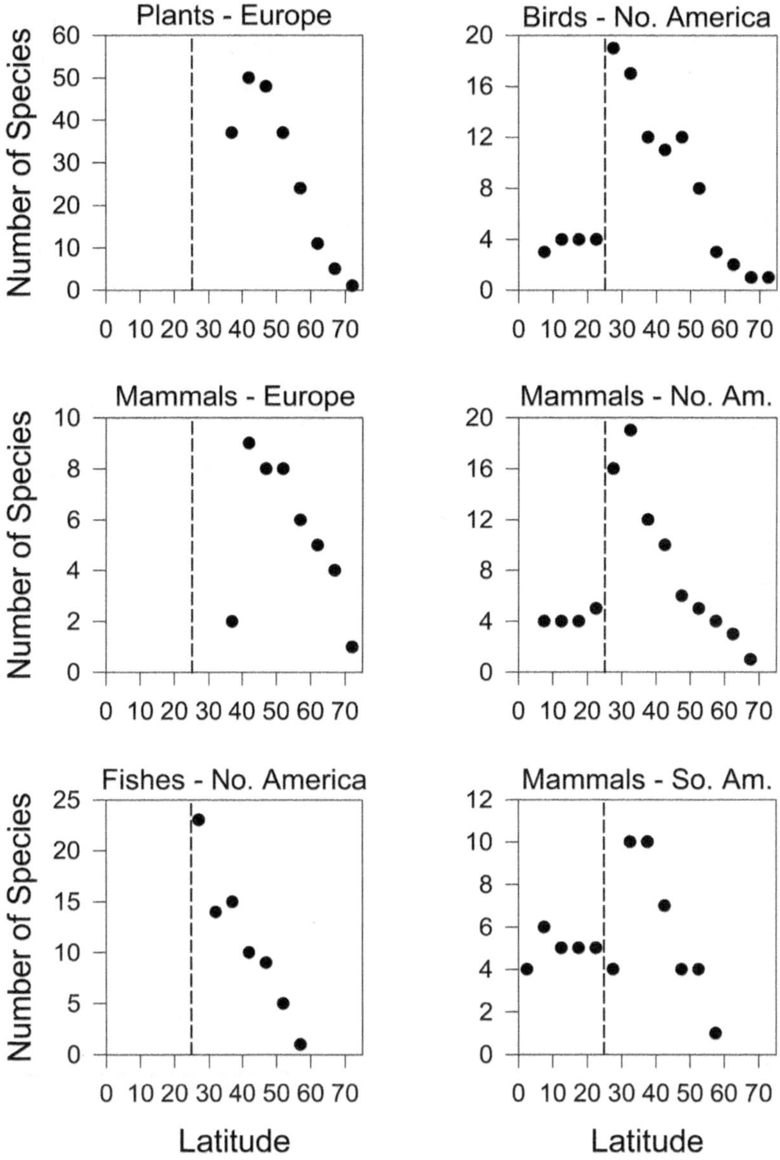

Fig. 1 Latitudinal gradients in richness of naturalized species. The species richness of naturalized exotic species in 5° latitudinal bands on continents. The dashed lines indicate the Tropic of Cancer or Capricorn, dividing tropical and temperate latitudes. Fish distributions were examined at temperate latitudes only. (*Reprinted with permission of Blackwell Publishing from Sax 2001*).

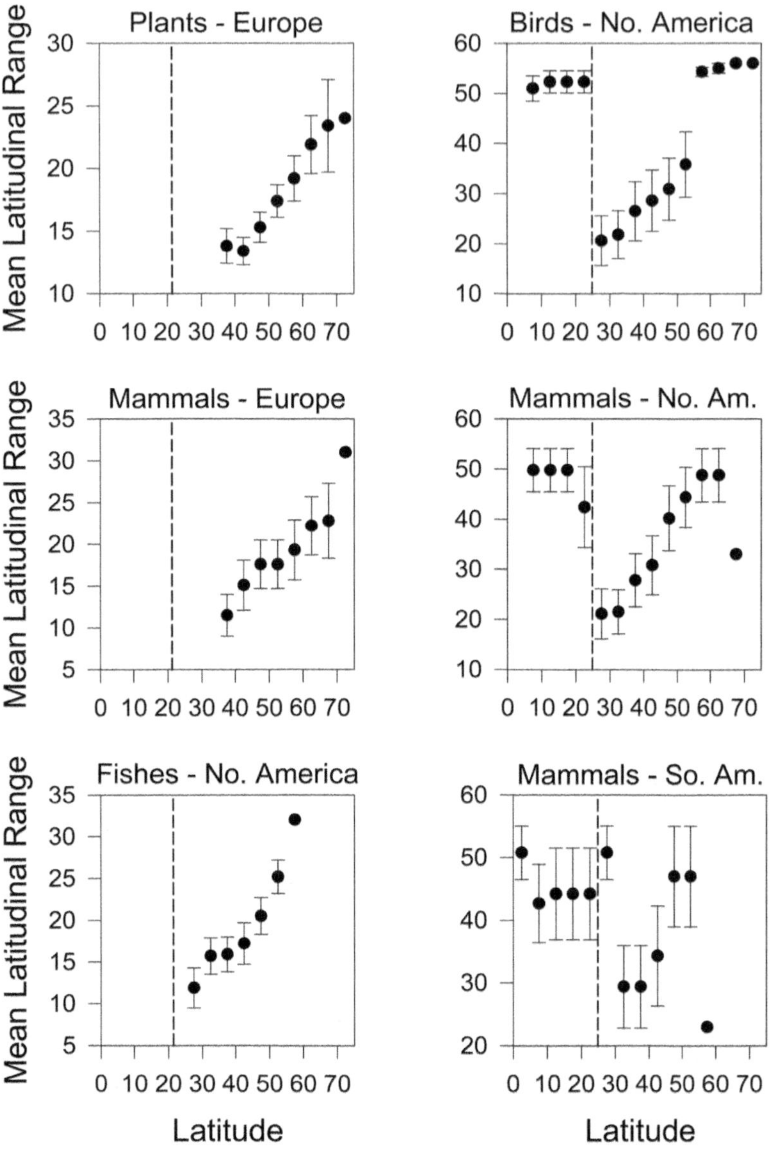

Fig. 2 Latitudinal gradients in geographic range size of naturalized species. The mean latitudinal extent of naturalized exotic species in 5° latitudinal bands on continents. Error bars represent + or – one standard error. Points without error bars are based on a single species. The dashed lines indicate the Tropic of Cancer or Capricorn, dividing tropical and temperate latitudes. (*Reprinted with permission of Blackwell Publishing from Sax 2001*).

This suggests a tradeoff in susceptibility of invasion between regions that is balanced between the absolute number of likely invaders and the absolute geographic extent those invaders are likely to achieve.

Many questions regarding these patterns remain unanswered. Are these patterns general to all continents and all taxonomic groups of naturalized species? Will these patterns change over time as more introduced species become established and as existing species continue to expand their geographic ranges? And, will future range collapses in the geographic distribution of some naturalized species affect these patterns? Answering these questions will involve much work, but doing so offers the promise of great intellectual contributions to our understanding of both biogeography and species invasions.

Invasions of hotspots of diversity

In contrast to experimental work done at small spatial scales and in contrast to the supposition of Elton (1958) that species-rich areas should be more difficult to invade than species-poor ones, work at biogeographic scales has generally shown that native and naturalized richness of species are positively correlated across landscapes and regions. Two papers that drew particular attention to this phenomenon were published in 1999.

Lonsdale (1999), examined 184 published studies of plant richness from around the world; the spatial extent covered by these studies varied from a few square kilometers to greater than a million square kilometers, and the regions covered included islands and continents, temperate and tropical areas. By examining the 177 floras that distinguished between native and exotic species he was able to show that native and exotic species richness were significantly positively correlated across these sites, even after controlling for differences in area among sites. The total amount of variance in the number of exotic species predicted by the number of native species was improved by considering whether a particular site was an island or a reserve; however, even without this information a relatively clear and robust pattern had been identified: native and exotic richness were positively correlated at biogeographic scales.

Stohlgren *et al.* (1999) collected field data on native and exotic plant species at local scales, across vegetation types, and within biomes in the western US. At the scale of vegetation types and biomes they found a positive relationship between native and exotic richness. Further, Stohlgren *et al.* (1999) indicated the obvious, but previously unarticulated observation that the areas richest in native species were in fact being invaded by the most exotic species.

This work by Lonsdale (1999) and Stohlgren *et al.* (1999) has subsequently been confirmed at biogeographic spatial scales by several other studies. Most of this work has also been done with plants; Stadler *et al.* (2000) have shown a positive correlation between native and exotic plant richness in ecogeographic zones of Kenya; McKinney (2002a) found this pattern in 77 protected areas of the US; Pysek *et al.* (2002) found this pattern in nature reserves of the Czech Republic;

Deutschewitz *et al.* (2003) found this pattern in regions of the Dessau district of Germany; and finally, Stohlgren *et al.* (2003) show that native and exotic richness are positively correlated across counties within US States, and across the states themselves. While the focus of these patterns has been on plants, an exploration of other groups (particularly at large spatial scales) shows similar patterns, particularly within the temperate zones. For example, a comparison of native and exotic mammalian richness within 5-degree bands of latitude in temperate North America reveals a strikingly strong positive relationship between native and exotic richness (Fig. 3a). Similarly, data from Sax *et al.* (2002) shows that native and exotic bird richness are positively correlated on oceanic islands (Fig. 3b). In contrast, however, to these positive correlations, at least one negative correlation between native and exotic species at biogeographic scales is known from the literature, i.e., for freshwater fishes in watersheds of North America (Gido and Brown 1999). There are some reasons, which we will return to later, that we might expect atypical patterns from freshwater fishes. Despite this exception, however, with the evidence available to date, it appears that the general pattern is for native and exotic richness to be correlated across large areas (at least within a given region, although comparisons of areas across regions may yield different results — a topic we return to in the section below on invasions of mainland tropical areas).

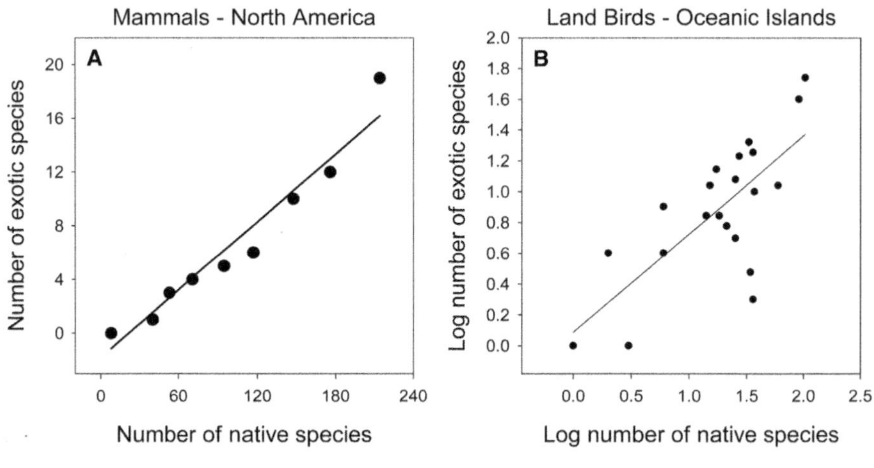

Fig. 3 Positive correlation between native and naturalized species richness. A, Richness of native and naturalized mammal species in 5 degree belts of latitude (from 35-80 degrees) in North America; $R^2 = 0.95$, $p < 0.0001$; data from Stevens (1989) and Sax (2001). B, Richness of historic number of native and current number of naturalized bird species on oceanic islands; $R^2 = 0.53$, $p < 0.001$; data from Sax *et al.* (2002).

This relatively simple observation (that native and exotic richness are positively correlated across large areas) comes in contrast with much work done at experimental scales that shows a negative relationship between native richness and exotic invasion (e.g., Tilman 1997, Stachowicz *et al.* 1999, Naeem *et al.* 2000). This experimental work is consistent with modeling work (e.g., Case 1991) and perhaps more significantly with the supposition of Elton (1958) that species rich areas should be more difficult to invade than species poor areas. A number of studies have attempted to reconcile the differences observed between experimental and observational studies (e.g., Wardle 2001, Fridley *et al.* 2004). Perhaps the most convincing explanation for these differences have been articulated by Levine (2000), who suggests that regardless of any impact native richness may have on species invasions (when all else is equal in controlled experimental settings), that these impacts will be outweighed by real-world spatial variation in environmental conditions (such as nutrient and energy availability, precipitation, and habitat heterogeneity) that promote covariance in the richness of both native and exotic species.

This explanation by Levine (2000) for reconciling differences observed in observational and experimental studies suggests that native and exotic richness should be expected to positively covary whenever both groups respond similarly to environmental variables, i.e., that site conditions that promote high native richness will often also promote high exotic richness. Other explanations for a positive association between native and exotic richness, however, are also possible. One alternative explanation is that introduction efforts by humans (i.e., propagule pressure) of exotic species are higher in species-rich than species-poor environments, such that regions with many native species are also the same regions where humans introduce the most species. For this to be a universal explanation for these patterns, however, the pattern of introduction effort by humans would have to consistently match native species richness across regions, which seems unlikely to be generally true and is definitely not the case in many places, e.g., bird introduction effort in North America does not match patterns of native species richness (see above). A second alternative explanation is that the positive association between native and exotic richness across regions is due to differences in the area of regions compared. Because of the species-area relationship, large regions should have more native and more exotic species. However, while species-area relationships certainly may be important in determining such patterns, positive relationships between native and exotic richness are observed even among equal size areas (e.g., Sax 2002) or when the effect of area is controlled (e.g., Lonsdale 1999). Given that these alternative explanations are unlikely to drive the positive relationship frequently observed between native and exotic richness, it seems likely that the observed relationships are indeed due to variation in environmental characteristics (such as energy and nutrient availability) that influence native and exotic species richness in similar ways.

If environmental characteristics are driving the positive association between native and exotic richness observed across sites then these patterns have important implications for conservation and environmental management. They suggest that species-rich native preserves cannot be expected to repel exotic species invasions and that appropriate management control efforts for exotic species must be maintained to prevent them from becoming established. Further, with respect to ecological theory, these patterns (and its likely explanation) suggest two things. First, they suggest that even diverse continental areas can often support many additional species. Second, they suggest that the richness of native species within an area may indicate the relative capacity of that area to support species, since sites rich in native species have presumably increased in richness the most, while sites poor in native species have presumably increased in richness the least.

Tropical mainlands: regions with few invaders

Because native and exotic richness are positively correlated across sites at large spatial scales one might suspect that tropical regions, where the global peaks in native richness exist, are also invaded by many exotic species. Nothing, however, could be further from the truth, especially in mainland tropical areas, where relatively few exotics have generally invaded (Rejmanek 1996, Sax 2001, Fine 2002). This means that the generally positive correlation between native and exotic richness often breaks down when comparing across areas in both temperate and tropical zones, e.g., the strong pattern depicted in Fig. 3A would break down if tropical points were included on the plot. Nevertheless, within tropical areas the positive correlation between native and exotic richness can still hold, e.g., in Mexican States native and exotic species of flowering plants are positively correlated, $R^2 = 0.54$, $p < 0.0001$, data from Villaseñor and Espinosa-Garcia 2004). Also, unlike tropical mainland environments, on tropical islands many exotic species have become naturalized (e.g., Sax *et al.* 2002). Why, however, so few exotic species have become established in mainland tropical areas in not clearly known.

Two principal studies to date have provided empirical data showing the contrast between the number of exotic species that have become naturalized in tropical mainland, tropical island and temperate areas. Rejmanek (1996) examined the numbers of native and exotic plants in various floras of the world, showing that the relative proportion of exotic species, as well as the absolute number of exotics (after controlling for area), declines precipitously within tropical mainland regions relative to tropical islands or temperate areas. Similarly, Sax (2001) shows that few species of birds or mammals have become naturalized on continents within the tropics (Fig. 1), but reports that many have become established on tropical islands and within temperate areas. Thus, both studies report qualitatively similar patterns: few species have invaded mainland tropical areas, while relatively many species have invaded tropical islands and temperate areas.

The mechanism responsible for this pattern remains elusive. One possibility, however, is that patterns of human disturbance or introduction effort of exotic species are different in mainland tropical areas then elsewhere. While each region does of course have its own unique history, such an explanation does not appear to be generally true. For example, mainland Mexico and islands of the Caribbean share similar histories of European colonization, disturbance, subsequent population growth and presumably introduction efforts, yet these islands harbor many more naturalized species of birds and mammals (> 40) than mainland Mexico (< 10, Sax 2001). Another possibility is that mainland tropical regions are resistant to invasion because of a combination of biotic and abiotic factors (Rejmanek 1996). The potential role of abiotic factors, however, is difficult to reconcile with the data provided in Sax (2001), which shows that coastal areas of tropical continents (which presumably have very similar abiotic conditions as tropical islands) have few exotic bird or mammal species. In light of this Sax (2001) argues that abiotic factors should not be the limiting variable in exotic species invasion of mainland tropical areas. Instead, he argues (based on the differential placement of naturalized species' low-latitude range boundaries on continents and islands) that biotic resistance, which may be strongly correlated with species richness, is the primary factor resisting species invasions of mainland tropical areas. Of course, if high species richness in mainland tropical areas is resulting in some sort of biotic resistance to invasion then why doesn't species richness correlate with invasion resistance in temperate latitudes? At the moment, there is no clear answer to this question. One possibility, however, is that some threshold in species richness must be reached (which may be met in mainland tropical areas) before biotic resistance is an effective force in resisting species invasions. Such an explanation would be consistent with the greater apparent invasibility of tropical islands and temperate areas, where species richness is reduced relative to mainland tropical areas.

The suggestion that biotic resistance is important in resisting species invasion of mainland tropical areas is also consistent with biogeographic hypotheses that postulate that low latitude range boundaries of species (or the ability of species to extend their geographic range closer to the equator) are set by biotic interactions with other species (e.g., Dobzhansky 1950, MacArthur 1972). These biotic interactions are assumed to include all negative pairwise species interactions (i.e., competition, predation and amensalism), but with respect to species invasion could also include missing mutualistic species (Richardson *et al.* 2000). Of course, research in ecology since the time of MacArthur has had a strong focus on the role of competition (e.g., Bruno *et al.* 2003). In spite of this, work in invasion biology suggests that competition itself may not be as important in affecting species richness as other types of biotic interactions. For example, recent work by Davis (2003) suggests that competition from exotic species should rarely have negative consequences on native species richness. Further, most of the negative consequences of exotic species on native richness have been associated with introduced predators and diseases, e.g., black rats, the brown tree snake, and avian malaria (e.g., Ebenhard 1988, Fritts and Rodda 1998; Scott *et al.* 2001). If this information can reciprocally be applied to understanding what prevents species invasions, then this work would suggest that predation (meant here in the broad sense of herbivory, predation, parasitism and disease) could likely explain why mainland tropical areas are so difficult to invade; in the tropics, particularly the mainland tropics, not only are there, for example, more mammalian predators and herbivores, but also greater numbers of parasites and diseases (e.g., Willig *et al.* 2003, Rodriguero and Gorla 2004).

If predation (*sensu lato*) is important in preventing species invasion of mainland tropical areas then this may suggest (depending on how thresholds of richness of predators operate) that predation may be important in preventing species invasion of other places as well; because, for example, even if predation cannot prevent species invasion at biogeographic scales in temperate areas, it may still be important at ecological spatial scales. If this is true, then it could have large implications for ecological theory and experimentation. For example, currently in ecology much effort is focused on understanding the role of richness in preventing species invasion. Much of this work, however, is focused on the role that richness of competitors may have in preventing species invasion (e.g., Kennedy *et al.* 2002); a compliment to this would be increased attention to the role that richness of predators (including herbivores, parasites and diseases) may play in preventing species invasion.

Patterns of invasion on islands

What Darwin (1859) wrote about islands almost 150 years ago couldn't be more true today; humans have been filling islands "far more fully" than has nature. Lifeforms that were entirely absent from many islands (particularly oceanic ones), such as terrestrial mammals and freshwater fishes, have been introduced and become established in abundance. Similarly, lifeforms that were present, but less diverse on islands than in similar sized areas on continents, such as vascular plants, have become much more speciose on islands today than in the past. In many cases, those species that have been introduced and become established have had devastating effects on the native flora and fauna, causing extinctions in some cases, and reductions in the abundance of native species in many (if not all) cases. Nonetheless, the net effect of these changes has been an increase in net richness for many groups of species on many islands (e.g., Wilson 1997, Chown *et al.* 1998, Sax *et al.* 2002). The conservation challenges posed by such patterns are severe and complex, but are not the focus of this discussion. Instead, here we report on the patterns themselves and on what these patterns can tell us about biogeography, invasive species, and what we do not yet understand.

Island floras and faunas are often described as being out of balance (or not harmonic) with mainland biotas (Brown and Lomolino 1998). This is because evenly representative subsets of the suite of organisms found on continents are not found on islands. Instead, a disproportionate number of species found on islands are from taxonomic groups that are likely to disperse across ocean waters. This includes: birds and insects, which can often fly long distances or be blown by storms to distant localities, plants, whose propagules can be blown by the wind or carried by birds (in their guts, on their feet, etc.), and reptiles, which are often able to survive extended voyages (with little food or water) on "rafts" of natural vegetation that have been washed out to sea by storms. By contrast, taxonomic groups that are not likely to make or survive such dispersal events (such as freshwater fish, terrestrial mammals, and amphibians) are likely to be absent or depauperate on isolated islands.

Over the past decade much has been learned about how humans are mediating changes on islands, for example the importance of introduced predators and pathogens in causing native species extinctions (e.g., Fritts and Rodda 1998, Atkinson *et al.* 2000), the relative role of different ecological correlates in the establishment success of introduced species (e.g., Duncan *et al.* 2003), and how the establishment of novel vegetation can facilitate the colonization of other species (e.g., Hutton 1990). Much has also been learned about how the composition and richness of different taxonomic groups are changing on islands. For example, there have been large changes in the composition of bird species, large increases in the total richness of plant species (coupled with substantial declines in native plant abundance), and large increases in the presence of taxa that were previously rare or entirely absent on islands, such as freshwater fishes and non-volant mammals (e.g., Wilson 1997, Sax *et al.* 2002). What we do not yet understand is how the species richness of these different taxonomic groups will change in the future, or whether islands are approaching any sort of maximum capacity for holding additional species. These issues require considerable additional work, but some key insights can be gleaned by examining species-area relationships.

SPECIES-AREA RELATIONSHIPS OF NATURALIZED SPECIES

The positive relationship between increasing area and increasing number of species has been known for some time (Whittaker 2004). Arrhenius (1921), however, was the first to plot species area against species richness on a log-log scale. Doing this laid the foundation for comparisons between species-area relationships (for different taxonomic groups and for different sets of localities), because species-area relationships on log-log plots commonly take a linear form, simplifying comparisons of slopes and intercepts (however, see Lomolino and Weiser 2001 for a discussion of when non-linear patterns are expected). The slopes and intercepts of log-log species-area relationships contain information about how species are distributed, how dispersal-limited different taxonomic groups are, and how isolation and limited area affect species richness (e.g., Rosenzweig 1995). In spite of the usefulness of these patterns in characterizing the distributions of species, the precise mechanisms for species-area relationships are still strongly contested (e.g., Matter *et al.* 2002). Species-area relationship are believed to be a consequence of, 1) patterns of species abundance, 2) the interplay of species abundance and geographic range distributions, 3) the number of habitats present (and consequently the variety of species supported), 4) passive sampling of a larger fraction of a region's species pool, 5) lower extinction risks occurring in larger populations, 6) a dynamic relationship between colonization and extinction, and 7) the dynamics of speciation, colonization and extinction (Matter *et al.* 2002). Part of the difficulty in teasing these different hypotheses apart is that they clearly are not mutually exclusive. Additionally, the mechanisms that are most important for the species-area relationships may vary across spatial scales; for example, at the small spatial scales of most ecological studies, mechanisms like speciation will generally be relatively unimportant, whereas at larger spatial scales, speciation may be a very important process. In this discussion, we restrict our focus to species-area relationships occurring at large spatial scales, i.e., at spatial scales that exceed several hectares.

Although there is a rich literature on species-area relationships (see reviews by Connor and McCoy 1979, Rosenzweig 1995, and Lomolino 2001), relatively little has been published on species-area relationships of naturalized (or otherwise exotic) species. Although it seems likely that someone has examined a species-area relationship for exotic species prior to 20 years ago, the first paper we are aware of that explicitly discusses such a relationship is by Crawley (1987), who compared the species-area relationships of native and exotic plant species. More recently, Pysek (1998) examined the species-area relationships of native and exotic species in cities of central Europe. Other subsequent studies that have considered the relationship between area and exotic species include work by Chown *et al.* (1998) on Southern Ocean Islands, by Lonsdale (1999) across 104 sites from around the world, by Moody (2000) on the Channel Islands of California, by McKinney (2002a) in protected areas of the U.S., by McKinney (2002b) and Rejmanek (2003) on exotic fishes and plants in U.S. States, by Pysek *et al.* (2002a and 2002b) on exotic plants in nature reserves of central Europe, and most recently by Russell *et al.* (2004) on the offshore islands of New Zealand. Here we build on this previous work to consider what examinations of exotic species-area relationships and their comparisons with native species-area relationships can tell us about species invasion and about the factors that limit and maintain species richness. In particular, we focus on two sets of issues. First, what do contrasts as well as similarities between the form of native and exotic species-area relationships tell us about invasion and biogeographic processes. Second, how do changes in species area-relationships following species invasions inform our understanding of how species richness has changed and will change in the future at different spatial scales and in locations that vary in their degree of isolation.

Contrasts and similarities between native and naturalized species-area relationships

Some previous work has considered the implications of differences in the slopes of species-area relationships between native and exotic species. Crawley (1987) found significant species-area relationships for native and exotic plant species when examining data from across regions in Britain, Ireland, eastern North America, and Western Australia. However, he found the slope of the species-area relationship to be steeper for native species, which he suggested was a consequence of the higher beta diversity of native species. In another study, Pysek (1998) found in cities of central Europe that neophytes (exotic species that have invaded the region since 1500 AD) show a significantly steeper species-area slope than do archaeophytes (exotics that invaded prior to 1500) or native species. He suggested that this might be due to the potentially inhospitable environment for neophytes that exists outside urban areas; if this were true, cities would serve as habitat "islands" for neophytes, thus explaining their steeper species-area slope, as islands typically exhibit much steeper slopes than comparable continental regions (Rosenzweig 1995). Russell *et al.* (2004) found the slope of the species-area relationship for exotic mammals on offshore islands of New Zealand to be 0.21. They suggest that this slope, which is shallower than those often seen on islands, may suggest that the interaction between area and isolation is reduced for these mammals, since their transport and introduction has been facilitated by humans. Of course, interpreting differences between the slopes of different species-area relationships is not a trivial matter and is an activity with many potential pitfalls (e.g., Martin 1981, Lomolino 1989). Nevertheless, we believe that making such comparisons between native and exotic relationships may be extremely valuable. As this chapter, however, is a first attempt to search for generalities from comparisons of many native and exotic species-area relationships, we here examine a topic that we believe is more readily tractable, namely a comparison of native and exotic species-area relationships that do and do not show significant linear relationships on a log-log scale.

Although the species-area relationship is a very general pattern, not all data sets show a significant relationship between these variables. Further, the native and exotic components of these relationships may not behave in similar ways. Four outcomes are possible in comparisons of native and exotic species-area relationships: 1) natives show a significant linear relationship, but exotics do not, 2) both natives and exotics show significant relationships, 3) exotics show a significant relationship, but natives do not, or 4) neither natives or exotics show significant relationships. As the occurrence of the fourth outcome, in which neither natives or exotics show significant species-area relationships, should be exceedingly rare, we restrict our discussion here to the first three outcomes listed above.

Species-area relationships for natives, but not for exotics

We begin with the situation where native species do and exotic species do not show a significant positive linear relationship (on a log-log scale) between area and species richness. We illustrate this with plant species in California Counties, US States, and Mexican States (Fig. 4). In each case, native species show a strong positive species-area relationship, but exotic species do not. It is important to note that not only are the relationships between exotic species richness and area not significant, but that there is no hint of a trend towards a positive relationship. Given the broad spatial range examined (nearly three orders of magnitude of area in each case) and large numbers of exotic species (more than 1000 exotic species in each case), it is somewhat surprising that exotic species do not show the expected relationship. We suggest, here, that exotics may not show the expected relationship for any one of the following three reasons: 1) exotic species are inherently different than native species and should not be expected to respond in similar ways to area as native species, 2) there are a limited number of exotic species within each of these regions, such that even small subregions (i.e., counties or states) could hold all or most of the exotic species in question, thereby making it unlikely for there to be a significant relationship at the scales analyzed, 3) exotic species are more strongly influenced by the distribution of human populations than by area — this could be true for any number of reasons such as a correspondence between human population size and introduction effort of exotic species, between human population density and disturbance (which may favor exotic species), or between human population density and climate matching that occurs from source areas for exotic species (where many humans have colonized from and exotics are locally adapted) and recipient areas for exotic species (where many humans now reside and exotics are preadapted for local conditions).

Discriminating among these hypotheses is not trivial, as they are neither mutually exclusive nor necessarily exhaustive. Nevertheless, with the data available to us we should be able to discriminate among at least some of our stated

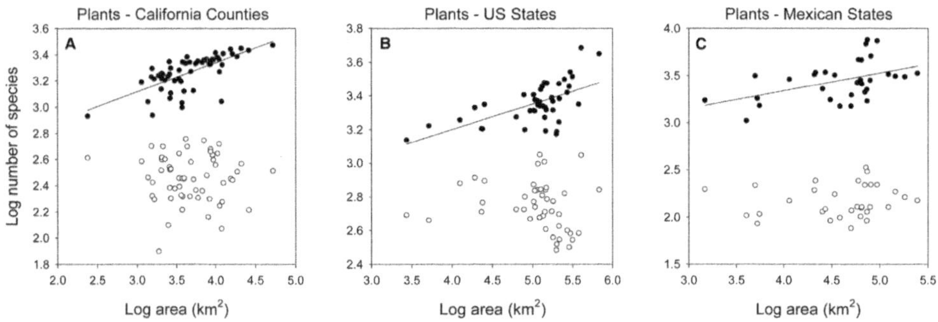

Fig. 4 Species-area relationships for native and exotic species of plants. Solid circles are native and empty circles are exotic species of vascular plants. A, Natives show a positive linear relationship, $R^2 = 0.48$, $z = 0.22$, $C = 2.44$, $p < 0.0001$, while exotics do not show a significant relationship; data from Calflora (www.calflora.org). B, Natives show a positive linear relationship, $R^2 = 0.35$, $z = 0.15$, $C = 2.61$, $p < 0.001$, while exotics do not show a significant relationship; see Appendix 1 for a description of data sources and methods. C, Natives show a positive linear relationship, $R^2 = 0.22$, $z = 0.19$, $C = 2.58$, $p < 0.01$, while exotics do not show a significant relationship; note that these data are for flowering plants only; data from Villaseñor and Espinosa-Garcia (2004).

hypotheses, with the goal of better understanding the curious absence of species-area relationships for these exotic species. The first hypothesis, that of inherent differences between native and exotic species, seems unlikely; this is because the characteristics of the native and exotic species in question are unlikely to be so different that they should respond in completely different ways to physical or biotic characteristics of the environment that promote richness with increasing area (such as increases in habitat diversity). Nevertheless, this is a possibility and one that is difficult to rule out. For Mexican States this does not appear to be the case, however, as native and exotic richness are strongly positively correlated across states ($R^2 = 0.54$, $p < 0.0001$). For California Counties and US States, however, no such positive relationship is apparent, leaving this hypothesis as a possibility for those areas. The second hypothesis, that of exotic species pool exhaustion within individual sample units, also seems unlikely, since a relatively small fraction of the total number of exotic species in these regions are present within individual sample areas. For example, between ca. 1/8 and 1/2 of the total number of exotic plant species in Mexico are found within individual states, suggesting that the number of exotic species present within individual states has not approached (or exhausted) the total pool of species available in Mexico; note that similar patterns exist for California Counties and US States. The third hypothesis, which suggests that human population density will influence exotic species richness, is strongly supported in California Counties, US and Mexican States, where in all cases human population density is strongly correlated with exotic species richness (Fig. 5). Differentiating the proximate cause of this pattern is difficult, however, as introduction effort, disturbance or some type of climate-matching explanation all seem credible. It is curious to note additionally, that native species either do not show a significant relationship with human population (in California Counties) or show relationships where less variance is explained relative to exotic species (in US and Mexican States), indicating that human population is a much better indicator of exotic than native richness.

Species-area relationships for natives and exotics

Native and exotic components of particular species-area relationships (i.e., for particular taxa across a particular set of areas) can both show significant positive linear relationships on plots of log area and log species richness. We illustrate this here with two examples, plants on the Channel Islands of California and plants on oceanic islands (Fig. 6B and C). Note also that plants in mainland sites of California show a non-significant, but still positive trend between area and exotic species richness and a significant positive relationship between native richness and area (Fig. 6A). Such patterns, i.e., where both native and exotic species show positive relationships between area and richness, can be argued to be consistent with the null hypothesis that species richness will always be influenced by area. As such, these results are not particularly surprising, but nevertheless of interest, because they suggest that native and exotic species are both responding in similar

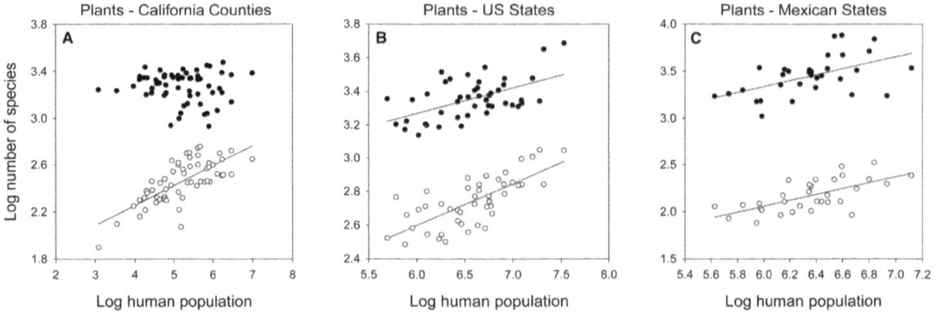

Fig. 5 Human population and richness of native and exotic species. Solid circles are native and empty circles are exotic species. A, Exotic species show a positive linear relationship, $R^2 = 0.53$, $p < 0.0001$, while natives do not show a significant relationship. B, Exotic and native species show positive linear relationships: naturalized species, $R^2 = 0.53$, $p < 0.0001$, natives, $R^2 = 0.30$, $p < 0.001$. C, Exotic and native species show positive linear relationships: naturalized species, $R^2 = 0.43$, $p < 0.0001$, natives, $R^2 = 0.28$, $p < 0.01$. Data sources as cited in Fig. 4 for plants; data for human population size are from California Counties (2003 estimates): http://www.dof.ca.gov/html/Demograp/repndat.htm, US States (2000 estimates): http://www.census.gov/main/www/cen2000.html, and Mexican states (2000 estimates): http://www.citypopulation.de/Mexico.html.

ways to differences in area. On the Channel Islands, however, the patterns are not identical; the slopes of the relationships are similar (natives: z = 0.26, exotics: z = 0.21), but the slope of the exotic relationship is shallower, which is consistent with Crawley's (1987) conclusion that beta diversity is reduced for exotic species relative to natives. This shallower slope is also consistent with the expectation that exotic species are not functionally as dispersal limited as natives, because steeper slopes are generally considered to be indicative of increased insularity and isolation (e.g., Rosenzweig 1995, Russell *et al.* 2004). This conclusion is strongly supported in this case by examining the relationship that native and exotic species have respectfully with isolation; by plotting the residuals of log area and log richness against distance of each island to the mainland it is clear that the richness of native species is strongly influenced by isolation, while the richness of exotic species is not (Fig. 7). In this case it appears, therefore, that both native and exotic species are responding in similar ways to area, but are differentially affected by isolation.

In contrast to this pattern on the Channel Islands, the pattern on oceanic islands is somewhat different. On oceanic islands, both the slopes and intercepts of the species-area relationships for native and exotic species are almost identical (natives: z = 0.31, C = 1.46, exotics: z = 0.31, C = 1.45). This result is a consequence of the extremely tight coupling between native and exotic species richness

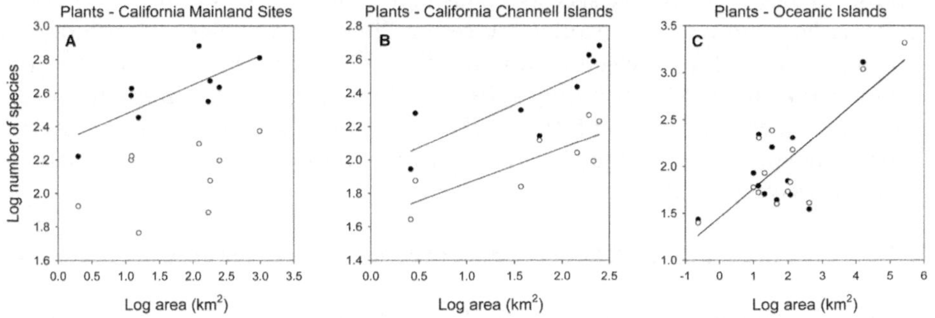

Fig. 6 Species-area relationships for native and exotic species of plants. Solid circles are native and empty circles are exotic species of vascular plants. A, Natives show a positive linear relationship, $R^2 = 0.59$, $z = 0.17$, $C = 2.30$, $p < 0.05$, while exotics do not show a significant relationship (although the trend is nevertheless positive); see Appendix 1 for a description of data sources and methods. B, Natives and exotics show positive linear relationships: natives, $R^2 = 0.66$, $z = 0.26$, $C = 1.95$, $p < 0.05$, exotics, $R^2 = 0.66$, $z = 0.21$, $C = 1.65$, $p < 0.05$; data from Junak *et al.* (1995). C, Natives and exotics show positive linear relationships: natives, $R^2 = 0.64$, $z = 0.31$, $C = 1.46$, $p < 0.01$, exotics, $R^2 = 0.66$, $z = 0.31$, $C = 1.45$, $p < 0.001$; data from Sax *et al.* (2002).

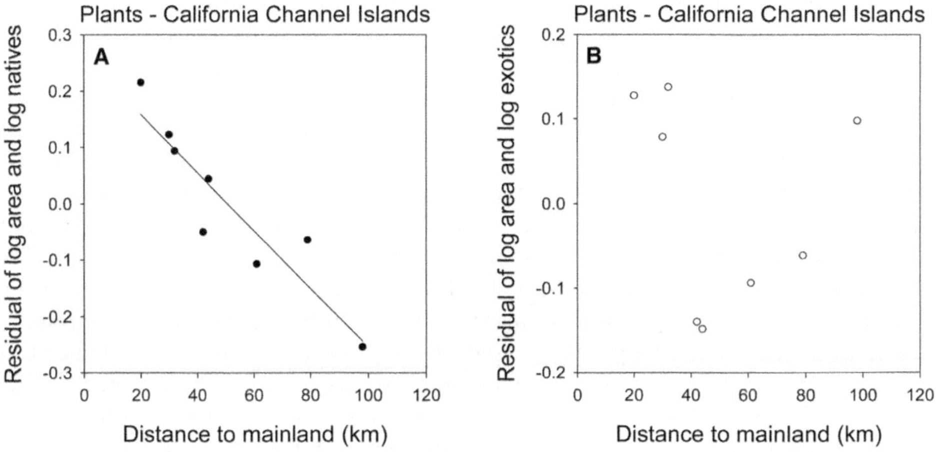

Fig. 7 Isolation and the richness of plant species on the Channel Islands of California. A, Significant linear relationship between the distance of islands to the mainland and the residuals of log area and log number of native species, $R^2 = 0.86$, $p < 0.001$. B, The same relationship for exotic species is not significant. Data from Junak *et al.* (1995). See Moody (2000) for a similar examination of these data.

on these islands (log native species vs. log exotic species: $R^2 = 0.96$, $p < 0.0001$). Clearly, both natives and exotics are responding to area, given the highly significant relationship between area and richness, but in this case it appears that other factors are also contributing to the tight coupling between native and exotic species richness. At this point, these other factors are unclear, but this pattern is intriguing and at present may be one of the more important unexplained patterns in species invasions and biogeography.

Species-area relationships for exotics, but not for natives

Finally, we consider scenarios in which exotic species show a species-area relationship and native species do not. We illustrate this here with freshwater fishes in watersheds of temperate North America and watersheds of California (Fig. 8). In both these examples exotic species (defined here as being non-native to the watersheds in question) show significant positive linear relationships in log-log space, while native species do not. We believe that there are two principal hypotheses that can explain these patterns. First, freshwater fishes are extremely dispersal limited and may have historically been unable to take advantage of large areas because of extreme barriers to dispersal. If the likelihood of fishes colonizing a watershed is driven more by the nature and severity of the dispersal barrier than by the size of the watershed then native richness may be poorly correlated with watershed area. Dispersal barriers, however, should not be as important for exotic freshwater fishes, because in most cases the spread of these species is facilitated by direct human introductions. A second potential explanation for the difference in native and exotic species-area relationships of freshwater fishes in North America and California may be due to recent anthropogenic changes in the environment. For example, many of the largest watersheds in the database of US watersheds are found in the western US. In the western US many dams have been built and lacustrian environments created in regions that historically had very few lakes (e.g., Marchetti *et al.* 2001). The creation of these lakes may have increased the capacity of these regions to support a larger variety of freshwater fish species. This pattern is therefore consistent with the hypothesis that exotic species are responding to current environmental conditions, while native species (which have not had sufficient time to speciate and fill these newly available habitats) are responding to conditions that were historically present. If this is true, then it would suggest that exotic species (when they are widely introduced across potential suitable habitats) may be a better indicator of an areas capacity to support species than the natives themselves, which may instead be in a dispersal-limited state of disequilibrium between environmental capacity and species richness.

Changes in species-area relationships

One of the most pertinent aspects of studying species-area relationships in today's changing world is examining how species-area relationships have changed with

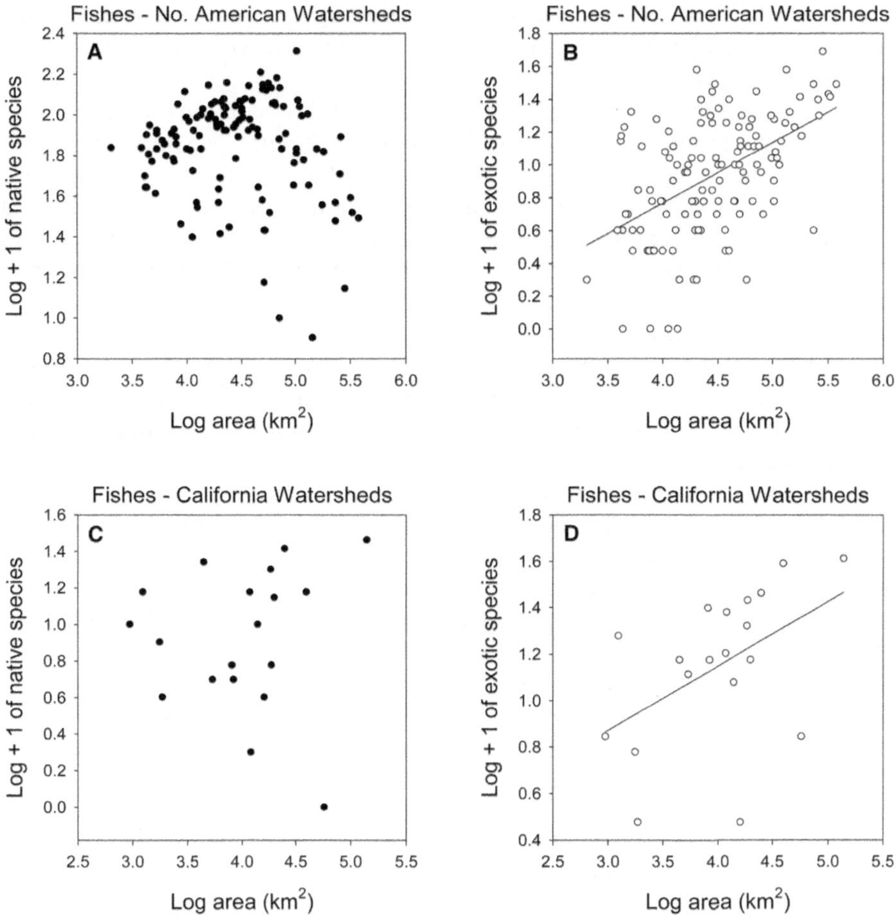

Fig. 8 Species-area relationships for native and exotic species of freshwater fish. A, Natives do not show a significant relationship; data from Gido and Brown (1999). B, Exotic species show a positive linear relationship, $R^2 = 0.23$, $z = 0.28$, $C = 0.04$, $p < 0.05$; data from Gido and Brown (1999). C, Natives do not show a significant relationship; data from Moyle (2002); area estimates for the 19 watersheds used in this analysis are from http://endeavor.des.ucdavis.edu/newcara. D, Exotic species show a positive linear relationship, $R^2 = 0.26$, $z = 0.37$, $C = -0.71$, $p < 0.0001$; data sources as cited above for panel C.

the addition of exotic species and the extinction of natives. Rosenzweig (2001) and Collins *et al.* (2002) consider this issue with thought experiments that vary the number of exotic species exchanged between regions to predict what the effects of exotic species introductions will be for local and global species richness.

Here we explore the closely related issue of how species-area relationships have actually changed overtime (using empirical data) and consider what insights these changes may provide.

In Fig. 9 we have plotted the trend-lines of historic and current species-area relationships of plant species for several regions; by 'historic' we mean the number of native species believed to have been present before any anthropogenic extinctions occurred and by 'current' we mean the number of species present today, which is the sum of extant native and naturalized species. Data on the number of native extirpations is available for all of the plotted trend lines except for floras from mainland sites in California. However, given the extremely low number of plant species that have been extirpated from comparable Channel Island sites (mean of fewer than four species per island) and from the other plant data sets considered here, it seems likely that the number of species lost from the California mainland sites should also be relatively low and should not significantly impact the observed patterns.

All of the patterns plotted show one overriding similarity, i.e., an increase in the number of species present across sites (Fig. 9). The magnitude of this increase, however, is not identical between regions, but is instead generally consistent with the degree of isolation and relative mean richness of each of these regions (average increase in richness on oceanic islands: 100%, Channel Islands: 44%, California mainland sites: 33%, US States: 23%, and California Counties: 17% — note that California Counties and US States have similar mean numbers of native species, 1925 and 2358 respectfully, but that the spatial extent and effective isolation between US States is much higher than that between California Counties). These increases in species richness, particularly the positive correlation between isolation and the magnitude of increased species richness, is consistent with the changes predicted to occur by Collins *et al.* (2002) and also with Darwin's observation that humans are filling up isolated regions far more fully than nature has done.

Another pattern evident in Fig. 9 is that with the establishment of naturalized species relatively large areas are 'catching up' in total plant richness with those regions that have historically been less isolated. Thus, the largest Channel Islands now have as many plant species as were historically found on equal sized areas of the California mainland (Fig. 9). Similarly, large oceanic islands now have as many plant species as were historically found within US States of comparable size (Fig. 9). While comparing the Channel Islands with California mainland sites seems reasonable, it is not clear, however, to which landmasses the largest oceanic islands should be compared. To explore this further, we consider the two largest oceanic islands in our database, Hawaii and New Zealand. Hawaii's closest continental land mass is North America, and in fact Hawaii's greatest number of natural colonists are believed to have come from the American continents (V. Funk, personal communication). Because Hawaii is present at high tropical latitudes, the most reasonable comparison in North America would seem to be with equal sized areas of Mexico. Comparing both the historic and current number

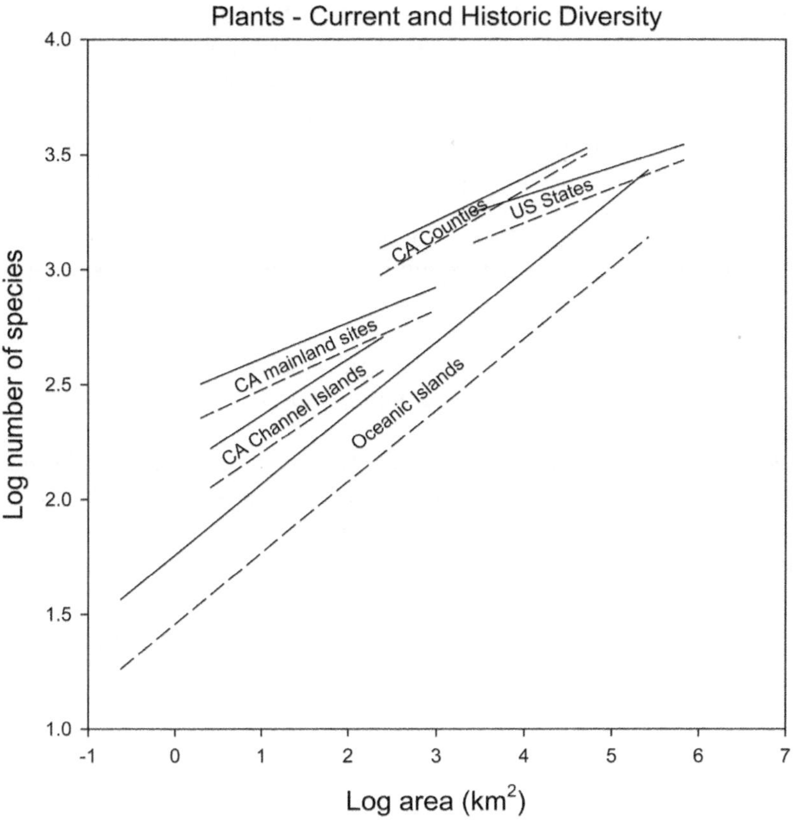

Fig. 9 Change in species-area relationships for plants. The dashed lines indicate the historic relationship between area and native species of vascular plants. The solid lines indicate the current relationship between area and all extant species of vascular plants (native and exotic). In all cases the intercept of these species-area relationships have shifted up and in some cases the slope of these relationships have become more shallow.

of plant species in Hawaii with the number of native species present in Mexican States, it is clear that the historic flora of Hawaii was depauperate compared with Mexican States. However, this discrepancy has now largely disappeared. Current plant richness is relatively close to what would be expected if Hawaii were a piece of mainland of North America (Fig. 10A). New Zealand's closest continent is Australia, and New Zealand's greatest number of natural colonists have come from Australia (McGlone *et al.* 2001). Comparing plant species in New Zealand with those in Australian States shows that historic species richness of plants in New Zealand was only minimally lower than in Australian States of comparable size and that current plant richness is now somewhat greater than would be expected (considering the role of area alone) were New Zealand present on the mainland of Australia (Fig. 10B). The long-term implications of these changes in richness are difficult to ascertain, as are the long-term trajectories of continuing changes in species richness on these islands. It is possible, that net richness will eventually decrease, if there is a long time lag before significant numbers of native extinctions occur. Alternatively, richness may continue to increase, if extinction rates remain low, and additional species become established (Rosenzweig 2001, Collins *et al.* 2002, Sax *et al.* 2002). Undoubtedly, additional study will help to clarify this issue. However, given the evidence available from the species-area relationships considered here, it appears that further effective decreases in isolation (mediated by the introduction of additional exotic species) should lead to continued increases in the intercept of species-area relationships for any region and to further decreases in the slope of those relationships, i.e., to continued increases in richness for individual areas.

It is important to stress that the increases in richness described here should not be construed as 'good', but instead as evidence of the types of differential impacts that humans are having on various regions around the world. Indeed, these data

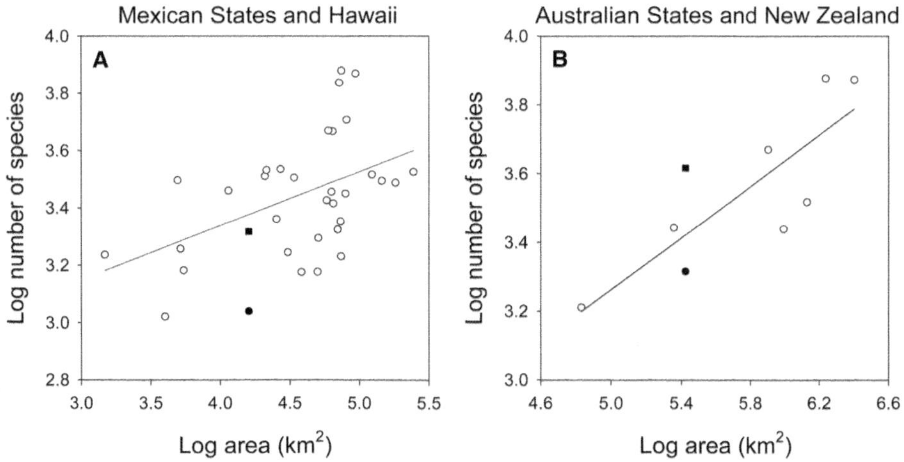

Fig. 10 Change in species richness of plants on Hawaii and New Zealand relative to species-area relationships of nearest continents. Empty circles represent historic number of native species on continents, solid circles represent historic number of native species on islands, and solid squares represent the current extant (native and naturalized) number of species on islands. A, The historic richness of flowering plant species on Hawaii was well below the line of the species-area relationship for Mexican states, but is currently very close to this line. B, The historic richness of vascular plant species in New Zealand was close to, but below, the line of the species-area relationship for Australian States, but is currently above this line.

suggest that the most isolated and distinctive biotas of the world are the ones being modified by the largest degree. Such modifications lead to at least two types of biotic homogenization. First, they lead to homogenization of species *composition* among regions, as unique species are lost and common ones are gained (e.g., Lockwood and McKinney 1999). Second, they lead to homogenization of species *richness* among regions, in which variation in species richness and species-area relationships is reduced. The long-term implications of biotic homogenization are still unknown, but it is clear that the loss of distinctive biotas and the modification of patterns of biodiversity across the globe are topics worthy of much additional study.

CONCLUDING THOUGHTS

We have attempted to review and present some of the basic patterns of biogeographic distribution of exotic species. Given the limited scope of a single book chapter we were not able to review all of the many patterns that have been described to date, but instead have focused on those patterns that we deemed to be of the greatest interest to ecologists. Had we instead chosen to explore those issues of greatest interest to evolutionary biologists, we would undoubtedly have chosen a slightly different set of topics to review. For example, one pattern that is extremely intriguing with naturalized species is the rapid evolution of clines in body size across latitude that quantitatively match, but qualitatively differ from those seen in the native range — suggesting that evolution can simultaneously operate in a deterministic and contingent manner (Huey *et al.* 2000, Gilchrist *et al.* 2004). Still, we hope that the work described here provides a good inroad to the growing literature on this topic, a literature that promises to provide reciprocal insights to ecological and biogeographical theory, as well as to our applied understanding of invasion biology. Finally, we believe that if we as a society are going to successfully learn to understand and mitigate the environmental threats posed by anthropogenic changes to the environment, then we must employ all available evidence to do so; undoubtedly studies of species invasion at biogeographic scales can do much to help achieve these goals.

ACKNOWLEDGEMENTS

We would like to thank the editors of this book for giving us the opportunity to write this chapter. Many of the ideas in this ms benefited from conversations with J. Brown. The quality of this ms has benefited significantly from comments by D. Perault, M. Beals and M. Lomolino. This is contribution number 165 from PISCO, the Partnership for Interdisciplinary Studies of Coastal Oceans: A Long-term Ecological Consortium funded by the David and Lucile Packard Foundation.

REFERENCES

Arrhenius, O. 1921. Species and area. Journal of Ecology **4**, 68-73.

Atkinson, C. T., R. J. Dusek, K. L. Woods and W. M. Iko. 2000. Pathogenicity of avian malaria in experimentally-infected Hawaii Amakihi. Journal of Wildlife Diseases **36**, 197-204.

Banks, D. L. 1999. A vascular flora of the Agua Tibia Mountains, southern California. Rancho Santa Ana Botanical Garden Occasional Publication number 4, Claremont, CA.

Bickford, C. and P. Rich. 1984. Vegetation and flora of the Landels-Hill Big Creek Reserve, Monterey County, California, Second Edition. Publication number 15, Environmental Field Program, University of California, Santa Cruz, CA.

Blackburn, T. M. and K. J. Gaston. 1996. Spatial patterns in the species richness of birds in the New World. Ecography **19**, 369-376.

Boyd, S. D. 1983. A flora of the Gavilan Hills, Western Riverside County, California. Masters of Science Thesis, University of California, Riverside, CA.

Brown, J. H. 1995. Macroecology. University of Chicago Press, Chicago, IL.

Brown, J. H. and M. V. Lomolino. 1998. Biogeography, Second Edition. Sinauer Associates, Sunderland, MA.

Brown, J. H. and D. F. Sax. 2004. Gradients in species diversity: why are there so many species in the tropics? Pages 1145-1154 in M. V. Lomolino, D. F. Sax and J. H. Brown, editors. Foundations of Biogeography: classic papers with commentaries. University of Chicago Press, Chicago, IL.

Bruno, J. F., J. J. Stachowicz and M. D. Bertness. 2003. Inclusion of facilitation into ecological theory. Trends in Ecology and Evolution **18**, 119-125.

Case, T. J. 1991. Invasion resistance, species build-up and community collapse in model competition communities. Pages 239-266 in M. E. Gilpin and I. Hanski, editors. Metapopulation dynamics. Academic Press, London, UK.

Chown, S. L., N. J. M. Gremmen and K. J. Gaston. 1998. Ecological biogeography of southern ocean islands: species-area relationships, human impacts, and conservation. American Naturalist **152**, 562-575.

Collins, M. D., D. P. Vazquez and N. J. Sanders. 2002. Species-area curves, homogenization and the loss of global diversity. Evolutionary Ecology Research **4**, 457-464.

Connor, E. F. and E. D. McCoy. 1979. The statistics and biology of the species-area relationship. American Naturalist **113**, 791-833.

Crawley, M. J. 1987. What makes a community invasible? Pages 429-453 in A. J. Gray, M. J. Crawley and P. J. Edwards, editors. Colonization, succession and stability Blackwell Scientific Publications, Oxford, UK.

Darwin, C. 1859. On the origin of species. Murray, London, UK.

Davis, M. A. 2003. Biotic globalization: does competition from introduced species threaten biodiversity? Bioscience **53**, 481-489.

De Nevers, G. 1985. Pepperwood Flora: Pepperwood Ranch Natural Preserve, Sonoma County, California. California Academy of Sciences, San Francisco, CA.

Deutschewitz, K., A. Lausch, I. Kuhn and S. Klotz. Native and alien plant species richness in relation to spatial heterogeneity on a regional scale in Germany. Global Ecology and Biogeography **12**, 299-311.

Dobzhansky, T. 1950. Evolution in the tropics. American Scientist **38**, 209-221.

Duncan, R. P., T. M. Blackburn and D. Sol. 2003. The ecology of bird introductions. Annual Review of Ecology, Evolution and Systematics **34**, 71-98.

Ebenhard, T. 1988. Introduced birds and mammals and their ecological effects. Swedish Wildlife Research **13**, 1-107.

Elton, C. S. 1958. The ecology of invasions by animals and plants. Methuen and Co. LTD, London, UK.

Fine, P. V. A. 2002. The invasibility of tropical forests by exotic plants. Journal of Tropical Ecology **18**, 687-705.

Fischer, A. G. 1960. Latitudinal variations in organic diversity. Evolution **14**, 64-81.

France, R. 1992. The North American Latitudinal gradient in species richness and geographic range of freshwater crayfish and amphipods. American Naturalist **139**, 342-354.

Fridley, J. D., R. L. Brown and J. F. Bruno. 2004. Null models of exotic invasion and scale-dependent patterns of native and exotic species richness. Ecology, in press.

Fritts, T. H. and G. H. Rodda. 1998. The role of introduced species in the degradation of island ecosystems: a case history of Guam. Annual Review of Ecology, Evolution and Systematics **29**, 113-140.

Gaines, S. D. and J. Lubchenco. 1982. A unified approach to marine plant-herbivore interactions. 2. Biogeography. Annual Review of Ecology and Systematics **13**, 111-138.

Gaston, K. J., T. M. Blackburn and J. I. Spicer. 1998. Rapoport's rule: time for an epitaph? Trends in Ecology and Evolution **13**, 70-74.

Gido, K. B. and J. H. Brown. 1999. Invasion of North American drainages by alien fish species. Freshwater Biology **42**, 387-399.

Gilchrist, G. W., R. B. Huey, J. Balanya, M. Pascual and L. Serra. 2004. A time series of evolution in action: a latitudinal cline in wing size in South American Drosophila subobscura. Evolution **58**, 768-780.

Groombridge, B., ed. 1992. Global biodiversity: status of the Earth's living resources. Chapman & Hall, London, UK.

Hawkins, B. A. 2001. Ecology's oldest pattern? Trends in Ecology and Evolution **16**, 470.

Huey, R. B., G. W. Gilchrist, M. L. Carlson, D. Berrigan and L. Serra. 2000. Rapid evolution of a geographic cline in size in an introduced fly. Science **287**, 308-309.

Hutton, I. 1990. Birds of Lord Howe Island: past and present. Lithocraft Graphics, Melbourne, Australia.

Junak, S., T. Ayers and R. Scott. 1995. A flora of Santa Cruz Island. Santa Barbara Botanic Gardens, Santa Barbara, CA.

Kennedy T. A., S. Naeem, K. M. Howe, J. M. H. Knops, D. Tilman and P. Reich. 2002. Biodiversity as a barrier to ecological invasion. Nature **417**, 636-638.

Lathrop, E. W. and R. F. Thorne. A flora of the Santa Rosa Plateau, southern California. Southern California Botanists Special Publication number 1, Claremont, CA.

Letcher, A. J. and P. H. Harvey. 1994. Variation in geographic range size among mammals of the Palearctic. American Naturalist **144**, 30-42.

Levine, J. M. 2000. Species diversity and biological invasions: relating local process to community pattern. Science **288**, 852-854.

Lonsdale, W. M. 1999. Global patterns of plant invasions and the concept of invasibility. Ecology **80**, 1522-1536.

Lomolino, M. V. 1989. Interpretations and comparisons of constants in the species-area relationship: an additional caution. American Naturalist **133**, 277-280.

Lomolino, M. V. 2001. The species-area relationship: new challenges for an old pattern. Progress in Physical Geography **25**, 1-21

Lomolino, M. V., D. F. Sax and J. H. Brown. 2004. Foundations of Biogeography: classic papers with commentaries. University of Chicago Press, Chicago, IL.

Lomolino, M. V. and M. D. Weiser. 2001. Towards a more general species-area relationship: diversity on all islands, great and small. Journal of Biogeography **28**, 431-445.

Luning, K. 1990. Seaweeds: their environment, biogeography, and ecophysiology. John Wiley and Sons, Inc., New York, NY.

MacArthur, R. H. and E. O. Wilson. 1967. The Theory of Island Biogeography. Princeton University Press, Princeton, NJ.

MacArthur, R. H. 1972. Geographic ecology: patterns in the distribution of species. Harper & Row, New York, NY.

Marchetti, M. P., T. Light, J. Feliciano, T. Armstrong, Z. Hogan, J. Viers and P. B. Moyle. 2001. Homogenization of California's fish fauna through abiotic change. Pages 259-278 in J. L. Lockwood and M. L. McKinney, editors. Biotic Homogenization, Kluwer Academic/Plenum Publishers, New York, NY.

Martin, T. E. 1981. Species-area slopes and coefficients: a caution on their interpretation. American Naturalist **118**, 823-837.

Matter, S. F., I. Hanski and M. Gyllenberg. 2002. A test of the metapopulation model of the species-area relationship. Journal of Biogeography **29**, 977-983.

McClintock, E., W. Knight and N. Fahy. 1968. A flora of the San Bruno Mountains, San Mateo County, California. Proceedings of the California Academy of Sciences, fourth series **32**, 587-677.

McGlone, M. S., R. P. Duncan and P. B. Heenan. 2001. Endemism, species selection and the origin and distribution of the vascular plant flora of New Zealand. Journal of Biogeography **28**, 199-216.

McKinney, M. L. 2002a. Influence of settlement time, human population, park shape and age, visitation and roads on the number of alien plant species in protected areas in the USA. Diversity and Distributions **8**, 311-318.

McKinney, M. L. 2002b. Do human activities raise species richness? Contrasting patterns in United States plants and fishes. Global Ecology and Biogeography **11**, 343-348.

McKinney, M. L. and Lockwood, J. L. 1999. Biotic homogenization: a few winners replacing many losers in the next mass extinction. Trends in Ecology and Evolution **14**, 450-453.

Moody, A. 2000. Analysis of plant species diversity with respect to island characteristics on the Channel Islands, California. Journal of Biogeography **27**, 711-723.

Moyle, P. B. 2002. Inland fishes of California, revised and expanded edition. University of California Press, Berkeley, CA.

Naeem, S., J. M. H. Knops, D. Tilman, K. M. Howe, T. Kennedy and S. Gale. 2000. Plant diversity increases resistance to invasion in the absence of covarying extrinsic factors. Oikos **91**, 97-108.

Pianka, E. R. 1966. Latitudinal gradients in species diversity: a review of concepts. American Naturalist **100**, 33-46.

Pysek, P. 1998. Alien and native species in central European urban floras: a quantitative comparison. Journal of Biogeography **25**, 155-163.

Pysek, P., V. Jarosik and T. Kucera. 2002. Patterns of invasion in temperate nature reserves. Biological Conservation **104**, 13-24.

Pysek, P., T. Kucera and V. Jarosik. 2002. Plant species richness of nature reserves: the interplay of area, climate and habitat in a central European landscape. Journal of Biogeography **11**, 279-289.

Rapoport, E. H. 1982 Areography: geographic strategies of species, translation by B. Drausel. Pergamon, Oxford, UK.

Raven, P. H., H. J. Thompson and B. A. Prigge. 1986. Flora of the Santa Monica Mountains, California, Second Edition. Southern California Botanists Special Publication No. 2.

Rejmanek, M. 1996. Species richness and resistance to invasions. Pages 153-172 in G. H. Orians, R. Dirzo & J. H. Cushman, editors. Biodiversity and ecosystem processes in tropical forests. Springer Verlag, Berlin, Germany.

Rejmanek, M. 2003. The rich get richer — responses. Frontiers in Ecology and the Environment **1**, 122-123.

Richardson, D. M. N. Allsopp, C. M. D'Antonio, S. J. Milton and M. Rejmanek. 2000. Plant invasions — the role of mutualisms. Biological Reviews **75**, 65-93.

Rodriguero, M. S. and D. E. Gorla. 2004. Latitudinal gradient in species richness of the New World Triatominae Reduviidae. Global Ecology and Biogeography **13**, 75-84.

Rohde, K. 1992. Latitudinal gradients in species diversity: the search for the primary cause. Oikos **65**, 514-527.

Rosenzweig, M. L. 1995. Species diversity in space and time. Cambridge University Press, Cambridge, UK.

Rosenzweig, M. L. 2001. The four questions: What does the introduction of exotic species do to diversity? Evolutionary Ecology Research **3**, 361-367.

Russell, J. C., M. N. Clout and B. H. McArdle. 2004. Island biogeography and the species richness of introduced mammals of the New Zealand offshore islands. Journal of Biogeography **31**, 653-664.

Sax, D. F. 2001. Latitudinal gradients and geographic ranges of exotic species: implications for biogeography. Journal of Biogeography **28**, 139-150.

Sax, D. F., S. D. Gaines and J. H. Brown. 2002. Species invasions exceed extinctions on islands worldwide: a comparative study of plants and birds. American Naturalist **160**, 766-783.

Scott, J. M., S. Conant and C. van Riper III, editors. 2001. Evolution, ecology, conservation, and management of Hawaiian birds: a vanishing avifauna. Cooper Ornithological Society, Camarillo, CA.

Simpson, G. G. 1964. Species density of North American Recent mammals. Systematic Zoology **13**, 57–73.

Stachowicz, J. J., R. B. Whitlatch and R. W. Osman. 1999. Species diversity and invasion resistance in a marine ecosystem. Science **286**, 1577-1579.

Stadler, J., A. Trefflich, S. Klotz and R. Brandl. 2000. Exotic plant species invade diversity hot spots, the naturalized flora of northwestern Kenya. Ecography **23**, 169-176.

Stevens, G. C. 1989. The latitudinal gradient in geographic range: how so many species coexist in the tropics. American Naturalist **133**, 240-256.

Stevens, G. C. and B. J. Enquist. 1998. Macroecological limits to the abundance and distribution of Pinus. Pages 183-190 in D. M. Richardson, editor. Ecology and biogeography of Pinus. Cambridge University Press, Cambridge, UK.

Stohlgren, T. J., D. Binkley, G. W. Chong, M. A. Kalkhan, L. D. Schell, K. A. Bull, Y. Otsuki, G. Newman, M. Bashkin and Y. Son. 1999. Exotic plant species invade hot spots of native plant diversity. Ecological Monographs **69**, 25-46.

Stohlgren, T. J., D. T. Barnett and J. T. Kartesz. 2003. The rich get richer: patterns of plant invasions in the United States. Frontiers in Ecology and the Environment **1**, 11-14.

Tilman, D. 1997. Community invasibility, recruitment limitation, and grassland biodiversity. Ecology **78**, 81-92.

Villaseñor, J. L. and F. J. Espinosa-Garcia. 2004. The alien flowering plants of Mexico. Diversity and Distributions **10**, 113-123.

Wardle, D. A. 2001. Experimental demonstration that plant diversity reduces invasibility — evidence of a biological mechanism or a consequence of sampling effect? Oikos **95**, 161-170.

Willig, M. R., D. M. Kaufman and R. D. Stevens. 2003. Latitudinal gradients of biodiversity: pattern, process, scale, and synthesis. Annual Review in Ecology, Evolution and Systematics **34**, 273-309.

Willoughby, J. W. 1981. A flora of the Vaca Mountains, California. Masters Thesis, California State University, Sacramento, CA.

Wilson, K. 1997. Extinct and introduced vertebrate species in New Zealand: a loss of biodistinctiveness and gain in biodiversity. Pacific Conservation Biology **3**, 301-305.

APPENDIX 1 — DETAILED METHODS

Plants in US States:

The number of native and naturalized species of vascular plants for each state are from Mac *et al.* (1998) and Stein *et al.* (2000). Data on number of extinct or extirpated plants for each state were collected from reports of private and public state agencies, predominately the Natural Heritage Programs of individual states; the data used in these analyses are available upon request from the authors. In each case, the number of species listed as "SX" (extinct or extirpated from a state) or "SH" (present historically in a state, but no longer known to occur) were summed to provide the total number of species that have been extirpated. Complete data for Massachusets, Minnesota and North Dakota were unavailable, so these states were excluded from these analyses. Alaska and Hawaii were also excluded from these analyses.

Plants in mainland areas of California:

Available floras of particular sites on the mainland of California were selected to represent the span of areas present on the California Channel Islands. In total nine sites were used. These included the floras of the Santa Monica Mountains (Raven *et al.* 1986), San Bruno Mountains (McClintock *et al.* 1968), Vaca Mountains (Willoughby 1981), Agua Tibia Mountains (Banks 1999), Santa Rosa Plateau (Lathrop and Thorne 1985), Gavian Hills (Boyd 1983), Pescadero Marsh (W. Anderson and R. Morgan, unpublished manuscript), Landels-Hill Big Creek Reserve (Bickford and Rich 1984), and Peperwood Reserve (De Nevers 1985). Data on the number of native and non-native species are from these sources. No data was available on number of extirpated species.

Synthesis

Chapter twenty-one

Linking scale dependent processes in invasions

M. W. Cadotte, T. Fukami and S. M. McMahon

INTRODUCTION

The overarching goal of this volume was to examine how conceptual ecology informs our understanding of species invasions, while simultaneously viewing species invasions as potential tests of ecological theory (see Chapter 1). As one half the purpose of the present volume attests to, ecological theory is useful for contextualizing when and how species invade (e.g., Shea and Chesson 2002). More than this, and as the second half of this volume's purpose, species invasions allow us to see hypothesized dynamics, and to test major ideas in ecology and evolution (e.g., Lodge 1993, Sax *et al.* 2005). Throughout this volume authors have explicitly explored the advances in knowledge gained by combining conceptual ecology with invasion ecology.

However, biological invasions have been studied for their own sake, as they are considered to be a major conservation risk (Vitousek *et al.* 1996), and an important component of invasions studies is the study of how non-indigenous species enter and proliferate in new communities. A multitude of hypotheses have been created to explain how species invade new habitats (Inderjit *et al.* 2005). Yet, these hypotheses can be considered as special cases of more general ecological phenomena, or as *ad hoc* explanations, filling the void where we lack a general theory (e.g., Lakatos 1970). Though the primary goal of this volume has been to combine conceptual ecology and invasion biology, many of the authors examined

M.W. Cadotte et al. (eds.), Conceptual ecology and invasion biology, 483–494.

ecological theory or particular invasions from different levels of organization. We organized the chapters into three general scales of organization: population, community and biogeographic scales.

Instead of viewing these scales as necessarily discrete or hierarchically related, it may be useful to view these processes as a continuum feeding back into one another (Fig. 1). In order to adequately describe and understand invasions we must look to processes allowing species to overcome: i) biogeographical barriers in the movement of propagules; ii) abiotic barriers in establishing new populations; and iii) biotic barriers in maintaining higher birth than death rates, and being able to increase numbers and spread to new locales. These are essentially filters subsampling potential colonists from larger species pools (Belyea and Lancaster 1999). The dispersal filter limits species that are unlikely to migrate to the local community, while those with certain characteristics (e.g., long-distance dispersal or common agricultural seed contaminants) will have a higher probability of colonizing, and abiotic filters select certain species or traits as those with a higher change of surviving in the new habitat (Belyea and Lancaster 1999).

Observations made at different scales often require mechanistic links to other scales of organization. For example, extinctions are ultimately observed at a biogeographic scale, but the mechanisms leading to extinction must be observed at the population level (e.g., what elevates death rates or lower birth rates).

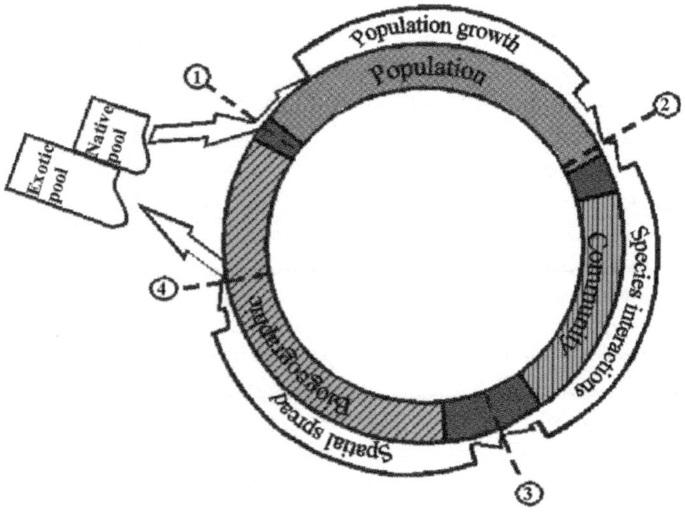

Fig. 1 The invasion process as a continuum across spatial scales. The ring represents the scale of organization (population, community and biogeographic). Directional boxes indicate dominant processes at each scale, which lead into one another. The numbered dashed lines refer to the four invasion steps outlined in this chapter: 1) immigration and establishment, 2) species interactions, 3) spatial spread, and 4) large scale patterns.

Explanatory mechanisms, such as the causes of extinction, can draw explicit links between scales, and this is where we see opportunity to better understand the processes driving invasions and how ecological theory informs us about these invasions.

The scales used in this volume do not necessarily translate into ecologically relevant spatial scales (e.g., 1 m^2 vs. 1,000 m^2), but reflect levels of ecological processes affecting species' abundance and distribution. Scale-explicit approaches can tell us much about the nature of ecological processes (e.g., Cadotte and Fukami 2005, Rahbek 2005). Further, a number of recent studies have shown that our understanding of the processes driving species invasions are affected by the spatial scale at which species invasions are examined (Collingham *et al.* 2000; Lloret *et al.* 2004, Hamilton *et al.* 2005). Not only do ecological processes appear scale-contingent, but so do temporal dynamics of invasions. For example, both Collingham *et al.* (2000) and Hamilton *et al.* (2005) show that species invasions appear to reach stasis or equilibrium (i.e., stop spreading) earlier at larger or coarser scales compared to fine spatial scales.

Is there a more general context within which we can conceive of invasions, especially one that links different scales? Many of the chapters in this volume discuss invasions in terms of small populations avoiding extinction. There are a number of ways by which populations avoid extinction, including prolonging life spans, protracted dormancy, and especially increasing reproductive rates and population growth rates. For example, Grotkopp *et al.* (2002) show that traits associated with rapid population increases (high relative growth rate, small seed mass and short generation time) were the best predictors of pine invasions, one of the most aggressive groups of plant invaders. In this chapter we look at a species' ability to increase population size (intrinsic rate of increase, r) as the portal to understanding patterns of invasions and linking spatial scales. Through the rest of this chapter we explore the processes affecting r at different scales of organization, and by understanding the nature of r and how it drives dynamics at different scales, we believe that light can be shed on the ultimate causes of species invasions. While understanding the ecological processes affecting r appear important for understanding the invasion process, we are not claiming that it alone can explain all invasions, nor that it even necessarily be the most important factor. Rather, we wish to show how a conceptual treatment of a species trait can help in systematically understanding different processes at different scales and stages of invasions.

Step 1: immigration and establishment

Species classified as invaders have one thing in common–they have overcome dispersal, and abiotic barriers and have arrived in new habitats. Chapters in this volume assume that species are able to disperse into new habitats through their own abilities or human vectors.

The first stage in any invasion is the establishment of individuals, commencing the new population. There are a multitude of factors that affect the likelihood of establishment success. These include abiotic conditions, composition of resident species, life-history traits promoting invasion (Baker 1974) and, the number of immigrants initiating the population (i.e., propogule pressure — see Carlton 1996; Lockwood *et al.* 2005).

Warren *et al.* (Chapter 16) explicitly consider the population processes involved with establishment. They view establishment as a probabilistic process where the probability of establishment ($P(n)$) is determined by the initial population size (n) and birth and death rates, b and d:

$$P(n) = 1 - \left(\frac{d}{b}\right)^n = 1 - \left(1 - \frac{r}{b}\right)^n$$

Here the intrinsic rate of increase, r, is simply $b - d$. Essentially populations overcome the risks of small population sizes either through increased propagule pressure (n) or by rapidly increasing population sizes by increasing r relative to d via higher birth rates.

The interaction between r and n is critical for establishment (see Chapter 16, their Fig. 1). Species with extremely low mortality and therefore high r have a high probability of success, largely independent of the number of initial progagules. However, species with low r or high mortality will need high numbers of propaglues to ensure successful establishment.

These two variables (n and r) are important because the establishment of a new population is determined by the processes affecting small populations. By increasing in size (either through immigration or birth) a population can better avoid some of the perils plaguing small populations. Chapters 5-8 deal with the ways in which populations of invaders establish and persist. Chapter 5 (Buckley and Metcalf) and Chapter 6 (Freckleton *et al.*) consider what happens to small populations. Stochasticity (demographic and environmental) as well as Allee effects can be strong forces affecting a new population's ability to persist. By understanding how stochasticity, as well as the abiotic environment influence r at different times and places, researchers can then make long-term probabilistic predictions regarding invader persistence and population growth.

Subsequent population growth also depends upon r, such as in the logistic growth model (see Chapter 5). In Chapter 7, Murrell shows that even though localized invader success partially depends upon dispersal rates, variation in r still has a large effect on success. However, populations are not isolated entities, solely interacting with their environments. Can we then incorporate multiple species into the notion of invader success?

Step 2: species interactions

Once species overcome abiotic and geographic barriers and establish in a new habitat, they are still faced with biotic barriers (the biological mechanisms affecting their invasion (e.g., Inderjit *et al.* 2005)), and will be faced with a suite of biotic interactions (competition, predation, pathogens, etc.). These interactions can lead to very complex dynamics. Taken on their own, Chapters 10-16, 18 and 19 appear to cover rather idiosyncratic interactions, which can lead to individual hypotheses explaining different types of invasion-promoting interactions. This would seem to give credence to the notion that we are unable to generate laws in community ecology (Lawton 1999), despite the fact that understanding the community-level is where our conservation and invasion control efforts are directed (Simberloff 2004).

However, basic ecological theory may be the best way for biologists understand the important interactions and ecological processes affecting invasions success (Shea and Chesson 2002). Most ecological theory that examines the outcomes of species interactions explicitly considers the role of negative interactions on individual population growth rates and population persistence. Whether using Lotka-Volterra-type equations (e.g., Courchamp and Caut, Chapter 11) where parameters approximate competitive interactions without specifying a mechanism, or resource-based mechanistic approaches (e.g., Tilman 1977, Chase and Leibold 2003), ecological theory has the potential to help ecologists understand and possibly predict the outcomes of multispecies interactions.

There may be a generalized way to think about an invader's success in lieu of negative interactions. Again we can look to the things that affect r. The processes affecting a population's intrinsic rate of increase can be very complex, yet for an invader to persist, its long-term growth rate, $\bar{r}$, must be ≥ 0. For successful invaders, r is much greater than 0 at certain times and places.

Chesson (2000) considers that the long-term growth rate ($\bar{r}$) is related to three mechanisms as: $\bar{r} \approx \bar{r}' - \Delta N + \Delta I$, where $\bar{r}'$ is the effect of fluctuation independent mechanisms on r (i.e., in the presence of competitors but without environmental changes), ΔN is the effect of the nonlinear competition responses to variance in resources, and ΔI is the storage effect (see Box 1 for elaborations of these three parameters). The storage effect represents a species' ability to diminish the effects of competition when the species is not favored by the environment (Chesson 2000). Species can reduce competitive interactions in a number of ways, such as migrating to regions with free resources or lacking dominant competitors, changing their behavior to reduce competitive overlap, or entering a dormant state and re-emerging under better conditions.

Box 1

THE THREE COMPONENTS OF LONG-TERM GROWTH (CHESSON 2000)

The fluctuation independent growth rate, $\bar{r}'$, depends upon the rate at which r declines in response to declining resources for the invader i, b_i, and the difference between the average fitness of the invader in isolation (k) compared to that in the presence of resident competitors ($\bar{k}_s$), where $k = u/b$ (where u is the per capita growth rate in the absence of resource limitation). So that:

$$\bar{r}' = b_i (k - \bar{k}_s)$$

Nonlinear competition responses, ΔN, depends upon two additional properties: the difference between the nonlinear responses to limiting resources, τ, for the invader in isolation and with competitors, and the variance of the limiting factor for both invader and resident, $V(F^{-i})$. So that:

$$\Delta N = b_i (\tau_i - \tau_s)\, V(F^{-i}).$$

Finally, the storage effect, ΔI, is the ability of a species to alter population growth to avoid less favorable environments where the invader would be at a competitive disadvantage:

$$\Delta I = \frac{b_i (1 - \rho)(-\gamma)\sigma^2}{n - 1}$$

Here, ρ is the correlation between the environmental responses between different species, $-\gamma$ is the buffered population growth, and σ^2 is the variance in environment response. See Chesson (2000) for further details.

The ability for species to exhibit behavioral or life history attributes that serve to reduce the negative effects of competition may be widespread. Successful invasive plants, for example, often grow rapidly in competition reduced (successional) habitats (Grotkopp *et al.* 2002) or have small seed sizes (Cadotte and Lovett-Doust 2001; Hamilton *et al.* 2005) which serve to help plants move further (Howe and Smallwood 1982) and establish longer seeds banks than large seeds (Groves 1992).

Of course species interactions can be very complex. Species interactions may produce dynamics that appear chaotic (May 1974) or unpredictable (Huisman and Weissing 2001). Yet, Courchamp and Caut (Chapter 11), Duncan and Forsyth (Chapter 18) and Smith and Shurin (Chapter 19) show that invasions can have regular and understandable dynamics in the presence of multiple species interactions. Courchamp and Caut (Chapter 11) go further to show how competition and predation can interact to promote or hinder animal invasions on islands.

Step 3: spatial spread

Dispersal and spread is a characteristic defining problematic invasions. Several chapters explicitly examine what happens to populations of invaders in a spatial context (Chapters 4 (Murphy *et al.*), 8 (Lewis *et al.*), and 17 (Harding *et al.*)).

Skellam's (1951) formulation of a population's spread according to a reaction-diffusion equation has been paramount to understanding how invaders move across a landscape (e.g., Kot *et al.* 1996, Kot 2001, With 2002, Cadotte Chapter 2, Lewis *et al.* Chapter 8). Skellam's original formulation can be summarized as follows (see Chapters 2 and 8 for more on Skellam's equations):

$$\frac{dx}{dt} = 2\sqrt{rD}.$$

This reaction-diffusion model reveals that the rate of spread is dependent upon two species' characteristics: the intrinsic rate of increase, *r*, and the diffusion coefficient, *D*, measured as distance over time. Figure 2 shows that species with low *r*, though they may have high movement rates, will not spread very fast. Similarly, a high *r* species with low movement will fail to spread rapidly.

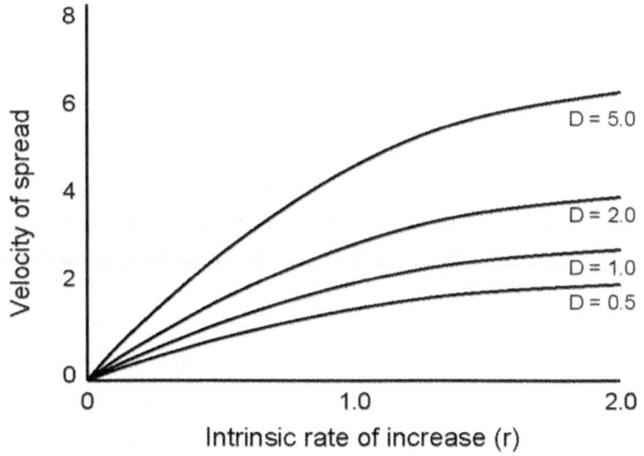

Fig. 2 The velocity of spread as influenced by a diffusion coefficient, D, and intrinsic rate of increase, r, from Skellam's (1951) reaction-diffusion equation (see text).

Spread models are the link between population and metapopulation dynamics. Skellam's model assumes a uniform environment, which may be a realistic assumption at larger spatial scales. At intermediate scales, however, habitats are often patchy, eliciting metapopulation dynamics onto species spread (e.g., Wolff 1980, Hanski 1999). Murphy *et al.* (Chapter 4) and Harding *et al.* (Chapter 17) use a metapopulation perspective to understand dynamics at landscape levels.

The rate of spread, using a metapopulation approach, will depend upon two parameters: the rates of migration and patch extinction. A higher rate of migration will mean that new patches have a higher probability of being colonized and lower patch extinction means that more local patches will act as sources, increasing the pool of potential colonizers. As pointed out earlier, a high r can help a colonizing population overcome inherent small population risks and thereby reduce extinction probabilities, and r is also likely important for immigration rate. If we assume that population size, N, at any given time is largely dependent upon r, as in a logistic equation:

$$N_{t+1} = N_t + N_t\left(1 - \frac{N_t}{K}\right),$$

and we assume that immigration into empty patches is some function of the local density in occupied patches, then the parameters that determine local density likely have a functional relationship with mean density and some distance function (τ) related to the spatial distribution of extant populations: $I_t = f(\overline{N}_t, \tau)$, and we assume that environmental variation affecting r is random and normally distributed about some mean and carrying capacity (K) is constant across patches, then $\bar{r}$ should characterize changes in density across a larger spatial scale, and consequently the rate at which new patches are colonized.

Step 4: large scale patterns

Spatial spread of invaders is a process occurring across spatial scales and exhibits patterns best understood by biogeography or macroecology. Sax and Gaines (Chapter 20) explore large scale patterns of species invasions. They examine how different taxonomic groups of invaders form range size and richness patterns through the lens of existing biogeographical theory. To understand the processes influencing range size, we can still apply our basic variable, r.

Through r we not only understand species spread, but also the final range of distribution (Holt *et al.* 1997; Freckleton *et al.* 2005). Using the well-founded idea that a species' distribution is correlated with mean local abundance (e.g., Brown 1984, Hanski *et al.* 1993), Holt *et al.* (1997) provides an important way to conceptualize these relationships. Local equilibrial population density (N^*) of species i at site x is: $N_i^*(x) = (b_0(x) - d_i)/u = r_i(x)/u$, so that species with higher abundances will have either lower death rates (d), or higher birth rates (b), or both. If we assume

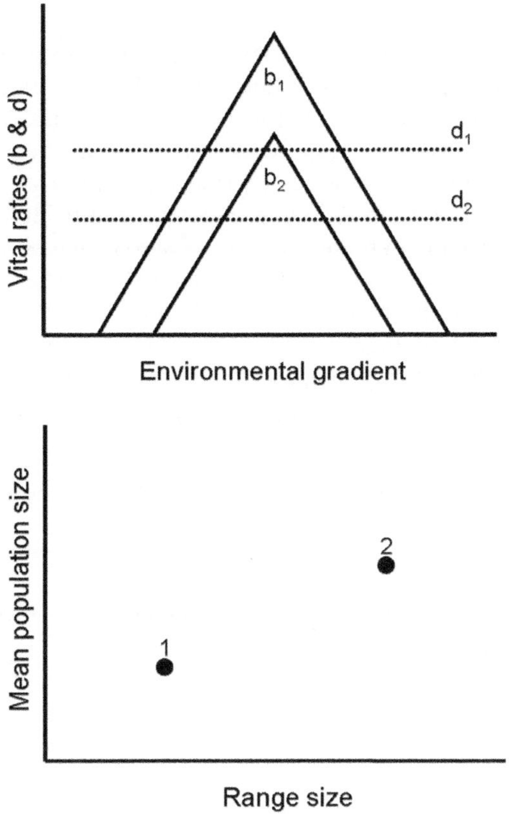

Fig. 3 Birth or death rate differences across and environment (top panel) can either lead to range size differences (bottom panel) between two species (adapted from Holt *et al.* 1997).

that birth rates decline from the center of an environmental gradient, then birth or death rate differences between species will predict range size (Fig. 3).

Freckleton *et al.* (2005) expanded this to include the more realistic scenario where habitat suitability is patchy, similar to that in metapopulation dynamics. Ultimately they show that patchiness had little effect on overall range size. It is still the demographic parameters (birth and death rates) within patches that critically determined range size. Further Holt and Keitt (2000) show that gradients in extinction and immigration rates, as well as patch availability in a metapopulation, can limit range size.

CONCLUSION

This cursory examination uses a population's intrinsic rate of increase (r) as the vehicle to understanding invasions at different scales. We show that r is not only important for local population establishment and size, but can also help us to understand the processes of species interactions, spatial spread and range sizes in invasions across multiple spatial scales. By understanding *r* we can understand the changes in local populations overtime, the factors lead to the success of invaders when interacting with other species, the rate of spread across larger spatial scales, and the ultimate limits to their inventive ranges. Of course other ecological and evolutionary factors will be important for individual invasions, but we show using *r* as an example that by taking a conceptual approach to the ecological factors, we can understand the dynamics of invasions at different scales of organization.

REFERENCES

Baker, H. G. 1974. The evolution of weeds. Annual Review of Ecology and Systematics **5**, 1-24.

Belyea, L. R. and Lancaster, J. 1999. Assembly rules within a contingent ecology. Oikos **86**, 402-416.

Brown, J. H. 1984. On the relationship between abundance and distribution of species. American Naturalist **124**, 255-279.

Cadotte, M. W. and Fukami, T. 2005. Dispersal, spatial scale, and species diversity in a hierarchically structured experimental landscape. Ecology Letters **8**, 548-557.

Carlton, J. T. 1996. Pattern, process and prediction in marine invasion ecology. Biological Conservation **78**, 97-106.

Chase, J. M. and Leibold M. A. 2003. Ecological Niches: Linking Classical and Contemporary Approaches. University of Chicago Press, Chicago, IL.

Chesson, P. 2000. Mechanisms of maintenance of species diversity. Annual Review of Ecology and Systematics **31**, 343-66.

Collingham, Y. C., Wadsworth, R. A., Willis, S. G., Huntley, B. and Hulme, P. E. (2000). Predicting the spatial distribution of alien riparian species: issue of spatial scale and extent. J. Appl. Ecol., **37** (Suppl. 1), 13-27.

Freckleton, R. P., Gill, J. A., Noble, D. and Watkinson, A. R. 2005. Large-scale population dynamics, abundance-occupancy relationships and the scaling from local to regional population size. Journal of Animal Ecology **74**, 353-364.

Grotkopp, E., Rejmánek, M. and Rost, T. L. 2002. Toward a causal explanation of plant invasiveness: seedling growth and life-history strategy of 29 pine (Pinus) species. American Naturalist **159**, 396-419.

Groves, R. H. 1992. Weed ecology, biology and spread. *Proceedings of the First International Weed Control Congress*. Melbourne, Australia, pp. 83-88

Hamilton, M. A., B. R. Murray., M. W. Cadotte, G. C. Hose, A. C. Baker, C. J. Harris and D. Licari. 2005. Life-history correlates of plant invasiveness at multiple spatial scales. Ecology Letters **8**, 1066-1074.

Hanski, I. 1999. Metapopulation Ecology. Oxford University press, Oxford, UK.

Hanski, I., Kouki, J. and Halkka, A. 1993. Three explanations of the positive relationship between distribution and abundance of species. In: Ricklefs, R. E. and Schluter, D. (eds), Species diversity in ecological communities: historical and geographical perspectives. University of Chicago Press, Chicago, pp. 108-116.

Holt, R. D. and Keitt, T. H. 2000. Alternative causes for range limits: a metapopulation perspective. Ecology Letters **3**, 41-47.

Holt, R. D., Lawton, J. H., Gaston, K. J. and Blackburn, T. M. 1997. On the relationship between range size and local abundance: back to basics. Oikos **78**, 183-190.

Howe, H. F. and Smallwood, J. 1982. Ecology of seed dispersal. Annual Review of Ecology and Systematics, **13**, 201-228.

Huisman, J. and Weissing, F. J. 2001. Fundamental unpredictability in multispecies coexistence. American Naturalist **157**, 488-494.

Inderjit, Cadotte, M. W. and Colautti, R. I. 2005. The ecology of biological invasions: past, present and future. In: *Invasive Plants: Ecological and Agricultural Aspects* (ed. Inderjit). Birkhauser, pp. 19-43.

Kot, M. 2001. Elements of Mathematical Ecology. Cambridge University Press, New York, NY.

Kot, M., Lewis, M. A. and van den Driessche, P. 1996. Dispersal data and the spread of invading organisms. Ecology **77**, 2027-2042.

Lakatos, I. 1978. Falsification and the methodology of scientific research programmes. In: *Criticism and the Growth of Knowledge* (ed. I. Lakatos and A. Musgrave). Cambridge University Press, pp 91-195.

Lawton, J. H. 1999. Are there general laws in ecology? Oikos, **84**, 177-192.

Lloret, F., Médail, F., Brundu, G. and Hulme, P. 2004. Local and regional abundance of exotic plant species on Mediterranean islands: are species traits important? Global Ecology and Biogeography **13**, 37-45.

Lockwood, J. L., Cassey, P. and Blackburn, T. 2005. The role of propagule pressure in explaining species invasions. Trends in Ecology and Evolution **20**, 223-228.

Lodge, D. M. 1993. Biological invasions: lessons for ecology. *Trends in Ecology and Evolution* **8**, 133-137.

May, R. M. 1974. Biological populations with nonoverlapping generations: stable points, stable cycles, chaos. Science **186**, 645-647.

Rahbek, C. 2005. The role of spatial scale and the perception of large-scale species-richness patterns. Ecology Letters **8**, 224-239.

Sax, D. F., Stachowicz, J. J. and Gaines, S. D. 2005. Species Invasions: Insights into Ecology, Evolution, and Biogeography. Sinauer Associates, Sunderland, MA.

Shea, K. and Chesson, P. 2002. Community ecology theory as a framework for biological invasions. Trends Ecol. Evol., **17**, 170-176.

Skellam, J. G. 1951. Random dispersal in theoretical populations. Biometrika **38**, 196-218.

Tilman, D. 1977. Resource competition between planktonic algae: an experimental and theoretical approach. Ecology **58**, 338-348.

Vitousek, P. M., C. M. D'Antonio, L. L. Loop and R. Westbrooks. 1996, Biological invasions as global environmental change: American Scientist. **84**, 468-478.

With, K. A. 2002. The landscape ecology of invasive spread. Conservation Biology **16**, 1192-1203.

Wolff, J. O. 1980. The role of habitat patchiness in the population dynamics of snowshoe hares. Ecological Monographs **50**, 111-130.

Index

D

N

O

P

GPSR Compliance

The European Union's (EU) General Product Safety Regulation (GPSR) is a set of rules that requires consumer products to be safe and our obligations to ensure this.

If you have any concerns about our products, you can contact us on ProductSafety@springernature.com

In case Publisher is established outside the EU, the EU authorized representative is:

Springer Nature Customer Service Center GmbH
Europaplatz 3
69115 Heidelberg, Germany

Batch number: 10809439

Printed by Printforce, the Netherlands